Periodic Table of the Elements with the Gmelin System Numbers

1 H 2																	2 He 1
3 Li 20	4 Be 26										5 B 13	6 C 14	7 N 4	8 O 3	9 F 5	1 H 2	10 Ne 1
11 Na 21	12 Mg 27										13 Al 35	14 Si 15	15 P 16	16 S 9	17 Cl 6	18 Ar 1
19 * K 22	20 Ca 28	21 Sc 39	22 Ti 41	23 V 48	24 Cr 52	25 Mn 56	26 Fe 59	27 Co 58	28 Ni 57	29 Cu 60	30 Zn 32	31 Ga 36	32 Ge 45	33 As 17	34 Se 10	35 Br 7	36 Kr 1
37 Rb 24	38 Sr 29	39 Y 39	40 Zr 42	41 Nb 49	42 Mo 53	43 Tc 69	44 Ru 63	45 Rh 64	46 Pd 65	47 Ag 61	48 Cd 33	49 In 37	50 Sn 46	51 Sb 18	52 Te 11	53 I 8	54 Xe 1
55 Cs 25	56 Ba 30	57** La 39	72 Hf 43	73 Ta 50	74 W 54	75 Re 70	76 Os 66	77 Ir 67	78 Pt 68	79 Au 62	80 Hg 34	81 Tl 38	82 Pb 47	83 Bi 19	84 Po 12	85 At 8a	86 Rn 1
87 Fr 25a	88 Ra 31	89*** Ac 40	104 71	105 71													

**Lanthanides 39	58 Ce	59 Pr	60 Nd	61 Pm	62 Sm	63 Eu	64 Gd	65 Tb	66 Dy	67 Ho	68 Er	69 Tm	70 Yb	71 Lu

***Actinides	90 Th 44	91 Pa 51	92 U 55	93 Np 71	94 Pu 71	95 Am 71	96 Cm 71	97 Bk 71	98 Cf 71	99 Es 71	100 Fm 71	101 Md 71	102 No 71	103 Lr 71

* NH₄ 23

A Key to the Gmelin System is given on the Inside Back Cover

Gmelin Handbook of Inorganic Chemistry

8th Edition

Gmelin Handbook
of Inorganic Chemistry

8th Edition

Gmelin Handbuch der Anorganischen Chemie

Achte, völlig neu bearbeitete Auflage

Prepared
and issued by

Gmelin-Institut für Anorganische Chemie
der Max-Planck-Gesellschaft
zur Förderung der Wissenschaften

Director: Ekkehard Fluck

Founded by

Leopold Gmelin

8th Edition

8th Edition begun under the auspices of the
Deutsche Chemische Gesellschaft by R. J. Meyer

Continued by

E. H. E. Pietsch and A. Kotowski, and by
Margot Becke-Goehring

Springer-Verlag Berlin Heidelberg GmbH 1990

Volumes published on "Rare Earth Elements" (Syst. No. 39)

* Completely or ° partly in English

Gmelin Handbook of Inorganic Chemistry

8th Edition

Sc, Y, La–Lu
RARE EARTH ELEMENTS

C 11a

Compounds with Boron

With 88 illustrations

AUTHORS

Hiltrud Hein, Claus Koeppel, Ursula Vetter,
Eberhard Warkentin

EDITORS

Hartmut Bergmann, Gerhard Czack, Hiltrud Hein,
Sigrid Ruprecht, Ursula Vetter

CHIEF EDITOR

Hartmut Bergmann

System Number 39

Springer-Verlag Berlin Heidelberg GmbH 1990

LITERATURE CLOSING DATE: 1988
IN MANY CASES MORE RECENT DATA HAVE BEEN CONSIDERED

Library of Congress Catalog Card Number: Agr 25-1383

ISBN 978-3-662-07505-0   ISBN 978-3-662-07503-6 (eBook)
DOI 10.1007/978-3-662-07503-6

© by Springer-Verlag Berlin Heidelberg 1989
Originally published by Springer-Verlag, Berlin · Heidelberg · New York · London · Paris · Tokyo in 1989
Softcover reprint of the hardcover 8th edition 1989

Typesetting

Preface

The volume "Rare Earth Elements" C 11 deals with the compounds and systems of the rare earth elements with boron, i.e., borides, borates, and associated alkali double compounds. As in all earlier volumes of "Rare Earth Elements" Series C ("Seltenerdelemente" Reihe C), comparative data are presented in sections preceding treatment of the individual compounds and systems.

Topics of the present volume C 11a are the comparative data on the borides and the individual sections on the systems and borides containing Sc, Y, and La. The individual sections for the systems and borides with Ce to Lu can be found in the following volume C 11b together with the borates, their alkali double compounds, and other compounds containing Ce to Lu, boron and elements related by the Gmelin system.

The most extensively studied borides treated in the comparative sections are of the type MB_6. These rare earth hexaborides are refractory compounds and some of them, especially LaB_6, are good thermionic emissive materials or exhibit other interesting physical properties. For example, many data on mechanical, thermal, or electrical properties are presented in tables and figures. In connection with these useful properties, the preparation of the samples and their chemical and thermal stability play an important role. Interesting crystallographic structures are presented in the comparative sections on MB_4 and MB_{66}, the latter formula gives only an approximate composition for the most boron-rich compound in the M–B systems. Main topics of these sections and those for the borides of the type MB_2 and MB_{12} are also physical properties. The most comprehensive chapter in the individual sections of volume C 11a deals with LaB_6, which is a well-known thermionic material of great practical importance with high electron emission efficiency and an excellent stability in the surface composition at high temperatures. The effects of preparation conditions, composition, temperature, surface structures, surface treatment, and of the atmosphere on the thermionic emission are studied thoroughly. Another important item of the LaB_6 chapter is the electronic structure, which often serves as an example for the other rare earth hexaborides. In the individual section on yttrium borides, the thermionic properties and the electronic structure of YB_4 and YB_6 are also remarkable features.

Frankfurt am Main
November 1989

Hartmut Bergmann

Table of Contents

32 Rare Earth Elements and Boron

32.1 Borides

General References:

Fisk, Z.; Remeika, J. P.; Growth of Single Crystals from Molten Metal Fluxes, in: Gschneidner, K. A., Jr.; Eyring, LeRoy; Hüfner, S.; Handbook on the Physics and Chemistry of Rare Earths, Vol. 12, North-Holland, Amsterdam 1989.

Eliseev, A. A.; Binary Inorganic Compounds of the Rare-Earth Elements with Intermediate Valences, Izv. Akad. Nauk SSSR Neorgan. Materialy 24 [1988] 181/92; Inorg. Materials [USSR] 24 [1988] 125/35.

Gschneidner, K. A., Jr.; Eyring, LeRoy; Hüfner, S.; Handbook on the Physics and Chemistry of Rare Earths, Vol. 10, High Energy Spectroscopy, North-Holland, Amsterdam 1987.

Villars, P.; Calvert, L. D.; Pearson's Handbook of Crystallographic Data for Intermetallic Phases, Vol. 1/3, ASM, Metals Park, Ohio, 1986.

Etourneau, J.; Critical Survey of Rare-Earth Borides: Occurrence, Crystal Chemistry and Physical Properties, J. Less-Common Metals 110 [1985] 267/81.

Etourneau, J.; Hagenmuller, P.; Structure and Physical Features of the Rare-Earth Borides, Phil. Mag. [8] B 52 [1985] 589/610.

Landolt-Börnstein New Ser. Group III 17 Pt. g [1984] 24/34, 356/67.

Kost, M. E.; Shilov, A. L.; Mikheeva, V. I.; et al.; Khimiya Redkikh Elementov [Chemistry of Rare Earth Elements], Soedineniya Redkozemel'nykh Elementov, Nauka, Moscow 1983, pp. 1/270, 40/68.

Gordienko, S. P.; Fenochka, B. V.; Viksman, G. Sh.; Termodinamika Soedinenii Lantanoidov, Nauka Dumka, Kiev 1979, pp. 1/373, 68/84.

Goryachev, Yu. M.; Kovenskaya, B. A.; The Physical Properties and the Electronic Structure of the Higher Borides, J. Less-Common Metals 67 [1979] 273/9.

Studies on Lanthanum Boride, Muki Zaishitsu Kenkyushu Kenkyu Hokokusho [Research Reports of the National Institute for Research in Inorganic Materials] No. 17 [1978] 1/139; C.A. 91 [1979] No. 101319.

Matkovich, V. I.; Boron and Refractory Borides, Springer, Berlin – Heidelberg – New York 1977.

Samsonov, G. V.; Vinitskii, I. M.; Tugoplavkie Soedineniya, 2nd Ed., Metallurgiya, Moscow 1976.

Spear, K. E.; Phase Behavior and Related Properties of Rare-Earth Borides, in: Alper, A. M.; Refractory Materials, Phase Diagrams, Vol. 4, Academic, New York 1976, pp. 91/159.

Fomenko, V. S.; Podchernyaeva, I. A.; Emissionnye i Adsorbtsionnye Svoistva Veshchestv i Materialov [Emissive and Adsorptive Properties of Materials], Atomizdat, Moscow 1975, pp. 1/319, 166/75.

Samsonov, G. V.; Serebryakova, T. I.; Neronov, V. A.; Boridy [Borides], Atomizdat, Moscow 1975, pp. 1/375, 9/132, 161/204.

Hoard, J. L.; Hughes, R. E.; Elementary Boron and Compounds of High Boron Content: Structure, Properties, and Polymorphism, in: Muetterties, E. L.; The Chemistry of Boron and Its Compounds, Wiley, New York – London – Sydney 1967, pp. 25/154.

Samsonov, G. V.; Tugoplavkie Soedineniya Redkozemel'nykh Metallov s Nemetallami, Metallurgiya, Moscow 1964; High-Temperature Compounds of Rare Earth Metals with Nonmetals, Consultants Bureau, New York 1965, pp. 1/280, 1/100.

32.1.1 Comparative Data for MB_2

Diborides are formed with M = Sc, Y, Sm, Gd to Lu. SmB_2 is only obtained by high-pressure synthesis. Earlier reported LaB_2 and CeB_2 seem questionable.

32.1.1.1 Preparation

Diborides with M = Tb to Er were prepared by arc-melting stoichiometric amounts of the elements, Will et al. [1], see also Buschow [2]. SmB_2, GdB_2, HoB_2, and TmB_2 were prepared by high-pressure synthesis at 60 to 70 kbar (6 to 7 GPa) and 1513 to 2053 K in a BN crucible, Cannon et al. [3, 4]; NdB_2 was not obtained by this method, Cannon, Hall [5].

References:

[1] Will, G.; Buschow, K. H. J.; Lehmann, V. (Conf. Ser. Inst. Phys. [London] No. 37 [1978] 255/61).
[2] Buschow, K. H. J. (in: Matkovich, V. I., Boron and Refractory Borides, Springer, Berlin 1977, pp. 494/515, 500).
[3] Cannon, J. F.; Cannon, D. M.; Hall, H. T. (J. Less-Common Metals **56** [1977] 83/90).
[4] Cannon, J. F.; Cannon, D. M.; Hall, H. T. (High Pressure Sci. Technol. 6th AIRAPT Conf., Boulder, Colo., 1977 [1979], Vol. 1, pp. 1000/6).
[5] Cannon, J. F.; Hall, H. T. (Rare Earths Mod. Sci. Technol. 1 [1977/78] 219/24).

32.1.1.2 Crystallographic Properties. Density

The diborides are hexagonal, AlB_2 type, space group $P6/mmm$-D_{6h}^1 (No. 191), Spear [1], or $P\bar{3}m1$-D_{3d}^3 (No. 164), Will et al. [2], Z=1, with the following lattice parameters:

MB_2	a in Å	c in Å	c/a	d_{M-B} in Å[1]	r_M in Å[1]	Ref.
ScB_2	3.148	3.516	1.117	2.529	1.620	[4]
YB_2	3.298	3.843	1.165	2.705	1.753	[5]
SmB_2[2]	3.310	4.019	1.214	—	—	[6, 7]
GdB_2	3.318	3.933	1.185	2.745	1.787	[8]
GdB_2[2]	3.315	3.936	1.187	—	—	[6]
TbB_2	3.290	3.878	1.179	~2.70	~1.75	[2, 9]

MB$_2$	a in Å	c in Å	c/a	d$_{M-B}$ in Å[1]	r$_M$ in Å[1]	Ref.
DyB$_2$	3.287	3.847	1.170	2.702	1.753	[10]
	3.287	3.845	1.170	—	—	[2, 9]
HoB$_2$	3.281	3.813	1.162	—	—	[2, 9]
	3.281	3.811	1.162	—	—	[11]
	3.273	3.814	1.165	2.685	1.740	[3]
HoB$_2$[2]	3.279	3.811	1.162	—	—	[6]
ErB$_2$	3.271	3.782	1.156	—	—	[2, 9]
	3.265	3.768	1.154	2.665	1.722	[12]
TmB$_2$	3.261	3.755	1.151	2.659	1.718	[3]
	3.250	3.739	1.150	—	—	[11]
TmB$_2$[2]	3.258	3.745	1.149	—	—	[6]
YbB$_2$	3.2503	3.7315	1.148	2.646	1.708	[13]
LuB$_2$	3.246	3.704	1.141	2.635	1.698	[14]

[1] Data listed in Spear [3], see below. – [2] Obtained by high-pressure synthesis.

For additional references see [1].

The structure consists of hexagonal layers normal to the c axis (see **Fig. 1** [1]): The M atoms in (000), with coordination number 6 in the plane, form hexagonal close-packed layers. The B atoms in ($\frac{1}{3}$ $\frac{2}{3}$ $\frac{1}{2}$) and ($\frac{2}{3}$ $\frac{1}{3}$ $\frac{1}{2}$) form graphite-like layers, exhibit sp^2 type bonding and coordination number 3 in the layer. Each M has twelve B neighbors which form a hexagonal prism, each B has six M neighbors which form a trigonal prism, see for instance [1, 2, 6, 7]. For SmB$_2$ the interplanar spacings and the intensities of the Debye-Scherrer diagram are listed in [6, 7]. Considered from a crystal chemical viewpoint, the chemical bonding in MB$_2$ seems to depend on the ability of M to deform from a spherical to an elliptic shape. The shortest distances d$_{M-B}$ and the M radius in the B direction r$_M$ = d$_{M-B}$ − r$_B$ are listed in the above table. Various crystal chemical parameters were calculated and used to compare relative bond strengths [3]. The results of Pauling's metallic radii and polyhedral atomic volumes calculations provide an estimate of effective charges, valences, atomic volumes, and nonintegral coordination number. These calculations, too, indicate for studied TbB$_2$ and LuB$_2$ an anisotropic metal radius, Carter [15]. The lattice constants depend linearly on the M ionic radii, showing trivalent M in MB$_2$. Implications of this relation concerning the boron sublattice are discussed [1]. The calculated densities of various MB$_2$ are summarized by Post [16]:

MB$_2$	ScB$_2$	YB$_2$	GdB$_2$	TbB$_2$	DyB$_2$	HoB$_2$	ErB$_2$	LuB$_2$
D in g/cm^3	3.67	5.54	7.96	8.34	8.53	8.80	8.89	9.76

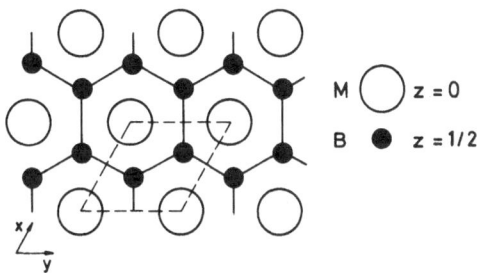

Fig. 1. Projection of the crystal structure of MB$_2$ along [001]. The unit cell is indicated by the dashed lines.

M ◯ z = 0
B ● z = 1/2

References:

[1] Spear, K. E. (in: Alper, A. M., Refractory Materials, Phase Diagrams, Vol. 4, Academic, New York 1976, pp. 91/159, 105/8, 120/6).
[2] Will, G.; Buschow, K. H. J.; Lehmann, V. (Conf. Ser. Inst. Phys. [London] No. 37 [1978] 255/61).
[3] Spear, K. E. (J. Less-Common Metals 47 [1976] 195/201).
[4] Peshev, P.; Etourneau, J.; Naslain, R. (Mater. Res. Bull. 5 [1970] 319/27).
[5] Lundin, C. E. (in: Spedding, F. H.; Daane, A. H., The Rare Earths, Wiley, New York 1961, pp. 224/385, 248).
[6] Cannon, J. F.; Cannon, D. M.; Hall, H. T. (J. Less-Common Metals 56 [1977] 83/90; High Pressure Sci. Technol. 6th AIRAPT Conf., Boulder, Colo., 1977 [1979], Vol. 1, pp. 1000/6).
[7] Cannon, J. F.; Hall, H. T. (Rare Earths Mod. Sci. Technol. 1 [1977/78] 219/24).
[8] Spear, K. E.; Petsinger, D. W. (in [1]).
[9] Buschow, K. H. J. (in: Matkovich, V. I., Boron and Refractory Borides, Springer, Berlin 1977, pp. 494/515, 500).
[10] Spear, K. E. (Proc. Intern. Symp. Boron Borides, Tbilisi, USSR, 1972 in [1]).

[11] Bauer, J.; Debuigne, J. (Compt. Rend. C 277 [1973] 851/3).
[12] Imperato, E. G.; Spear, K. E. (in [1]).
[13] Bauer, J. (Compt. Rend. C 279 [1974] 501/4).
[14] Przybylska, M.; Reddoch, A. H.; Ritter, G. J. (J. Am. Chem. Soc. 85 [1963] 407/11).
[15] Carter, F. L. (J. Less-Common Metals 47 [1976] 157/63).
[16] Post, B. (in: Adams, R. M., Boron, Metallo-Boron Compounds, and Boranes, Wiley, New York 1964, pp. 301/71, 311).

32.1.1.3 Thermal and Magnetic Properties. Electronic Structure

Melting temperatures T$_f$ and behavior were estimated, or originate from unpublished data and are summarized as follows:

M in MB$_2$	Sc	Y	Gd	Tb	Dy	Ho	Er	Tm	Yb	Lu
T$_f$ in K	2523	2373	2323	2373	2373	2473	2458	2523	2523	2523
mode	C	C	C	C	P	P	P	P	P	P

C indicates congruent, P peritectic melting, Spear [8].

Heats of formation ΔH_f in kcal/mol (kJ/mol in parentheses) were calculated, applying Pauling's bond distance-bond order relation, and compared with predicted ΔH_f values:

MB$_2$	ScB$_2$	YB$_2$	SmB$_2$	GdB$_2$	LuB$_2$	Ref.
$-\Delta H_f$(calc)	19.3(80.8)	19.4(81.2)	11.6(48.5)	19.5(81.6)	19.4(81.2)	[1]
$-\Delta H_f$(pred)	20(83.7)	20(83.7)	—	—	—	[2]

For Gd and Lu, f bonding is ignored. The electronic structure of M is decisive in determining the stability and occurrence of MB$_2$. The energy contributions from electron populations in various energy levels are discussed, Mulokozi [1], see also [3]. For "LaB$_2$" (see the phase diagram La–B, p. 162) $\Delta H_f = -17$ kcal/g atom (-71 kJ/g atom) is predicted, Miedema [2].

The standard entropy S°_{298} in $J \cdot mol^{-1} \cdot K^{-1}$ of MB$_2$ in which B forms nets was evaluated by comparison of S° of the metals and by additive methods:

MB$_2$	....	ScB$_2$	YB$_2$	DyB$_2$	LuB$_2$
S$^\circ_{298}$	....	38.5 ± 2.1	50.0 ± 2.1	82.8 ± 1.7	56.1 ± 1.2

Borovikova, Fesenko [4].

From magnetic measurements on polycrystalline MB$_2$ (M = Tb, Dy, Ho, and Er), using Faraday's method, the following ferromagnetic ordering temperatures T_C, extrapolated paramagnetic Curie temperatures Θ_p, effective magnetic moments μ_{eff} (values of M^{3+} in parentheses), and moments μ at 18 kOe field strength at 4.2 K are obtained:

MB$_2$	T_C in K	Θ_p in K	μ_{eff} in μ_B	μ in μ_B
TbB$_2$	151	151	9.96 (9.72)	4.0
DyB$_2$	55	33	10.7 (10.64)	5.4
HoB$_2$	15	25	10.2 (10.60)	7.5
ErB$_2$	16	9	9.47 (9.58)	5.1

Above T_C, the magnetic susceptibility displays Curie-Weiss behavior with μ_{eff} close to theoretical values. Except for TbB$_2$, considerable differences between T_C and Θ_p are observed. It is suggested that the magnetic ordering phenomena are more complicated than simple ferromagnetic ordering. This follows also from the shape of the magnetization vs. temperature curves of TbB$_2$ and DyB$_2$, which involve two steps. Studies of the field dependence of the magnetization at 4.2 K suggest that relatively strong magneto-crystalline anisotropies are involved. For all studied MB$_2$, saturation was still not reached at 18 kOe (experimental limit) at 4.2 K. This anisotropy is very probably induced by crystal fields, Buschow [5], Will et al. [6]. However, neutron-diffraction measurements on TbB$_2$ between 4.2 and 300 K do not support the assumption of a complex magnetic structure [6]. For details see the corresponding sections on individual borides in "Rare Earth Elements" C 11b.

The interaction of M orbitals with those of a planar graphite-like B net and with those of other M controls the electronic structure in AlB$_2$ type borides like ScB$_2$. Electronic density of states, overlap populations, dispersion of some energy bands along Δ and Γ of the hexagonal zone, all based on parameters for TiB$_2$, and further details are given in the paper, Burdett et al. [7]. For band structure calculations on ScB$_2$, see p. 122.

References:

[1] Mulokozi, A. M. (J. Less-Common Metals **71** [1980] 105/11).
[2] Miedema, A. R. (J. Less-Common Metals **46** [1976] 67/83, 79).
[3] Mulokozi, A. M. (J. Less-Common Metals **64** [1979] 145/53).
[4] Borovikova, M. S.; Fesenko, V. V. (J. Less-Common Metals **117** [1986] 287/91).
[5] Buschow, K. H. J. (in: Matkovich, V. I., Boron and Refractory Borides, Springer, Berlin 1977, pp. 494/515, 500).
[6] Will, G.; Buschow, K. H. J.; Lehmann, V. (Conf. Ser. Inst. Phys. [London] No. 37 [1978] 255/61).
[7] Burdett, J. K.; Canadell, E.; Miller, G. J. (J. Am. Chem. Soc. **108** [1986] 2651/8).
[8] Spear, K. E. (in: Alper, A. M., Refractory Materials, Phase Diagrams, Vol. 4, Academic, New York 1976, pp. 91/159, 134).

32.1.2 Comparative Data for M_2B_5

M_2B_5 have monoclinic symmetry and space group $P2_1/c-C_{2h}^5$ (No. 14). They are obtained only with M = Nd, Sm, and Gd, see for example the review by Etourneau [1]. The structure contains parallel layers that are virtually identical to layers found in the xy plane of MB_4. These layers are separated by intervening B_2 units that connect the B_6 octahedra lying in adjacent layer planes. In the MB_4 layers the B atoms are of the sp^2 and B_6 type. The intervening B_2 units are similar to sp^2 type, each B is coordinated by 6 M arranged in a distorted trigonal prism. However, each of these B is bonded to only 2 other B rather than 3, and the site of the third B is empty, Spear [2], see also Etourneau, Hagenmuller [3]. According to unpublished observations of Bucher et al. [4] in Buschow [5], Pr_2B_5 exists also with silvery aspects like the other M_2B_5 with M = Nd, Sm, and Gd. Néel temperatures T_N = 15.1, 32.7, 23.5, and ~50 K for M = Pr, Nd, Sm, and Gd, respectively (less accurate for the latter due to metallic Gd traces) and paramagnetic Curie temperatures Θ_p = 14 K for Pr_2B_5 and Nd_2B_5 were determined, Buschow [5].

References:

[1] Etourneau, J. (J. Less-Common Metals **110** [1985] 267/81, 268).
[2] Spear, K. E. (in: Alper, A. M., Refractory Materials, Phase Diagrams, Vol. 4, Academic, New York 1976, pp. 91/159, 108).
[3] Etourneau, J.; Hagenmuller, P. (Phil. Mag. [8] B **52** [1985] 589/610, 593).
[4] Bucher, E.; Schmidt, P. H.; Castellano, R. N.; Dernier, P. D.; Cooper, A. S. (from [5]).
[5] Buschow, K. H. J. (in: Matkovich, V. I., Boron and Refractory Borides, Springer, Berlin – New York 1977, pp. 494/515, 501).

32.1.3 Comparative Data for "MB_x" (x = 3 to 4)

The earlier reported phases MB_x with x = 3 to < 4 for M = Y, La, Pr, Sm, Gd, and Yb were shown later to be ternary phases containing C, see for example Spear, K. E. (in: Alper, A. M., Refractory Materials, Phase Diagrams, Vol. 4, Academic, New York 1976, pp. 91/159, 101).

32.1.4 Comparative Data for MB_4

32.1.4.1 Formation, Preparation (Color)

Tetraborides of the elements Sc, Y, La through Lu are known for all except Sc, Pm, Eu. The formation of PmB_4 has been predicted, Spear [1, p. 99]. Attempts to obtain ScB_4, see p. 125, and EuB_4 failed, Felten et al. [2], Fisk et al. [3], Savitskii et al. [4], Kaznoff et al. [5]. The instability of a hypothetical phase "EuB_4" was discussed in terms of the Eu^{2+}/Eu^{3+} stability relation and ion size effects, Etourneau [6], or was related to the high vapor pressure of Eu, Etourneau et al. [7].

There are mainly two methods of synthesis: directly from the elements above 1273 K and reduction of the oxides under vacuum by boron (borothermal method), by boron carbide, or by carbon in the presence of boron above 1773 K [6].

Borothermal reduction in high vacuum apparently is most convenient for the congruently melting MB_4 compounds with M = Y, Gd to Er, Eick, Gilles [8], Etourneau et al. [7], and M = Y, Gd to Tm, Severyanina et al. [9]. The conditions applied by the various authors are described for GdB_4, see "Rare Earth Elements" C 11b. The method is apparently not appropriate for the synthesis of the incongruently melting tetraborides with M = La to Sm because of small quantities of hexaboride or rare earth metal in the products [7]. However, also by this method the synthesis of samples of NdB_4 and SmB_4 free from diffraction lines of other phases was reported [8], see also Galloway, Eick [10]; subsequent arc-melting under Ar gave pure samples of PrB_4 and NdB_4, Paderno et al. [11].

Direct synthesis from the elements was preferred generally for tetraborides with M = La to Sm, Berrada et al. [12], Buschow, Creyghton [13], Etourneau et al. [14], for M = Yb [13, 14], and also for M = Y, Gd, Dy to Er [13]. The procedures applied by [13] and [14] are described for the synthesis of CeB_4, see "Rare Earth Elements" C 11b, and LaB_4 (see p. 165), respectively; a different procedure was used for M = Tb [13].

Crystals with dimensions up to 1 mm were obtained by growth in a metal flux. An excess of the rare earth metal M served as flux for M = La to Nd with initial compositions $MB_{0.65}$ (M = La, Ce) and MB_3 (M = Pr, Nd). An aluminium metal flux was used for M = Sm, Gd to Lu, with 5% of the mixture of M plus B and 95% (presumably atomic %) of Al, see the crystal growth of SmB_4 in "Rare Earth Elements" C 11b. The Al flux was leached in saturated NaOH solution, Fisk et al. [3]. The same method was also described for M = Sm, Gd to Er by [14].

The floating-zone method, Johnson, Daane [15], Tanaka et al. [16], and the Czochralski method, Bressel et al. [17] were used in the growth of large single crystals of YB_4.

The tetraboride of Y has a gold-metallic color [18], the tetraborides of La, Pr, Ho, Yb, and Lu form black crystals, those of Sm and Gd form bluish colored crystals, whereas MB_4 with M = Nd, Tb, Dy, Er, and Tm form golden colored crystals. CeB_4, unlike the other tetraborides, forms distinctive bronze-colored crystals, Fisk et al. [3].

References:

[1] Spear, K. E. (in: Alper, A. M., Refractory Materials, Phase Diagrams, Vol. 4, Academic, New York 1976, pp. 91/159).
[2] Felten, E. J.; Binder, I.; Post, B. (J. Am. Chem. Soc. **80** [1958] 3479).
[3] Fisk, Z.; Cooper, A. S.; Schmidt, P. H.; Castellano, R. N. (Mater. Res. Bull. **7** [1972] 285/8).
[4] Savitskii, E. M.; Arabei, B. G.; Bakarinova, V. I.; Salibekov, S. E.; Romashov, V. M.; Timofeeva, N. I. (Izv. Akad. Nauk SSSR Neorgan. Materialy **7** [1971] 617/9; Inorg. Materials [USSR] **7** [1971] 539/41).
[5] Kaznoff, A. I.; Hoyt, E. W.; Grossman, L. N. (Advan. Energy Convers. **3** [1963] 167/73).
[6] Etourneau, J. (J. Less-Common Metals **110** [1985] 267/81).
[7] Etourneau, J.; Mercurio, J.-P.; Naslain, R.; Hagenmuller, P. (Compt. Rend. C **274** [1972] 1688/91).
[8] Eick, H. A.; Gilles, P. W. (J. Am. Chem. Soc. **81** [1959] 5030/2).
[9] Severyanina, E. N.; Dudnik, E. M.; Paderno, Yu. B. (Poroshkovaya Metal. **1974** No. 10, pp. 83/5; Soviet Powder Met. Metal Ceram. **13** [1974] 843/5).
[10] Galloway, G. L.; Eick, H. A. (J. Inorg. Nucl. Chem. **27** [1965] 293/6).

[11] Paderno, Yu. B.; Severyanina, E. N.; Dudnik, E. M.; Lazorenko, V. (Splavy Redk. Metall. Osobymi Fiz. Khim. Svoistvami Mater. 2nd Vses. Soveshch., Moscow 1974 [1975], pp. 118/21; C.A. **85** [1976] No. 12590).
[12] Berrada, A.; Mercurio, J.-P.; Chevalier, B.; Etourneau, J.; Hagenmuller, P.; Lalanne, M.; et al. (Mater. Res. Bull. **11** [1976] 1519/26).
[13] Buschow, K. H. J.; Creyghton, J. H. N. (J. Chem. Phys. **57** [1972] 3910/4).
[14] Etourneau, J.; Mercurio, J.-P.; Berrada, A.; Hagenmuller, P.; Georges, R.; Bourezg, R.; Gianduzzo, J. C. (J. Less-Common Metals **67** [1979] 531/9).

[15] Johnson, R. W.; Daane, A. H. (J. Chem. Phys. **38** [1963] 425/32).

[16] Tanaka, T.; Otani, T.; Ishizawa, Y. (J. Less-Common Metals **102** [1984] 281/7).

[17] Bressel, B.; Chevalier, B.; Etourneau, J.; Hagenmuller, P. (J. Cryst. Growth **47** [1979] 429/33).

[18] Okada, S.; Atoda, T. (Yogyo Kyokaishi **89** [1981] 339/45; C.A. **95** [1981] No. 89085).

32.1.4.2 Crystallographic Properties

The tetraborides of the rare earth elements M = Y, La through Lu crystallize at room temperature isomorphous with the tetragonal UB_4 type, as was shown by Guette et al. [1], Giese et al. [2], Kato et al. [3], Zavalii et al. [4], Will et al. [5 to 7], McCarthy [8], Elf et al. [9], Schäfer, Will [10] by powder diffraction studies and single-crystal (SC) data using both X-rays and neutrons: Space group $P4/mbm-D_{4h}^5$ (No. 127), $Z = 4$, atomic positions: M in 4g $(x, \frac{1}{2} + x, 0)$, B(1) in 4e $(0, 0, z)$, B(2) in 4h $(x, \frac{1}{2} + x, \frac{1}{2})$, B(3) in 8j $(x, y, \frac{1}{2})$ with:

MB_4	YB_4	LaB_4	SmB_4	TbB_4	DyB_4	ErB_4
x(M)	0.3179(1)	0.31661(6)	0.31675(9)	0.3175(11)	0.319(2)	0.3183(9)
z(B(1))	0.2027(20)	0.2088(30)	0.214(4)	0.2017(17)	0.196(7)	0.2031(13)
x(B(2))	0.0871(9)	0.0884(12)	0.0894(18)	0.0875(16)	0.086(2)	0.0859(18)
x(B(3))	0.1757(9)	0.1743(12)	0.1799(23)	0.1758(7)	0.175(2)	0.1767(4)
y(B(3))	0.0389(8)	0.0394(11)	0.0392(16)	0.0387(7)	0.039(2)	0.0382(5)
type of data ..	X-ray, SC	X-ray, SC	X-ray, SC	n, SC	n, powd.	n, SC
Ref.	[1]	[3]	[4]	[5]	[6]	[5]
additional data in	[2]			[8, 9]		[7, 10]

These data result in the following interatomic distances in Å:

distance		rel. mult.	YB_4	LaB_4	SmB_4	TbB_4	DyB_4	ErB_4
B(1)–B(1)′	along c	1	1.629	1.746(25)	1.73(2)	1.630(10)	1.56[1)]	1.625(1)
B(1)–B(3)	octah.	8	1.750	1.787(11)	1.76(2)	1.760(6)	1.74	1.745(4)
B(3)–B(3)′		4	1.809	1.850(12)	1.86(2)	1.813(7)	1.78	1.808(5)
B(2)–B(2)′	7-ring	1	1.752	1.831(24)	1.81(1)	1.762(16)	1.71	1.718(18)
B(2)–B(3)		4	1.721	1.775(11)	1.69(2)	1.721(12)	1.71	1.712(13)
M–B(1)		4	2.729	2.818(4)	2.765(5)	2.732	2.690	2.72[2)]
M–B(2)		2	3.069	3.155(9)	3.074(7)	3.074	3.049	3.07
M–B(2)′		4	2.856	2.969(4)	2.900(3)	2.870	2.811	2.84
M–B(3)		4	2.823	2.915(6)	2.845(8)	2.833	2.795	2.81
M–B(3)′		4	2.743	2.849(6)	2.764(9)	2.754	2.711	2.72
			[1]	[3]	[4]	[5]	[6]	[5]

[1)] From [10], error limit ±0.01 Å. – [2)] From [7].

Atomic positions marked by ′ are generated by symmetry; rel. mult. = relative multiplicity; { } not listed in the papers [5] and [6], respectively, but for consistency calculated at the Gmelin Institute from the data given.

The MB_4 structure type is shown (projection along the c axis) in **Fig. 2** from Spear [14, p. 109], a three-dimensional view is given in **Fig. 3** from Elf et al. [9]. It was described to be a hybrid of the hexagonal AlB_2 type and the cubic CaB_6 type, Zalkin, Templeton [11, 12]. Octahedra of six boron atoms are linked by boron-boron bonds to linear chains along the c axis equivalent to the chains along the cubic axes of the CaB_6 type. Perpendicular to the c axis, in the (001) plane these chains are linked by bonds to pairs of boron atoms, the structural element from the AlB_2 type, Etourneau et al. [13]. With the metal atoms M in the voids of this three-dimensional framework of boron atoms, the coordination by M for both structural elements is nearly identical to that of the parent structures [14, pp. 109/10].

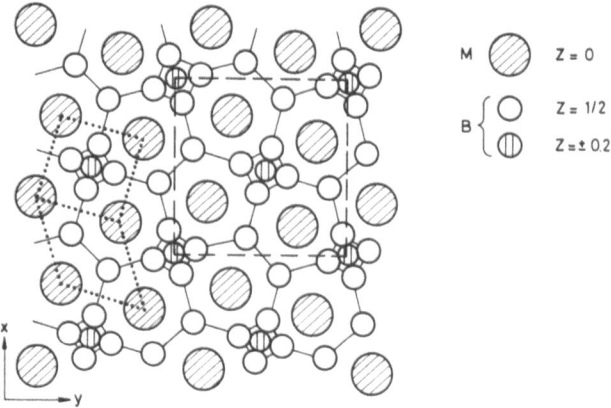

Fig. 2.   Projection of the MB_4 structure along the c axis. The unit cell is indicated by the dashed lines. The formation of the structure from MB_2 and MB_6 units is illustrated by the dotted lines; the heights are shifted by $z = \frac{1}{2}$ relative to the setting used by [14].

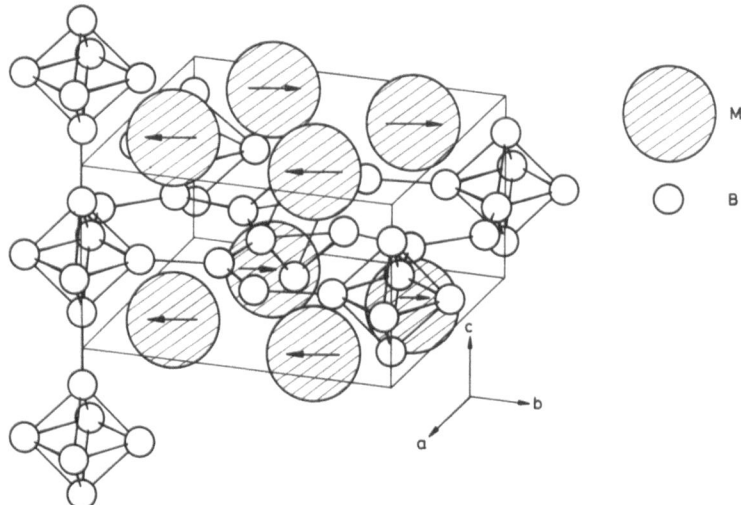

Fig. 3.   Three-dimensional view of the MB_4 structure; the arrows on M indicate the relative orientation of the magnetic moments, which are perpendicular to c, in antiferromagnetic TbB_4.

The coordination polyhedron around M is made of 18 B atoms: two planar seven-membered rings at ±c/2 off M and four B on the fourfold axes, from the boron octahedra; this coordination was also referred to as a 16 + 2 coordination [13, p. 123] because 16 of the M–B distances fall in a rather narrow range ($\Delta < 0.15$ Å), two distances, M–B(2) on the mirror plane parallel to (110), are considerably longer, see the table on p. 8.

The mean of the B–B bond lengths within the boron octahedra is larger than the other B–B bond lengths in YB₄, Guette et al. [1]; the same holds also for TbB₄, ErB₄, Will et al. [5], DyB₄, Will, Schäfer [6], and at the given precision of data there are hardly significant differences between corresponding B–B bond lengths within these four tetraborides.

In LaB₄ the B–B bond lengths are generally larger and the difference between the distances within the octahedra and outside is less obvious. Compared to LaB₆, the B–B distances in and between the octahedra are larger in LaB₄, whereas on the other hand, the La–B distances are shorter in LaB₄, Kato et al. [3]. The variation of the bond lengths of the intermediate SmB₄ is not uniform.

The following lattice parameters for the tetraborides of La through Lu were determined by Fisk et al. [15] on metal flux grown crystals at room temperature (accuracy ±0.001 Å):

MB₄	LaB₄	CeB₄	PrB₄	NdB₄	SmB₄	GdB₄	TbB₄
a in Å	7.324	7.208	7.241	7.220	7.178	7.145	7.119
c in Å	4.181	4.091	4.119	4.102	4.071	4.048	4.029

MB₄	DyB₄	HoB₄	ErB₄	TmB₄	YbB₄	LuB₄
a in Å	7.102	7.087	7.071	7.057	7.064	7.036
c in Å	4.017	4.008	3.997	3.987	3.989	3.974

For YB₄: a = 7.111 Å, c = 4.017 Å, both ±0.002 Å, Guette et al. [1]. The unit cell dimensions have been related to the atomic number of M, Fisk et al. [15], Gschneidner [16], Etourneau et al. [17, 18] or as shown in **Fig. 4** to the ionic radii, Spear [14, p. 120]. Except for CeB₄ and YbB₄, the linear correlation is very good. The deviations were discussed in terms of a cation valence differing from + 3. Parallel to the change of size from La to Lu there is also a slight, but monotonic decrease of the ratio c/a, except for Y, Ce, and Yb, Spear [14, p. 110].

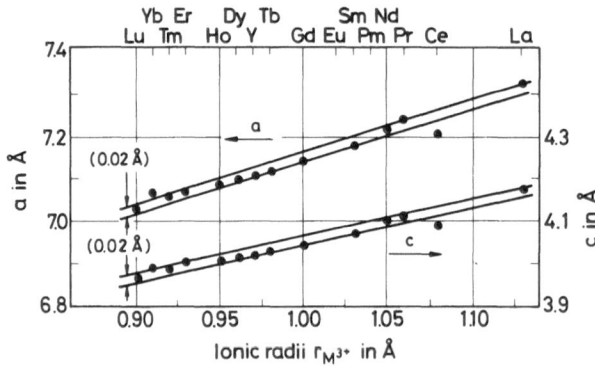

Fig. 4. Plot of the MB₄ lattice constants versus M³⁺ ionic radii.

For the several rare earth boride structures, a discussion of the relative change in lattice dimensions per change in M³⁺ ionic radii indicates that the MB₄ structure is much less rigid along [001] than the MB₆ structure along the cubic axes. In the (001) plane the rigidity of the MB₄ structure is the average of that of the two parent structures MB₂ and MB₆ [14, pp. 120/5];

for discussions of the relative stability of the boride structures see Etourneau et al. [13, p. 123], Hoard, Hughes [19, pp. 105/6]; see also Etourneau [20].

The temperature dependence of the lattice parameters of four of the tetraborides is shown in **Fig. 5**, which indicates discontinuities only in the parameter a at the magnetic phase transitions, Berrada et al. [21]. Deviation from tetragonal symmetry in the low-temperature phases have so far been proved only for TbB$_4$ and ErB$_4$, Heiba et al. [22], see "Rare Earth Elements" C 11b.

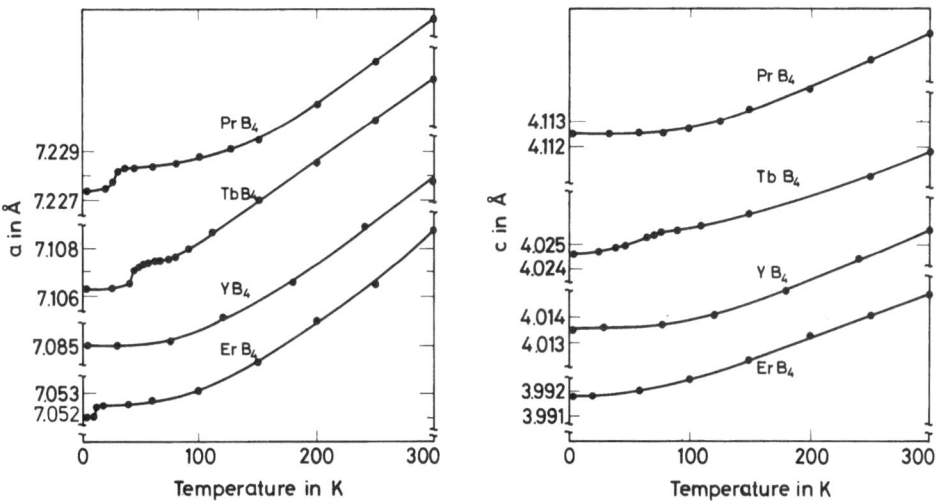

Fig. 5.   Temperature dependence of the lattice parameters a and c of some MB$_4$ compounds.

References:

[1] Guette, A.; Vlasse, M.; Etourneau, J.; Naslain, R. (Compt. Rend. C **291** [1980] 145/8).
[2] Giese, R. F.; Matkovich, V. I.; Economy, J. (Z. Krist. **122** [1965] 423/32).
[3] Kato, K.; Kawada, I.; Oshima, C.; Kawai, S. (Acta Cryst. B **30** [1974] 2933/4).
[4] Zavalii, L. V.; Bruskov, V. A.; Kuz'ma, Yu. B. (Izv. Akad. Nauk SSSR Neorgan. Materialy **24** [1988] 1576/7; Inorg. Materials [USSR] **24** [1988] 1350/1).
[5] Will, G.; Schäfer, W.; Pfeiffer, F.; Elf, F.; Etourneau, J. (J. Less-Common Metals **82** [1981] 349/55).
[6] Will, G.; Schäfer, W. (J. Less-Common Metals **67** [1979] 31/9).
[7] Will, G.; Pfeiffer, F.; Schäfer, W.; Etourneau, J.; Georges, R. (Rev. Chim. Minerale **17** [1980] 533/40).
[8] McCarthy, C. M. (Diss. Univ. Missouri, Columbia 1981, pp. 1/120; Diss. Abstr. Intern. B **42** [1982] No. 3743).
[9] Elf, F.; Schäfer, W.; Will, G.; Etourneau, J. (Solid State Commun. **40** [1981] 579/81).
[10] Schäfer, W.; Will, G (Z. Krist. **144** [1977] 217/25).

[11] Zalkin, A.; Templeton, D. H. (J. Chem. Phys. **18** [1950] 391).
[12] Zalkin, A.; Templeton, D. H. (Acta Cryst. **6** [1953] 269/72).
[13] Etourneau, J.; Mercurio, J.-P.; Hagenmuller, P. (in: Matkovich, V. I., Boron and Refractory Borides, Springer, Berlin 1977, pp. 115/38).

[14] Spear, K. E. (in: Alper, A. M., Refractory Materials, Phase Diagrams, Vol. 4, Academic, New York 1976, pp. 91/159).

[15] Fisk, Z.; Cooper, A. S.; Schmidt, P. H.; Castellano, R. N. (Mater. Res. Bull. **7** [1972] 285/8).

[16] Gschneidner, K. A., Jr. (Rare Earth Alloys, Van Nostrand, Princeton, N.J., 1961, pp. 116/30, 116/8).

[17] Etourneau, J.; Mercurio, J.-P.; Naslain, R.; Hagenmuller, P. (Compt. Rend. C **274** [1972] 1688/91).

[18] Etourneau, J.; Mercurio, J.-P.; Naslain, R.; Hagenmuller, P. (Bor. Poluchenie Strukt. Svoistva Mater. 4th Mezhdunar. Simp. Boru, Tbilisi 1972 [1974], pp. 228/32).

[19] Hoard, J. L.; Hughes, R. E. (in: Muetterties, E. L., The Chemistry of Boron and its Compounds, Wiley, New York 1967, pp. 25/154).

[20] Etourneau, J. (J. Less-Common Metals **110** [1985] 267/81).

[21] Berrada, A.; Mercurio, J.-P.; Chevalier, B.; Etourneau, J.; Hagenmuller, P.; et al. (Mater. Res. Bull. **11** [1976] 1519/26).

[22] Heiba, Z.; Schäfer, W.; Jansen, E.; Will, G. (J. Phys. Chem. Solids **47** [1986] 651/8).

32.1.4.3 Valency

With few exceptions, the rare earth tetraborides have the cations in the oxidation state $+3$, Buschow, Creyghton [1], Etourneau [2]. The exceptions are CeB$_4$ with an oxidation state near Ce^{4+} deduced from the abnormal magnetic susceptibility and the relatively small unit-cell volume [1], and YbB$_4$ where an abnormal, intermediate valence state between 2 and 3 was inferred [1 to 3], see "Rare Earth Elements" C 11b. The valence state in SmB$_4$ may be poorly defined, Buschow [3]. Contrary to other conclusions, but in agreement with experimental magnetic moments, a tetravalent state was assumed by Goryachev et al. [4] for Tb in TbB$_4$ in calculations on the electronic structure of MB$_4$.

References:

[1] Buschow, K. H. J.; Creyghton, J. H. N. (J. Chem. Phys. **57** [1972] 3910/4).

[2] Etourneau, J. (J. Less-Common Metals **110** [1985] 267/81).

[3] Buschow, K. H. J. (in: Matkovich, V. I., Boron and Refractory Borides, Springer, Berlin 1977, pp. 494/515, 501/5).

[4] Goryachev, Yu. M.; Kovenskaya, B. A.; Dudnik, E. M.; Severyanina, E. N.; Arabei, B. G. (Zh. Strukt. Khim. **16** [1975] 1036/40; J. Struct. Chem. [USSR] **16** [1975] 951/4).

32.1.4.4 Mechanical and Thermal Properties

Density

The following X-ray densities, in g/cm^3, were calculated by Kost et al. [1] from data of [2 to 7]; because of some inconsistencies, recalculated (at the Gmelin Institute) values based on lattice parameters by [8] are also given.

M in MB$_4$	Y	La	Ce	Pr	Nd	Sm	Gd
D$_{calc}$ [8]	—	5.39	5.73	5.66	5.82	6.13	6.44
D$_{calc}$ [1]	4.36	5.44	5.74	5.74	5.83	6.14	6.52

M in MB$_4$	Tb	Dy	Ho	Er	Tm	Yb	Lu
D$_{calc}$ [8]	6.576	6.745	6.87	6.996	7.10	7.22	7.37
D$_{calc}$ [1]	6.579	6.74	6.88	7.26	7.09	7.31	7.52

Thermal Expansion, Debye Temperature

The thermal expansion is anisotropic, for example $\alpha_{\perp c} > \alpha_{\parallel c}$ for YB$_4$, Zhuravlev et al. [9], see p. 135. This is also evident from the temperature dependence of the lattice parameters, see p. 11.

The quasi-isotropic, linear thermal expansion coefficients α for the tetraborides of Y, Gd, Ho, and Er were measured on pressed and sintered samples in the range between 293 and 1273 K. The error in the determination of mean values and values for a certain temperature did not exceed 1.5 to 2%. The plots of relative strain vs. temperature were almost straight lines, Severyanina et al. [10]. The values of α for the tetraborides MB$_4$ with M = Nd, Gd, Tb, and Ho were reported by Goryachev et al. [11]. All these values are also given by Paderno et al. [12], completed by those for PrB$_4$, DyB$_4$, and TmB$_4$. The Debye temperatures Θ_D were calculated by Severyanina et al. [13] from the thermal expansion coefficients (for M = Y, Gd, Ho, Er see also [10]):

M in MB$_4$	Y	Pr	Nd	Gd	Tb	Dy	Ho	Er	Tm
α in $10^{-6} \cdot K^{-1}$	7.68	5.0	5.84	7.0	6.55	5.93	7.85	7.6	6.5
Θ_D in K [13]	670	784 [14]	733 $\pm$ 22 [11]	632	661	698	517	543	657

Melting

In the following, experimental melting temperatures T$_f$ are given for the congruent melting (C) or peritectic decomposition (P) of the tetraborides of Y, Lundin [15], Tanaka et al. [16], La, Johnson, Daane [17], and Ce, Stecher et al. [18] and values estimated by Spear [19]; the estimates of T$_f$ are higher or lower, respectively, than the melting temperatures of the hexaborides:

M in MB$_4$	Y	Y	La	Ce	Pr	Nd	Pm	Sm
T$_f$ in K	3073	2883 $\pm$ 10	2073	2653	2623	2623	2623	2673
mode	C [15]	C [16]	P	P	P	P	P	P

M in MB$_4$	Gd	Tb	Dy	Ho	Er	Tm	Yb	Lu
T$_f$ in K	2923	2873	2773	2773	2773	2823	2823	2823
mode	C	C	C	C	C	C	C	C

The congruent melting of the tetraborides of Gd, Dy, and Er was observed by Manelis et al. [5], Spear [20], and Imperato, Spear [21].

The peritectically melting tetraborides of La to Sm decompose according to $3 MB_4 \rightarrow 2 MB_6 + M(\uparrow)$ with $\Delta G_0^T = - RT \ln p_M$. The dissociation (or incipient decomposition) temperatures are 2123, 2473, 2223, 2123, and 1923 K for the tetraborides of La, Ce, Pr, Nd, and Sm, respectively. The tetraborides of Y and Gd to Er sublime without decomposition, Etourneau et al. [22].

The trend in the thermal stability of the tetraborides of La to Sm was explained by the counteracting effects of increasing metal volatility and decreasing unit cell volume. The first effect accounts for a reduced stability of LaB$_4$, the second for a maximum at CeB$_4$. The high thermal stability of the tetraborides of Y and Gd through Er is the result of the combination of both a relatively low metal volatility and a small unit cell volume, Etourneau et al. [23].

Formation Enthalpy, Standard Entropy, Vaporization Energy

Experimental results on the Gibbs free energy of formation at 1700 K for LaB_4, NdB_4, and GdB_4 are treated in the individual sections.

Estimates for the standard entropy S°_{298} were derived by the method of comparative calculation. The value for YB_4 was interpolated from the YB_2 and YB_6 data and from this, the data for the other rare earth tetraborides MB_4 were derived. Allowance was made for magnetic contributions; the uncertainty is $\pm 6.3\ J \cdot mol^{-1} \cdot K^{-1}$:

M in MB_4	Y	La	Ce	Pr	Nd	Sm	Gd
S°_{298} in $J \cdot mol^{-1} \cdot K^{-1}$	58.6	75.3	89.9	92.0	96.2	89.9	96.2

Borovikova, Fesenko [24].

Free energies of vaporization ΔG°_{2200} of MB_4 for congruent vaporization and for metal or boron evaporation were estimated from compatibility studies of the hexaborides and the tetraborides with the metals W, Zr, and Ta. The actual behavior could only be reproduced by restricting the heat of formation to between -55 and -40 kcal/mol ($\triangleq -230$ to -167 kJ/mol) for the MB_4 and to between -65 to -45 kcal/mol ($\triangleq -272$ to -188 kJ/mol) for the MB_6. This showed CeB_4 and LuB_4 to be the most stable and SmB_4 the least stable; for further details see the paper, Smith [25].

Thermal Conductivity

The thermal conductivities λ of the tetraborides MB_4 with M = Y, Gd, Tb, Dy, Ho, Er, and Tm were determined at room temperature by a steady-state method on pressed and sintered samples with a final porosity of between 22 and 30%. The following values for λ and the electronic fraction λ_{el} were obtained after conversion to zero porosity:

M in MB_4	Y	Gd	Tb	Dy	Ho	Tm
λ in $W \cdot m^{-1} \cdot K^{-1}$	29.16	148.65	126.44	118.25	27.69	158.22
λ_{el} in $W \cdot m^{-1} \cdot K^{-1}$	21.1	23.63	22.97	20.88	23.48	21.18

Both the electronic and the phononic components of the thermal conductivity are larger for the tetraborides than for the metals, Severyanina et al. [13].

References:

[1] Kost, M. E.; Shilov, A. L.; Mikheeva, V. I.; Uspenskaya, S. U.; et al. (Soedineniya Redkoze-mel'nikh Elementov: Gidridy, Boridy, Karbidy, Fosfidy, Pniktidy, Khal'kogenidy, Psevdoga-logenidy, Nauka, Moscow 1983, pp. 40/68).

[2] Eick, H. A.; Gilles, P. W. (J. Am. Chem. Soc. **81** [1959] 5030/2).

[3] Post, B.; Moskowitz, D.; Glaser, F. W. (J. Am. Chem. Soc. **78** [1956] 1800/2).

[4] Zalkin, A.; Templeton, D. H. (Acta Cryst. **6** [1953] 269/72).

[5] Manelis, R. M.; Meerson, G. A.; Zhuravlev, N. N.; Telyukova, T. M.; Stepanova, A. A.; Gramm, N. V. (Poroshkovaya Metal. **1966** No. 6, pp. 77/84; Soviet Powder Met. Metal Ceram. **1966** 904/9).

[6] Paderno, Yu B.; Samsonov, G. V. (Zh. Strukt. Khim. **2** [1961] 213/4; J. Struct. Chem. [USSR] **2** [1961] 202/3).

[7] Rudman, R.; LaPlaca, S.; Post, B. (Acta Cryst. **16** [1963] A29).

[8] Fisk, Z.; Cooper, A. S.; Schmidt, P. H.; Castellano, R. N. (Mater. Res. Bull. **7** [1972] 285/8).

[9] Zhuravlev, N. N.; Belousova, I. A.; Manelis, R. M.; Belousova, N. A. (Kristallografiya **15** [1970] 836/8; Soviet Phys.-Cryst. **15** [1970] 723/4).

[10] Severyanina, E. N.; Dudnik, E. M.; Paderno, Yu. B. (Poroshkovaya Metal. **1973** No. 3, pp. 72/4; Soviet Powder Met. Metal Ceram. **12** [1973] 1001/2).

[11] Goryachev, Yu. M.; Kovenskaya, B. A.; Dudnik, E. M.; Severyanina, E. N.; Arabei, B. G. (Zh. Strukt. Khim. **16** [1975] 1036/40; J. Struct. Chem. [USSR] **16** [1975] 951/4).

[12] Paderno, Yu. B.; Severyanina, E. N.; Dudnik, E. M.; Val'ovka, I. P.; Klochkov, L. A. (Tezisy Dokl. 2nd Vses. Konf. Kristallokhim. Intermetal. Soedin., Lvov 1974, pp. 148/9; C.A. **86** [1977] No. 8924).

[13] Severyanina, E. N.; Dudnik, E. M.; Paderno, Yu. B. (Poroshkovaya Metal. **1974** No. 10, pp. 83/5; Soviet Powder Met. Metal Ceram. **13** [1974] 843/5).

[14] Severyanina, E. N. (Autoref. Kand. Diss., Kiev 1974 from Samsonov, G. V.; Vinitskii, I. M., Tugoplavkie Soedineniya, Metallurgiya, Moskva 1976, pp. 202/25).

[15] Lundin, C. E. (in: Spedding, F. H.; Daane, A. H., The Rare Earths, Wiley, New York 1961, pp. 224/385, 247/8).

[16] Tanaka, T.; Otani, S.; Ishizawa, Y. (J. Less-Common Metals **102** [1984] 281/7).

[17] Johnson, R. W.; Daane, A. H. (J. Phys. Chem. **65** [1961] 909/15).

[18] Stecher, P.; Benesovsky, F.; Novotny, H. (Planseeber. Pulvermet. **13** [1965] 37/46).

[19] Spear, K. E. (in: Alper, A. M., Refractory Materials, Phase Diagrams, Vol. 4, Academic, New York 1976, pp. 91/159).

[20] Spear, K. E. (Bor. Poluchenie Strukt. Svoistva Mater. 4th Mezhdunar. Simp. Boru, Tbilisi 1972 [1974], pp. 207/15).

[21] Imperato, E. G.; Spear, K. E. (74th and 76th Ann. Meeting Am. Ceram. Soc. [1972] and [1974] from [19]).

[22] Etourneau, J.; Mercurio, J.-P.; Naslain, R.; Hagenmuller, P. (Compt. Rend. C **274** [1972] 1688/91).

[23] Etourneau, J.; Mercurio, J.-P.; Hagenmuller, P. (in: Matkovich, V. I., Boron and Refractory Borides, Springer, Berlin – New York 1977, pp. 115/38).

[24] Borovikova, M. S.; Fesenko, V. V. (J. Less-Common Metals **117** [1986] 287/91).

[25] Smith, P. K. (Diss. Univ. Kansas 1964, pp. 1/482; Diss. Abstr. **25** [1964/65] No. 5591).

32.1.4.5 Magnetic Properties

The tetraborides of Ce to Yb are paramagnetic and, except for Ce and perhaps Yb, order magnetically at low temperature. The magnetic susceptibility of CeB_4 is temperature independent in a large temperature range (~50 to 300 K), indicating a possible tetravalency of Ce. The tetraborides with M = Y, La, Lu are diamagnetic; this indicates complete charge transfer from M to B, leaving the respective 4d or 5d electron states empty, Buschow [1]. A pronounced anisotropy of the magnetic properties of PrB_4 and TmB_4 was found by Fisk et al. [2], see Fig. 6a on p. 16, and also for TbB_4 by Gianduzzo et al. [3] (favoring the a or b axis as easy direction), and ErB_4 and HoB_4 (favoring the c axis). The relative orientation of the magnetic moments in antiferromagnetic TbB_4 is shown in Fig. 3 on p. 9.

Magnetic Ordering

At low temperature, PrB_4 orders ferromagnetically (at T_C) whereas the tetraborides with M = Nd to Tm order antiferromagnetically (at T_N); ordering temperatures [1]:

M in MB_4	Pr	Nd	Sm	Gd	Tb	Dy	Ho	Er	Tm
T_C, T_N in K	25	7	57	42	43	21	7	13	11.8

Except for SmB$_4$, where also different values near 30 K were reported by Berrada et al. [4], Bucher et al. [5], the listed values are consistent with those of other studies. Additional magnetic transitions at lower temperature were reported for DyB$_4$, HoB$_4$, and TmB$_4$, see the respective sections in "Rare Earth Elements" C 11 b. ErB$_4$ is of special interest because at low temperature it is antiferromagnetic in zero and weak magnetic fields (||c), but shows two successive magnetic transitions with increasing field to a pure ferromagnetic structure, Etourneau et al. [6]. The situation for YbB$_4$ is controversial because several studies did not confirm the magnetic transition at T=10 K given by Buschow, Creyghton [7], but rather stated the absence of any magnetic ordering [4, 5], down to, for example, 0.4 K, Sales, Wohlleben [8]. The discrepancies were tentatively attributed to a poorly defined valence state in SmB$_4$ and YbB$_4$, which might be susceptible to deviations from ideal stoichiometry or to lattice defects [1]. The magnetic symmetries, Shubnikov groups, possible for the magnetic structures of the tetraborides were discussed by Schäfer et al. [9]. The essential magnetic coupling between the localized moments (4f shells) is thought to be indirect through the free electrons and thus, a strong correlation between magnetic and electrical properties is expected [6].

Paramagnetic Range

The temperature dependence of the paramagnetic susceptibility follows the Curie-Weiss law for M = Pr to Tm except Sm [1], see also **Fig. 6**a, b. This shows the susceptibility parallel and perpendicular to the c axis of single crystals of PrB$_4$, GdB$_4$, TbB$_4$, and TmB$_4$ [2]. The effective magnetic moments μ_{eff} and the paramagnetic Curie temperatures Θ_p, as given by Buschow, Creyghton [7], cf. [1] for powders were compared with calculated values (see p. 17) $\Theta_p^{(1)}$, $\Theta_p^{(4)}$:

M in MB$_4$	Pr	Nd	Gd	Tb	Dy	Ho	Er	Tm
μ_{eff} in μ_B	3.53	3.61	7.98	9.30	10.26	10.50	9.52	7.68
$g[J(J+1)]^{\frac{1}{2}}$	3.58	3.62	7.94	9.72	10.63	10.60	9.59	7.57
Θ_p in K	+5	−12	−66	−43	−22	−14	−10	+3[1], −27[2]
$\Theta_p^{(1)}$ in K	−3.4	−7.6	−66	−44	−29.5	−18.8	−10.7	—
$\Theta_p^{(4)}$ in K	+5	−10.4	−66	−37	−19.6	−14.4	−13.9	—

[1] From [5]. − [2] From [10].

For the abnormal variation of χ with the temperature of SmB$_4$ and YbB$_4$, see "Rare Earth Elements" C 11 b. The inconsistent experimental results on Θ_p for TmB$_4$ were tentatively attributed to the magnetic anisotropy or to differences in stoichiometry [1].

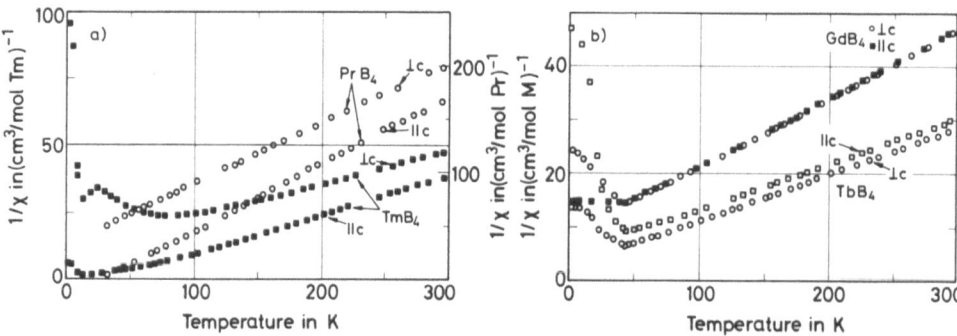

Fig. 6.   Inverse molar magnetic susceptibility 1/χ versus temperature of MB$_4$ compounds measured both parallel and perpendicular to the c axis.

The values $\Theta_p^{(1)}$ were calculated within the RKKY model for the indirect exchange and were normalized to the experimental value of GdB$_4$. Within the group of antiferromagnetic compounds, the approximate agreement supports the applicability of the indirect exchange model. The poor result for the ferromagnetic PrB$_4$, however, required a modification. By inclusion of the effects of the orbital momentum to the exchange interaction between the localized moments, a better agreement of the $\Theta_p^{(4)}$ for PrB$_4$ was obtained after adjusting four parameters to the experimental data of the other MB$_4$. A superexchange contribution was thought less likely [7], see also [1].

A deviation of observed effective moments from the free ion values to higher ones in the first subgroup, and to lower moments in the second subgroup of the rare earth tetraborides was attributed to different participation of the 4 f states in the valence band, Paderno et al. [11]. This kind of systematic deviation is not present in the data of other authors, see e.g. the table on p. 16.

References:

  [1] Buschow, K. H. J. (in: Matkovich, V. I., Boron and Refractory Borides, Springer, Berlin — New York 1977, pp. 494/515, 501/5).
  [2] Fisk, Z.; Maple, M. B.; Johnston, D. C.; Woolf, L. D. (Solid State Commun. **39** [1981] 1189/92).
  [3] Gianduzzo, J. C.; Georges, R.; Chevalier, B.; Etourneau, J.; Hagenmuller, P.; Will, G.; Schäfer, W. (J. Less-Common Metals **82** [1981] 29/35).
  [4] Berrada, A.; Mercurio, J.-P.; Chevalier, B.; Etourneau, J.; Hagenmuller, P.; et al. (Mater. Res. Bull. **11** [1976] 1519/26).
  [5] Bucher, E.; Schmidt, P. H.; Castellano, R. N.; Dernier, P. D.; Cooper, A. S. (unpublished, from [1]).
  [6] Etourneau, J.; Mercurio, J.-P.; Berrada, A.; Hagenmuller, P.; Georges, R.; Bourezg, R.; Gianduzzo, J. C. (J. Less-Common Metals **67** [1979] 531/9).
  [7] Buschow, K. H. J.; Creyghton, J. H. N. (J. Chem. Phys. **57** [1972] 3910/4).
  [8] Sales, B. C.; Wohlleben, D. K. (Phys. Rev. Letters **35** [1975] 1240/4).
  [9] Schäfer, W.; Will, G.; Buschow, K. H. J. (J. Magn. Magn. Mater. **3** [1976] 61/6).
 [10] Paderno, Yu. B.; Pokrzywnicki, S. (Phys. Status Solidi **24** [1967] K 11/K 12).

 [11] Paderno, Yu. B.; Severyanina, E. N.; Fedchenko, R. G.; Dudnik, E. M. (Elektron. Str. Fiz.-Khim. Svoistva Splavov Soedin. Osn. Perekhodnykh Met. Dokl. 8th Simp., Kiev 1974 [1976], pp. 117/20; C. A. **87** [1977] No. 160804).

32.1.4.6 Electrical Properties

Electronic Structure. Bonding

Because of the more complex crystal structure of the rare earth tetraborides compared to the hexaborides, no quantitative band structure calculations by the augmented plane wave method have so far been made, see for example the review of Etourneau [1]. In addition to the early qualitative concept by Lipscomb, Britton [2], calculations by the group orbital LCAO method (GO-LCAO) have been conducted by Goryachev et al. [3 to 8].

A tetraboride unit cell was treated to contain two octahedral B$_6$ groups and two boron pairs B$_2$ with trigonal sp^2 hybridized boron atoms. Each B$_6$ group requires 14 electrons for the seven internal bonding molecular orbitals and six more as its share in the six external bonds. A boron pair needs four electrons to bind the group, as in ethylene, and four more for the bonds to the octahedra. This makes a total of 56 electrons per unit cell. Thus in addition to the boron valence electrons (16 × 3), two electrons are required from each of the metal atoms [2] to ensure the stability of the structure (cf. Etourneau et al. [9]). This leaves, in the case of trivalent M, one valence electron per metal atom to account for the observed metallic conductivity [1]. This estimate of the conduction electron concentration, however, is in disagreement with (at least) the experimental results for YB$_4$, Johnson, Daane [10], which seems to be a semimetal with ~0.03 electrons/holes per unit cell, Tanaka, Ishizawa [11].

Using the "group-orbital-LCAO method" with inclusion of 20 atoms (i.e., 100 atomic orbitals of s and p type for B, and s, p, and d type for M) the spectrum of electron energies was evaluated. Neighbors up to the fifth and sixth order were considered, for more details on the tetraborides of Gd, Tb, and Ho, see for example Abdusalamova et al. [8]. The following parameters were obtained by Goryachev et al. [5] from these calculations: charge of M, contribution to the conduction band from the metal M (c.b.) and from boron B (c.b.), filling of d states of the metal M (d states), filling of s, p states of boron B (s, p states), and the dissociation energy D for the reaction MB$_4 \rightarrow$ M + 4B:

M in MB$_4$	Ce	Nd	Gd	Tb	Ho
valence of M	4	3	3	4	3
charge of M	+2.12	+1.14	+1.14	+1.58	+1.07
M (c.b.) in %	—	78.6	86.3	68.6	74.5
B (c.b.) in %*)	—	21.4	13.7	31.4	25.5
M (d states)	0.703	0.783	0.180	0.450	0.120
B (s, p states)	3.527	3.295	3.284	3.396	3.264
D in eV	378	362.8	359.6	370.7	292.0

*) Estimated in conjunction with the experimental data on the Hall coefficients.

Goryachev et al. [5]; for group-orbital-LCAO calculation see also [3, 4] and Samsonov, Kovenskaya [12, p. 13].

The calculations confirmed that the basic role in the tetraboride structure is played by strong covalent B–B bonds. It also follows that the conduction band is formed mainly by metal states with only slight participation of boron states [5].

A qualitative model of the band structure was discussed to explain the strong sensitivity of the electrical properties to magnetic fields as observed in PrB$_4$, TbB$_4$, and ErB$_4$. The model was based on the different interaction of the metal d orbitals. The crystal field strongly stabilizes the $d_{c^2-u^2}$ orbital (the u axis was along [110]), Etourneau et al. [13].

The B K$_\alpha$ emission bands of MB$_4$, studied for M = Y, La to Nd, Gd, Tb, and Ho, were similar to each other and to that in amorphous boron. This indicated mainly covalent bonding. From the shift to lower energies, a negative charge on the B atom was inferred. The width of the K$_\alpha$ bands indicated a width of at least 15 eV for the valence bands in MB$_4$, Bondarenko et al. [14], see also [15].

Electrical Conductivity

All rare earth tetraborides are metallic conductors [1] with room temperature resistivities in the range 30 to 50 μΩ·cm, Paderno et al. [16]. The resistivities are lower by about 20 to 60% than those of the respective metals [16, 17]. This was attributed to an increase in the mobility which was said to over-compensate the reduced carrier concentration by Severyanina et al. [17]. The following values for the room temperature resistivity and the mean temperature coefficient α in the range 300 to 1025 K were evaluated from resistivities of sintered samples after allowing for the porosity, which ranged from 22 to 30% [16, 17]:

M in MB_4	Y	Pr	Nd	Gd	Tb	Dy	Ho	Er	Tm
ϱ in $\mu\Omega\cdot cm$	34.85	40.4	39.2	31.1	32.0	35.2	30.0	49.5	34.7
α in $10^{-3}\cdot K^{-1}$	5.4	—	—	2.9	2.9	3.4	4.03	2.9	2.6

The low-temperature resistivities of single crystals with M = Gd, Tb, Dy, Ho, Er, and Tm, Fisk et al. [18] are shown in **Fig. 7**; the singularities are related to magnetic phase transitions; for a discussion see the respective sections in "Rare Earth Elements" C 11b.

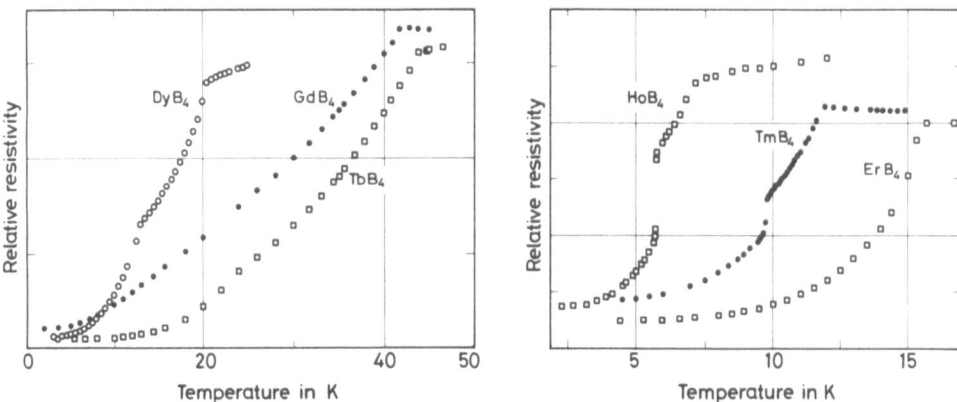

Fig. 7. Temperature dependence of the relative electrical resistivity at low temperatures of rare earth tetraborides MB_4 parallel to the c axis, except for GdB_4 for which the data are perpendicular to c.

Hall Effect. Carrier Concentration and Mobility

The simple single-band theory of the Hall effect has been shown to be inadequate for YB_4, Johnson, Daane [10], see p. 136.

The single-band approximation, in contrast, was thought appropriate and was used in the interpretation of the experimental Hall constants R_H at 293 K, corrected for sample porosity for the rare earth tetraborides with M = Pr, Nd, Gd to Tm. From these, together with the resistivity data ($\varrho \approx 30$ to $50\ \mu\Omega\cdot cm$), the following values for the concentrations n of the effective charge carriers and their mobilities μ were derived:

M in MB_4	Y	Pr	Nd	Gd	Tb	Dy	Ho	Er	Tm
$-R_H$ in $10^{-10}\ m^3/C$	7.02	3.91	5.80	8.41	6.60	4.37	5.19	5.71	4.06
n in $10^{22}\ cm^{-3}$	0.9	1.6	1.1	0.8	1.0	1.5	1.2	1.1	1.5
μ in $cm^2\cdot V^{-1}\cdot s^{-1}$	20.0	9.7	14.8	27.0	20.0	12.4	17.3	11.5	12.1

The results correspond to a carrier concentration of 0.4 to 0.8 electrons per M atom and an effective mass of 2 to 3 m_e. Besides this, the data confirm, that phonon scattering is predominant in MB_4 compounds at sufficiently high temperature, Paderno et al. [16]. At low temperature, the main contribution to the conduction electron scattering is the spin disorder, Etourneau et al. [19].

Thermionic Emission

The work function was evaluated for YB$_4$ and GdB$_4$ from the thermionic properties: $\Phi_{eff} \approx 3.45$ eV for YB$_4$ at 1890 K, Meerson et al. [20], and $\Phi_{eff} = 3.27$ eV for GdB$_{4.03}$ at 1700 K, Storms, Mueller [21]. Both materials are inferior by far to LaB$_6$ as cathode material, see p. 73. During operation at high temperature, the surface composition and the work function of any rare earth boride cathode shift to that of the congruently vaporizing phase, Jaskie, Jacobson [22]. For Y, Gd, Tb, and Er these are the tetraborides, namely YB$_{4.02}$ (estimated), GdB$_{4.03}$ (measured), TbB$_{4.05}$ (estimated), ErB$_{4.1}$ (estimated), for La, Ce, Pr, Nd these are the hexaborides [21].

References:

[1] Etourneau, J. (J. Less-Common Metals **110** [1985] 267/81).

[2] Lipscomb, W. N.; Britton, D. (J. Chem. Phys. **33** [1960] 275/80).

[3] Goryachev, Yu. M.; Dudnik, E. M.; Kovenskaya, B. A.; Severyanina, E. N.; Arabei, B. G. (Elektron. Str. Fiz.-Khim. Svoistva Splavov Soedin. Osn. Perekhodnykh Met. Dokl. 8th Simp., Kiev 1974 [1976], pp. 114/7; C.A. **87** [1977] No. 73650).

[4] Goryachev, Yu. M.; Kovenskaya, B. A.; Arabei, B. G. (Izv. Vysshikh. Uchebn. Zavedenii Fiz. **17** No. 10 [1974] 157; Soviet Phys. J. **17** [1974] 1476).

[5] Goryachev, Yu. M.; Kovenskaya, B. A.; Dudnik, E. M.; Severyanina, E. N.; Arabei, B. G. (Zh. Strukt. Khim. **16** [1975] 1036/40; J. Struct. Chem. [USSR] **16** [1975] 951/4).

[6] Goryachev, Yu. M.; Kovenskaya, B. A. (J. Less-Common Metals **67** [1979] 273/9).

[7] Arabei, B. G.; Goryachev, Yu. M.; Kovenskaya, B. A. (Tezisy Dokl. 5th Vses. Konf. Khim. Svyazi Poluprovodn. Polumetallakh, Minsk 1974, p. 33; C.A. **83** [1975] No. 183718).

[8] Abdusalamova, M. N.; Arabei, B. G.; Goryachev, Yu. M.; Kovenskaya, B. A. (Khim. Svyaz Krist. Ikh Fiz. Svoistva **1976** No. 2, pp. 33/43; C.A. **86** [1977] No. 10343).

[9] Etourneau, J.; Mercurio, J.-P.; Hagenmuller, P. (in: Matkovich, V. I., Boron and Refractory Borides, Springer, Berlin – New York 1977, pp. 115/38).

[10] Johnson, R. W.; Daane, A. H. (J. Chem. Phys. **38** [1963] 425/32).

[11] Tanaka, T.; Ishizawa, Y. (J. Phys. C **18** [1985] 4933/40).

[12] Samsonov, G. V.; Kovenskaya, B. A. (in: Matkovich, V. I., Boron and Refractory Borides, Springer, Berlin – New York 1977, pp. 5/18).

[13] Etourneau, J.; Chevalier, B.; Georges, R.; Gianduzzo, J. C.; Will, G.; Schäfer, W. (Rare Earths Mod. Sci. Technol. **3** [1982] 423/6).

[14] Bondarenko, T. N.; Zhurakovskii, E. A.; Paderno, Yu. B.; Severyanina, E. N.; Dudnik, E. M. (Ukr. Fiz. Zh. **21** [1976] 146/8).

[15] Zhurakovskii, E. A.; Bondarenko, T. N.; Paderno, Yu. B.; Dudnik, E. M.; Severyanina, E. N. (Teor. Eksperim. Khim. **12** [1976] 274/6; Theor. Exptl. Chem. [USSR] **12** [1976] 219/21).

[16] Paderno, Yu. B.; Severyanina, E. N.; Dudnik, E. M.; Lazorenko, V. (Splavy Redk. Metall. Osobymi Fiz. Khim. Svoistvami Mater 2nd Vses. Soveshch., Moscow 1974 [1975], pp. 118/21; C.A. **85** [1976] No. 12590).

[17] Severyanina, E. N.; Dudnik, E. M.; Paderno, Yu. B. (Poroshkovaya Metal. **1974** No. 10, pp. 83/5; Soviet Powder Met. Metal Ceram. **13** [1974] 843/5).

[18] Fisk, Z.; Maple, M. B.; Johnston, D. C.; Woolf, L. D. (Solid State Commun. **39** [1981] 1189/92).

[19] Etourneau, J.; Mercurio, J.-P.; Berrada, A.; Hagenmuller, P.; Georges, R.; Bourezg, R.; Gianduzzo, J. C. (J. Less-Common Metals **67** [1979] 531/9).

[20] Meerson, G. A.; Zhuravlev, N. N.; Manelis, R. M.; Runov, A. D.; Stepanova, A. A.; Grishina, L. P.; Gramm, N. V. (Izv. Akad. Nauk SSSR Neorgan. Materialy **2** [1966] 608/16; Inorg. Materials [USSR] **2** [1966] 527/33).

[21] Storms, E. K.; Mueller, B. A. (J. Appl. Phys. **52** [1981] 2966/70).

[22] Jaskie, J.; Jacobson, D. (Proc. Intersoc. Energy Convers. Eng. Conf. **15** No. 3 [1980] 2331/3; C.A. **94** [1981] No. 142615).

32.1.4.7 Chemical and Electrochemical Behavior

The oxidation behavior in air of Al-flux grown single crystals of YB_4, SmB_4, GdB_4, and TbB_4 was studied by Okada, Atoda [1, 2, 3]: Oxidation of YB_4 started at 1033 K [1], that of GdB_4 and TbB_4 at 1003 to 1023 K [3] giving MBO_3 and noncrystalline B_2O_3. In the case of SmB_4, oxidation starts at 1013 to 1023 K. After intermediate formation of SmB_6 and $SmBO_3$, the final products were monoclinic $Sm(BO_2)_3$ and noncrystalline B_2O_3 [2].

From the temperature dependence of the weight gain during oxidation, apparent activation energies in kcal/mol (kJ/mol in parentheses) of 95.6 (400) for YB_4 [1], 144.4 $\pm$ 4.2 (604) for GdB_4, and 91.8 $\pm$ 6.1 (384) for TbB_4 were evaluated [3].

Rare earth tetraborides were shown to give clearly defined maxima of the anodic oxidation currents when present in paste electrodes made from carbon powder with a Si-organic binder. The anodic-oxidation curves of the electrodes were traced in aqueous 1 M $NaClO_4$ at pH 1.0 by a standard polarographic circuit at a sweep rate of 0.5 mV/s. A saturated silver chloride electrode was used as reference in the three-electrode electrochemical cell. The respective potentials of all the studied rare earth borides are within the range 0.08 to 0.82 V. The tetraborides oxidize at much lower positive potentials than the hexa- and higher borides. The oxidation current is proportional to the respective boride concentration over a wide range, allowing quantitative phase analysis of mixtures. The following potentials φ_{max}, in V, of the current maxima were observed for tetraborides MB_4:

MB_4	LaB_4	GdB_4	DyB_4	HoB_4	ErB_4	LuB_4
φ_{max}	0.08	0.12	0.13	0.11	0.11	0.10

The basic process during anodic polarization is the ionization of the metal, Tkach, Paderno [4], data on only Gd borides in [5].

References:

[1] Okada, S.; Atoda, T. (Yogyo Kyokaishi **89** [1981] 339/45; C.A. **95** [1981] No. 89085).

[2] Okada, S.; Atoda, T. (Yogyo Kyokaishi **91** [1983] 136/47; C.A. **98** [1983] No. 152917).

[3] Okada, S.; Atoda, T. (Yogyo Kyokaishi **93** [1985] 301/10; C.A. **103** [1985] No. 46050).

[4] Tkach, A. V.; Paderno, Yu. B. (Dopov. Akad. Nauk Ukr. RSR Ser. B Geol. Khim. Biol. **1984** No. 9, pp. 52/4; C.A. **102** [1985] No. 35077).

[5] Tkach, A. V.; Paderno, Yu. B.; Masyuk, T. V. (Poroshkovaya Metal. **1984** No. 12, pp. 56/8; Soviet Powder Met. Metal Ceram. **23** [1984] 949/50).

32.1.5 Comparative Data for MB_6

32.1.5.1 Existence

Hexaborides MB_6 exist for M = Y, La to Dy, and Yb. The small M = Sc, Ho, Er, Tm, and Lu do not form hexaborides, see the review by Etourneau [1]. Originally it was assumed that all elements of the series can form this type of compound, see for example, the early reviews by Samsonov [2], Gschneidner [3], and Post [4, 5]. However, it was shown that this assumption was based on erroneous interpretations of X-ray diffraction powder data for MB$_6$ with M = Sc, Er, Tm, and Lu. Attempts to prepare these compounds resulted in MB$_4$, MB$_4$/MB$_{12}$, or MB$_2$/MB$_{12}$ mixtures; for M = Sc, see for example, Etourneau et al. [6], for M = Sc, Lu, Przybylska et al. [7], for M = Er, Tm, Lu, Sturgeon, Eick [8], for M = Er also Eick, Gilles [9, 10], M = Tm and Lu, Sturgeon, Eick [11, 12], and for M = Er and Tm, Geballe et al. [13]. For controversial data regarding the existence of HoB$_6$, see "Borothermal Reduction of Metal Oxides", p. 23, and Section 32.1.8.13 in "Rare Earth Elements" C 11 b.

References:

[1] Etourneau, J. (J. Less-Common Metals **110** [1985] 267/81, 270).
[2] Samsonov, G. V. (Usp. Khim. **28** [1959] 189/217; AERE-Trans-849 [1960] 1/37).
[3] Gschneidner, K. A., Jr. (Rare Earth Alloys, van Nostrand, Princeton, N.J., 1961, pp. 116/30).
[4] Post, B. (in: Adams, R. M., Boron, Metallo-Boron Compounds and Boranes, Interscience, New York 1964, pp. 301/71).
[5] Post, B. (NP-12652 [1962] 1/64; N.S.A. **17** [1963] No. 18761).
[6] Etourneau, J.; Mercurio, J.-P.; Naslain, R.; Hagenmuller, P. (Colloq. Intern. Centre Natl. Rech. Sci. [Paris] No. 205 [1972] 429/38; C.A. **78** [1973] No. 167937).
[7] Przybylska, M.; Reddoch, A. H.; Ritter, G. J. (J. Am. Chem. Soc. **85** [1963] 407/11).
[8] Sturgeon, G. D.; Eick, H. A. (Inorg. Chem. **2** [1963] 430/1).
[9] Eick, H. A.; Gilles, P. W. (J. Am. Chem. Soc. **81** [1959] 5030/2).
[10] Eick, H. A.; Gilles, P. W. (AECU-4087 [1958] 1/12; N.S.A. **13** [1959] No. 10826).

[11] Sturgeon, G. D.; Eick, H. A. (TID-17968 [1961] 1/10; N.S.A. **17** [1963] No. 14216).
[12] Sturgeon, G. D.; Eick, H. A. (Proc. 3rd Conf. Rare Earth Res., Clearwater, Fla., 1963 [1964], pp. 87/97).
[13] Geballe, T. H.; Matthias, B. T.; Andres, K.; Maita, J. P.; Cooper, A. S.; Corenzwit, E. (Science **160** [1968] 1443/4).

32.1.5.2 Homogeneity Range

In view of the changes in lattice parameter and density, rare earth hexaborides exhibit a range of homogeneity by the formation of metal vacancies, M$_x$B$_6$, which extends from the stoichiometric composition (x = 1.0) to x ≈ 0.7 in the case of M = Ce, Pr, Sm, Gd, Tb, Yajima, Niihara [1]. Similar large ranges were assumed for M = Nd and Dy, whereas for M = Y, La, and Yb the lower limit was given as x ≈ 0.8, and for M = Eu as x = 0.9, see the review by Binder [2]. For SmB$_6$ the existence of two distinct phases with the approximate compositions Sm$_{0.8}$B$_6$ and SmB$_6$ and nearly no homogeneity range is assumed in some papers, see Section 32.1.8.8 in "Rare Earth Elements" C 11 b. On the basis of chemical analyses, boron may also be deficient, see "Single-Crystal Growth" (p. 28) for LaB$_6$ and PrB$_6$ and the individual sections. Crystal structure analyses indicated severe local boron deficits in LaB$_6$ and CeB$_6$, see p. 33.

References:

[1] Yajima, S.; Niihara, K. (Proc. 9th Rare Earth Res. Conf., Blacksburg, Va., 1971, Vol. 2, pp. 598/609).
[2] Binder, F. (Radex Rundschau **1977** 52/71, 57; C.A. **87** [1977] No. 27597).

32.1.5.3 Preparation

Preparation of Polycrystalline Samples

Preparation from the Elements

Direct synthesis of MB_6 in the presence of a vapor phase is utilized in the case of the relatively volatile metals M = Sm, Eu, and Yb at 973 to 1473 K. It requires sealed containers that resist the corrosive action of boron and of metal gases under pressure, Etourneau et al. [1], see also Etourneau et al. [2]. The reaction of boron with the other rare earth elements, which are less volatile at high temperatures, are carried out in a vacuum or ultradry Ar (and He [1]). To facilitate solid-state diffusion, the boron and the metal (often as a finely divided hydride) are shaped into tablets under a pressure of ca. 1.5 GPa [1, 2]. Compounds MB_6 with M = Ce to Dy, and Yb were prepared in the arc furnace by heating the elements in an inert atmosphere. However, for M = Eu free boron in addition to EuB_6 was present, owing to the evaporation of some Eu. Samples of TbB_6 and DyB_6 contained small quantities of other boride phases. The HoB_6 was obtained in a similar way, however, it could not be prepared as a dominant phase, Geballe et al. [3]. Attempts to prepare MB_6 with M = Er, Tm, Lu by arc melting resulted in the formation of products containing MB_4 and MB_{12} but no MB_6, Sturgeon, Eick [4 to 6].

Preparation by Thermal Decomposition

MB_6, contaminated by amorphous boron, formed on thermal decomposition of $M_2(B_{10}H_{10})_3$ in a vacuum of ca. 10^{-3} Pa above 1273 K (M = Ce) or 1473 K (M = Gd). Pure MB_6 was prepared from the product obtained at 1073 K, 1 h, ca. 10^{-3} Pa by mixing with $MH_{\sim 2}$ in a N_2 atmosphere at a mole ratio of B/M = 6, followed by a heat treatment at 1373 K (M = Ce) or 1473 to 1673 K (M = Gd) for 0.5 h in an Ar flow of ca. 100 mL/min, and a final acid treatment in concentrated HCl for 15 min at room temperature, in order to remove inclusions such as oxides and MBO_3, Itoh et al. [11].

Borothermal Reduction of Metal Oxides

This preferred method of synthesis proceeds in an inert atmosphere at high temperatures via the overall reaction

$$M_mO_n + (6m+n)B \rightarrow m\,MB_6 + 0.5\,n\,B_2O_2$$

The nature of the volatile boron oxide is not well defined, Bliznakov, Peshev [7, 8], Etourneau et al. [1, 2]. It appears that X-ray amorphous B_2O_3 forms first, which reacts with the excess boron via $B + B_2O_3$ (liquid) $\rightarrow 3\,BO$ (gas) $\rightarrow 1.5\,B_2O_2$ (gas), the first part being an equilibrium process [7, 8]. Metal oxide-boron mixtures, pressed (1.5 GPa) into cylindrical tablets, are treated either by arc melting or high-frequency induction heating in Ta, Mo (or W [2]) crucibles lined with BN or the respective MB_6 in order to prevent any contamination. Starting with M_2O_3, M = La, Ce, Nd, and Sm and stoichiometric amounts of B (i.e., $2\,M_2O_3 + 30\,B$) in vacuum, single-phase hexaboride was formed only above 1873 K, and this from the intermediate product MB_4. Temperatures between 1773 and 1873 K lead to MB_4 or to $MB_6 + MBO_3$ at 1373 to 1773 K [1]. DTA and X-ray studies of the reduction of M_2O_3 in He between room temperature and 1973 K were interpreted to show that MB_6 (M = La, Sm) formed in a first step along with MBO_3 plus excess boron, and in a second step from MBO_3. The latter one occurred at 1448 K in the case of M = La and at 1543 K (diffuse SmB_6 diffraction pattern) and 1723 K (sharp diffraction pattern) for SmB_6, if the heating rate was 80 K/min, Dudnik et al. [12]. The MB_6 with M = Y, Gd, Tb, and Dy, free from MB_4, but perhaps with traces of free boron and borate, was obtained from stoichiometric M_2O_3–B mixtures in Ar or He at 1773 to 1973 K. In vacuum, MB_6 coexisted with the more stable MB_4 at these temperatures. In the case of M = Y and Dy in a vacuum, MB_6 contaminated with boron oxide was observed between 1573 and 1773 K, whereas below ca. 1573 K, a mixture of $MB_6 + MB_{12} + MBO_3$ resulted. Pure YbB_6 was obtained in a vacuum slightly

            References for 32.1.5.3 on p. 31

below 1873 K; above 1873 K, YbB$_{12}$, and below 1773 K, YbBO$_3$ occurred as by-product [1]. The same author group at the same time stated that MB$_6$ with M = La to Eu, and Yb were prepared with high purity above 1773 K at 10^{-6} Torr ($\triangleq$ 133.3 µPa). The HoB$_6$ and ErB$_6$ could not be prepared borothermally, Etourneau et al. [9], Hagenmuller et al. [10]. Stoichiometric quantities of CeO$_2$, Pr$_6$O$_{11}$, or Nd$_2$O$_3$ and amorphous B, all of considerable dispersity, were pressed into pellets (250 to 500 kg/cm^2 $\triangleq$ 24.5 to 49.1 MPa) and were heated in fused alumina crucibles in a Mo-vacuum furnace (10^{-4} Torr $\triangleq$ 13.3 mPa) for 1 h at 2073 K to yield almost pure stoichiometric MB$_6$. Samples prepared at lower temperatures, i.e., $\geqq$ 1373 K for CeB$_6$ and NdB$_6$, and $\geqq$ 1473 K for PrB$_6$ were assumed to contain some amorphous B$_2$O$_3$ [7, 8]. Samples of SmB$_6$ and EuB$_6$ obtained within 1 h at 2073 K, and of GdB$_6$ obtained at 2173 K, in a BN-lined vacuum furnace with graphite heater suffered from a deficit in the components and severe contamination by nitrogen. However, SmB$_6$ and GdB$_6$ prepared at 1823 to 1873 K and 1923 to 1973 K, respectively, had only a small B deficit and were contaminated by < 0.1 wt% carbon, Karasev et al. [13].

Most of the following studies performed with stoichiometric mixtures in a vacuum agree with the former results with respect to the light M = La to Sm, however, they mostly disagree for M = Y, Gd, etc.: Mixtures of preroasted (1073 K, 2 h) M$_2$O$_3$ with M = Y, La, or Gd and amorphous B, pressed into compacts under 20 to 50 MPa (2 to 5 kN/cm^2), were heated by a cylindrical electron beam from a W wire to give MB$_6$ free from carbon. Optimum conditions to obtain nearly stoichiometric compositions were 1 h at 1973 K (Y), 1873 K (La), and 2073 K (Gd), from studies in the ranges 1773 to 2073 K (Y), 1273 to 2073 K (La), and 1873 to 2073 K (Gd). The mean particle size of the powders (1.4 µm) was smaller by 50% than that of MB$_6$ obtained in a vacuum furnace with a graphite heating element, Derenovskii et al. [14]. The reduction of preroasted M$_2$O$_3$ with amorphous boron (99.5% B), studied at 1173 to 2173 K, yielded single-phase LaB$_6$ and SmB$_6$ contaminated by $\leqq$ 0.1% C at 2073 and 1973 K, respectively, in 1 h. With Gd$_2$O$_3$ at 1973 to 2073 K, GdB$_6$ plus GdB$_4$ formed, whereas on reduction of Y$_2$O$_3$, nearly stoichiometric YB$_6$ formed at 1973 K with repeated heating of the reaction product after grinding. Products of YB$_6$ obtained from crystalline boron (98.5% B) contained up to 0.2 wt% C, Bondarenko et al. [15]. Recommended temperatures for the reduction of M$_2$O$_3$ and Tb$_4$O$_7$ in a Mo sheath (vacuum furnace, tungsten or graphite heater) to yield MB$_6$ within 45 min were: ca. 1973 K for M = La, Eu, Tb; ca. 1873 K for M = Pr, Nd, Sm, Dy, Ho, Yb; 2073 to 2123 K for M = Gd; ca. 1823 K for M = Er, Lu (but see p. 21), Mordovin, Timofeeva [16]. Blue TbB$_6$ was obtained in 1 h at 1923 K, intense blue ErB$_6$ (but see p. 21) at 2073 K, from studies in an evacuated resistance furnace from 1773 to 2273 K, Samsonov et al. [17].

MB$_6$ was prepared from compacts of CeO$_2$, Pr$_6$O$_{11}$, Sm$_2$O$_3$, Gd$_2$O$_3$, or Tb$_4$O$_7$ with elemental B by heating in a vacuum of 13.3 mPa to 133.3 µPa at 2123 K (Ce), 2073 K (Pr) for 3 h, or 1923 K (Sm, Gd, Tb) for 2 h, and a final anneal for 100 h at 1073 K, Yajima, Niihara [18]. The use of pure boron was emphasized by Sobczak, Sienko [19] who started from mixtures containing 15% B in excess over the stoichiometric composition, in order to compensate for boron losses through evaporation. The reaction mixture containing 99.9999% La$_2$O$_3$ and 99.999% boron or 99.99% Y$_2$O$_3$ and 99.98% boron powder was heated under 13.3 mPa in an induction furnace in a ZrB$_2$ crucible for ca. 30 min at ca. 1273 K and then about 1 h at 1673 K. After cooling, the sample was reground and the second step was repeated, sometimes twice. Even this LaB$_6$ was contaminated by up to 0.02 at% Fe and this YB$_6$ by up to 0.3 at% Fe [19]. Stoichiometric powder mixtures with M = Y, La to Sm, Gd, Tb, Dy or Yb, pressed into pellets under 13.8 to 68.9 MPa (2000 to 10 000 lb/in^2), were mounted on a copper hearth and were repeatedly heated in an arc-melting furnace under 0.1 MPa of Ar with intermediate grinding and compacting. When the desired MB$_6$ did not form as a single phase after repeated meltings, additional boron or M$_2$O$_3$ was added and melting repeated. Finally the ground MB$_6$ was washed several times with warm 50% HCl followed by distilled H$_2$O to remove excess M$_2$O$_3$. An attempt to prepare HoB$_6$ produced an HoB$_4$–HoB$_6$–HoB$_{12}$ mixture. Boron-deficient starting mixtures corresponding to the reaction $M_2O_3 + 14\,B \rightarrow 2\,MB_6 + B_2O_3$, treated in this way, produced principally MB_6 plus MB_4 for all M, Smith [20]. According to Eick, Gilles [21, 22] products free from MB_4 (except for M = Ho, Tb) resulted, when mixtures of $M_2O_3 + 14\,B$, pressed under 3.4 MPa ($\triangleq 500$ psi) into wafers, were heated in a high vacuum in Ta, Mo, or graphite crucibles at 2073 to 2173 K for 1 to 6 h (M = Nd, Sm, Yb) or < 1773 K for 15 to 60 min (M = Gd to Ho). The product was ground in an alundum mortar and washed twice with concentrated HCl. The ErB_6 could not be prepared in this manner [21, 22]. Ready formation of MB_6' (M = La, Pr, Sm, Gd, Yb) from similar starting mixtures in a H_2 atmosphere in ZrB_2 crucibles at 1773 K (2 h) or 2073 K (1 h) was also observed earlier. On use of C crucibles, samples were contaminated by 1 to 2% C. Originally, mixtures of M_2O_3 (M = La, Sm, Gd, Yb) or Pr_4O_7, boron, and carbon black powders were used as starting materials. It was shown, however, that the boron rather than the carbon acted as the reducing agent, Post et al. [23], see also Post et al. [24]; for single-phase EuB_6 and for TmB_6 (but see p. 21) admixed with other boride phase, see Tvorogov [25].

MB_6 material used for zone melting (floating-zone melting) was prepared from stoichiometric mixtures: These (M = La to Nd) were cold pressed in steel molds and were heated in an induction oven in a boron-conditioned Ta crucible in a vacuum (< 133 mPa $\triangleq 10^{-3}$ Torr) at 1873 to 1973 K for ca. 1 h, Niemyski et al. [26]. The oxide CeO_2, Sm_2O_3, or Gd_2O_3 and amorphous B were mixed in a plastic ball mill (15 h), pressed under 29.4 MPa ($\triangleq 300$ kg/cm^2), and heated in a vacuum at ca. 1973 K for 1 h in an induction heated graphite susceptor. The reaction product was crushed in an agate mortar, ground by a ball mill of stainless steel, and the powder so obtained was boiled in dilute HCl solution (to free from stainless steel contaminations). To get rods, the material was pressed into a tablet at 29.4 MPa under a friction-free condition, then repressed hydrostatically at 98.1 MPa (1 ton/cm^2) in order to obtain uniform density. Camphor may be added as a binder. After sintering in an induction heated graphite susceptor under 0.1 MPa (1 atm) of Ar for 0.5 h at 2073 K, the rod had a density of ca. 65% of theoretical, Tanaka et al. [27]. For densification, see also p. 27.

Keeping a mixture of La_2O_3 and amorphous B (mole ratio 1:2) in flat ampules under a pressure of ca. 10 GPa ($\triangleq 100$ kbar) did not lead to even partial reaction of the reagents. However, shock waves acting on mixtures of M_2O_3 (M = La, Nd, Gd) and amorphous B in the mole ratios 1:2 or 1:15 produced a core of X-ray-pure MB_6, which was surrounded by a phase mixture also containing MB_6. The formation of the inner zone MB_6 obviously took place at temperatures close to the melting point. A repeated compression did not significantly increase the yield ($\sim 40\%$), Adadurov et al. [28].

Boron Carbide Method

This method can be used technically for the preparation of relative pure MB_6 with M = Y, La to Gd, and Yb at a yield of almost 100%, via:

$$M_2O_3 + 3\,B_4C \rightarrow 2\,MB_6 + 3\,CO \text{ (for M = Y, La, Nd to Gd, Yb)}$$
$$2\,CeO_2 + 3\,B_4C + C \rightarrow 2\,CeB_6 + 4\,CO$$
$$Pr_6O_{11} + 9\,B_4C + 2\,C \rightarrow 6\,PrB_6 + 11\,CO$$

The reaction is generally performed in a vacuum induction furnace at temperatures between 1873 and 2473 K with use of a C, Mo, Ta, BN, or ZrB_2 crucible. When using C crucibles, the sintered material exhibits a small carbidic outer zone which can be readily taken off. The hexaborides with M = Tb, Dy, and Ho are not obtained as a single phase; they show an increasing readiness for decomposition into $MB_4 + MB_{12}$, see the review by Binder [29]. The C content of MB_6 (M = Y, La, Ce) did not exceed 0.08 wt% and could be further reduced severely by supplementary heating of the MB_6 with excess metal oxide, followed by washing with a weak solution of HCl, Samsonov [46]. For example, single-phase MB_6 with M = La to Sm and Yb was prepared from appropriate mixtures of metal oxides + B_4C (plus C), pressed at

References for 32.1.5.3 on p. 31

room temperature under 98 MPa ($\triangleq$ 1000 kg/cm^2) into rods, by heating in a Ta crucible first at 1573 to 1773 K (evolution of CO) and finally for 1.5 to 2 h at 1823 to 2023 K; initial vacuum of 1.33 mPa ($\triangleq$ 10^{-5} Torr). Similar treatments with M = Gd, Tb, and Ho gave MB$_6$ plus MB$_4$, whereas those with M = Y and Er gave multiphase mixtures containing MB$_6$ (but see p. 21 for Er), Tvorogov [25]. For PrB$_6$, the best temperature for preparation within 1 h was 1923 K, the highest temperature studied. Under these conditions, no TbB$_6$ (but instead TbB$_4$) was formed, Samsonov et al. [17]. Attempts to prepare MB$_6$ with M = Gd, Dy, Ho (and Er, Lu, but see p. 21) at temperatures between 1673 and 1873 K resulted in products of MB$_6$ admixed with other boride phases, Neshpor, Samsonov [30].

Mixtures of B$_2$O$_3$ and C can be used instead of B$_4$C in accordance with the equation 2 B$_2$O$_3$ + 7 C $\rightarrow$ B$_4$C + 6 CO [29]. However, deviations from these ratios, for example, mixtures of La$_2$O$_3$ + 12 H$_3$BO$_3$ + 10.5 C or CeO$_2$ + 6 H$_3$BO$_3$ + 11 C roasted at 2173 to 2273 K led to LaB$_6$ or CeB$_6$ plus ternary M–C–B compounds. These latter can be largely removed (on boiling in HCl, washing in acetone, heating in air), leaving MB$_6$ contaminated by up to 0.5 wt% C, see Vekshina et al. [39].

Magnesothermic Method

Hexaborides MB$_6$ (M = La, Ce, Gd, Eu) of ca. 99.9% purity were obtained via M$_n$O$_m$ + 3n B$_2$O$_3$ + (m + 9n) Mg $\rightarrow$ nMB$_6$ + (m + 9n) MgO as follows: Appropriate mixtures of boric acid (H$_3$BO$_3$) and M$_2$O$_3$ (or MO$_2$ for M = Ce), dehydrated by slow heating to 1073 K, were mixed with the stoichiometric amount of Mg. The mixture was briquetted and heated at 1573 K (M = La, Ce), 1473 K (M = Gd), or 1373 K (M = Eu) in streaming Ar in a steel reactor. The product, cooled outside the furnace, was treated with concentrated HCl at the boil (removal of MgO), washed with H$_2$O, dried, and finally calcined in a vacuum furnace at 1873 to 1973 K for 1 h; yield 85 to 90% of the theoretical, Markovskii, Vekshina [31].

Flux Methods

For the preparation of microcrystalline CeB$_6$ and perhaps other MB$_6$ from a Na flux containing rare earth oxides or chlorides plus B$_2$O$_3$, or from a Na$_2$B$_4$O$_7$ flux containing rare earth oxides, see Tepper et al. [32].

Electrolysis

Very small crystals of MB$_6$ (M = Y, La, Ce, Nd, Gd, Yb, and Er but see p. 21), admixed with amorphous boron were originally obtained by electrolyzing fused mixtures of M$_2$O$_3$ (or CeO$_2$), Li$_2$B$_4$O$_7$, and LiF (for M = La, Ce, Nd) or MgB$_4$O$_7$ and MgF$_2$ (for M = Y, Ce, Gd, Er, Yb) near 1273 K, 20 to 25 A at $\leqq$ 14 V (depending on M) within a few hours, using a C crucible as anode and a C rod as cathode. Detailed conditions are given in the paper. The product was boiled in dilute HCl, afterwards in concentrated HCl. The separation from impurities such as B and C was troublesome and was done by mechanical and physical means, Andrieux [33]; see also von Stackelberg, Neumann [34]. Now, TiB$_2$ is mostly used as the electrode material, although ZrB$_2$ can also serve as an anode. The MB$_6$ prepared from the baths mentioned above may also be contaminated by other boride phases, see the review by Binder [29]. Other, less favorable electrolyte baths for MB$_6$ (M = Y, Ce, Gd) were given by Andrieux [33], for CeB$_6$ by Andrieux [35, 36]. Use of a LiCl flux yielded good LaB$_6$ crystals at low current densities, but SmB$_6$ crystals of poor quality. Experiments involving M = Y, Eu to Lu yielded only amorphous boron, Kunnmann [37]. Crystals of CeB$_6$, NdB$_6$, and SmB$_6$ free of Li were prepared from proper portions of Li$_2$B$_4$O$_7$ and MCl$_3$ in a LiCl flux (C electrodes, 923 K), Wold [40]. As shown for M = Ce, baths containing only CeO$_2$ plus B$_2$O$_3$ or CeO$_2$, B$_2$O$_3$, and CeF$_3$ are unsuitable or disadvantageous, respectively [35].

Production of MB_6 by electrolyzing a mixture of A_3AlF_6 (A = alkali) and 5 to 30% rare earth metal borate with a C rod as cathode and C crucible as anode at 1073 to 1323 K was reported recently. For example, a mixture of 320 g Li_3AlF_6 and 70 g lanthanum tetraborate was heated at 1223 K under N_2, electrolyzed at 3.0 V, 20 A or ~3.9 V, 40 A for 1 h, and the deposit was leached with hot aqueous $AlCl_3$–HCl or $Al_2(SO_4)_3$–H_2SO_4, yielding 15 and 21 g LaB_6, respectively, Uchida [38]. For electrolytes containing essentially A_3AlF_6, $Na_2B_2O_4$, and M_2O_3 (M = Y, La, Gd, Dy) or CeO_2 in a cell open to the air, see Gomes, Uchida [41].

Purification of Commercial Samples

Impurities such as free metal and carbon can be removed by digesting the samples in hot concentrated H_2SO_4, Ames, McGrath [42]. The carbon and oxygen content can be decreased significantly by adding enough M_2O_3 to MB_6 powder, sintering the mixture (CO↑), and washing the product in concentrated HCl, Westrum [43]. For other treatments, see "General Remarks", p. 72. Contamination by MB_4 can be almost eliminated by heating the light hexaborides with an admixture of ca. 10 mol% elemental B to 1773 to 1973 K for several hours. The solid residues acquired the distinct dark blue color of MB_6 and were single phase according to X-ray analyses. Dodecaboride impurities in HoB_6 disappeared on heating from 1973 to 2123 K for several hours, however, not those contained in TbB_6 and in ErB_6 (but see p. 21) [42].

Densification

The MB_6 powders are refractory, brittle, and nonelastic and cannot be transformed into dense products by sintering of prepressed briquettes. Stresses arising at the grain boundaries, and which are relieved during sintering, would cause a sharp increase in the porosity and sometimes even the disintegration of the sintered compacts, Samsonov, Neshpor [44]. For the preparation of MB_6 with M = Ce, Sm, Gd of ca. 35% porosity, see p. 25. Hot pressing in a vacuum is the most suitable method of producing dense material, see Hoyt, Chorné [45]. It is performed in graphite molds [44], Samsonov [46], Arabei et al. [47], Markov et al. [48]. In order to minimize MB_6–C reaction, the inner cavity of the mold and the surface of the dies in contact with MB_6 were wet with a paste of BN in distilled water [47, 48]. Because BN dissociates at high temperatures, a coating of pyrolytic graphite was preferred by [45]. Recommended conditions of temperature T, pressure p, and sintering time t to obtain dense material of low residual porosity (res. por.) were (1 kg/cm² $\triangleq 0.9807 \times 10^5$ Pa):

sample	T in K	p in MPa	t in min	res. por. in %	Ref.
YB_6	2423	12.3	5	1.8	[46]
	2423	19.7	15	—	[48]
	2500	15.6	60	7	[45]
LaB_6	2273	12.3	10	1.0	[46]
	2223 to 2323	34.3/49.0	5	≧3	[47]
CeB_6	2423	12.3	5	2.0	[46]
NdB_6	2473	14.7	15	3.0	[46]
SmB_6	2273	14.7	20	1.2	[46]
	2273 to 2373	34.3/49.0	5	≧3	[47]
	2373	49.0	5	—	[48]
	2500	22.6	60	13	[45]
EuB_6	2223 to 2243	34.3/49.0	5	≧4	[47]
	2500	25.2	90	10	[45]

References for 32.1.5.3 on p. 31

sample	T in K	p in MPa	t in min	res. por. in %	Ref.
GdB$_6$	2173	14.7	10	2.8	[46]
	2273	19.7	15	—	[48]
DyB$_6$	2273 to 2373	34.3/49.0	5	≧5	[47]
YbB$_6$	2273	12.3	10	1.1	[46]

In the pressure data from [47], the first value gives the initial pressure which, after the appearance of shrinkage, should be rapidly raised to the second value (in the text and the tables of this paper, pressure values are given as by one order of magnitude higher). For the effect of temperature (1973 to 2373 K), pressure (9.8 to 56.84 MPa) and holding time (1 to 25 min) on the residual porosity, see figures in the paper [47]. Similar discrepancies as to the order of pressure exist also for the data from [46], see [44]. According to Markov et al. [48], the dense material was single phase. For the observation of MB$_4$ in hot-pressed specimens, especially of YB$_6$, see the individual sections.

Treatment of MB$_6$ (M = La, Nd, Gd) material with a short shock wave, which avoids the heating-out of defects formed in the shock wave front, lowers the temperature for later hot pressing, in the case of LaB$_6$ to 673 K with the simultaneous reduction of the residual porosity from 15 to 5%, Adadurov et al. [28].

Specimens of required dimension can be obtained by electric erosion [44] or ultrasonic machining [47].

Powder compacts with a residual porosity of ≳45%, when irradiated with neutrons, exhibited pronounced "fission" sintering associated with the energy release of the ^{10}B fission event, see for example, Hoyt [49], and p. 87.

Single-Crystal Growth

For the preparation of single crystals of SmB$_6$, EuB$_6$, and YbB$_6$, see also the review of Korsukova, Gurin [61].

The rare earth hexaborides are highly reactive at elevated temperature. Therefore, crystal growth should be carried out in an inert atmosphere. A considerable overpressure of inert gas is required, where a large melt surface is involved, in order to minimize material evaporation, Davis et al. [50]. For the same reason, a vacuum is unsuited, Etourneau et al. [1, p. 434].

Floating-zone melting is suited to the growth of high-purity crystals of congruently melting MB$_6$ (M = La to Sm, see p. 49), whereas in the case of incongruently melting MB$_6$, e.g. GdB$_6$, the zone-leveling technique is used preferably in order to prevent MB$_4$ inclusions (see Section 32.1.8.10 in "Rare Earth Elements" C 11 b), Tanaka et al. [27]. In a critique of the various single-crystal growth methods used for MB$_6$, Davis et al. [50] note that use of r.f. induction heating to achieve the molten zone results in randomly oriented crystal rods of largest diameter, however, these rods do not appear to be truly monocrystalline, but consist of subgrains (see p. 29). Variously oriented single crystals can be grown by use of a single-crystal seed at the initiation zone. From the standpoint of simplicity, low cost and achieving high purity and oriented single-crystal rods suitable for field and thermionic emission cathodes (especially of LaB$_6$), the arc floating-zone method is regarded as preferable to other floating-zone techniques [50].

For single-crystal growth of LaB$_6$, CeB$_6$, and PrB$_6$ by the arc floating-zone refining method in Ar of ≧0.1 MPa (≧1 atm), a pointed, hot-pressed rod of the relevant MB$_6$ is regarded as the best counter electrode. Tantalum can also be used in the case of LaB$_6$ and CeB$_6$, and tungsten performs satisfactorily in the case of PrB$_6$. A low zone pass rate (5 cm/h) is used in order to prevent second-phase precipitation. The prepared seeded single crystals of selected directions such as $\langle 100 \rangle$, $\langle 211 \rangle$, and $\langle 310 \rangle$ have all been aligned along the rod axis to within 2°. A two-zone pass carried out with commercially available sintered feedstocks of $LaB_{6.2}$ (which is about the congruently melting composition) resulted in $LaB_{6.09}$ crystals, whereas starting compositions poorer in boron produced compositions of $B/La < 6$, a composition gradient, and an observable La-rich second phase. Similarly, starting from $PrB_{6.0}$ gave an average composition $PrB_{5.87 \pm 0.05}$, whereas a $CeB_{6.0}$ feedstock gave a zone-refined rod of average composition $CeB_{6.2 \pm 0.1}$. The $LaB_{6.09}$ had altogether < 100 ppm by weight of 27 main impurities (except C and O) and carbon of the order of 50 ppm, to be compared to contents in the starting material of $> 3.2 \times 10^4$ and 2160 ppm, respectively. Residual oxygen of 240 ppm in the zone-refined rod, caused by the starting material (13 100 ppm), produced second-phase inclusions of La–B–O (– and possibly C) compounds which rapidly evaporate from heated surfaces, for details see the paper. Similar inclusions were also observed in the two-pass $CeB_{6.2}$, the initial feedstock of which contained 7900 ppm of oxygen and 950 ppm carbon [50]. For an early synthesis of LaB_6 and CeB_6 by zone heating pressed rods of M_2O_3/B mixtures (mole ratio 1:15) in vertical position with an electron beam in a vacuum of 1.33 mPa (10^{-5} Torr) at start, see Ban, Sikirica [51].

Floating-zone melting in a r.f. induction furnace in an Ar atmosphere of 1.47 MPa (15 kg/cm^2) produced crystals, ca. 8 mm in diameter, 30 to 40 mm long, of bluish CeB_6, bluish black SmB_6, and violet GdB_6 containing golden GdB_4 inclusions, by a threefold (M = Ce, Gd) or fourfold (M = Sm) zone pass; zone pass rate 1 cm/h. Feedstocks had been prepared borothermally from stoichiometric amounts of metal oxide and boron. In all cases, the chemical compositions of the zone-passed crystals had a tendency to boron deficiency. The growth directions of CeB_6 crystals were near [100] or [110] in the core region, whereas subgrains formed the peripheral regions. The SmB_6 crystal had a rough surface and facets. The growth direction was far from low-index directions. The GdB_6 crystal consisted of several large grains. The authors note that the molten zones of CeB_6 and SmB_6 are considerably more stable than that of LaB_6, which makes the zone-refining process with CeB_6 and SmB_6 much easier [27]. For similar preparation of LaB_6 and SmB_6, see Tanaka et al. [52], and for MB_6 with M = La, Ce, Nd, Sm, Eu, and Gd, see Yamauchi et al. [53]. Data on the experimental set-up in order to obtain large-diameter rods of LaB_6 and GdB_6 having small number of dislocations were given by [54]. Early studies with LaB_6 revealed that a rapid pure Ar stream (3 to 4 L Ar/min) led to loss of material, Curtis, Graffenberger [55]. Therefore, zone melting was sometimes carried out in an Ar/H$_2$-gas mixture. A device for both the synthesis and melting of MB_6 consisted of a silver-plated double-walled copper boat, water cooled internally, which was situated in an inductively heated quartz tube. Raw material MB_6 with M = La to Sm and Gd was prepared from pressed stoichiometric metal oxide/boron mixtures, plus an added piece of previously fused MB_6, by heating the material at 1873 K under vacuum. After MB_6 formation the vacuum was replaced by a gas mixture of 3 parts of argon and 1 part of hydrogen, and purification by zone melting was carried out several times to give long, uniform, and sufficiently pure rods, Niemyski et al. [26]. Nearly stoichiometric LaB_6, NdB_6, and GdB_6, contaminated more or less by carbon, was zone melted in vertical position in an Ar-5% H$_2$ atmosphere. So-obtained $GdB_{5.97}$ (with 0.001 wt% O$_2$ and 0.061 wt% C) and LaB_6 contained slight traces of a brassy yellow phase, probably MB_4. The zone-refined NdB_6 rod was pure at the end where zone refining was initiated, but was contaminated with a gray phase (average content: 2 wt% "B_5C") toward the final end, Westrum [43].

A vapor-liquid-solid growth mechanism produced small single crystals of LaB_6, EuB_6, and YbB_6 besides YbB_{12}, but not of GdB_6. For this, mixtures of M_2O_3 and 5 wt% excess boron over the stoichiometric amount were slowly (4 h) heated in an open ZrB_2 crucible in flowing Ar to 2083 K, kept at that temperature for 4 h, and finally cooled during 4 h to room temperature, Rea, Kostiner [56].

The very common Al-flux method produces large single crystals, which often have some Al and Al$_2$O$_3$ inclusions, Olsen, Cafiero [57]. Crystal growth is performed in Al$_2$O$_3$ crucibles in an Ar atmosphere. After growth, the Al flux was dissolved in dilute HCl, Futamoto et al. [58], Gurin et al. [59], in concentrated HCl [57], or in a saturated solution of NaOH, Fisk et al. [60]. Crystal-growth procedures were as follows: A stoichiometric mixture of rare earth metal (M = La to Eu) and B powder and Al (ratio B:Al≈1wt%) was heated at a temperature between 1523 and 1573 K for several hours and cooled at 70 K/h [58]. Crystals of MB$_6$ with M = La to Nd, ca. 5 mm on a side, formed also from a solution of 95 at% Al + 5 at% (M + 4B, tetraboride ratio), held at 1823 K for 10 min, cooled to 1273 K in 10 min, and finally quenched to room temperature. All the steps were performed under 66.66 kPa (500 Torr) Ar in a r.f. induction furnace [60]. Rods of M = La, Sm, Eu, Yb, and amorphous B powder at the ratio M:B=1:6 or 1:12 and a ratio (M + B)/Al of ca. 5 to 10 wt% in a current of Ar were used by [59], see also Korsukova, Gurin [61]. After homogenization at 1573 to 1673 K for 2 to 10 h, the solution was slowly cooled to room temperature. The compositions of the crystals were LaB$_6$, SmB$_6$, EuB$_{6.02}$, and YbB$_{6.1}$ in the case of M:B=1:6 at start and were said to be LaB$_{5.86}$ and EuB$_{5.93}$ in the case of M:B=1:12 at start. Contaminations by Al ranged from 0.02 wt% for LaB$_6$ to 0.3 wt% in both SmB$_6$ and YbB$_{6.1}$ [59]. The oxide powders M$_2$O$_3$ instead of the metals were used by Olsen, Cafiero [57, 62] to prepare single crystals of MB$_6$ with M = La, Ce, and Eu, of (La,Y)B$_6$, (Eu,Y)B$_6$, and (Eu,La)B$_6$, as well as of (La,Cs)B$_6$ using CsNO$_3$. The charge contained stoichiometric amounts of M$_2$O$_3$ and B and a ratio B:Al of ca. 3 wt%. For a design of the furnace, see the paper. After soaking the charge for 5 d at 1773 K, the furnace was cooled, either to 873 K at a linear rate of 5 to 20 K/h and finally to room temperature or exponentially at an initial rate of 15 K/h. Attempts to grow YB$_6$ produced only the tetraboride. Ternary mixtures containing Y$_2$O$_3$ always produced a mixture of hexaboride and metallic brown YB$_4$ crystals. The mixed hexaboride crystals (La,Y)B$_6$, (Eu,Y)B$_6$, and (Eu,La)B$_6$ had a single composition that was within 15 mol% of the powder composition. Furthermore, in any ternary series, the product crystal tended to be richer in one metal than the starting powder according to the descending order La, Eu, Y [57]. These binary and ternary hexaborides, and also quaternary (Eu,La,Y)B$_6$ were prepared in the same way by [62]. Binary MB$_6$ crystals were grown by melting Al at ca. 973 K, adding an Al foil-wrapped rare earth metal and a Ni–(12 to 18%)B alloy as B source, heating to 1373 K at 300 K/h, homogenizing at ca. 1573 K for 3 h, and finally cooling at ca. 100 K/h, Hitachi, Ltd. [63].

MB$_6$ crystals grow from an Al flux as idiomorphic needles, plates, and isomeric crystals: plates exhibit a well-developed (100) face [58, 59, 64], as do bar-like crystals with M = La to Eu [58]. Needles of LaB$_6$ and EuB$_6$ are stretched either in the [110] direction [59] or [100] direction [64]. Cubes are observed for M = La to Eu [58] or M = La and Eu, whereas SmB$_6$ and YbB$_6$ formed cubooctahedra. The ratio N of the number of plates to that of needles crystallizing from a solution of LaB$_6$ in Al, homogenized at 1623 K, increased linearly with increasing starting concentration of LaB$_6$ in the solution and amounted to N≈0.45 at 6 wt% LaB$_6$ (upper limit studied) [64]. Cooling charges with M = La, Ce, Eu from 1773 to 873 K at a linear rate of 5 to 20 K/h resulted in cubes, plates, and polyhedra, with no obvious correlation between morphology and temperature gradient. Longer, rod-like shapes with (100) faces were produced by exponential cooling, initial rate of 15 K/h [57].

Crystal Quality

Apart from some Al/Al$_2$O$_3$ inclusions, Al flux-grown crystals are relatively perfect, with etch pit densities of ca. 10^4 to 10^5 cm^{-2} in LaB$_6$ and of ca. 10^4 cm^{-2} in EuB$_6$, Gurin et al. [59]. Arc zone-melted LaB$_6$ crystals revealed, on cleaved (100) surfaces, <10^6 cm^{-2} of square pits (ca. 0.2 μm on a side) with sharp corners. This relatively large defect concentration, also observed in laser zone-refined samples (see p. 191), appears to be inherent in the zone-refinement process, but may depend somewhat on the presence of impurities. The second-phase inclusions observed in these crystals cause large square or rectangular pits with well-rounded corners, ca. 1×10^4 cm^{-2} in LaB$_{6.09}$ and CeB$_{6.2}$, Davis et al. [50].

Preparation of Films

Films with various degrees of crystallinity were prepared by electron beam evaporation of MB$_6$ (M = La, Pr, Sm) at 1.3 mPa ($\triangleq 10^{-5}$ Torr) onto substrates of glass, silicon, quartz, or NaCl, heated to a temperature of 500 to 900 K, Bessaraba et al. [65].

References:

[1] Etourneau, J.; Mercurio, J.-P.; Naslain, R.; Hagenmuller, P. (Colloq. Intern. Centre Natl. Rech. Sci. [Paris] No. 205 [1972] 429/38).

[2] Etourneau, J.; Mercurio, J.-P.; Hagenmuller, P. (in: Matkovich, V. I., Boron and Refractory Borides, Springer, Berlin – Heidelberg – New York 1977, pp. 115/38, 118/9).

[3] Geballe, T. H.; Matthias, B. T.; Andres, K.; Maita, J. P.; Cooper, A. S.; Corenzwit, E. (Science **160** [1968] 1443/4).

[4] Sturgeon, G. D.; Eick, H. A. (TID-17968 [1961] 1/10; N.S.A. **17** [1963] No. 14216).

[5] Sturgeon, G. D.; Eick, H. A. (Inorg. Chem. **2** [1963] 430/1).

[6] Sturgeon, G. D.; Eick, H. A. (Proc. 3rd Conf. Rare Earth Res., Clearwater, Fla., 1963 [1964], pp. 87/97).

[7] Bliznakov, G.; Peshev, P. (J. Less-Common Metals **7** [1964] 441/6).

[8] Bliznakov, G.; Peshev, P. (Izv. Inst. Obshta Neorgan. Khim. Bulg. Akad. Nauk. **3** [1965] 5/12; C.A. **64** [1966] 13723).

[9] Etourneau, J.; Mercurio, J.-P.; Naslain, R. (Compt. Rend. C **275** [1972] 273/6).

[10] Hagenmuller, P.; Etourneau, J.; Mercurio, J.-P.; Naslain, R. N. (Redkozemel. Metal. Splavy Soedin. Mater. 7th Soveshch., Moscow 1972 [1973], pp. 229/36; C.A. **80** [1974] No. 152305).

[11] Itoh, H.; Tsuzuki, Y.; Yogo, T.; Naka, S. (Mater. Res. Bull. **22** [1987] 1259/66).

[12] Dudnik, E. M.; Kocherzhinskii, Yu. A.; Shishkin, E. A.; Paderno, Yu. B.; Zadvornyi, L. I.; Timofeeva, I. I.; Zaletilo, L. S. (Tugoplavkie Soedin. Redkozemel. Met. Mater. 3rd Vses. Semin., Novosibirsk 1977 [1979], pp. 26/9; C.A. **93** [1980] No. 60063).

[13] Karasev, A. I.; Dudnik, E. M.; Chekhovich, V. A. (Tugoplavkie Soedin. Redkozemel. Met. Mater. 3rd Vses. Semin., Novosibirsk 1977 [1979], pp. 32/4; C.A. **93** [1980] No. 206964).

[14] Derenovskii, M. V.; Shlyuko, V. Ya.; Chernyak, L. V. (Poroshkovaya Metal. **1967** No. 2, pp. 8/12; Soviet Powder Met. Metal Ceram. **1967** 93/6).

[15] Bondarenko, V. N.; Selivanova, N. F.; Shlyuko, V. Ya. (Vestn. Kiev. Politekhn. Inst. Mashinostr. No. 3 [1967] 174/9; C.A. **68** [1968] No. 118872).

[16] Mordovin, O. A.; Timofeeva, E. N. (Zh. Neorgan. Khim. **13** [1968] 3155/8; Russ. J. Inorg. Chem. **13** [1968] 1627/9).

[17] Samsonov, G. V.; Paderno, Yu. B.; Serebryakova, T. I. (Kristallografiya **4** [1959] 542/4; Soviet Phys.-Cryst. **4** [1959] 510/2).

[18] Yajima, S.; Niihara, K. (Proc. 9th Rare Earth Res. Conf., Blacksburg, Va., 1971, Vol. 2, pp. 598/609).

[19] Sobczak, R. J.; Sienko, M. J. (J. Less-Common Metals **67** [1979] 167/71).

[20] Smith, P. K. (COO-1140-103 [1964] 1/485, 149/55; N.S.A. **18** [1964] No. 25319; Diss. Univ. Kansas 1964, pp. 1/482; Diss. Abstr. **25** [1964/65] No. 5591).

[21] Eick, H. A.; Gilles, P. W. (J. Am. Chem. Soc. **81** [1959] 5030/2).

[22] Eick, H. A.; Gilles, P. W. (AECU-4087 [1958] 1/12; N.S.A. **13** [1959] No. 10826).

[23] Post, B.; Moskowitz, D.; Glaser, F. W. (J. Am. Chem. Soc. 78 [1956] 1800/2).

[24] Post, B.; Moskowitz, D.; Glaser, F. W. (Plansee Proc. 2nd Semin., Reutte/Tyrol, Austria, 1955 [1956], pp. 173/86; Met. Abstr. [2] 24 [1956/57] 350).

[25] Tvorogov, N. N. (Zh. Neorgan. Khim. 4 [1959] 1961/6; Russ. J. Inorg. Chem. 4 [1959] 890/3).

[26] Niemyski, T.; Pracka, I.; Jun, J.; Paderno, J. (J. Less-Common Metals 15 [1968] 97/9).

[27] Tanaka, T.; Nishitani, R.; Oshima, C.; Bannai, E.; Kawai, S. (J. Appl. Phys. 51 [1980] 3877/83).

[28] Adadurov, G. A.; Breusov, O. N.; Dremin, A. N.; Drobyshev, V. N. (Zh. Neorgan. Khim. 15 [1970] 2887/9; Russ. J. Inorg. Chem. 15 [1970] 1503/4).

[29] Binder, F. (Radex Rundschau 1977 52/71, 64).

[30] Neshpor, V. S.; Samsonov, G. V. (Zh. Fiz. Khim. 32 [1958] 1328/32; C.A. 1959 953).

[31] Markovskii, L. Ya.; Vekshina, N. V. (Zh. Prikl. Khim. 40 [1967] 1824/6; J. Appl. Chem. [USSR] 40 [1967] 1752/3).

[32] Tepper, F.; Mausteller, J. W.; Kabi, S. K., Mine Safety Appliances Co. (Ger. 1189055; Brit. 955730 [1960/64] 1/6; C.A. 61 [1964] 1537).

[33] Andrieux, L. (Ann. Chim. [Paris] [10] 12 [1929] 423/507, 458/76).

[34] von Stackelberg, M.; Neumann, F. (Z. Physik. Chem. B 19 [1932] 314/20).

[35] Andrieux, L. (Compt. Rend. 186 [1928] 1736/8).

[36] Andrieux, J.-L. (Rev. Met. 45 [1948] 49/59, 54; C.A. 1948 8089).

[37] Kunnmann, W. (in: Lefever, R. A., Preparation and Properties of Solid State Materials, Vol. 1, Dekker, New York 1971, pp. 1/36, 25/9).

[38] Uchida, K. (Japan. Kokai 77-78799 [1975/77] 1/4 from C.A. 87 [1977] No. 143367).

[39] Vekshina, N. V.; Markovskii, L. Ya.; Pron, G. F. (Vysokotemp. Neorgan. Soedin. 1965 404/14; C.A. 64 [1966] 13722).

[40] Wold, A. (AD-660615 [1967] 1/41, 5; C.A. 68 [1968] No. 90511).

[41] Gomes, J. M.; Uchida, K. (U.S. 3902973 [1973/75] 1/4; C.A. 83 [1975] No. 185521).

[42] Ames, L. L.; McGrath, L. (High Temp. Sci. 7 [1975] 44/54; C.A. 83 [1975] No. 66462).

[43] Westrum, E. F., Jr. (Colloq. Intern. Centre Natl. Rech. Sci. [Paris] No. 180 [1970] 443/50; C.A. 78 [1973] No. 89350).

[44] Samsonov, G. V.; Neshpor, V. S. (Redk. Metally Splavy Tr. 1st Vses. Soveshch., Moscow 1957 [1960], pp. 392/417; C.A. 1961 4314; JPRS-17750 [1963] 1/31, 17/8).

[45] Hoyt, E. W.; Chorné, J. (GEAP-3332 [1960] 1/15; C.A. 1961 1338).

[46] Samsonov, G. V. (Tugoplavkie Soedineniya Redkozemel'nykh Metallov s Nemetallami, Metallurgiya, Moscow 1964; High-Temperature Compounds of Rare Earth Metals with Nonmetals, Consultants Bureau, New York 1965, pp. 1/280, 50/60).

[47] Arabei, B. G.; Shtrom, E. N.; Lapitskii, Yu. A. (Poroshkovaya Metal. 1964 No. 5, pp. 65/70; Soviet Powder Met. Metal Ceram. 1964 406/9).

[48] Markov, Yu. M.; Trokhina, G. N.; Zernova, E. E.; Maksimovskii, V. V. (Izv. Akad. Nauk SSSR Neorgan. Materialy 14 [1978] 79/81; Inorg. Materials [USSR] 14 [1978] 61/3).

[49] Hoyt, E. W. (Proc. 2nd Conf. Rare Earth Res., Glenwood Springs, Colo., 1961 [1962], pp. 287/99; N.S.A. 16 [1962] No. 32072).

[50] Davis, P. R.; Swanson, L. W.; Hutta, J. J.; Jones, D. L. (J. Mater. Sci. 21 [1986] 825/36).

[51] Ban, Z.; Sikirica, M. (New Nucl. Mater. Incl. Non-Metal. Fuels Proc. Conf., Prague 1963, Vol. 2, pp. 175/82; C.A. 60 [1964] 10192).

[52] Tanaka, T.; Bannai, E.; Kawai, S. (Seramikkusu 11 [1976] 1083/92 from C.A. 86 [1977] No. 198128).

[53] Yamauchi, T.; et al. (Muki Zaishitsu Kenkyusho Kenkyu Hokokusho No. 17 [1978] 22/32 from C.A. 91 [1979] No. 184994).

[54] Denki Kagaku Kogyo K.K. (Japan. Kokai Tokkyo Koho 81-59691 [1979/81] 1/4 from C.A. **95** [1981] No. 142398).

[55] Curtis, B. J.; Graffenberger, H. (Mater. Res. Bull. **1** [1966] 27/31).

[56] Rea, J. R.; Kostiner, E. (J. Cryst. Growth **11** [1971] 110/2).

[57] Olsen, G. H.; Cafiero, A. V. (J. Cryst. Growth **44** [1978] 287/90).

[58] Futamoto, M.; Aita, T.; Kawabe, U. (Mater. Res. Bull. **14** [1979] 1329/34).

[59] Gurin, V. N.; Korsukova, M. M.; Nikanorov, S. P.; Smirnov, I. A.; Stepanov, N. N.; Shul'man, S. G. (J. Less-Common Metals **67** [1979] 115/23).

[60] Fisk, Z.; Cooper, A. S.; Schmidt, P. H.; Castellano, R. N. (Mater. Res. Bull. **7** [1972] 285/8).

[61] Korsukova, M. M.; Gurin, V. N. (Zh. Vses. Khim. Obshchestva **26** No. 6 [1981] 79/88; Mendeleev Chem. J. [USSR] **26** No. 6 [1981] 114/26).

[62] Olsen, G. H.; Cafiero, A. V. (U.S. 4260525 [1978/81] 1/5; C.A. **94** [1981] No. 218072).

[63] Hitachi, Ltd. (Japan. Kokai Tokkyo Koho 84-69497 [1982/84] 1/6 from C.A. **101** [1984] No. 141556).

[64] Korsukova, M. M.; Nardov, A. V.; Gurin, V. N. (in: Sb. Rassh. Tez. 6th Mezhdunar. Konf. Rost. Kristall. [Expanded Abstracts 6th Intern. Conf. Cryst. Growth], Moscow 1980, Vol. 3, pp. 259/61).

[65] Bessaraba, V. I.; Ivanchenko, L. A.; Paderno, Yu. B. (J. Less-Common Metals **67** [1979] 505/9).

32.1.5.4 Crystallographic Properties

Crystal Structure

The cubic rare earth hexaborides MB_6 crystallize in the CaB_6 type proposed by von Stackelberg, Neumann [1]; see also Allard [2]. $Z = 1$, space group $Pm\overline{3}m-O_h^1$ (No. 221) with M in (0,0,0) and B in $(x, \frac{1}{2}, \frac{1}{2})$ with $x \approx 0.2$ [1]. X-ray diffraction studies were performed on zone-melted single crystals of LaB_6, CeB_6, and SmB_6, the mean compositions of which were calculated from the lattice parameters and the experimental density D_{exp} on the assumption of either metal defects or boron defects. Anisotropic refinement yielded the following values of x and the refined (local) defect formula at specified R_w:

| mean composition | | a | D_{exp} | x | local | R_w |
M-defect	B-defect	in Å	in g/cm³		composition	
LaB_6	LaB_6	4.1570(1)	4.7096(8)	0.1993(10)	$LaB_{5.63(10)}$	0.024
$Ce_{0.99}B_6$	$CeB_{5.84}$	4.1408(1)	4.7538(15)	0.1985(6)	$CeB_{5.77(7)}$	0.018
$Sm_{0.99}B_6$	$SmB_{5.92}$	4.1347(1)	5.0348(15)	0.1986(5)	$Sm_{0.77(2)}B_6$	0.034

Calculated bond lengths M–B, $(B–B)_e$, and $(B–B)_i$, where e = edge or intraoctahedral and i = interoctahedral, all in Å, amount to:

MB_6	M–B	$(B–B)_e$	$(B–B)_i$
LaB_6	3.051(1)	1.766(2)	1.656(2)
CeB_6	3.036(1)	1.762(4)	1.642(5)
SmB_6	3.032(1)	1.760(2)	1.639(3)

For the bond angles see the paper, Eliseev et al. [3]. For the effect of variation of x on the intensity of various X-ray diffraction lines of LaB$_6$ and YbB$_6$ powders, see Barantseva, Paderno [4]. Values for x for all MB$_6$ (M = La to Lu) were calculated by means of the equation x = 0.3228 − 0.5122/a, using values for a from literature, see the paper and for M = La, Ho, Eu, and Yb also the table below, Kuz'micheva et al. [5]. Generally, the atom positions are given reversed, i.e., M in (½,½,½) and B in (x',0,0) with x' = (½ − x) ≈ 0.3, see for example Blum, Bertaut [6] and **Fig. 8** from the review by Post [7]. In this case the intraoctahedral bond length is given by (B–B)$_e$ = $\sqrt{2} \cdot$ x' · a and the interoctahedral one by (B–B)$_i$ = (1 − 2x') · a, Hulliger [8]. In order to show the change of the various bond lengths with ionic radius of the metal, values are assembled below for MB$_6$ with trivalent La and Ho and with divalent Eu and Yb. Also given are corresponding bond lengths calculated from the estimated metallic radius R$_M$(6) or R$_M$(24) for coordination number 6 or 24, respectively, and the boron "bond radius" R$_B$ = 0.88 Å (all values in Å):

MB$_6$	M–M distance obs.	2R$_M$(6)	M–B distance obs.	R$_M$(24) + R$_B$	Ref.	x	(B–B)$_e$	(B–B)$_i$	Ref.
LaB$_6$	4.154	3.57	3.05	2.85	[9]	0.202[1]	1.755	1.676	[10]
LaB$_6$	4.1561		3.054		[5]	0.1995	1.7662	1.6584	[5]
HoB$_6$	4.096	3.35	3.01	2.74	[9]	0.202[1]	1.723	1.654	[10]
HoB$_6$	4.096		3.007		[5]	0.1977	1.7511	1.6196	[5]
EuB$_6$	4.178	3.89	3.07	3.01	[9]	0.202[1]	1.758	1.687	[10]
EuB$_6$	4.1780		3.070		[5]	0.2002	1.7714	1.6726	[5]
YbB$_6$	4.147	3.70	3.04	2.91	[9]	0.202[1]	1.746	1.675	[10]
YbB$_6$	4.1478		3.047		[5]	0.1993	1.7639	1.6531	[5]

[1] x' = 0.298 ± 0.004 [10].

Bond lengths for the whole series are listed in Binder [10] (M = Y, La to Er, and Yb) and Kuz'micheva et al. [5] (M = La to Lu), i.e., inclusive the instable compounds with M = Er, Tm, Lu. A similar small change in both B–B bond lengths on going from LaB$_6$ to HoB$_6$, as listed above was also predicted earlier by Hoard, Hughes [9].

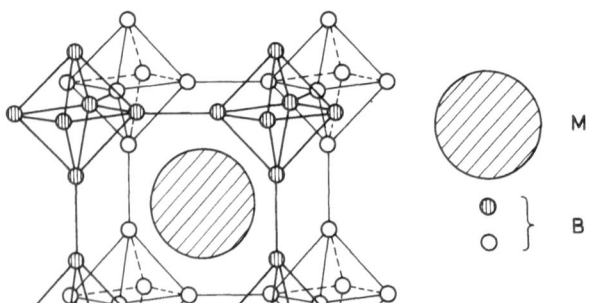

Fig. 8.   Crystal structure of MB$_6$.

The MB$_6$ structure can be regarded as a CsCl type with B$_6$ octahedra as anions [1]. It is built up of a three-dimensional framework of interlocked B$_6$ octahedra with M situated in the large holes between the octahedra. Each B has 5 B + 4 M neighbors, each M has 24 B + 6 M neighbors [6]. Each B has 4 B neighbors in the same octahedron and a fifth less-distant B neighbor belonging to the adjacent B_6 octahedron [9]. Earlier, the five B–B bond lengths were assumed to be equidistant, see for example [6], Kiessling [11], Neshpor, Samsonov [12], and the review by Samsonov [13]. Each M is enclosed by 24 equidistant B, three from each of the eight surrounding B_6 octahedra, thus forming an octahedrally truncated cube, see **Fig. 9** [7, 9]. The covalently bonded boron framework is very rigid [6, 11], which accounts for the small changes of all observed bond lengths along the MB_6 series [9]. This leads to the conclusion that the contributions of the M–B bonding to the stability of the structure would be very sensitive to the size of the metal, Spear [14, p. 125]. From the lower limit of space available for M it was concluded that metals as small as Ho (but see Section 32.1.8.13 in "Rare Earth Elements" C 11b), Er, Tm, and Lu cannot form hexaborides, Etourneau et al. [15, p. 123], for Er, Tm, and Lu, also Sturgeon, Eick [16], and for Sc, Er, Tm, and Lu, Kuz'micheva et al. [5] according to which HoB_6 might exist.

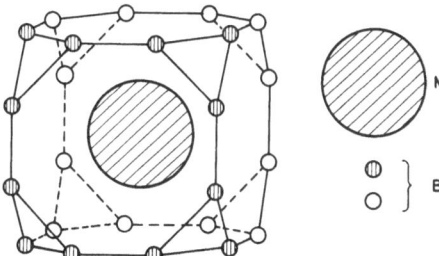

Fig. 9.   Coordination of M in MB_6.

Lattice Constants

The lattice constants of the hexaborides decrease with atomic number for MB_6 with trivalent cations and show maxima for EuB_6 and YbB_6 with divalent cations, see **Fig. 10**, p. 37, from Binder [10], which also includes values for the pseudohexaborides of Er, Tm, and Lu (MB_6 with M = Sc, Ho(?), Er, Tm, and Lu do not exist, see p. 21). For similar earlier presentations, see for example, Post et al. [17], Samsonov, Paderno [18], Samsonov et al. [19], Tvorogov [20]. The following assemblage gives lattice constants (in Å) observed for single crystals or their powders:

YB_6	4.102 [21]
LaB_6	4.153 [22], 4.156 [21, 23 to 25], 4.1556(1) [26], 4.1568 [27], 4.1570(1) [3]
CeB_6	4.1406 [27], 4.1408(1) [3], 4.140 [25], 4.141 [21], 4.144 [23, 24]
PrB_6	4.133 [21, 28], 4.135 [23, 24]
NdB_6	4.126 [21, 28], 4.129 [23, 24]
SmB_6	4.131 [23, 24], 4.132 [25], 4.133 [21], 4.1347(1) [3], 4.1414(1) [26]
EuB_6	4.1843 [27], 4.186 [24], 4.187(2) [26]
GdB_6	4.109 [21]
YbB_6	4.145 [22]

Lattice constants for polycrystalline samples prepared by borothermal reduction are:

YB_6	4.1022(5) [29], 4.1025(5) [10]
LaB_6	4.153 [28], 4.156 [30], 4.1561(5) [10], 4.1562(5) [29]
CeB_6	4.1399(5) [10], 4.1399(8) [31, 32], 4.1418(5) [29], 4.145 [30]
PrB_6	4.1316(6) [31, 32], 4.1336(5) [10], 4.14 [30]
NdB_6	4.1250(4) [31, 32], 4.1252(5) [10], 4.1260(5) [33, 34], 4.1265(5) [29]
PmB_6	4.128 (extrapolated value) [19]
SmB_6	4.130 [30], 4.1333(5) [33, 34], 4.1352(5) [29], 4.134 [22]

EuB$_6$ 4.176 [30], 4.1789(5) [10], 4.184 [22], 4.185(5) [29]
GdB$_6$ 4.1066(5) [10], 4.1078(5) [33, 34], 4.109(5) [29]
TbB$_6$ 4.1020(5) [33, 34], 4.1036(5) [29]
DyB$_6$ 4.0976(5) [33, 34], 4.1002(5) [29]
HoB$_6$ 4.096(5) [33, 34]
YbB$_6$ 4.1468(5) [10, 33, 34], 4.1478(5) [29], 4.148 [22], 4.155 [30]

Additional values are given, for example, in the following papers, for samples prepared mostly borothermally: for M = La to Lu except Ce, Tm, Mordovin, Timofeeva [35]; for M = La, Pr, Sm, Gd, Yb, Post et al. [17, 36]; for M = Y, La to Tb, Yb, Zhuravlev et al. [37]; for M = Y, La to Yb except Dy, Tm, Tvorogov [20]; for ScB$_6$ (a = 4.35 kX $\cong$ 4.36 Å) and M = Gd, Dy, Ho, Er, Yb, Lu, Neshpor, Samsonov [12]; for M = Dy, Ho, Lu, also Neshpor, Samsonov [38]; for M = Pr, Tb, Samsonov et al. [39]; for M = Eu, Tb, Tm, Samsonov et al. [19]; for M = La, Ce, Pr, Stepanova, Umanskii [40]. Values for samples prepared by flux electrolysis: for M = Y, La, Ce, Nd, Gd, Yb, Bertaut, Blum [41], Blum, Bertaut [6]; for M = La to Nd, Er, von Stackelberg, Neumann [1]; for M = La, Ce, von Stackelberg [42]; for M = Y, La, Ce, Nd, Gd, Er, Yb, Allard [2]; see also Kiessling [11].

In the homogeneity range M$_x$B$_6$ with x $\leqq$ 1, the lattice constants were found to decrease with the increase of B/M ratio (decrease of x) for M = Sm whereas they increase for M = Ce, Pr, Gd, Tb, Yajima, Niihara [43], and Dy, Spear [14, pp. 126/30] and they are constant for LaB$_6$, Johnson, Daane [44]; see below and the reviews of Spear [14, pp. 126/30], Binder [10, p. 57], and Etourneau et al. [15, p. 116]. For M = Sm no homogeneity range, but two forms with x $\approx$ 1 and $\approx$ 0.8 ($\cong$ β-SmB$_6$) with different lattice constants are assumed, see Section 32.1.8.8 in "Rare Earth Elements" C 11 b. Additional limiting lattice constants for the homogeneity regions of YB$_6$, NdB$_6$, and YbB$_6$, reported in [10, p. 57] appear to be the so-called "best" high and "best" low values reported by Spear [14, p. 112], which are based on a critical study of early literature data on isolated lattice constants for a given M. Limiting values of x and lattice constants of M$_x$B$_6$ are:

M	x	a in Å	Ref.	M	x	a in Å	Ref.	M	x	a in Å	Ref.
Y	1.00	4.1035[a]	[10]	Nd	1.00	4.1265[a]	[15, 29]	Tb	1.00	4.1008	[43]
	0.82	4.1132[a]	[10]		1.00	4.1250[a]	[10]		0.75	4.1052	[43]
La	1.00	4.1561	[14, 44]		0.75	4.1260[a]	[10]	Dy	?	4.0969	[14]
	0.77	4.1561	[14, 44]	Sm	1.00	4.1304	[43]		?	4.1008	[14]
Ce	1.00	4.1396	[43]		0.68	4.1278[b]	[43]	Yb	1.00	4.1478[a]	[10, 15]
	0.70	4.1415	[43]	Eu	1.00	4.1855[a]	[45]		0.84	4.1422[a]	[10]
Pr	1.00	4.1329	[43]	Gd	1.00	4.1065	[43]				
	0.69	4.1355	[43]		0.70	4.1113	[43]				

[a] Not from studies of the homogeneity range. – [b] In some papers attributed to β-SmB$_6$, see Section 32.1.8.8 in "Rare Earth Elements" C 11b.

Impurities, especially C, change the lattice constants. Especially in the case of EuB$_6$, this change possibly simulates a large metal-deficient homogeneity range, see Section 32.1.8.9 in "Rare Earth Elements" C 11b. For overall compositions derived from lattice constants and densities on the assumption of defects in one of the sublattices and refined (local) compositions derived from crystal structure analysis of powdered single crystals of LaB$_6$, CeB$_6$, and SmB$_6$, all prepared by zone melting in Ar, see p. 33. Lattice parameters and compositions (from chemical analysis) of single crystals grown from an Al flux were reported as:

MB$_6$	LaB$_6$	LaB$_{5.86}$	SmB$_6$	EuB$_{6.02}$	EuB$_{5.93}$	YbB$_{6.1}$	YbB$_6$*)
a in Å	4.157(2)	4.158(2)	4.135(2)	4.187(2)	4.187(2)	4.148(2)	4.147(2)

*) Composition not studied.

Gurin et al. [46]; for EuB$_6$, SmB$_6$, and YbB$_6$, see also the review by Korsukova, Gurin [47].

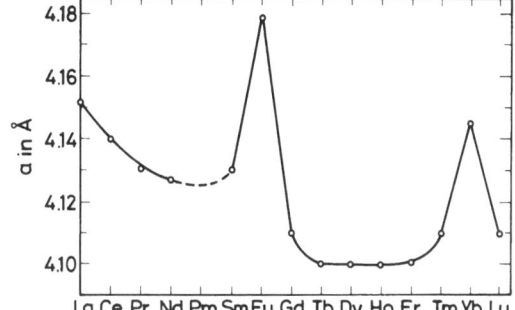

Fig. 10. Lattice constants of MB$_6$ for M = La to Lu (but see text on p. 35).

On the basis of a statistical model, Frenkel defects were predicted for NdB$_6$ and mutual migration of the Frenkel and Schottky defects for YB$_6$ and CeB$_6$, Kutolin et al. [48].

The temperature dependence of the lattice constants of pressed MB$_6$ (M = Y, La, Pr, Nd, and Yb) between 272 and 972 K, measured by X-rays in pure He, deviates from linearity (see **Fig. 11**) owing to the appearance of higher degrees of anharmonicity of thermal vibrations of the lattice atoms and the increase of anharmonicity with temperature, Dutchak et al. [49]. Earlier studies on MB$_6$ (M = Y, La to Tb, Yb) at 293, 773 to 1073 K indicated a linear variation a = f(T), Zhuravlev et al. [37]. The behavior below room temperature is shown in **Fig. 12** for MB$_6$, M = La, Ce, Sm. The lattice constants have been related to the CeB$_6$ constant at 300 K. The change observed for intermediate valent SmB$_6$ could be due to competition between delocalization of electrons

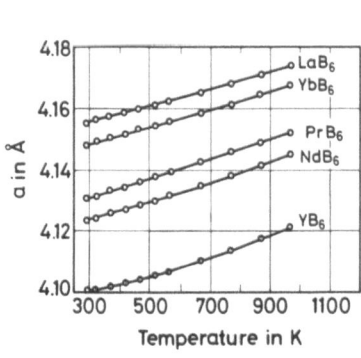

Fig. 11. Lattice constants of rare earth hexaborides as a function of temperature above room temperature (the curves are erroneously labeled in the original paper [49]).

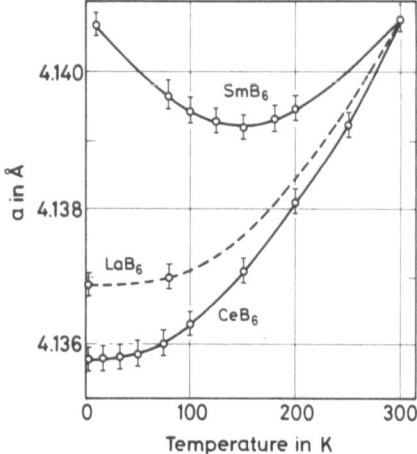

Fig. 12. Lattice constants of rare earth hexaborides as a function of temperature below room temperature.

(5d^1→5d6s), which leads to a reduction of the Sm–Sm distances, and the normal lattice dilatation with temperature, which becomes predominant above about 70 K [15], Mercurio et al. [50].

References:

[1] von Stackelberg, M.; Neumann, F. (Z. Physik. Chem. B **19** [1932] 314/20).

[2] Allard, G. (Bull. Soc. Chim. France [2] **51** [1932] 1213/5).

[3] Eliseev, A. A.; Efremov, V. A.; Kuz'micheva, G. M.; Konovalova, E. S.; Lazorenko, V. I.; Paderno, Yu. B.; Khlyustova, S. Yu. (Kristallografiya **31** [1986] 803/5; Soviet Phys.-Cryst. **31** [1986] 476/7).

[4] Barantseva, I. G.; Paderno, Yu. B. (Poroshkovaya Metal. **1982** No. 7, pp. 83/7; Soviet Powder Met. Metal Ceram. **21** [1982] 585/8).

[5] Kuz'micheva, G. M.; Khlyustova, S. Yu.; Tolstova, V. A.; Eliseev, A. A. (Zh. Neorgan. Khim. **33** [1988] 2205/10; Russ. J. Inorg. Chem. **33** [1988] 1259/62).

[6] Blum, P.; Bertaut, F. (Acta Cryst. **7** [1954] 81/6).

[7] Post, B. (in: Adams, R. M., Boron, Metallo-Boron Compounds and Boranes, Interscience, New York – London – Sydney 1964, pp. 301/71, 322/6).

[8] Hulliger, F. (Solid State Commun. **15** [1974] 933/6).

[9] Hoard, J. L.; Hughes, R. E. (in: Muetterties, E. L., The Chemistry of Boron and its Compounds, Wiley, New York – London – Sydney 1967, pp. 25/154, 117/29).

[10] Binder, F. (Radex Rundschau **1977** 52/71, 54/57; C.A. **87** [1977] No. 27597).

[11] Kiessling, R. (Acta Chem. Scand. **4** [1950] 209/27, 218/9).

[12] Neshpor, V. S.; Samsonov, G. V. (Zh. Fiz. Khim. **32** [1958] 1328/32; C.A. **1959** 953).

[13] Samsonov, G. V. (Tugoplavkie Soedineniya Redkozemel'nykh Metallov s Nemetallami, Metallurgiya, Moscow 1964; High-Temperature Compounds of Rare Earth Metals with Nonmetals, Consultants Bureau, New York 1965, pp. 1/280, 1/15).

[14] Spear, K. E. (in: Alper, A. M., Refractory Materials, Phase Diagrams, Vol. 4, Academic, New York 1976, pp. 91/159).

[15] Etourneau, J.; Mercurio, J.-P.; Hagenmuller, P. (in: Matkovich, V. I., Boron and Refractory Borides, Springer, Berlin – Heidelberg – New York 1977, pp. 115/38).

[16] Sturgeon, G. D.; Eick, H. A. (Inorg. Chem. **2** [1963] 430/1).

[17] Post, B.; Moskowitz, D.; Glaser, F. W. (J. Am. Chem. Soc. **78** [1956] 1800/2).

[18] Samsonov, G. V.; Paderno, Yu. B. (Boridy Redkozemel'nykh Metallov, Izdatel. Akad. Nauk Ukr.SSR, Kiev 1961; Borides of the Rare Earth Metals, AEC-tr-5264 [1962] 1/99, 9; N.S.A. **17** [1963] No. 2959).

[19] Samsonov, G. V.; Dzeganovskii, V. P.; Semashko, I. A. (Dokl. Akad. Nauk SSSR **119** [1958] 506/7; Proc. Acad. Sci. USSR Chem. Sect. **118/123** [1958] 237/8).

[20] Tvorogov, N. N. (Zh. Neorgan. Khim. **4** [1959] 1961/6; Russ. J. Inorg. Chem. **4** [1959] 890/3).

[21] Fisk, Z.; Lawson, A. C.; Fitzgerald, R. W. (Mater. Res. Bull. **9** [1974] 633/6).

[22] Leger, J. M.; Percheron-Guegan, A.; Loriers, C. (Phys. Status Solidi A **60** [1980] K23/K26).

[23] Okano, H.; Futamoto, M.; Hosoki, S.; Kawabe, U. (Shinku **20** No. 4 [1977] 127/35; C.A. **87** [1977] No. 94150).

[24] Futamoto, M.; Aita, T.; Kawabe, U. (Mater. Res. Bull. **14** [1979] 1329/34).

[25] Swanson, L. W.; McNeely, D. R. (Surf. Sci. **83** [1979] 11/28).

[26] Shelykh, A. I.; Sidorin, K. K.; Karin, M. G.; Bobrikov, V. N.; Korsukova, M. M.; Gurin, V. N.; Smirnov, I. A. (J. Less-Common Metals **82** [1981] 291/6).

[27] Olsen, G. H.; Cafiero, A. V. (J. Cryst. Growth **44** [1978] 287/90).

[28] Ali, N.; Woods, S. B. (Solid State Commun. **46** [1983] 33/5).

[29] Etourneau, J.; Mercurio, J.-P.; Naslain, R.; Hagenmuller, P. (Colloq. Intern. Centre Natl. Rech. Sci. [Paris] No. 205 [1972] 429/38; C.A. **78** [1973] No. 167937).

[30] Dudnik, E. M.; Bessaraba, V. I.; Paderno, Yu. B. (Poroshkovaya Metal. **1976** No. 12, pp. 60/2; Soviet Powder Met. Metal Ceram. **15** [1976] 945/7).

[31] Bliznakov, G.; Peshev, P. (J. Less-Common Metals **7** [1964] 441/6).

[32] Bliznakov, G.; Peshev, P. (Izv. Inst. Obshta Neorg. Khim. Bulg. Akad. Nauk. **3** [1965] 5/12; C.A. **64** [1966] 13723).

[33] Eick, H. A.; Gilles, P. W. (J. Am. Chem. Soc. **81** [1959] 5030/2).

[34] Eick, H. A.; Gilles, P. W. (AECU-4087 [1958] 1/12; N.S.A. **13** [1959] No. 10826).

[35] Mordovin, O. A.; Timofeeva, E. N. (Zh. Neorgan. Khim. **13** [1968] 3155/8; Russ. J. Inorg. Chem. **13** [1968] 1627/9).

[36] Post, B.; Moskowitz, D.; Glaser, F. W. (Plansee Proc. 2nd Semin., Reutte/Tyrol, Austria, 1955 [1956], pp. 173/86; Metal. Abstr. [2] **24** [1956/57] 350).

[37] Zhuravlev, N. N.; Stepanova, A. A.; Paderno, Yu. B.; Samsonov, G. V. (Kristallografiya **6** [1961] 791/4; Soviet Phys.-Cryst. **6** [1961] 636/8).

[38] Neshpor, V. S.; Samsonov, G. V. (Dopov. Akad. Nauk Ukr. RSR **1957** 478/9; C.A. **1958** 4370).

[39] Samsonov, G. V.; Paderno, Yu. B.; Serebryakova, T. I. (Kristallografiya **4** [1959] 542/4; Soviet Phys.-Cryst. **4** [1959] 510/2).

[40] Stepanova, A. A.; Umanskii, M. M. (Bor. Tr. Konf. Khim. Bora Ego Soedin., Moscow 1955 [1956], pp. 102/5; AEC-tr-4270 [1960] 1/4; N.S.A. **15** [1961] No. 1838).

[41] Bertaut, F.; Blum, P. (Compt. Rend. **234** [1952] 2621/3).

[42] von Stackelberg, M. (Z. Elektrochem. **37** [1931] 542/5).

[43] Yajima, S.; Niihara, K. (Proc. 9th Rare Earth Res. Conf., Blacksburg, Va., 1971, Vol. 2, pp. 598/609).

[44] Johnson, R. W.; Daane, A. H. (J. Phys. Chem. **65** [1961] 909/15).

[45] Kasaya, M.; Tarascon, J. M.; Etourneau, J.; Hagenmuller, P. (Mater. Res. Bull. **13** [1978] 751/6).

[46] Gurin, V. N.; Korsukova, M. M.; Nikanorov, S. P.; Smirnov, I. A.; Stepanov, N. N.; Shul'man, S. G. (J. Less-Common Metals **67** [1979] 115/23).

[47] Korsukova, M. M.; Gurin, V. N. (Zh. Vses. Khim. Obshchestva **26** No. 6 [1981] 79/88; Mendeleev Chem. J. [USSR] **26** No. 6 [1981] 114/26).

[48] Kutolin, S. A.; Komarova, S. N.; Frolov, Yu. A. (Zh. Fiz. Khim. **56** [1982] 996/9; Russ. J. Phys. Chem. **56** [1982] 606/8).

[49] Dutchak, Ya. I.; Fedyshin, Ya. I.; Paderno, Yu. B. (Izv. Akad. Nauk SSSR Neorgan. Materialy **8** [1972] 2134/7; Inorg. Materials [USSR] **8** [1972] 1877/80).

[50] Mercurio, J.-P.; Etourneau, J.; Naslain, R.; Hagenmuller, P. (J. Less-Common Metals **47** [1976] 175/80).

32.1.5.5 Lattice Vibrations

Group theoretical analysis yields for MB_6 (space group $Pm\bar{3}m\text{-}O_h^1$) the following optical modes at Γ points: $A_{1g} + E_g + F_{2g} + F_{1g} + 2F_{1u} + F_{2u}$. Of these, 3 modes of symmetries A_{1g}, E_g, and F_{2g} are Raman active, Ishii et al. [1]. An acoustic mode of F_{1u} symmetry exists in addition, see the individual sections. In the Raman spectra of MB_6 the 3 expected modes are observed together with a first-order symmetry forbidden scattering intensity near 200 cm^{-1}. The latter is assigned to a F_{1u} optical mode, Zirngiebl et al. [2], as was done in the case of LaB_6 by Smith et al. [3]. **Fig. 13** shows the frequencies of these 4 phonon symmetry modes as a function of lattice parameter to vary linearly for MB$_6$ with trivalent M = Y, La, Ce, Nd, and Gd. The allowed modes of intermediate valent SmB$_6$ have their frequencies between those of trivalent MB$_6$ and divalent EuB$_6$, whereas the F$_{1u}$ mode of SmB$_6$ exhibits softening [2]. In earlier studies of the Raman spectra of pressed powders of MB$_6$ (M = La, Pr, Nd, Sm, Gd, Eu, and Yb) only 3 modes (A$_{1g}$, E$_g$, and F$_{2g}$) were observed. The frequencies of compounds with divalent cations, EuB$_6$ (at 1238, 1101, and 767 cm^{-1}, respectively), YbB$_6$, and alkaline earth hexaborides, as a function of lattice parameter, also showed a linear dependence. With YbB$_6$, the frequencies of A$_{1g}$ and E$_g$ are about that for NdB$_6$, whereas the frequency of F$_{2g}$ is only a little higher than that of EuB$_6$, see a figure in the paper. The decrease of the frequencies of A$_{1g}$ and E$_g$ with the increase of the lattice constant was attributed to a decrease in the strength of interoctahedral B–B bond. The behavior of F$_{2g}$ suggests that the M–B interaction in MB$_6$ is ionic rather than covalent [1].

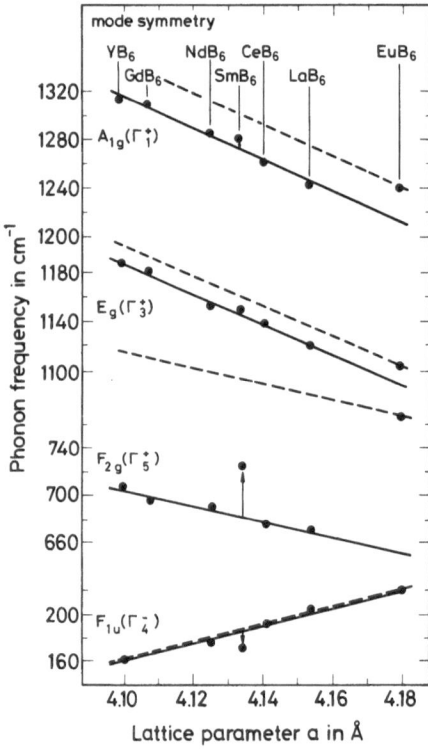

Fig. 13.   Frequencies of the phonon symmetry modes A$_{1g}$, E$_g$, F$_{2g}$, and F$_{1u}$ as a function of the lattice parameter for divalent EuB$_6$, intermediate valent SmB$_6$, and trivalent YB$_6$, LaB$_6$, CeB$_6$, NdB$_6$, and GdB$_6$.

Vibrational frequencies for MB$_6$ (M = La, Nd, Gd, Tb, Dy, Yb) observed in recent infrared (range 400 to 4000 cm^{-1}) and Raman studies are given in the table on p. 41, assigned to modes at Γ points. However, comparison of observed frequencies with published theoretical dispersion-curve frequencies for LaB$_6$ and YbB$_6$ (from Takegahara, Kasuya [4]) leads to the conclusion that the observed spectra are due mainly to defect-induced activity of phonons from X points, Turrell et al. [5].

mode	LaB$_6$ Raman	LaB$_6$ IR	NdB$_6$ Raman	NdB$_6$ IR	GdB$_6$ Raman	GdB$_6$ IR	TbB$_6$ Raman	TbB$_6$ IR	DyB$_6$ Raman	DyB$_6$ IR	YbB$_6$ Raman	YbB$_6$ IR
F$_{1g}$	210 w				242 w		500 br, w		262 w			650 vw, br
F$_{2g}$	670 m; 690 sh; 782 vw		682 m		680 w; 842 vw; 940 vw		692 m; 715 sh; 845 vw; 918 vw		715 w; 730 m; 860 vw; 935 vw		732 sh; 760 m	
F$_{2u}$		700; 790		690; 790		680 w; 795		790		790		600 to 740; 810 to 960
E$_g$	1112 w; 1152 w		1136 mw; 1145 mw		1142 w; 1156 w		1168 m		1180 mw		1122 m; 1150 sh	
F$_{1u}$		1095 to 1170		1095 to 1165		1100 to 1160		1085 to 1155		1100 to 1160		1090 to 1170
A$_{1g}$	1242 s		1274 s		1279 s		1298 s		1314 s		1258 vs	
$\nu_5 + \nu_T$		1340 to 1430		1320 to 1480		1330 to 1430		1375 to 1410		1330 to 1470		1310 to 1480

s = strong, m = medium, mw = medium to weak, w = weak, vw = very weak, vs = very strong, sh = shoulder, br = broad.

References:

[1] Ishii, M.; Aono, M.; Muranaka, S.; Kawai, S. (Solid State Commun. **20** [1976] 437/40).

[2] Zirngiebl, E.; Blumenröder, S.; Mock, R.; Güntherodt, G. (J. Magn. Magn. Mater. **54/57** [1986] 359/60).

[3] Smith, H. G.; Dolling, G.; Kunii, S.; Kasaya, M.; Liu, B.; Takegahara, K.; Kasuya, T.; Goto, T. (Solid State Commun. **53** [1985] 15/9).

[4] Takegahara, K.; Kasuya, T. (Solid State Commun. **53** [1985] 21/5).

[5] Turrell, S.; Yahia, Z.; Huvenne, J. P.; Lacroix, B.; Turrell, G. (J. Mol. Struct. **174** [1988] 455/60).

32.1.5.6 Nuclear Magnetic Resonance (NMR)

The ^{11}B NMR spectra of MB$_6$ (point symmetry m$\bar{3}$m at the M sites, 4 mm at the B sites) show a central line corresponding to transitions between nuclear Zeeman levels $m_I = +\frac{1}{2} \leftrightarrow m_I = -\frac{1}{2}$ and satellites due to nuclear electric quadrupole splitting of the nuclear resonance, characteristic of the nuclear spin $I = \frac{3}{2}$ of the ^{11}B nucleus, Gossard, Jaccarino [1]. Studies were performed on powder samples at room temperature at fixed frequencies of 9.5 and 16 MHz in the case of MB$_6$ with M = trivalent La to Nd, Gd to Ho, and intermediate valent Sm [1] and at 14.5 MHz for M = La, Sm, divalent Eu and Yb. The magnitude of the ^{11}B nuclear electric quadrupole interaction $|e^2qQ|$, derived from the separation of the first-order satellite peaks, and the corresponding values of the magnitude of the electric field gradient $|eq|$, based on the value $Q = 0.03 \times 10^{-24}$ cm^2 for ^{11}B, decrease strongly with increasing lattice parameter, see **Fig. 14**, Aono, Kawai [2]. Both values of $|e^2qQ|$ reported by Kushida et al. [3] for M = La, Ce, Pr, Sm, and Gd and values of v_Q ($= 3e^2qQ/\{2I(2I-1) \cdot h\}$, see Jones et al. [4]) reported by Bose et al. [5] for M = La, Pr, and Sm (see the individual sections) agree with this trend. For YB$_6$, $|e^2qQ| = 0.91$ MHz, from the second-order broadening of the ^{11}B central resonance see [3]. The trend of $|eq|$ in the divalent MB$_6$ can be explained by a change in the electronic structure in the boron framework due to the increase in the lattice parameter, see the broken curve in Fig. 14, which gives eq[M^{2+}(B$_6$)$^{2-}$], regarding M^{2+} as point charges and neglecting the interaction between B and M; for details, see the paper. At a given lattice parameter, the experimental $|eq|$ value for the trivalent MB$_6$ is distinctly smaller than that for divalent MB$_6$. Analysis of this difference as a mere size effect (dotted curve in Fig. 14) leads to the conclusion that the distribution of conduction electrons in the trivalent MB$_6$ deviates considerably from a uniform contribution [2].

The central line in the ^{11}B NMR spectra (studies from 2 to 15 MHz) exhibits second-order quadrupolar splitting below ca. 5 MHz for LaB$_6$ and below 6 MHz for PrB$_6$ and SmB$_6$. At higher frequencies a single line is observed for all three borides, indicating that quadrupolar and magnetic interactions are interwoven, Bose et al. [6]. Studies at 9.5 and 16 MHz at temperatures between 4 and 300 K show that the nonmagnetic compounds "ScB$_6$" (but see p. 21), YB$_6$, and LaB$_6$ have a relatively narrow absorption line, a temperature-independent line width of 17.8, 17.4, and 18.3 Oe, respectively, at 16 MHz at 20 and 295 K, and a temperature-independent shift in field for resonance (referred to ^{11}B in Na$_2$B$_2$O$_4$). The magnetic compounds MB$_6$ with M = Ce, Pr, Nd, Gd to Ho display appreciable NMR shifts and line widths, which depend on both field and temperature [1]. In addition, the latter compounds reveal pronounced asymmetric lines, McNiff, Shapiro [7], for GdB$_6$ also [3]. On employing the method of Jones et al. [4] for interwoven interactions, the following values were obtained for the isotropic (K_{iso}) and axial (K_{ax}) components of the chemical (Knight) shift tensor and the ratio $a = K_{ax}/(1 + K_{iso})$ at room temperature:

MB$_6$	K$_{iso}$ in %	K$_{ax}$ in %	a
LaB$_6$	+ 0.028	+ 0.157	+ 0.153
PrB$_6$	− 0.04	− 0.02	− 0.02
SmB$_6$	− 0.03	+ 0.1067	+ 0.111

Bose et al. [5]. Earlier, the observed shift of the ^{11}B NMR absorption derivative of the central line was analyzed in considering the magnetic and the quadrupolar interaction to act independently of each other. After correcting the resonance shift (at 7.911 MHz, referred to ^{11}B in Na$_2$B$_2$O$_4$) for quadrupolar interaction, the chemical shifts shown in **Fig. 15** for room temperature were obtained [7]. Similar analyses of data at 9.5 and 16 MHz led to a positive chemical shift for MB$_6$ with M = Y (0.04%), La (0.02%), and Ce to Sm and to a negative one for M = Gd to Ho. The shifts of the paramagnetic compounds were explained by the exchange interaction between localized 4f moments and conduction electrons, which results in a magnetic polarization of the conduction electrons [1]. Metal-deficient samples obtained by reaction of commercial samples with excess boron had K = + 0.018% (LaB$_6$) and − 0.133% (EuB$_6$) compared to + 0.031% and − 0.059% for the starting samples, respectively. For a discussion, see the paper [7]. On cooling PrB$_6$ and GdB$_6$, the magnitudes of the shifts increased in proportion to the susceptibility, reaching at 20 K values of + 0.14% in the former and − 0.58% in the latter (still paramagnetic) sample [1].

Fig. 14. Magnitudes of the ^{11}B nuclear electric quadrupole interaction $|e^2qQ|$ and electric field gradient $|eq|$ at the boron nucleus as a function of the lattice constants for MB$_6$ with divalent M = Eu, Yb, and Ca, intermediate valent Sm, and trivalent M (all others). Squares from Aono, Kawai [2], circles from Gossard, Jaccarino [1].

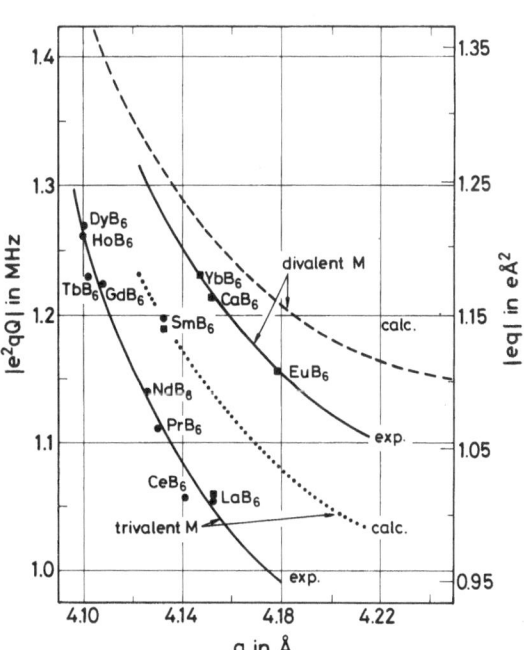

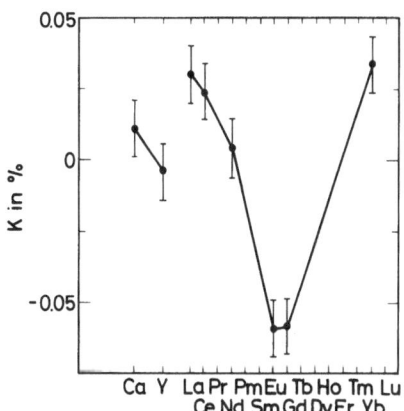

Fig. 15. Chemical shift of the ^{11}B resonance in rare earth hexaborides.

The chemical shift of the ^{139}La resonance in LaB₆ amounts to -0.038% [5] or $-0.04 \pm 0.02\%$. The same shift was observed for ^{45}Sc in "ScB₆" [1].

References:

[1] Gossard, A. C.; Jaccarino, V. (Proc. Phys. Soc. [London] **80** [1962] 877/81).
[2] Aono, M.; Kawai, S. (J. Phys. Chem. Solids **40** [1979] 797/802).
[3] Kushida, T.; Laurance, N.; Silver, A. H. (Bull. Am. Phys. Soc. [2] **7** [1962] 226).
[4] Jones, W. H., Jr.; Graham, T. P.; Barnes, R. G. (Phys. Rev. [2] **132** [1963] 1898/909).
[5] Bose, M.; Roy, K.; Basu, A. (Proc. Nucl. Phys. Solid State Phys. Symp. **21c** [1978] 688; C.A. **92** [1980] No. 155577).
[6] Bose, M.; Roy, K.; Basu, A. (J. Phys. C **13** [1980] 3951/9).
[7] McNiff, E. J., Jr.; Shapiro, S. (J. Phys. Chem. Solids **24** [1963] 939/45).

32.1.5.7 Crystal Field Splitting

The large crystal field splitting (CFS) observed in CeB₆ can be explained by the p(B)–f(Ce) mixing mechanism. As the B₆ bonding states have much stronger p–f mixing than the unoccupied antibonding states, the c–f² mechanism is dominant, indicating the similar p–f mixing effect on the crystal field splitting in PrB₆ and NdB₆, Kasuya et al. [1]. Earlier calculations on the CFS of these three compounds were performed on the basis of the s–f and p–f mixing model by taking into account the following points: the energy levels of the molecular orbitals of the B₆ octahedron (see p. 57), the occupied f levels derived from the photoemission spectrum, the unoccupied f levels determined by means of the correlation energy $U_{ff} = 6$ eV for CeB₆ and 8 eV for PrB₆ and NdB₆, and the mixing parameters (pfσ) $= -0.34$ eV for CeB₆ and -0.287 eV for PrB₆ and NdB₆. The calculated CFS agreed fairly well with recent experimental results (from Zirngiebl et al. [2] and Loewenhaupt et al. [3]) only in the case of CeB₆, see **Fig. 16.** Discrepancies concerning the character of the excited levels in PrB₆ and the overall splitting in NdB₆ might be reduced when considering different mixing matrices for the occupied and unoccupied f states, Ikeda et al. [4].

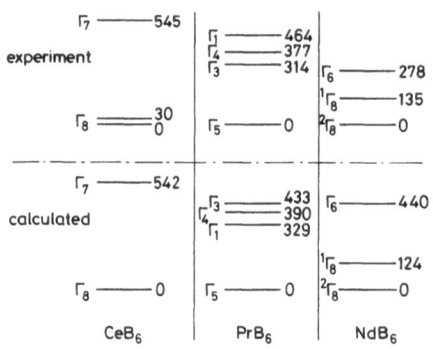

Fig. 16. Experimental and calculated crystal field splitting of CeB₆, PrB₆, and NdB₆ (values in K).

References:

[1] Kasuya, T.; Sakai, O.; Harima, H.; Ikeda, M. (J. Magn. Magn. Mater. **76/77** [1988] 46/52).
[2] Zirngiebl, E.; Hillebrands, B.; Blumenröder, S.; Güntherodt, G.; Loewenhaupt, M.; Carpenter, J. M.; Winzer, K.; Fisk, Z. (Phys. Rev. [3] B **30** [1984] 4052/4).
[3] Loewenhaupt, M.; Carpenter, J. M.; Loong, C.-K. (J. Magn. Magn. Mater. **52** [1985] 245/9).
[4] Ikeda, M.; Aoki, Y.; Kasuya, T. (J. Magn. Magn. Mater. **52** [1985] 264/6).

32.1.5.8 Mechanical Properties

Density

According to Binder [1] values for the density reported in the literature deviate strongly in most cases, because they were not corrected for sample porosity and stoichiometry. Along the homogeneity range M_xB_6 ($x \leqq 1$), the density decreases with the increase of B/M (decrease of x) for all M, Yajima, Niihara [2]. Values for D_{calc} derived from lattice constants and for D_{exp} determined pycnometrically for nominal MB_6 or specified M_xB_6, prepared by borothermal reduction, are (both in g/cm^3):

M	nominal MB_6				M_xB_6		
	D_{calc}*)	D_{calc}	D_{exp}*)	D_{exp}*)	D_{exp} (at x=1) to	D_{exp} (at x_{low})	x_{low}
Y	3.70	3.695	—	3.71	—	—	—
La	4.71	4.711	4.68	4.717	—	4.71 to 3.98	0.77
Ce	4.79	4.798	4.67	4.80	4.80 to 4.00	4.80 to 4.00	0.70
Pr	—	4.840	—	4.83	4.65 to 3.95	4.84 to 3.95	0.69
Nd	4.94	4.943	—	4.94	—	4.94 to 4.14	0.72
Sm	5.05	5.071	5.03	5.06	5.02 to 3.90	5.05 to 3.90	0.68
Eu	4.91	4.936	4.83	4.88	—	4.88 to 4.38	0.87
Gd	5.32	5.325	5.20	5.31	5.17 to 4.20	5.35 to 4.68	0.70
Tb	5.38	5.386	—	—	5.22 to 4.68	5.35 to 4.68	0.75
Dy	5.48	5.487	—	5.48	—	—	—
Yb	5.53	5.538	5.47	5.52	—	—	—
Ref.	[3]	[1]	[3]	[1]	[2]	[1]	[1]

*) Error limits of ± 0.02 g/cm^3.

$D_{calc} = 5.554$ g/cm^3 for HoB_6 and 5.602 g/cm^3 for "ErB_6" (but see p. 21) [1]. Values other than D_{exp} for MB_6, taken from the review by Binder [1], are based on studies from Tvorogov [4] for M = Y to Yb; from Mordovin, Timofeeva [5] for M = La, Pr, Nd, Sm to Er, Yb; from Yajima, Niihara [2] for the homogeneity ranges of M = Ce, Pr, Sm, Gd, and Tb, and on the review from Spear [6]. For recent D_{exp} values for zone-melted crystals with M = La, Ce, and Sm, see p. 33. Early studies on MB_6 prepared by flux electrolysis led to D_{calc} values similar to those given above for M = La, Ce, Pr, Nd (and Er), see von Stackelberg, Neumann [7], but resulted for the most part in much lower D_{exp} values (M = Y, La, Ce, Nd, Gd, Er, Yb), see Andrieux [8]. Exceptionally large values were calculated for borothermally prepared samples with M = Y and La, but not Gd, by Derenovskii et al. [9].

Microhardness. Brittleness

All values were converted from kp/mm^2 into Pa; $1\,kp/mm^2 = 9.80665$ MPa.

The rare earth hexaborides MB_6 are hard compounds. The Knoop microhardness H_K at the (100) faces of single crystals increases with decreasing load, with (100) LaB_6 in the [010] direction from 16.87 GPa to 24.03 GPa when the load P was decreased from 3923 to 490 mN. The microhardness of (100) planes exhibits anisotropy, in that H_K ($\parallel$[011]) < H_K ($\parallel$[010]). Average Knoop hardness values measured on the (100) plane along [010] under P = 1960 mN are:

MB_6	LaB_6	CeB_6	PrB_6	NdB_6	SmB_6	EuB_6
H_K in GPa ...	19.42±1.27	18.98±1.37	18.44±1.27	18.98±0.59	17.46±0.64	17.80±0.64

Futamoto et al. [10]. $H_K = 19.96$ GPa for LaB$_6$, but the same values as given by [10] for M = Ce, Pr, Nd, and Sm were earlier reported by Okano et al. [11]. The Vickers microhardness H_V of flux-grown needles on the (1$\bar{1}$0) plane was found to be the same (within the error of measurements) as that of plates on the (100) plane. Mean values observed for crystals of different compositions under P = 980 mN (M = La, Eu) or 490 mN (M = Sm, Yb) are:

MB$_6$	LaB$_6$, LaB$_{5.86}$	SmB$_6$	EuB$_{6.02}$, EuB$_{5.83}$	YbB$_{6.1}$
H_V in GPa	25.69 ± 0.78	24.52 ± 1.47	24.03 ± 1.18	24.03 ± 1.57

Gurin et al. [12]. The Knoop hardness of polycrystalline samples was measured under P = 980 mN by Binder [1], see table below, and was compared with data from the literature, both H_K values measured under the same load and H_V values measured under 490 mN. The latter values (H_V) ranged from 23.04 GPa for DyB$_6$ and HoB$_6 \approx$ TbB$_6$ to 25.50 GPa for LaB$_6$ and YbB$_6$ [1]. Values for H_K (error limit of ± 0.98 GPa) measured by [1] under P = 980 mN and values for H_V measured by Ballowe et al. [13] are given below, together with nonspecified values from other authors, presumably measured under P = 490 mN (all in GPa):

YB$_6$	LaB$_6$	CeB$_6$	PrB$_6$	NdB$_6$	SmB$_6$	EuB$_6$	GdB$_6$	TbB$_6$	DyB$_6$	YbB$_6$	Ref.
19.12	20.79	20.10	20.59	19.12	18.83	18.73	18.73	17.95	—	18.63	[1]
15.69	—	—	—	—	21.57*)	22.55	—	—	18.14	—	[13]
—	25.99	—	26.28	25.01	25.50	23.73	—	27.85	23.14	26.48	[5]
—	24.52	—	—	—	25.50	24.32	24.03	—	24.32	25.30	[14]
32.01	—	—	—	—	24.52	—	24.03	—	—	—	[15]

*) Under P = 980 mN, Hoyt, Chorné [16].

For data on MB$_6$ (M = La, Pr, Nd, Sm, Eu, Yb) under P = 137 to 235 mN, see Paderno et al. [17]. Irradiating hot-pressed YB$_6$ and SmB$_6$ specimens up to 226 d with neutrons from the reactor core led to a pronounced softening, which was more significant at the sample surface than in the center; for H_V values, see the paper, Hoyt, Zimmerman [18].

Rare earth hexaborides MB$_6$ are brittle, Binder [1]. An electron microscopic investigation of the structure of the fracture surface of zone-refined crystals revealed that the brittleness increases (plasticity decreases) with relative increase in the distances between the metal atoms of the compounds, i.e. in the order EuB$_6$ > YbB$_6$ > LaB$_6$ > SmB$_6$ > PrB$_6 \approx$ NdB$_6$ > GdB$_6$, Paderno et al. [19, 20]. Later, the microelasticity was found to decrease in a similar sequence, except for SmB$_6$. This compound was found to have the largest microelasticity. Micropressure diagrams for LaB$_6$ revealed hysteresis (for details, see the paper), Paderno et al. [17].

Elastic Properties

The bending strength σ amounts to 126.52 MPa for LaB$_6$, 148.09 MPa for SmB$_6$, and 183.40 MPa for EuB$_6$, Arabei et al. [21]. Ranges for σ from 0.2 to 0.4 GPa for each MB$_6$ (M = Y, La to Er, Yb) are listed in the review by [1]. Experimental values for the Young modulus E are: 393 GPa, Paderno et al. [17], and 402 GPa, both for LaB$_6$; 383 GPa for YB$_6$, and 393 GPa for CeB$_6$. Calculated values for the other compounds range from E = 373 GPa for DyB$_6$ and HoB$_6$ to E = 412 GPa for GdB$_6$, with an error limit of ± 39 GPa for all [1].

Pressure-volume diagrams derived from lattice constants of single crystals at room temperature are shown in **Fig. 17**. The volume V_0 for zero pressure, in the case of LaB$_6$ and GdB$_6$ with trivalent cations is the measured one. The curves for SmB$_6$ with intermediate valent Sm and EuB$_6$ with divalent Eu have been shifted upwards by $(V_0 - V_0^{3+})/V_0$, the difference between their true zero-pressure volume and their hypothetical trivalent volume at zero pressure. The curvature in the p–V data for SmB_6 is due to a smooth valence change of Sm from 2.8 at p = 0 to 2.9 at p = 6 GPa. Measured data were fit to the Birch-Murnaghan equation to give the bulk modulus K_0 and its pressure derivative K_0':

MB_6	LaB_6	SmB_6	EuB_6	GdB_6
K_0 in GPa	191	139	157	190
K_0'	−3	5	0	−3

Plots of $K_0 \cdot V_0$ versus valence, from observed data, deviate from data predicted for an ionic crystal with a Born-Mayer repulsive potential and reveal that K_0 of the integral-valent compounds has both a valence dependent and a large valence independent contribution. The latter, being largest for EuB_6, probably arises from the boron-boron repulsion, which is ignored in the ionic model. The product $K_0 \cdot V_0$ of intermediate valent SmB_6 lies below that predicted from the ionic model (see a figure in the paper), King et al. [22]. For crystals with an intermediate bond type like MB_6, the compressibility $\varkappa_0$ obeys the rule of inverse additivity of the ionic component (index 1) and the covalent component (index 2) of the lattice energy and thus of the compressibility, $1/\varkappa_0 = \varepsilon^2/\varkappa_1 + (1 - \varepsilon^2)/\varkappa_2$. On use of $\varepsilon = 0.22$ as mean value for the ionicity of all MB_6, a reduced Madelung constant of 1.67, and a repulsive potential of boron of 0.345, values for $\varkappa_0$, $\varkappa_1$, and $\varkappa_2$ (see **Fig. 18**) were calculated for the whole MB_6 series, i.e., inclusive of the instable compounds with M = Er, Tm, and Lu, Kuz'micheva et al. [23]. (For example, K_0 (= $1/\varkappa_0$) ≈ 170 GPa for LaB_6, ~125 GPa for SmB_6, and ~120 GPa for EuB_6 according to this figure.)

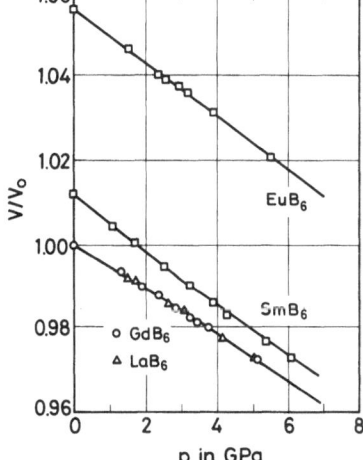

Fig. 17. Pressure-volume diagrams for MB_6 (M = La, Sm, Eu, and Gd); see text.

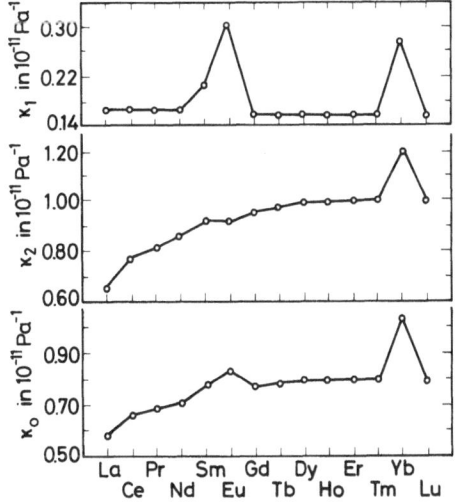

Fig. 18. Calculated compressibility $\varkappa_0$ of MB_6 (M = La to Lu) and the ionic ($\varkappa_1$) and covalent ($\varkappa_2$) contributions.

References:

[1] Binder, F. (Radex Rundschau **1977** 52/71; C.A. **87** [1977] No. 27597).

[2] Yajima, S.; Niihara, K. (Proc. 9th Rare Earth Res. Conf., Blacksburg, Va., 1971, Vol. 2, pp. 598/609).

[3] Etourneau, J.; Mercurio, J.-P.; Naslain, R.; Hagenmuller, P. (Colloq. Intern. Centre Natl. Rech. Sci. [Paris] No. 5 [1972] 429/38; C.A. **78** [1973] No. 167937).

[4] Tvorogov, N. N. (Zh. Neorgan. Khim. **4** [1959] 1961/6; Russ. J. Inorg. Chem. **4** [1959] 890/3).

[5] Mordovin, O. A.; Timofeeva, E. N. (Zh. Neorgan. Khim. **13** [1968] 3155/8; Russ. J. Inorg. Chem. **13** [1968] 1627/9).

[6] Spear, K. E. (in: Alper, A. M., Refractory Materials, Phase Diagrams, Vol. 4, Academic, New York 1976, pp. 91/159, 112, 127).

[7] von Stackelberg, M.; Neumann, F. (Z. Physik. Chem. B **19** [1932] 314/20).

[8] Andrieux, L. (Ann. Chim. [Paris] [10] **12** [1929] 423/507, 476).

[9] Derenovskii, M. V.; Shlyuko, V. Ya.; Chernyak, L. V. (Poroshkovaya Metal. **1967** No. 2, pp. 8/12; Soviet Powder Met. Metal Ceram. **1967** 93/6).

[10] Futamoto, M.; Aita, T.; Kawabe, U. (Mater. Res. Bull. **14** [1979] 1329/34).

[11] Okano, H.; Futamoto, M.; Hosoki, S.; Kawabe, U. (Shinku **20** No. 4 [1977] 127/35; C.A. **87** [1977] No. 94150).

[12] Gurin, V. N.; Korsukova, M. M.; Nikanorov, S. P.; Smirnov, I. A.; Stepanov, N. N.; Shul'man, S. G. (J. Less-Common Metals **67** [1979] 115/23).

[13] Ballowe, W. C.; Holden, A. N.; Ozeroff, W. J.; Williamson, H. E.; Weidenbaum, B. (GEAP-3355-Pt. 2 [1960] 1/132, 1/31, 17; N.S.A. **15** [1961] No. 26541).

[14] Timofeeva, N. I.; Timofeeva, E. N. (Izv. Akad. Nauk SSSR Neorgan. Materialy **4** [1968] 1789/91; Inorg. Materials [USSR] **4** [1968] 1559/61).

[15] Markov, Yu. M.; Trokhina, G. N.; Zernova, E. E.; Maksimovskii, V. V. (Izv. Akad. Nauk SSSR Neorgan. Materialy **14** [1978] 79/81; Inorg. Materials [USSR] **14** [1978] 61/3).

[16] Hoyt, E. W.; Chorné, J. (GEAP-3332 [1960] 1/15; C.A. **1961** 1338).

[17] Paderno, V. N.; Paderno, Yu. B.; Pilyankevich, A. N.; Lazorenko, V. I.; Bulychev, S. I. (J. Less-Common Metals **67** [1979] 431/6).

[18] Hoyt, E. W.; Zimmerman, D. L. (GEAP-3743-Pt. 1 [1962] 1/41, 27; C.A. **59** [1963] 12331).

[19] Paderno, Yu. B.; Pilyankevich, A. N.; Paderno, V. N. (Bor: Poluch. Strukt. Svoistva Mater. 4th Mezhdunar. Simp. Boru, Tbilisi 1972 [1974], Vol. 2, pp. 124/32 from C.A. **83** [1975] No. 140041).

[20] Paderno, Yu. B.; Pilyankevich, A. N.; Paderno, V. N. (Izv. Akad. Nauk SSSR Neorgan. Materialy **11** [1975] 856/8; Inorg. Materials [USSR] **11** [1975] 733/5).

[21] Arabei, B. G.; Shtrom, E. N.; Lapitskii, Yu. A. (Poroshkovaya Metal. **1964** No. 5, pp. 65/70; Soviet Powder Met. Metal Ceram. **1964** 406/9).

[22] King, H. E., Jr.; La Placa, S. J.; Penney, T.; Fisk, Z. (Valence Fluctuations Solids St. Barbara Inst. Theor. Phys. Conf., Santa Barbara, Calif., 1981, pp. 333/6; C.A. **95** [1981] No. 157022).

[23] Kuz'micheva, G. M.; Khlyustova, S. Yu.; Tolstova, V. A.; Eliseev, A. A. (Zh. Neorgan. Khim. **33** [1988] 2205/10; Russ. J. Inorg. Chem. **33** [1988] 1259/62).

32.1.5.9 Thermal Properties

Thermal Expansion

The temperature dependence of the true thermal expansion coefficient, $\alpha = \alpha_0 + \alpha_1 \cdot T$ for various hexaborides was derived from X-ray intensity studies on the (410) and (411) reflections between 273 and 973 K. Selected values for α and the percentage of the temperature-independent contribution α_0 are:

MB_6	T in K	273	293	373	473	573	673	773	873	973
YB_6	$10^6\alpha$ in K^{-1}	5.82	5.91	6.29	6.76	7.22	7.69	8.15	8.62	9.08
	α_0/α in %	100	98.40	92.49	86.04	80.43	75.51	71.15	67.27	63.79
LaB_6	$10^6\alpha$ in K^{-1}	5.48	5.55	5.83	6.19	6.54	6.89	7.25	7.60	7.95
	α_0/α in %	100	98.71	93.87	88.44	83.61	79.28	75.38	71.84	68.62
PrB_6	$10^6\alpha$ in K^{-1}	6.93	6.97	7.13	7.33	7.52	7.72	7.92	8.11	8.31
	α_0/α in %	100	99.42	97.15	94.49	91.96	89.56	87.28	85.11	83.06
NdB_6	$10^6\alpha$ in K^{-1}	6.37	6.44	6.72	7.06	7.41	7.75	8.10	8.44	8.78
	α_0/α in %	100	98.90	94.76	90.05	85.78	81.90	78.35	75.10	72.11
YbB_6	$10^6\alpha$ in K^{-1}	5.78	5.85	6.12	6.45	6.79	7.12	7.46	7.79	8.12
	α_0/α in %	100	98.83	94.45	89.47	85.00	80.95	77.27	73.92	70.83

In view of the proportional dependences $\alpha_0 \sim \Phi_3$ and $\alpha_1 \sim \Phi_5$ existing between the components of the coefficients of thermal expansion and the odd terms in the expansion of the lattice potential energy in powers of the displacements of the atoms from the equilibrium positions, the coefficient of linear expansion is mainly determined by Φ_3, in the whole temperature range studied. The effect of Φ_5 increases with temperature, minimal for PrB_6, Dutchak et al. [1].

The mean linear thermal expansion coefficient was derived from early lattice constant measurements at 293 K and in the range from 773 to 1073 K:

MB_6	YB_6	LaB_6	CeB_6	PrB_6	NdB_6	SmB_6	EuB_6	GdB_6	TbB_6	YbB_6
$10^6\alpha$ in K^{-1}	6.2	6.4	7.3	7.5	7.3	6.8	6.9	8.7	7.8	5.8

Zhuravlev et al. [2], for M = La, Ce, and Yb also Zhdanov et al. [3], and for M = La and Ce also Stepanova, Umanskii [4]. Similar values were listed by Binder [5]. Values determined by Goryachev, Kovenskaya [6] for MB_6 with M = Ce to Nd, Gd to Er increase from (in 10^{-6} K^{-1}) ca. 7.2 for CeB_6 to 7.9 for HoB_6, whereby GdB_6 has about the same value (ca. 7.7) as NdB_6. For "ErB_6" (but see p. 21), $\alpha \approx 8.5 \times 10^{-6}$ K^{-1}, see a figure in the paper [6].

Melting Point

MB_6 with M = La to Sm melts congruently, whereas MB_6 with M = Y, Gd to Ho melts peritectically, see the review by Spear [7, pp. 141/55]. For the complex melting behavior of EuB_6 and YbB_6, see the individual sections. Melting points are (in K):

YB_6	2573 [8], 2873 [7, 9]	GdB_6	2783 [8, 10]
LaB_6	2988 [10], >2773 [11]	TbB_6	2613 [10]
CeB_6	2823 [7], >2463 [11]	DyB_6	2473 [10]
PrB_6	2883 [10]	HoB_6	2453 [10]
NdB_6	2883 [10], 2813 [11]	ErB_6[2)]	2458 [10]
PmB_6	2873[1)] [7]	YbB_6	2643 [10]
SmB_6	2853 [10], 2813 [11], 2800 [9], 2673 [8]	LuB_6[2)]	2443 [10]
EuB_6	2933 [10], 2800 [9]		

[1)] Estimated. – [2)] Instable compounds, see p. 21.

Values determined by Mordovin, Timofeeva [10], which have an error of ± 30 K, were adopted by [7]. Similar values were determined by Goryachev, Kovenskaya [6] for MB$_6$ with M = Ce to Nd, Gd to Er, see a figure in the paper. For melting points of YB$_6$, LaB$_6$ to TbB$_6$, and YbB$_6$, derived from the mean thermal expansion coefficient, see Zhuravlev et al. [2].

Debye Temperature

Values for the Debye temperature at room temperature, as derived from published values of the elastic constants by Smith et al. [12], from the Lindemann formula by Dutchak et al. [13], and from the mean linear thermal expansion coefficient by Zhuravlev et al. [2] are:

MB$_6$	YB$_6$	LaB$_6$	CeB$_6$	PrB$_6$	NdB$_6$	SmB$_6$	EuB$_6$	GdB$_6$	TbB$_6$	YbB$_6$	Ref.
Θ_D in K	—	404	381	—	—	373	—	—	—	265	[12]
Θ_D in K	811	690	687	730	730	—	—	—	—	625	[13]
Θ_D in K	922	885	747	730	752	755	735	745	690	763	[2]

Values from [2] were listed by Samsonov [14]; for earlier different values, see Samsonov, Paderno [15].

Heat Capacity

In **Fig. 19** from Kasuya et al. [16] is shown the low-temperature molar heat capacity C$_p$ for zone-refined LaB$_6$ and SmB$_6$ single crystals, measured by these authors (solid curves) and for zone-refined LaB$_6$ and sintered YbB$_6$ measured by Etourneau et al. [17] (dashed curves) and is compared to the calculated Debye heat capacity of LaB$_6$ with $\Theta_D = 404$ K. With SmB$_6$, the small hump at the lowest temperature may be due to some impurities, whereas the strong rise near 10 K originates in the magnetic contribution [16]. Recent measurements above 10 K on purer samples of LaB$_6$ and YbB$_6$, however, do agree with the heat capacity predicted from the acoustic modes (Debye plus Einstein modes), see Smith et al. [12] and the individual sections. The heat capacity of zone-refined MB$_6$ with M = La, Nd, and Gd and of less purified CeB$_6$ and PrB$_6$ was measured in an adiabatic vacuum cryostat between 5 and 350 K. Diamagnetic LaB$_6$ exhibits a normal sigmate behavior, however, with an unusual curvature near 50 K. The other MB$_6$ (M = Ce, Pr, Nd, Gd) all showed cooperative, antiferromagnetic to paramagnetic λ-type transitions, CeB$_6$ near 7 K, NdB$_6$ at 7.5 K, and GdB$_6$ at 16 K. These were followed by complex Schottky transitions, a C$_p$ maximum at ca. 35 K for CeB$_6$ and at 50 K for NdB$_6$, and a shoulder near 30 K for GdB$_6$, Westrum [18, 19]. For the possible use of MB$_6$ (M = Ce, Pr, Nd) and of EuB$_{5.97}$C$_{0.03}$ in low-temperature cryocooler regenerators, see Barbisch et al. [20].

The heat capacity of hot-pressed samples of YB$_6$, SmB$_6$, and GdB$_6$ with given porosity P was measured by the method of plane temperature waves at 700 to 1200 K and 667 µPa. Values for c$_p$ in kJ·kg^{-1}·K^{-1} are (overall error $\pm 4\%$):

MB$_6$	P	700 K	800 K	900 K	1000 K	1200 K
YB$_6$	9.9%	0.97	1.00	1.02	1.04	1.10
	0%	1.03	1.07	1.08	1.11	1.16
SmB$_6$	8.0%	0.89	0.92	0.92	0.93	0.97
	4.3%	0.94	0.95	0.97	0.97	1.00
GdB$_6$	2.6%	0.63	0.65	0.67	0.68	0.7
	0.4%	0.69	0.68	0.71	0.71	0.71

Values for intermediate porosities, see the paper, Markov et al. [8]. Enthalpy measurements on samples prepared by the borothermal method lead to the following temperature dependences of c_p (in $kJ \cdot kg^{-1} \cdot K^{-1}$):

NdB_6 (298 to 2000 K): $c_p = 0.4812 + 0.4356 \times 10^{-3} \, T - 0.1135 \times 10^5 \, T^{-2}$

SmB_6 (298 to 1300 K): $c_p = 0.5841 + 0.3378 \times 10^{-3} \, T - 0.62283 \times 10^5 \, T^{-2}$

EuB_6 (298 to 1300 K): $c_p = 0.4686 + 0.4624 \times 10^{-3} \, T - 0.1799 \times 10^5 \, T^{-2}$

Prilepskii et al. [21]. With SmB_6 between 700 and 1200 K, values from [8] deviate by less than 14% from those by [21]. For $C_p = f(T)$, with C_p in $cal \cdot mol^{-1} \cdot K^{-1}$, for LaB_6 between 1000 and 2000 K and for NdB_6, SmB_6, and EuB_6 in the above given temperature ranges, see the review by Gordienko et al. [22].

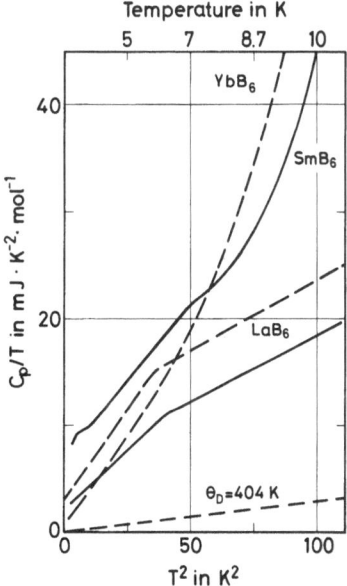

Fig. 19. Experimental molar heat capacity of MB_6 single crystals (M = La, Sm) and sintered YbB_6 as well as Debye heat capacity of LaB_6 calculated with $\Theta_D = 404$ K (solid curves of the C_p/T vs. T^2 plot are from Kasuya et al. [16] and dashed curves from Etourneau et al. [17]).

Thermodynamic Data of Formation

The enthalpies of formation $\Delta H_{f,298}$ for the congruently vaporizing hexaborides MB_6 with M = La to Eu were calculated by Ames, McGrath [24] from the average enthalpies of vaporization (see p. 90) and the enthalpies of sublimation of the elements (taken from literature). These values are compiled below, together with those $\Delta H_{f,298}$ values estimated by Smith [23, p. 413], based on reactions in the system M–B–C and M–B–W plus the stability of the various rare earth borides, and those $\Delta H_{f,298}$ values listed in the review by Gordienko et al. [22], all in kJ/mol (converted from kcal/mol):

MB_6	LaB_6	CeB_6	PrB_6	NdB_6	SmB_6	EuB_6	Ref.
$-\Delta H^\circ_{f,298}$	259.4	255.2	246.9	240.6	230.1	223.8	[23]
$-\Delta H^\circ_{f,298}$	117.2 ±20.9	—	121.3 ±20.9	165.3 ±20.9	225.9[1] ±12.6	—	[22]
$-\Delta H^\circ_{f,298}{}^{3)}$	189.1	214.2	55.6	—	52.7	110.5	[24]

MB$_6$	GdB$_6$	TbB$_6$	DyB$_6$	HoB$_6$	YbB$_6$	Ref.
$-\Delta H^\circ_{f,298}$	217.6	213.4	205.0	198.7	179.9	[23]
$-\Delta H^\circ_{f,298}$	129.7 ± 16.7	142.3 [2] ± 20.9	—	—	—	[22]

[1] At 2210 K. – [2] At 2162 K. – [3] Error limit 167 to 209 kJ/mol (± 40 to 50 kcal/mol) for all.

The (instable compounds) ErB$_6$, TmB$_6$, and LuB$_6$ have $-\Delta H^\circ_{f,298} = 46.0$, 44.5, and 41.5 kcal/mol (192.5, 186.2, and 173.6 kJ/mol), respectively [23]. Values ranging from -99.6 kcal/mol for PrB$_6$ to -134.0 kcal/mol (from -416.7 to -560.7 kJ/mol) for LuB$_6$ were extrapolated on the basis of literature data for YB$_6$ (-24 kcal/mol or -100.4 kJ/mol) and CeB$_6$ (-81 kcal/mol or -338.9 kJ/mol), Portnoi et al. [25], Timofeev, Timofeeva [26], and were used for thermodynamic calculations of the beginning formation of MB$_6$ from M$_2$O$_3$+15B, M$_2$O$_3$+3B$_4$C, and M$_2$O$_3$+12B+3C, see the paper [25]; for YB$_6$ and LaB$_6$, see also Peshev [27]. The values from [25, 26] are listed in the handbook by Samsonov et al. [28]; see also a figure in the handbook by Kost et al. [29].

Enthalpy and Entropy

Enthalpy increments as a function of temperature, determined calorimetrically on NdB$_6$, SmB$_6$, and EuB$_6$ prepared by the borothermal method are (H$_T$ – H$_{298.15}$ in kJ/kg):

NdB$_6$ (298 to 2000 K):   H$_T$ – H$_{298}$ = 0.4812 T + 0.2178$\times 10^{-3}$ T^2 + 0.1135$\times 10^5$ T^{-1} – 200.9

SmB$_6$ (298 to 1300 K):   H$_T$ – H$_{298}$ = 0.5814 T + 0.1689$\times 10^{-3}$ T^2 + 0.2283$\times 10^5$ T^{-1} – 265.7

EuB$_6$ (298 to 1300 K):   H$_T$ – H$_{298}$ = 0.4686 T + 0.2312$\times 10^{-3}$ T^2 + 0.1799$\times 10^5$ T^{-1} – 220.6

The root mean square deviation of the experimental data (see paper) and the data calculated by these formulas is less than ± 0.5 to 0.6%, Prilepskii et al. [21]. In the review of Gordienko et al. [22], formulas for H$_T$ – H$_{298}$ = f(T) in units of cal/mol are given for LaB$_6$ between 1000 and 2200 K and for NdB$_6$, SmB$_6$, and EuB$_6$ in the above given temperature ranges.

Standard entropies S$^\circ_{298}$ for several MB$_6$ have been calculated by the method of comparative calculation on use of values for LaB$_6$, NdB$_6$, and GdB$_6$, measured by Westrum [18]. The lattice entropy was set equal to the nonmagnetic contribution of 19.83 J·mol^{-1}·K^{-1} and the magnetic contribution was S$_{mag}$ = R ln (2J+1), where J is the total quantum number and R is the universal gas constant. The difference between the electronic contributions to the entropy of the borides and that of the metals was neglected, Borovikova, Fesenko [30]. Experimental values (exp) from [18] and calculated values from [30] are:

MB$_6$	YB$_6$	LaB$_6$	CeB$_6$	PrB$_6$
S$^\circ_{298}$ in J·mol^{-1}·K^{-1} ..	67.8 ± 4.2	83.16 ± 0.21(exp)	97.9 ± 4.2	101.6 ± 4.2

MB$_6$	NdB$_6$	SmB$_6$	GdB$_6$
S$^\circ_{298}$ in J·mol^{-1}·K^{-1} ..	103.45 ± 0.21(exp)	97.9 ± 4.2	126.37 ± 0.21(exp), 104.6 ± 6

For estimated values of S$^\circ$ at 2200 K for all MB$_6$, see [23, p. 390].

Values for the Gibbs function $-(G_T - H_{298})/T$, in cal·mol^{-1}·K^{-1} (in J·mol^{-1}·K^{-1} in parentheses), are:

MB$_6$	2000 K	2100 K	2200 K	2300 K
LaB$_6$	56.19 (235.1)	57.94 (242.4)	59.66 (249.6)	61.33 (256.6)
CeB$_6$	60.38 (252.6)	62.54 (261.7)	64.67 (270.6)	66.78 (279.4)

MB$_6$	2000 K	2100 K	2200 K	2300 K
PrB$_6$	61.12 (255.7)	63.28 (264.8)	65.41 (273.7)	67.52 (282.5)
SmB$_6$	63.73 (266.7)	65.89 (275.7)	68.02 (284.6)	70.12 (293.4)
EuB$_6$	64.18 (268.5)	66.55 (278.5)	68.90 (288.3)	71.23 (298.0)

Behrens, Gilles [31].

Thermal Diffusivity

The thermal diffusivity α of hot-pressed samples of YB$_6$, SmB$_6$, and GdB$_6$ of different porosity P was measured by the method of plane temperature waves at 700 to 1200 K and 667 μPa. Values for α (in 10^{-2} m^2/h) are (overall error $\pm 7\%$):

MB$_6$	P	700 K	800 K	900 K	1000 K	1100 K	1200 K
YB$_6$	9.9%	2.30	2.27	2.32	2.33	2.45	2.45
	0%	2.54	2.54	2.52	2.48	2.54	2.56
SmB$_6$	8.0%	1.42	1.47	1.47	1.48	1.49	1.54
	4.3%	1.72	1.71	1.68	1.71	1.73	1.78
GdB$_6$	2.6%	2.76	2.82	2.91	2.99	3.01	3.08
	0.4%	3.10	3.15	3.16	3.17	3.24	3.25

Values for intermediate porosities, see the paper, Markov et al. [8].

Thermal Conductivity

Values at room temperature of the measured thermal conductivity λ_{total}, the electronic contribution λ_{el} obtained from the Wiedemann-Franz law, and the phonon contribution $\lambda_{ph}(=\lambda_{total}-\lambda_{el})$ for hot-pressed samples are (in 10^{-3} cal$\cdot$cm$^{-1}\cdot$s$^{-1}\cdot$K^{-1}):

MB$_6$	YB$_6$	LaB$_6$	CeB$_6$	PrB$_6$	NdB$_6$	SmB$_6$	EuB$_6$	GdB$_6$	TbB$_6$	YbB$_6$
λ_{total}	70$\pm$10	114$\pm$10	81$\pm$2	98$\pm$5	113$\pm$8	33$\pm$4	55$\pm$2	49$\pm$3	48$\pm$3	60$\pm$4
λ_{el}	40	106	54	81	79	8	19	35	42	34
λ_{ph}	30	8	27	17	34	25	36	14	6	26

The relatively high values of λ_{total}, λ_{el}, and λ_{ph} of the hexaborides compared to the metals are explained with the strong covalent B–B bonds in MB$_6$, L'vov et al. [32, 33]. Differently, λ_{total} was found to decrease from ca. 80 to 70 W$\cdot$m$^{-1}\cdot$K^{-1} ($\sim$(191 to 167)$\times 10^{-3}$ cal$\cdot$cm$^{-1}\cdot$s$^{-1}\cdot$K^{-1}) on going from CeB$_6$ to NdB$_6$ and from ca. 100 to 50 W$\cdot$m$^{-1}\cdot$K^{-1} ($\sim$(239 to 119)$\times 10^{-3}$ cal$\cdot$cm$^{-1}\cdot$s$^{-1}\cdot$K^{-1}) between GdB$_6$ and (instable, see p. 21) ErB$_6$, see a figure in Goryachev, Kovenskaya [6]. The thermal conductivity of hot-pressed samples of YB$_6$, SmB$_6$, and GdB$_6$ of different porosity P between 700 and 1200 K was calculated from $\lambda = D\cdot c_p\cdot\alpha$ (with D = density at 293 K, c_p = measured heat capacity, and α = measured thermal diffusivity). Values for λ in W$\cdot$m$^{-1}\cdot$K^{-1} are:

MB$_6$	P	700 K	800 K	900 K	1000 K	1100 K	1200 K
YB$_6$	9.9%	20.33	20.68	21.56	22.08	23.66	24.55
	0%	25.84	26.60	27.14	27.20	28.62	29.87
SmB$_6$	8.0%	15.80	16.91	16.90	17.21	17.69	18.11
	4.3%	21.02	21.12	21.18	21.56	22.27	23.14

MB$_6$	P	700 K	800 K	900 K	1000 K	1100 K	1200 K
GdB$_6$	2.6%	23.52	24.79	26.38	27.50	27.69	29.16
	0.4%	29.59	29.63	31.04	31.13	32.27	31.92

For values of samples with intermediate porosities, see the paper, Markov et al. [8].

References:

[1] Dutchak, Ya. I.; Fedyshin, Ya. I.; Paderno, Yu. B. (Izv. Akad. Nauk SSSR Neorgan. Materialy 8 [1972] 2134/7; Inorg. Materials [USSR] 8 [1972] 1877/80).

[2] Zhuravlev, N. N.; Stepanova, A. A.; Paderno, Yu. B.; Samsonov, G. V. (Kristallografiya 6 [1961] 791/4; Soviet Phys.-Cryst. 6 [1961] 636/8).

[3] Zhdanov, G. S.; Zhuravlev, N. N.; Stepanova, A. A.; Umanskii, M. M. (Redk. Metally Splavy Tr. Pervogo Vses. Soveshch. Splavam Redkikh Metal., Moscow 1957 [1960], pp. 366/71; C.A. **1961** 3366).

[4] Stepanova, A. A.; Umanskii, M. M. (Bor. Tr. Konf. Khim. Bora Ego Soedin., Moscow 1955 [1956], pp. 102/5, AEC-tr-4270 [1960] 1/4; N.S.A. **15** [1961] No. 1838).

[5] Binder, F. (Radex Rundschau **1977** 52/71; C.A. **87** [1977] No. 27597).

[6] Goryachev, Yu. M.; Kovenskaya, B. A. (J. Less-Common Metals **67** [1979] 273/9).

[7] Spear, K. E. (in: Alper, A. M., Refractory Materials, Phase Diagrams, Vol. 4, Academic, New York 1976, pp. 91/159).

[8] Markov, Yu. M.; Trokhina, G. N.; Zernova, E. E.; Maksimovskii, V. V. (Izv. Akad. Nauk SSSR Neorgan. Materialy **14** [1978] 79/81; Inorg. Materials [USSR] **14** [1978] 61/3).

[9] Hoyt, E. W.; Chorné, J. (GEAP-3332 [1960] 1/15; C.A. **1961** 1338).

[10] Mordovin, O. A.; Timofeeva, E. N. (Zh. Neorgan. Khim. **13** [1968] 3155/8; Russ. J. Inorg. Chem. **13** [1968] 1627/9).

[11] Okano, H.; Futamoto, M.; Hosoki, S.; Kawabe, U. (Shinku **20** No. 4 [1977] 127/35; C.A. **87** [1977] No. 94150).

[12] Smith, H. G.; Dolling, G.; Kunii, S.; Kasaya, M.; Liu, B.; Takegahara, K.; Kasuya, T.; Goto, T. (Solid State Commun. **53** [1985] 15/9).

[13] Dutchak, Ya. I.; Fedyshin, Ya. I.; Paderno, Yu. B.; Vadets, D. I.; Odintsov, V. V. (Tezisy Dokl. 2nd Vses. Konf. Kristallokhim. Intermetal. Soedin., Lvov, USSR, 1974, p. 149; C.A. **86** [1977] No. 10825).

[14] Samsonov, G. V. (Tugoplavkie Soedineniya Redkozemel'nykh Metallov s Nemetallami, Metallurgiya, Moscow 1964; High Temperature Compounds of Rare Earth Metals with Nonmetals, Consultants Bureau, New York 1965, pp. 1/280, 18).

[15] Samsonov, G. V.; Paderno, Yu. B. (Vysokotemp. Metallokeram. Mater. **1962** 102/8; C.A. **58** [1963] 8483).

[16] Kasuya, T.; Takegahara, K.; Fujita, T. (J. Phys. Colloq. [Paris] **40** [1979] C5-308/C5-313).

[17] Etourneau, J.; Mercurio, J.-P.; Naslain, R.; Hagenmuller, P. (J. Solid State Chem. **2** [1970] 332/42).

[18] Westrum, E. F., Jr. (Colloq. Intern. Centre Natl. Rech. Sci. [Paris] No. 180 [1970] 443/50; C.A. **78** [1973] No. 89350).

[19] Westrum, E. F., Jr. (U.S. At. Energy Comm. COO-1149-149 [1969] 1/15; N.S.A. **23** [1969] No. 32578).

[20] Barbisch, B.; Olsen, J. L.; Kwasnitza, K. (Advan. Cryog. Eng. **32** [1986] 271/8; C.A. **106** [1987] No. 140186).

[21] Prilepskii, V. N.; Timofeeva, E. N.; Timofeev, V. A.; Trubitsyn, A. Ya. (Izv. Akad. Nauk SSSR Neorgan. Materialy **6** [1970] 2069/70; Inorg. Materials [USSR] **6** [1970] 1816/7).

[22] Gordienko, S. P.; Fenochka, B. V.; Viksman, G. Sh. (Termodinamika Soedinenii Lanta-noidov, Naukova Dumka, Kiev 1979, pp. 1/373, 68/84).

[23] Smith, P. K. (COO-1140-103 [1964] 1/485, 390, 413; N.S.A. 18 [1964] No. 25319, Diss. Univ. Kansas 1964, 1/485, 390, 413; Diss. Abstr. 25 [1964/65] 5591).

[24] Ames, L. L.; McGrath, L. (High. Temp. Sci. 7 [1975] 44/54; C.A. 83 [1975] No. 66462).

[25] Portnoi, K. I.; Timofeev, V. A.; Timofeeva, E. N. (Izv. Akad. Nauk SSSR Neorgan. Materialy 1 [1965] 1513/20; Inorg. Materials [USSR] 1 [1965] 1378/85).

[26] Timofeev, V. A.; Timofeeva, E. N. (Zh. Neorgan. Khim. 11 [1966] 1233/5; Russ. J. Inorg. Chem. 11 [1966] 659/61).

[27] Peshev, P. (Rev. Intern. Hautes Temp. Refract. 4 No. 4 [1967] 289/96; C.A. 69 [1968] No. 5740).

[28] Samsonov, G. V.; Serebryakova, T. I.; Neronov, V. A. (Boridy, Atomizdat, Moscow 1975, pp. 1/375, 171/7).

[29] Kost, M. E.; Shilov, A. L.; Mikheeva, V. I.; et al. (Khimiya Redkikh Elementov [Chemistry of Rare Earth Elements], Soedineniya Redkozemel'nykh Elementov, Nauka, Moscow 1983, pp. 40/68, 61).

[30] Borovikova, M. S.; Fesenko, V. V. (J. Less-Common Metals 117 [1986] 287/91).

[31] Behrens, R.; Gilles, P. (from [24]).

[32] L'vov, S. M.; Nemchenko, V. F.; Paderno, Yu. B. (Dokl. Akad. Nauk SSSR 149 [1963] 1371/2; Dokl. Phys. Chem. Proc. Acad. Sci. USSR 148/153 [1963] 353/4).

[33] L'vov, S. M.; Nemchenko, V. F.; Paderno, Yu. B. (Vysokotemp. Neorgan. Soedin. 1965 445/50; C.A. 64 [1966] 13538).

32.1.5.10 Magnetic Properties

LaB_6 and YbB_6 are diamagnetic, see the review by Etourneau et al. [1]. The observed weak paramagnetism, see a figure in Paderno et al. [2], may be due to the presence of impurities [1]. The other MB_6 compounds are paramagnetic and, except SmB_6 and EuB_6, become antiferro-magnetic at low temperatures [2]. With EuB_6, a transition to the antiferromagnetic state has also been observed, however, it turned out that pure EuB_6 becomes ferromagnetic (see Section 32.1.8.9 in "Rare Earth Elements" C 11b) as was found by Matthias et al. [3] and Hacker et al. [4]. The following values for the Néel temperature T_N and, in the case of EuB_6, the Curie temperature T_C (all in K) were derived from resistivity studies [3], Ali, Woods [5] and the magnetic susceptibility [3, 4]:

CeB_6	PrB_6	NdB_6	EuB_6	GdB_6	TbB_6	DyB_6	HoB_6	Ref.
3.0	7	8.6	8	~17.5	23	~21	9	[3]
—	8.3*)	—	8.8*)	16.4*)	—	—	—	[4]
—	6.99	7.74	—	15.2	—	20.3	—	[5]

*) Error limit of ± 0.2 K.

The samples of TbB_6, DyB_6, and especially HoB_6 used by [3] were contaminated with other boride phases, see Geballe et al. [6]. The measured T_N values of CeB_6, PrB_6, and NdB_6 are by a factor of more than two larger than those expected from a de Gennes scaling of the measured T_N values of MB_6 (M = Tb, Dy, Ho), see a figure in the paper. In the case of NdB_6, coinciding ferroquadrupolar and antiferromagnetic ordering may be responsible for the relatively high T_N, Pofahl et al. [7]. The specific magnetic susceptibility χ_g for MB_6 with M = Ce, Pr, Nd, Gd, Tb, and

Dy below room temperature is shown in **Fig. 20** [1]. The Curie-Weiss law was valid above 150 K [2] or 200 K for CeB_6, above 100 K for PrB_6, above 30 K for EuB_6, and above 70 K for GdB_6 (studies from 7 to 300 K) [4]. The law was found to be valid in the whole range studied, between 80 and 300 K for M = Pr to Dy except Sm [2] and between 80 and 1100 K for M = Pr, Eu, and Gd, Aivazov et al. [8]. The effective magnetic moments μ_{eff} of EuB_6 and YbB_6 correspond to the divalent metals, those of the others except SmB_6 to the trivalent metals. The μ_{eff} of SmB_6 increased with increasing temperature and its values, lying between those calculated for Sm^{2+} and Sm^{3+}, indicated an Sm^{2+} content of 40% in the sample studied [2]. Values for μ_{eff} (in μ_B) in comparison to the free ion values (in parentheses), the paramagnetic Curie temperature Θ_p (in K), for χ_g (in cm^3/g) at 293 K, and the molar susceptibility χ_{mol} in cm^3/mol at room temperature are:

MB_6	μ_{eff}		μ_{eff}	μ_{eff}	μ_{eff}[2]	Θ_p	Θ_p	Θ_p	$10^6\,\chi_g$	$10^6\,\chi_{mol}$
LaB_6	0	(0)	—	—	0.64	—	—	—	—	172
CeB_6	2.49	(2.56)	2.51	—	2.64	−76	−82	—	10.21	2800
PrB_6	3.59	(3.62)	3.64	3.6	3.58	−68	−41	−50	24.12	5300
NdB_6	3.54	(3.68)	—	—	3.68	−42	—	—	—	5600
SmB_6	~1.5 to 2.3[1]		—	—	2.61	—	—	—	10.3[3]	2830
EuB_6	8.1	(7.94)[4]	7.90	7.6	8.35	+9	−0.5	+15	121.7	29300
GdB_6	8.01	(7.94)	7.98	7.8	7.90	−55	−66.5	−50	99.8	26100
TbB_6	9.43	(9.7)	—	—	9.92	−35	—	—	—	41000
Ref.	[2]		[4]	[8]	[9]	[2]	[4]	[8]	[4]	[9]

[1] Between 80 and 300 K. – [2] Calculated from χ_{mol} at room temperature. – [3] Value from [8]. –
[4] Divalent M.

Values for DyB_6: $\mu_{eff} = 10.63\ \mu_B$ (in comparison to 10.6 μ_B for Dy^{3+}), and $\Theta_p = -21$ K. The YbB_6 with divalent Yb has $\mu_{eff} = 0\ \mu_B$ [2], however, $\mu_{eff} = 0.55\ \mu_B$ and $\chi_{mol} = 126 \times 10^{-6}\ cm^3/mol$ at room temperature according to [9]. For early values of μ_{eff} and Θ_p for rather impure samples of MB_6 with M = La, Ce, Pr, Nd, and Sm, studied between 293 and 713 K, see Klemm et al. [10], with M = Y, La, Ce, Nd, Gd, and (pretended trivalent) Yb, studied between ca. 300 and 1300 K, see Benoit, Blum [11], Benoit [12].

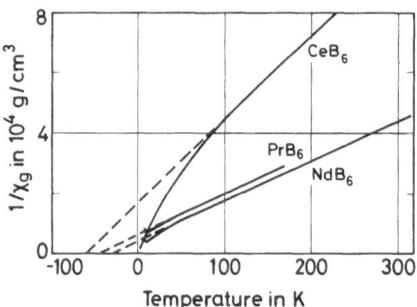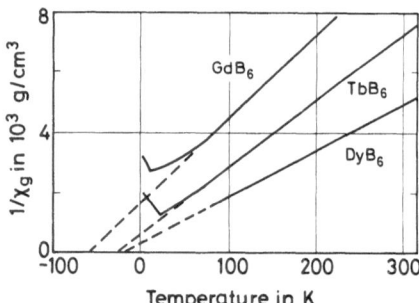

Fig. 20.   Reciprocal magnetic susceptibility $1/\chi_g$ of rare earth hexaborides.

References:

[1] Etourneau, J.; Mercurio, J.-P.; Hagenmuller, P. (in: Matkovich, V. I., Boron and Refractory Borides, Springer, Berlin – Heidelberg – New York 1977, pp. 115/38, 129/33).

[2] Paderno, Yu. B.; Pokrzywnicki, S.; Staliński, B. (Phys. Status Solidi **24** [1967] K73/K76).

[3] Matthias, B. T.; Geballe, T. H.; Andres, K.; Corenzwit, E.; Hull, G. W.; Maita, J. P. (Science **159** [1968] 530).

[4] Hacker, H., Jr.; Shimada, Y.; Chung, K. S. (Phys. Status Solidi A **4** [1971] 459/65).

[5] Ali, N.; Woods, S. B. (Phys. Letters A **104** [1984] 212/4).

[6] Geballe, T. H.; Matthias, B. T.; Andres, K.; Maita, J. P.; Cooper, A. S.; Corenzwit, E. (Science **160** [1968] 1443/4).

[7] Pofahl, G.; Zirngiebl, E.; Blumenröder, S.; Brenten, H.; Güntherodt, G.; Winzer, K. (Z. Physik B **66** [1987] 339/43).

[8] Aivazov, M. I.; Aleksandrovich, S. V.; Evseev, B. A.; Tsarev, O. M. (Fiz. Metal. Metalloved. **56** [1983] 249/52; Phys. Metals Metallog. [USSR] **56** No. 2 [1983] 33/6; C.A. **99** [1983] No. 132467).

[9] L'vov, S. M.; Nemchenko, V. F.; Paderno, Yu. B. (Vysokotemp. Neorgan. Soedin. **1965** 445/50; C.A. **64** [1966] 13538).

[10] Klemm, W.; Schüth, W.; von Stackelberg, M. (Z. Physik. Chem. B **19** [1932] 321/7).

[11] Benoit, R.; Blum, P. (Compt. Rend. **234** [1952] 2428/30).

[12] Benoit, R. (J. Chim. Phys. Physicochim. Biol. **52** [1955] 119/32).

32.1.5.11 Electrical Properties

32.1.5.11.1 Electronic Structure

For recent band structure calculations, see the individual sections, particularly LaB$_6$, p. 214.

Early investigations of the electronic structure of MB$_6$ concerned the boron framework and its stabilization: A tight-binding approximation with the 2s and 2p atomic orbitals of boron led to the conclusion that each B$_6$ octahedron requires 20 electrons. Of these, 6 are involved in the formation of 2-electron covalent bonds to neighboring octahedra leaving 14 to form the 12 bonds within the octahedra. There is a finite gap between this valence band, composed by various symmetry types of bonding states with different energies, and the conduction band; for details, see the paper. Since the 6 B atoms in the unit cell provide 18 electrons, a transfer of 2 electons from each metal is required. Thus MB$_6$ should be insulating or semiconducting if M is divalent (like Eu and Yb) and be metallic if M has a valency greater than two, Longuet-Higgins, Roberts [1]. A similar result was obtained also by Flodmark [2] and Yamazaki [3]. The latter author calculated the energy band structure of CaB$_6$ based on the atomic orbitals 2s and 2p of boron and one ns of the metal [3]. A different treatment by Flodmark [4], considering hybrids of 2s, 2p, 3s, 3p, and 3d orbitals, led to a different closed-shell structure for MB$_6$ (also divalent MB$_6$ should be metallic) and to a different sequence of energy bands [4]; for earlier studies, see Flodmark [5] and Fischer-Hjalmars, Flodmark [6]. A more recent calculation of the molecular orbitals (MO) of each B$_6$ octahedron from s, p orbitals on each boron performed by Ikeda et al. [7], essentially agrees with that of [1]. The MO's are represented by the irreducible representation of point group O$_h$ and are classified into bonding, antibonding, and non-bonding orbitals. The center of each level was evaluated by comparing with the APW (augmented plane wave) band calculation results for LaB$_6$ (from Hasegawa, Yanase [8]), and is listed relative to the Fermi energy E$_F$:

MO	energy level in eV	MO	energy level in eV
E$_g$	6.0	T$_{2g}$	−3.5
T$_{1u}$	6.0	T$_{1u}$	−8.0
T$_{1g}$	4.0	B$_6$–B$_6$ bonding	−8.0
B$_6$–B$_6$ antibonding	3.0	(A$_{1g}''$, E$_g''$, T$_{1u}''$) nonbonding	
(A$_{1g}''$, E$_g''$, T$_{1u}''$) nonbonding		A$_{1g}$	−16.0
T$_{2u}$ { unoccupied[*]	2.0		
T$_{2u}$ { occupied[*]	−1.0		

[*] This bond corresponds to the band that cuts the Fermi energy.

The nonbonding orbitals are extended outside of the B$_6$ molecule and thus form bonding and antibonding orbitals between B$_6$ molecules. Bonding orbitals of each B$_6$ and of intermolecules are occupied, whereas antibonding orbitals are unoccupied, except the T$_{2u}$ state of B$_6$. This state forms the lowest conduction levels as the bonding orbital with the 5d(e$_g$) states [7]. The magnitude of the electric field gradient at the boron nucleus derived by means of NMR can be explained in terms of this model (considering only 2s and 2p electrons of boron) and a weak metal-boron interaction in the case of divalent MB$_6$ (M = Eu, Yb). For the trivalent MB$_6$, data indicate in addition that the distribution of conduction electrons deviates considerably from a uniform distribution, which is in agreement with Fermi surface studies (see p. 59), Aono, Kawai [9].

Peaks observed in the photoemission (XPS) and X-ray spectra of MB$_6$ at 15, 10, and 5 eV below E$_F$ were attributed to arise mainly from B 2s and 2p contributions. Peak positions corresponding to the localized 4f states of trivalent MB$_6$ range from the upper edge of the valence band in CeB$_6$ (4f at 2.5 eV; however, the hybridization peaks at 4.5 eV [7]) to deep inside the valence band in TbB$_6$ (4f at 11 eV). In intermediate valent SmB$_6$ the 4f states of divalent Sm are situated in the vicinity of the Fermi level, and in EuB$_6$ and YbB$_6$ they are slightly below it, see "Photoemission and X-Ray Spectra", p. 85. For the 4f position see also a figure in Kawai et al. [10]. From NMR data on LaB$_6$, SmB$_6$, and EuB$_6$ it was concluded that 4f electrons participate in the formation of covalent interatomic bonding in SmB$_6$ and, increasingly, in EuB$_6$, whereas in LaB$_6$ (no 4f electrons), 5d electrons participate, Paderno et al. [11]. Recent band structure calculations by Harima et al. [12] for LaB$_6$ (see p. 216) revealed that the position of the unoccupied 4f level is of great importance in order to explain experimental facts on the detailed Fermi surface (see p. 59). Based on these calculations, the p(B)–f(La) mixing was defined to be (pfσ) = 0.25 eV and (pfπ) = −0.17 eV, i.e. (pfπ)/(pfσ) ≈ −0.7. In the case of MB$_6$ (M = Ce, Pr, and Nd) having occupied 4f levels, the B$_6$ bonding states have much stronger p(B)–f(M) mixing than the unoccupied antibonding states. The bonding band p(B)–f(M) mixing is large, because it acts as the sum of the 24 nearest neighbor borons, Kasuya et al. [13]. Values of p(B)–f(M) and s(B)–f(M) mixings, used earlier in order to explain the photoemission spectra (PES) in the 4f region, amounted to (pfσ) = −0.231 eV and (sfσ) = −0.6 (pfσ) for CeB$_6$ and to (pfσ) = −0.162 eV for PrB$_6$; i.e., (pfσ) in PrB$_6$ is ca. 70% of CeB$_6$. The amount of mixing decreases on going from CeB$_6$ to PrB$_6$ and NdB$_6$ owing to both the dynamical d–f Coulomb screening effect and the contraction of the 4f wave function. In calculating the crystal field splitting due to c–f mixing, agreement with experimental data was obtained in the case of CeB$_6$ on use of (pfσ) = −0.34 eV and a correlation energy U$_{ff}$ = 6 eV for the determination of the unoccupied f levels. In the case of PrB$_6$ and NdB$_6$ the values (pfσ) = −0.287 eV, U$_{ff}$ = 8 eV led to a different character of the excited levels (PrB$_6$) or to larger overall splitting (NdB$_6$) than that of the experiment, Ikeda et al. [7]. More precisely, according to the recent discussion, the c–f^2

mechanism is dominant in CeB_6, indicating the similar p(B)–f(M) mixing effect on the crystal field splitting in PrB_6 and NdB_6 [13].

The GO-LCAO method (group orbitals, linear combination of atomic orbitals) was used earlier to calculate some parameters of the electronic structure of MB_6 by considering a fragment consisting of two formula units. The 66 GO taken into account were derived from 6s, 6p, 5d metal orbitals and 2s and 2p boron orbitals: As the number of f electrons of M increases, the so-called charge q of M (i. e., the amount of valence electrons given up from M to the B sublattice in order to strengthen the B–B and B–M bonds) decreases, and consequently the B–B bond weakens, see a figure in the paper for M = Ce, Nd, Gd, Tb, and Dy, Goryachev, Kovenskaya [14]. Different values for q were given by Goryachev et al. [15] (M = Ce, Tb, and Sm^{II} and Sm^{III}) and still different ones in the figures of Goryachev et al. [16, 17] (M = La, Ce, Nd, Sm, Gd, Tb, Ho, Tm). Also shown in [16] is the occupation density of $sp^3(B)$ states. All these papers also give data on the energy of atomization. For additional data, see also Goryachev et al. [18] (M = La, Ce, Sm, Gd, Tb) and Shvartsman [19] (same M plus Ho). The role of d(M) orbitals and their relative position to s(M) orbitals in the formation of MB_6 ($\Delta E < 2.76$ eV) was emphasized by Garf et al. [20], from MO-LCAO calculations in L approximation; see also Garf et al. [21]. Originally, the possibility of forming hexaborides was assumed to be connected with the values of the first and second ionization potentials of the metals, see for example [2, 3], Neshpor, Samsonov [22].

The Fermi surface of LaB_6 most probably consists of nearly spherical ellipsoid-bellies at X points connected by short necks and small ellipsoid-electron pockets (ϱ branches of de Haas-van Alphen (dHvA) signals) which are elongated along the Σ axes and overlap on the necks. This model was obtained by displacing the unoccupied 4f levels upward from those of the self-consistent band calculation, which leads to a larger hybridization of the 4f states at the Fermi level, Harima et al. [12]; for details see p. 218. The Fermi surfaces of PrB_6 and CeB_6 appear to be essentially equal to that of LaB_6, Kasuya et al. [13], see also the individual sections, and for the ellipsoid-bellies (dHvA studies), van Deursen et al. [23] in contrast to ten Cate, de Groot [24]. Two-dimensional angular correlation of positron annihilation radiations yielded the absolute locations of the ellipsoids and the following dimensions d_1 and d_2 of the ellipsoid-bellies in the direction of X-M and X-R, respectively:

	LaB_6	CeB_6	PrB_6	NdB_6	Ref.
d_1 in π/a	0.95	1.16	1.08	0.61	[25]
in mrad	2.8 ± 0.1	3.4 ± 0.1	3.2 ± 0.1	1.8 ± 0.1	[26]
in a.u.	0.38 ± 0.02	0.47 ± 0.02	0.44 ± 0.01	0.24 ± 0.02	[26]
d_2 in π/a	0.95	0.95	0.88	0.61	[25]
in mrad	2.8 ± 0.1	2.8 ± 0.1	2.6 ± 0.1	1.8 ± 0.1	[26]
in a.u.	0.38 ± 0.02	0.38 ± 0.02	0.36 ± 0.02	0.24 ± 0.02	[26]

For cross sections of the Lock-Crisp-West k space density along the three principal axes Γ-X, X-M, and M-Γ, see figures in the papers, Tanigawa et al. [25, 26], and for a discussion, Tanigawa [30]. According to dHvA data the ellipsoid-bellies in PrB_6 are almost the same size as those of LaB_6 and the ϱ, ϱ' branches may be ascribed to the similar pockets as in LaB_6, namely to electron pockets with up and down spins in PrB_6, Ōnuki et al. [27]. The observed pair structure of the dHvA signals in this compound has been analyzed as due to different nonlinear behaviors of the up and down spin Fermi surfaces in two successive magnetic phases. It indicates that the effective exchange field on the conduction electron spin is of opposite sign in the two successive phases. The cyclotron mass enhancement in PrB_6 referred to LaB_6 (see p. 60) is thought to be due to the electron-magnon effect. In CeB_6 which is a dense Kondo system, the magnetic interaction between the 4f moments on different sites seems to be not so much affected by the Kondo state. The pair behavior for the neck Fermi surface is similar to that in PrB$_6$, but with a large mass enhancement due to the Kondo-like magnetic fluctuation of kT$_K$ (where T$_K$ = Kondo temperature) with a large degeneracy; i.e., the Coqblin-Schrieffer model for the trivalent Ce is applicable and the Fermi surface in the Kondo regime is essentially equal to that of LaB$_6$ [13]. The main Fermi surface of NdB$_6$, however, appears to be somewhat different from that of LaB$_6$, regarding dHvA data [27], see also table on p. 59 [25, 26]. The magnetoresistance of NdB$_6$ shows the existence of open orbits as in LaB$_6$ and PrB$_6$, implying a multiply connected Fermi surface [27].

The cyclotron masses m*/m$_0$ of the various orbits, given as the corresponding dHvA frequencies, are shown in **Fig. 21** for MB$_6$ with M = La, Ce, Pr, and Nd, based on data from the literature. The mass of α orbit on the ellipsoid-belly in CeB$_6$ is field-dependent, ranging from m* = 6 m$_0$ to 17 m$_0$. The mass enhancement in CeB$_6$ and PrB$_6$ relative to LaB$_6$ was interpreted so as to reflect the hybridization effects of 4f electrons with conduction electrons [27]. A similar interpretation was also given earlier, for example by van Deursen et al. [28] and Crabtree [29] on the basis of the following masses for the $\chi_1(\equiv\alpha_1)$ orbit observed at ca. 8×10^3 T ($\triangleq 8 \times 10^7$ Oe) for magnetic field parallel to [100]: m* = 0.6 m$_0$ for both LaB$_6$ and NdB$_6$ (in contrast to Fig. 21 [27]), 6 m$_0$ for CeB$_6$, and 1.6 m$_0$ for PrB$_6$ [28]. The corresponding average value of the renormalized Fermi velocity (in 10^8 cm/s) which is a qualitative measure of the degree of f character in the conduction state, amounted to 0.94 (LaB$_6$), 0.10 (CeB$_6$), 0.36 (PrB$_6$), and to 0.95 (NdB$_6$). It was concluded that, despite the existence of a local moment, there is significant hybridization of the f electrons with the conduction band in CeB$_6$, slight f hybridization in PrB$_6$, and nearly no f hybridization for NdB$_6$, like for LaB$_6$ [29].

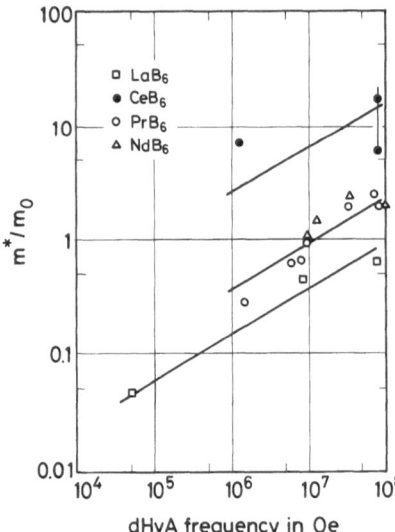

Fig. 21.   Cyclotron masses vs. de Haas-van Alphen frequencies of rare earth hexaborides.

For the effective mass and other properties of charge carriers, derived from optical studies and the Hall effect, see p. 70.

References:

[1] Longuet-Higgins, H.C.; Roberts, M. de V. (Proc. Roy. Soc. [London] A **224** [1954] 336/47).

[2] Flodmark, S. (Arkiv Fysik **9** [1955] 357/76).

[3] Yamazaki, M. (J. Phys. Soc. Japan **12** [1957] 1/6).

[4] Flodmark, S. (Arkiv Fysik **18** [1960] 49/64).

[5] Flodmark, S. (Arkiv Fysik **14** [1959] 513/50).

[6] Fischer-Hjalmars, I.; Flodmark, S. (J. Chem. Phys. **22** [1954] 1950).

[7] Ikeda, M.; Aoki, Y.; Kasuya, T. (J. Magn. Magn. Mater. **52** [1985] 264/6).

[8] Hasegawa, A.; Yanase, A. (J. Phys. F **7** [1977] 1245/60).

[9] Aono, M.; Kawai, S. (J. Phys. Chem. Solids **40** [1979] 797/802).

[10] Kawai, S.; Tanaka, T.; Muranaka, S.; Aono, M.; Oshima, C. (Process. Kinet. Prop. Electron. Magn. Ceram. Proc. U.S.-Japan Semin. Basic Sci. Ceram., Hakone, Japan, 1975 [1976], pp. 125/34; C.A. **88** [1978] No. 113454).

[11] Paderno, Yu.V.; Afanas'ev, M.L.; Dudnik, E.M.; Ermolaev, V.K. (Tezisy Dokl. Vses. 5th Konf. Khim. Svyazi Poluprovodn. Polumet., Minsk, USSR, 1974, p. 100; C.A. **83** [1975] No. 139490).

[12] Harima, H.; Sakai, O.; Kasuya, T.; Yanase, A. (Solid State Commun. **66** [1988] 603/7).

[13] Kasuya, T.; Sakai, O.; Harima, H.; Ikeda, M. (J. Magn. Magn. Mater. **76/77** [1988] 46/52).

[14] Goryachev, Yu.M.; Kovenskaya, B.A. (J. Less-Common Metals **67** [1979] 273/9).

[15] Goryachev, Yu. M.; Kovenskaya, B. A.; Shvartsman, E. I. (Vysokotemp. Boridy Silitsidy **1978** 85/9; C.A. **91** [1979] No. 95481).

[16] Goryachev, Yu. M.; Kovenskaya, B. A.; Shvartsman, E. I. (Elektron. Strukt. Perekhodnykh Met. Ikh Splavov Intermet. Soedin. Material. 2nd Mezhdunar. Simp. ISESTM, Kiev 1977 [1979], pp. 124/7; C.A. **92** [1980] No. 169484).

[17] Goryachev, Yu. M.; Kovenskaya, B. A.; Shvartsman, E. I. (Tugoplavkie Soedin. Redkozemel. Met. Mater. 3rd Semin., Novosibirsk 1977 [1979], pp. 42/5; C.A. **93** [1980] No. 54205).

[18] Goryachev, Yu. M.; Kovenskaya, B. A.; Shvartsman, E. I. (Konfiguratsionnye Predstavleniya Elektron. Str. Fiz. Materialoved Mater. 2nd Nauchn. Semin. Konfiguratsionnoi Modeli Kondens. Sostoyaniya Veshchestva, Lvov, USSR, 1976 [1977], pp. 73/6; C.A. **89** [1978] No. 169313).

[19] Shvartsman, E.I. (Poluch. Issled. Svoistv Nov. Mater. Mater. 10th Nauchn. Konf. Aspir. Molodykh Issled. Inst. Probl. Materialoved. Akad. Nauk Ukr. SSR, Kiev, 1976 [1978], pp. 7/13; C.A. **90** [1979] No. 92798).

[20] Garf, E.S.; Paderno, Yu. B.; Goryachev, Yu. M. (Kristallokhim. Tugoplavkikh Soedin. **1972** 77/84; C.A. **78** [1973] No. 75971).

[21] Garf, E.S.; Paderno, Yu. B.; Goryachev, Yu. M. (Redkozemel. Metal. Ikh Soedin. **1970** 101/9; C.A. **77** [1972] No. 143934).

[22] Neshpor, V.S.; Samsonov, G.V. (Zh. Neorgan. Khim. **4** [1959] 1967/9; Russ. J. Inorg. Chem. **4** [1959] 893/4).

[23] van Deursen, A.P.J.; Pols, R.E.; de Vroomen, A.R.; Fisk, Z. (J. Less-Common Metals **111** [1985] 331/4).

[24] ten Cate, R. W.; de Groot, R. A. (from [23]).

[25] Tanigawa, S.; Terakado, S.; Iwase, Y.; Suzuki, R.; Komatsubara, T.; Ōnuki, Y. (J. Magn. Magn. Mater. **52** [1985] 313/6).

[26] Tanigawa, S.; Terakado, S.; Ito, K.; Morisue, A.; Komatsubara, T.; Ōnuki, Y.; Shiotani, N. (Positron Annihilation Proc. 7th Intern. Conf., New Delhi 1985, pp. 285/7; C.A. **105** [1986] No. 121085).

[27] Ōnuki, Y.; Kurosawa, Y.; Omi, T.; Komatsubara, T.; Yoshizaki, R.; Ikeda, H.; Maezawa, K.;
     Wakabayashi, S.; Umezawa, A.; Kwok, W. K.; Crabtree, G. W. (J. Magn. Magn. Mater. **76/77**
     [1988] 37/9).
[28] van Deursen, A. P. J.; Fisk, Z.; de Vroomen, A. R. (Solid State Commun. **44** [1982] 609/12).
[29] Crabtree, G. W. (J. Magn. Magn. Mater. **52** [1985] 169/73).
[30] Tanigawa, S. (Mater. Sci. Forum **37** [1987/89] 111/22; from C. A. **110** [1989] No. 237310).

32.1.5.11.2 Electrical Resistivity. Magnetoresistance

The rare earth hexaborides MB$_6$ with trivalent M = Y, La to Nd, Gd to Ho have metallic
character. Intermediate valent SmB$_6$ behaves as a very narrow gap semiconductor and EuB$_6$
and YbB$_6$ with divalent cations are semiconductors. Details for the latter (SmB$_6$, EuB$_6$, YbB$_6$),
see in the individual sections and the review from Korsukova, Gurin [1]. Among the hexaborides, only YB$_6$ is a reasonable superconductor (see p. 65).

Room Temperature Resistivity

At Ambient Pressure

Measured values of the electrical resistivity ϱ (in $\mu\Omega \cdot cm$) are assembled below for zone-
melted (z.m.) or melted (m) specimens, single crystals grown from an Al flux (fl) and hot-
pressed samples with specified porosity p. Also presented are ϱ values extrapolated to zero
porosity and the temperature coefficient α (in K^{-1}) around room temperature:

MB$_6$ sample	ϱ z.m.	ϱ z.m.	ϱ m[2]	ϱ fl	ϱ p=0	ϱ p=0	ϱ p=5 to 15%	10$^3\alpha$ p=0[3]	10$^3\alpha$
YB$_6$	—	—	45 to 80	—	40	20	40 to 60	1.24	1.23
LaB$_6$	9.1	—	8.5	95	15.0	6.8	11 to 25	2.68	2.88
CeB$_6$	—	—	18	240	29.4	14	20 to 55	1.00	1.05
PrB$_6$	24	15	20	190	19.5	16	22 to 50	1.92	1.99
NdB$_6$	19.6	14.6	16	150	20.0	13	18 to 40	1.93	1.97
SmB$_6$	249	—	250	1800	207.0	168	250 to 800	−0.42[4]	−0.47
EuB$_6$	2×10^4	—	(10 to 40)×10^3	—	84.7	80	140 to 370	0.90	0.90
GdB$_6$	—	31[1]	55 to 105	—	44.7	27	45 to 95	1.40	1.40
TbB$_6$	—	—	38 to 78	—	37.4	19	28 to 60	1.31	1.36
DyB$_6$	—	40[1]	45 to 100	—	—	19	30 to 70	—	1.82
HoB$_6$	—	—	55 to 150	—	—	18	30 to 70	—	1.77
YbB$_6$	4×10^3	—	(2 to 4)×10^3	—	46.6	33	40 to 96	2.34	2.44
Ref.	[2]	[3]	[4]	[5]	[6]	[4]	[4]	[6]	[4]

[1] Arc-melted samples. – [2] Single-phase material in the case of M = La to Sm and multiphase
material, due to incongruent melting in the case of M = Eu to Yb. Multiphase material with
M = Sm has ϱ = 400 to 800 $\mu\Omega \cdot cm$. – [3] Temperature range from 273 to 373 K. – [4] α becomes
positive abcve 573 K.

Values for ϱ and α from Paderno et al. [6] were also listed by Samsonov et al. [7 to 10]. The
proportionality existing between resistivity and porosity is shown in a figure in [9] for LaB$_6$.
Powders have ϱ (in $\mu\Omega \cdot cm$) = 9.1 (LaB$_6$), 29.4 (CeB$_6$), 24.0 (PrB$_6$), 19.6 (NdB$_6$), and 14.7 (GdB$_6$),

Ivanchenko et al. [11]. For early ϱ data on MB_6, $M = La$ to Sm, Gd, Tb, Yb, see Samsonov, Paderno [12].

Along the homogeneity range ϱ varied from 0.8 m$\Omega \cdot$cm at $Sm_{1.0}B_6$ to 2.6 m$\Omega \cdot$cm at $Sm_{0.68}B_6$, see a figure in the paper. The variation of ϱ is also small for M_xB_6 with $M =$ Ce, Pr, Gd, and Tb in their homogeneity ranges $x = 1.0$ to ca. 0.7. All the samples were prepared by borothermal reduction in a vacuum of ca. 0.1 to 13 mPa, followed by a 100 h-annealing at 1073 K. The relatively large conductivity at the metal-deficient boundary was assumed to be due to impurities and the effect of metal d band. Additionally, an electron-deficient boron framework requiring less than the calculated two electrons from M (see p. 57) was considered for the rare earth hexaborides, similar to that apparently existing in $Ba_{0.57}Na_{0.43}B_6$ and $Th_{0.23}Na_{0.77}B_6$, Yajima, Niihara [13].

Analysis of the reflection spectra led to the following high-frequency resistivities ϱ_{opt} in $\mu\Omega \cdot$cm:

melted polycrystalline samples: LaB_6: 40; PrB_6: 44; SmB_6: 142 [14]

powders   : LaB_6: 21 [15], 16.9; CeB_6: 20; PrB_6: 17.2 [11], 36 [15];
                                     NdB_6: 17.5; GdB_6: 16.1 [11]

polycrystalline films   : LaB_6: 90; PrB_6: 79; SmB_6: 188 [14]

At Elevated Pressure

Under hydrostatic pressure the resistivities of powder samples of SmB_6, EuB_6, and YbB_6, in contrast to LaB_6, decrease rapidly up to ca. 2 GPa, remain essentially constant above 4 GPa, and do not reach values typical of metallic resistivities up to 10 GPa (range studied), see a figure in the paper. Characteristics of the pressure effect (applied pressure P in Pa) are:

MB_6	ϱ(0 Pa) in m$\Omega \cdot$cm	(d log ϱ/dP)$_{P=0}$ in 10^{-11} Pa^{-1}	ϱ(10 GPa) in m$\Omega \cdot$cm	ϱ(10 GPa)/ϱ(0 Pa)
SmB_6	1.2	-25	0.35	0.29
EuB_6	4.0	-27	1.1	0.27
YbB_6	72.0	-65	2.8	0.039

The resistivity of an Al-flux grown YbB_6 crystal, with $\varrho = 13.0$ m$\Omega \cdot$cm at P = 0 was reduced by pressure in the same ratio as that of the powder sample, Leger et al. [16]. **Fig. 22**, p. 64, shows ϱ(P)/ϱ(P = 0) up to 1.4 GPa for single crystals grown from an Al flux using atomic ratios of M : B = 1 : 6 in the case of M = La, Eu, and Yb, however, of 1 : 7, 1 : 9, and 1 : 12 for SmB_6. The slope of the curves for MB_6, M = Sm, Eu, Yb, decreases as the lattice parameter of MB_6 decreases, Korsukova et al. [17]. The observed pressure effect may be attributed to a slight change in the valence state in the case of SmB_6 and to the presence of a level unlike 4f inside the band gap in the case of EuB_6 and YbB_6 [16].

Temperature Dependence of Resistivity. Superconductivity

Fig. 23, p. 64, shows the resistivity of zone-melted MB_6 specimens with M = La, Pr, Nd, Sm, Eu, and Yb between 1.39 and ~1000 K, Paderno et al. [2]. Hot-pressed samples have a positive temperature coefficient α between 273 and 373 K for all M except Sm (see table on p. 62). The SmB_6 has a negative α up to ca. 573 K and a positive one at higher temperatures, Paderno, Samsonov [6]. The resistivity of hot-pressed samples with M = La, Ce, Nd, Sm, and Gd below room temperature to 77 K is shown in **Fig. 24**, p. 65, from Samsonov et al. [18]. In the case of EuB_6, ϱ decreases nonlinearly in this temperature range, see a figure in Paderno et al. [19]. For deviating preliminary data on MB_6 with M = La to Gd except Pr, see Garf, Shcherbina [20].

Recent studies in the range 2 to 20 K show the resistivity of nonmagnetic LaB$_6$ to be nearly independent of temperature, Ali, Woods [21]. Other MB$_6$ reveal just below their Néel temperature T$_N$ a T^4 (M = Pr, Nd) or T^3 temperature dependence (M = Gd, Dy), which was associated with electron-spin wave scattering. At lower temperatures (above 2 K), the resistivity of these compounds (such as that of CeB$_6$, see Section 32.1.8.4 in "Rare Earth Elements" C 11 b) varies as T^2, which may be due to Baber type electron-electron scattering. Residual resistivities ϱ(293 K)/ϱ(4.2 K) and extents of T^n ranges (in K) are:

MB$_6$	ϱ(293 K)/ϱ(4.2 K)	T$_N$ in K	T^4 or T^3 range	T^2 range
PrB$_6$	18	6.99	6.8 to 5.1	$\leqq$4.2
NdB$_6$	16	7.74	7.6 to 5.9	$\leqq$5.8
GdB$_6$	8	15.2	14.4 to 6.7	$\leqq$5.7
DyB$_6$	9	20.3	19.5 to 12.0	7.3 to 3.8

Ali, Woods [3]. Abrupt ϱ drops at T$_N$ were also observed for MB$_6$ with M = Ce, Pr, Nd, Gd, Tb, Dy, and Ho by Matthias et al. [22] and Geballe et al. [23].

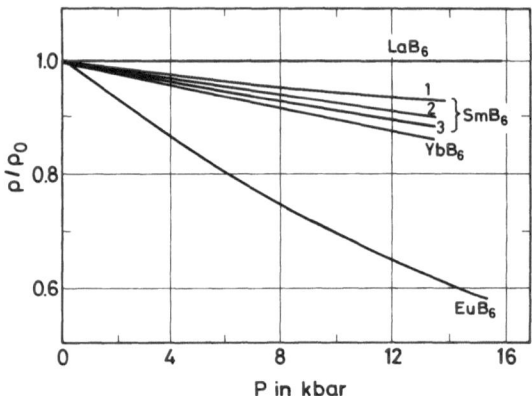

Fig. 22. Electrical resistivity ratios ϱ(P)/ϱ(0) of rare earth hexaborides as a function of the hydrostatic pressure P up to ca. 14 kbar (1.4 GPa) at room temperature. Curves 1, 2, and 3 were obtained on SmB$_6$ prepared from starting Sm:B ratios of 1:7, 1:9, and 1:12, respectively.

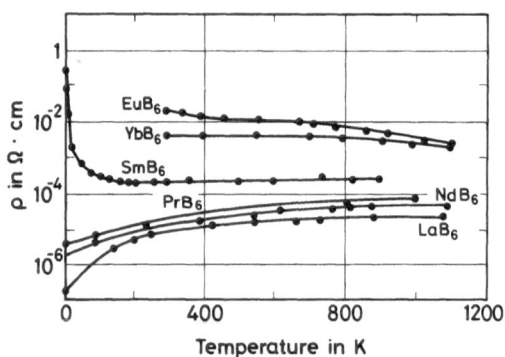

Fig. 23. Electrical resistivity ϱ of zone-melted rare earth hexaborides as a function of temperature.

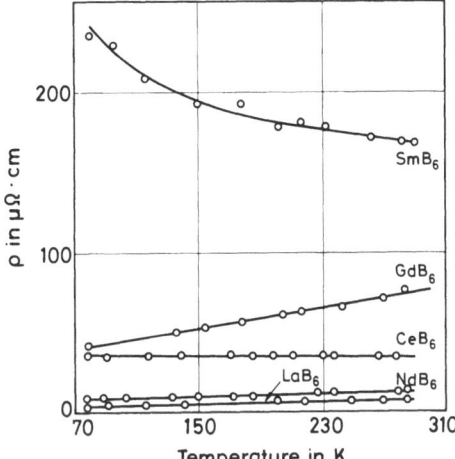

Fig. 24. Electrical resistivity ϱ of hot-pressed rare earth hexaborides as a function of temperature below room temperature.

Superconductivity is observed for YB_6 below ca. 7 K (see p. 149) and for LaB_6 below 0.1 K (see p. 232). The other rare earth hexaborides do not become superconducting, see for example [22] for M = Ce, Pr, Nd, Eu to Ho and [19] for M = La (studied above 1.28 K), Pr, Nd, and Sm. For an observation of superconductivity on some samples of NdB_6 below ca. 3 K, but on none of the other MB_6, see Shulishova, Shcherbak [24]. The depression of the superconducting transition temperature T_c of YB_6 by 1 at% addition of rare earth impurity (Ce to Yb) for Y is shown in **Fig. 25**, p. 66. For the most part the depressions follow the trend of $(g-1)^2 J(J+1)$, where J is the total angular momentum of the localized 4f spins and g is the Landé g factor. Additions of Ce, probably owing to the presence of a virtual 4f level, of divalent Eu and Yb, and of Tm do not follow either this trend or that of the spin factor $S(S+1)$, Fisk et al. [25]. Values for T_c of powder samples of YB_6, high-purity LaB_6, and $Y_{1-x}La_xB_6$ are assembled below, together with selected values for YB_6 contaminated by various amounts of Ce, Nd, and Yb, and/or Fe, which latter may originate from the boron used for sample preparation (amount of impurity in at%):

| sample | impurity | | T_c in K | sample | impurity | | | T_c in K |
	Fe	Yb			Fe	Ce	Nd	
YB_6	—	—	6.0 ± 0.1	YB_6	0.3	0.1	—	3.1
YB_6	0.2	—	4.1	"$Y_{0.95}Ce_{0.05}B_6$"	0.2	0.5	—	1.6
YB_6	0.3	—	3.1	YB_6	0.2	—	0.2	3.3
YB_6	—	0.005	3.9	$Y_{0.6}La_{0.4}B_6$	?	—	—	4.4 ± 0.1
YB_6	0.2	0.01	3.6; 3.3	$Y_{0.5}La_{0.5}B_6$	0.2	—	—	3.3
YB_6	0.2	0.1	3.0	LaB_6	—	—	—	0.1

It appears that the very low T_c of LaB_6 is due to a small magnetic moment from the lanthanum itself. From the influence of the impurities on the transition temperature of YB_6 it was concluded that rare earth impurities with localized f electrons are more effective in lowering T_c than is Fe having localized d electrons, Sobczak, Sienko [26]. The T_c vs. x data for $Y_{1-x}M_xB_6$ with M = nonmagnetic La, Yb, or Th and Ca exhibit an identical linear decrease of T_c in the concentration range $0 \leq x \leq 0.2$, which is coincident with the range of nonstoichiometry assumed for YB_6 (see p. 22). So it might be that the observed decrease of T_c in this region is a stoichiometry effect. The divergence of the curves at higher concentrations originates from the rising magnetic moment; in the case of Yb, owing to the presence of some trivalent Yb^{3+}, Hiebl, Sienko [27].

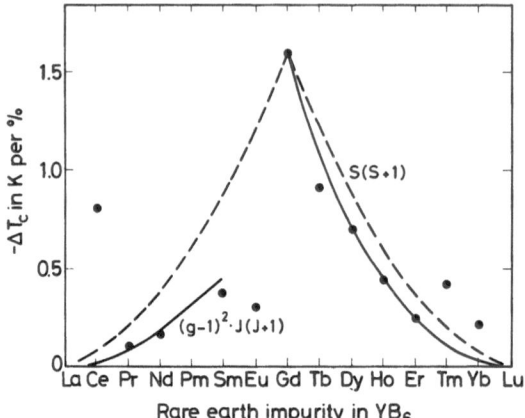

Fig. 25.   Depression of the superconducting transition temperature T$_c$ of YB$_6$ by substitution of 1 at% Y by rare earth elements.

Magnetoresistance

The magnetoresistance $\Delta\varrho/\varrho_0$ of MB$_6$ (M = Pr, Nd, Gd, Dy) was studied in applied magnetic fields B up to 3 T between 4 and 30 K. The main features of the longitudinal magnetoresistance, $(\Delta\varrho/\varrho_0)_{\|}$ where B III, are as follows: In the antiferromagnetic state $(\Delta\varrho/\varrho_0)_{\|}$ is positive and increases with increasing field. In addition, it decreases with increasing temperature at a constant field according to $(\Delta\varrho/\varrho_0)_{\|} = A(B/T)^2$, where A is a constant. At the Néel temperature, $(\Delta\varrho/\varrho_0)_{\|}$ drops severely, is very small in the paramagnetic phase, and becomes negative at higher fields. Studies of the transverse magnetoresistance, $(\Delta\varrho/\varrho_0)_{\perp}$ where B ⊥ I, reveal an anisotropy in the magnetoresistance $\Delta\varrho_{an}[=(\Delta\varrho/\varrho_0)_{\|}-(\Delta\varrho/\varrho_0)_{\perp}]$ for all MB$_6$ studied except GdB$_6$. In general $\Delta\varrho_{an}$ increases with increasing field and decreases with increasing temperatures. Values of $\Delta\varrho_{an}$ are negative for PrB$_6$ and NdB$_6$, positive for DyB$_6$, and zero for GdB$_6$. This behavior may be caused by anisotropic conduction electron f-electron interactions, Ali, Woods [28].

References:

[1] Korsukova, M. M.; Gurin, V. N. (Zh. Vses. Khim. Obshchestva **26** [1981] 679/88; C.A. **96** [1982] No. 44359).

[2] Paderno, Yu. B.; Garf, E. S.; Niemyskii, T.; Pračka, I. (Poroshkovaya Metal. **1969** No. 10, pp. 55/8; Soviet Powder Met. Metal Ceram. **1969** 821/3).

[3] Ali, N.; Woods, S. B. (Phys. Letters A **104** [1984] 212/4).

[4] Binder, F. (Radex Rundschau **1977** 52/71, 62; C.A. **87** [1977] No. 27597).

[5] Okano, H.; Futamoto, M.; Hosoki, S.; Kawabe, U. (Shinku **20** No. 4 [1977] 127/35; C.A. **87** [1977] No. 94150).

[6] Paderno, Yu. B.; Samsonov, G. V. (Dokl. Akad. Nauk SSSR **137** [1961] 646/7; Proc. Acad. Sci. USSR Phys. Chem. Sect. **136/141** [1961] 293/4).

[7] Samsonov, G. V.; Serebryakova, T. I.; Neronov, V. A. (Boridy [Borides], Atomizdat, Moscow 1975, pp. 1/375, 154).

[8] Samsonov, G. V.; Paderno, Yu. B. (Vysokotemp. Metallokeram. Mater. 1962 102/8; C.A. 58 [1963] 8483).

[9] Samsonov, G. V.; Vainshtein, E. E.; Paderno, Yu. B. (Fiz. Metal. Metalloved. 13 [1962] 744/9; Phys. Metals Metallog. [USSR] 13 No. 5 [1962] 100/4).

[10] Samsonov, G. V.; Paderno, Yu. B.; Vainshtein, E. E. (Izv. Sibirsk. Otd. Akad. Nauk SSSR Ser. Khim. Nauk 1964 No. 3, pp. 78/84; C.A. 63 [1965] 2401).

[11] Ivanchenko, L. A.; Bessaraba, V. I.; Paderno, Yu. B.; Vereshchak, V. M. (Electron. Str. Fiz. Khim. Svoistva Splavov Soedin. Osn. Perekhodnykh Met. Dokl. 8th Simp., Kiev 1974 [1976], pp. 111/4; C.A. 87 [1977] No. 125852).

[12] Samsonov, G. V.; Paderno, Yu. B. (Dopov. Akad. Nauk Ukr. RSR 1959 1215/8; C.A. 1961 6329).

[13] Yajima, S.; Niihara, K. (Proc. 9th Rare Earth Res. Conf., Blacksburg, Va., 1971, Vol. 2, pp. 598/609).

[14] Bessaraba, V. I.; Ivanchenko, L. A.; Paderno, Yu. B. (J. Less-Common Metals 67 [1979] 505/9).

[15] Ivanchenko, L. A.; Paderno, Yu. B.; Pilyankevich, A. N. (Poroshkovaya Metal. 1978 No. 8, pp. 38/48; Soviet Powder Met. Metal Ceram. 17 [1978] 602/9).

[16] Leger, J. M.; Percheron-Guegan, A.; Loriers, C. (Phys. Status Solidi A 60 [1980] K23/K26).

[17] Korsukova, M. M.; Stepanov, N. N.; Goncharova, E. V.; Gurin, V. N.; Nikanorov, S. P.; Smirnov, I. A. (J. Less-Common Metals 82 [1981] 211/7).

[18] Samsonov, G. V.; Sorin, L. A.; Vlasova, M. V.; Shcherbina, V. I. (Dokl. Akad. Nauk SSSR 178 [1968] 1346/7; Dokl. Chem. Proc. Acad. Sci. USSR 178/183 [1968] 176/7).

[19] Paderno, Yu. B.; Novikov, V. I.; Garf, E. S. (Poroshkovaya Metal. 1969 No. 11, pp. 70/3; Soviet Powder Met. Metal Ceram. 1969 921/3).

[20] Garf, E. S.; Shcherbina, V. I. (Tr. 1st Nauchn. Konf. Aspir. Inst. Probl. Materialoved. Akad. Nauk Ukr. SSR, Kiev 1967 [1968], pp. 110/6; C.A. 72 [1970] No. 26184).

[21] Ali, N.; Woods, S. B. (Solid State Commun. 46 [1983] 33/5).

[22] Matthias, B. T.; Geballe, T. H.; Andres, K.; Corenzwit, E.; Hull, G. W.; Maita, J. P. (Science 159 [1968] 530).

[23] Geballe, T. H.; Matthias, B. T.; Andres, K.; Maita, J. P.; Cooper, A. S.; Corenzwit, E. (Science 160 [1968] 1443/4).

[24] Shulishova, O. I.; Shcherbak, I. A. (Izv. Akad. Nauk SSSR Neorgan. Materialy 3 [1967] 1495/7; Inorg. Materials [USSR] 3 [1967] 1304/6).

[25] Fisk, Z.; Matthias, B. T.; Corenzwit, E. (Proc. Natl. Acad. Sci. USA 64 [1969] 1151/4).

[26] Sobczak, R. J.; Sienko, M. J. (J. Less-Common Metals 67 [1979] 167/71).

[27] Hiebl, K.; Sienko, M. J. (Inorg. Chem. 19 [1980] 2179/80).

[28] Ali, N.; Woods, S. B. (J. Appl. Phys. 61 [1987] 4393/4).

32.1.5.11.3 Hall Effect. Thermoelectric Power

Hall effect data of MB_6 with trivalent M can be interpreted on the basis of electronic conduction alone. This is not the case with intermediate valent M = Sm and divalent Eu and Yb, Paderno et al. [1]. These three compounds have relatively high absolute values of the thermoelectric power, Paderno, Samsonov [2]. Values at room temperature for the Hall coefficient R_H, the quantity $\delta = R_H/e \cdot \varrho^2$ where ϱ is the electrical resistivity, and the thermoelectric power S are assembled on p. 68 for zone-melted samples from [1] and hot-pressed samples with zero porosity from [2]:

MB$_6$	R$_H$×10^4 in cm^3/C		$|\delta|$×10^{-23} in cm·V^{-2}·s^{-2}		S in µV/K	
YB$_6$	—	−4.56	—	17.8	—	−0.5
LaB$_6$	−4.68	−4.96	353	137.8	—	0.1
CeB$_6$	—	−4.18	—	30.2	—	2.8
PrB$_6$	−5.08	−4.33	55	71.1	—	−0.6
NdB$_6$	−4.98	−4.39	81	68.5	—	0.4
SmB$_6$	2.68	1.54	16.9×10^{-2}	0.2	8.4	7.6
EuB$_6$	—	−50.2	—	43.7	−38.5	−17.7
GdB$_6$	—	−4.39	—	13.7	—	0.1
TbB$_6$	—	−4.57	—	20.4	—	−1.1
YbB$_6$	—	−83.6	—	240.3	18.4	−25.5
Ref.	[1]	[2]	[1]	[2]	[1]	[2]

Above given values for hot-pressed samples were also listed by Samsonov et al. [3 to 5]. For early data on S for MB$_6$ with M = La, Ce, Pr, Nd, Sm, and Gd, see Samsonov, Paderno [6] and for M = La to Nd in addition Samsonov, Strel'nikova [7]. **Fig. 26** shows the thermal variation of S for zone-melted samples of SmB$_6$ between 1.39 and 1100 K and of EuB$_6$ and YbB$_6$ above ca. 350 K [1]. Hot-pressed samples with M = La, Ce, Nd, Sm, and Eu behave between 323 and 123 K as shown in **Fig. 27** from Paderno et al. [8]. The thermoelectric power of GdB$_6$, positive throughout, is less dependent on temperature in this range than that of NdB$_6$, with S(GdB$_6$) > S(NdB$_6$) above ca. 200 K, see a figure in Samsonov et al. [9] for M = La, Ce, Nd, Sm, and Gd. The thermoelectric power of polycrystalline LaB$_6$ (obtained by melting) and zone-melted single crystals of PrB$_6$ and NdB$_6$ along [110] was studied between 2 and 20 K, see **Fig. 28**. It exhibits a negative peak at ca. 5.5 K, with S = −0.28 µV/K for LaB$_6$, −0.55 µV/K for PrB$_6$, and −1.15 µV/K for NdB$_6$. In the case of nonmagnetic LaB$_6$, this peak most probably is due to the competing effects of the phonon drag contributions S$_g$, which are proportional to T^3 and originate from normal electron-phonon scattering processes below ca. 5.5 K, and from Umklapp processes above this temperature. The increased height of the peak in PrB$_6$ and NdB$_6$ is attributed to magnon drag contribution S$_m$ which has the same temperature dependence as S$_g$. The sharp increase in S near the Néel temperature (T$_N$ ≈ 6.9 K for PrB$_6$, ~7.7 K for NdB$_6$), similar to that in the resistivity, may be due to electron-spin wave scattering. Well above T$_N$ a linear contribution to S is expected due to spin disorder scattering (S$_{spd}$) along with electron

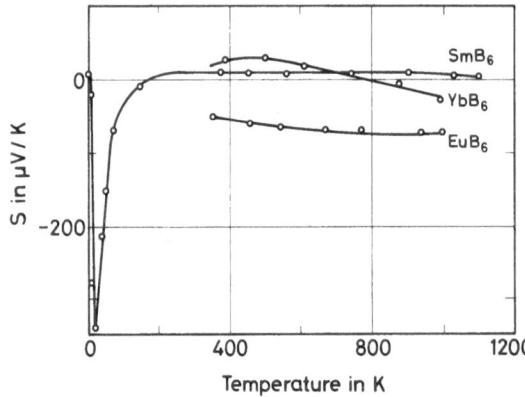

Fig. 26. Thermoelectric power S of zone-melted MB$_6$ (M = Sm, Eu, and Yb) as a function of temperature.

diffusion (S_e). This latter (S_e) presumably also determines the behavior of LaB_6 at the lowest temperatures studied. The positive contribution to S in PrB_6 at ca. 3.5 K is associated with the appearance of the low-temperature commensurate phase, Ali, Woods [10].

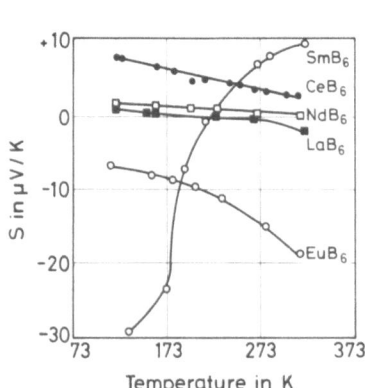

Fig. 27. Thermoelectric power S of hot-pressed MB_6 (M = La, Ce, Nd, Sm, and Eu) as a function of temperature.

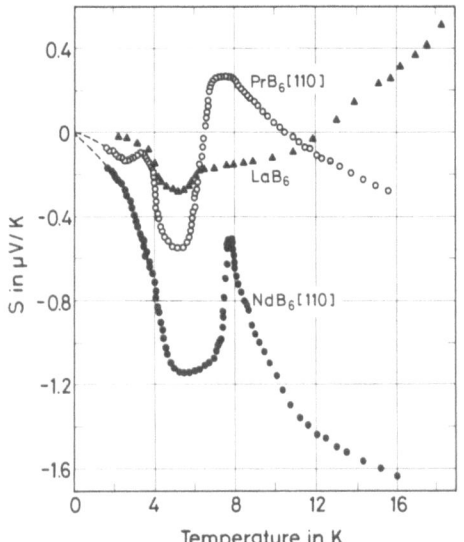

Fig. 28. Thermoelectric power S of melted LaB_6 and NdB_6 or PrB_6 single crystals along the [110] direction as a function of temperature at low temperatures.

References:

[1] Paderno, Yu. B.; Garf, E. S.; Niemyskii, T.; Pračka, I. (Poroshkovaya Metal. **1969** No. 10, pp. 55/8; Soviet Powder Met. Metal Ceram. **1969** 821/3).

[2] Paderno, Yu. B.; Samsonov, G. V. (Dokl. Akad. Nauk SSSR **137** [1961] 646/7; Proc. Acad. Sci. USSR Phys. Chem. Sect. **136/141** [1961] 293/4).

[3] Samsonov, G. V.; Serebryakova, T. I.; Neronov, V. A. (Boridy [Borides], Atomizdat, Moscow 1975, pp. 1/375, 154).

[4] Samsonov, G. V.; Vainshtein, E. E.; Paderno, Yu. B. (Fiz. Metal. Metalloved. **13** [1962] 744/9; Phys. Metals Metallog. [USSR] **13** No. 5 [1962] 100/4).

[5] Samsonov, G. V.; Paderno, Yu. B.; Vainshtein, E. E. (Izv. Sibirsk. Otd. Akad. Nauk SSSR Ser. Khim. Nauk **1964** No. 3, pp. 78/84; C.A. **63** [1965] 2401).

[6] Samsonov, G. V.; Paderno, Yu. B. (Dopov. Akad. Nauk Ukr. RSR **1959** 1215/8; C.A. **1961** 6329).

[7] Samsonov, G. V.; Strel'nikova, N. S. (Ukr. Fiz. Zh. **3** [1958] 135/8; C.A. **1958** 19309).

[8] Paderno, Yu. B.; Novikov, V. I.; Garf, E. S. (Poroshkovaya Metal. **1969** No. 11, pp. 70/3; Soviet Powder Met. Metal Ceram. **1969** 921/3).

[9] Samsonov, G. V.; Sorin, L. A.; Vlasova, M. V.; Shcherbina, V. I. (Dokl. Akad. Nauk SSSR **178** [1968] 1346/7; Dokl. Chem. Proc. Acad. Sci. USSR **178/183** [1968] 176/7).

[10] Ali, N.; Woods, S. B. (Solid State Commun. **46** [1983] 33/5).

32.1.5.11.4 Properties of Charge Carriers

Values at room temperature for the carrier concentration n_H derived from Hall measurements assuming a single-band model and the Hall mobility μ_H are given below, together with the number of 5d electrons n_d per metal atom as derived from the K_{α_1} line shift and values listed for n and μ (z.m. = zone-melted sample, h.p. = hot-pressed sample with zero porosity, pd = pressed powder):

MB$_6$	n_H per metal atom	n_H	n_d	n in 10^{22} cm^{-3}	μ_H in cm$^2 \cdot$V$^{-1} \cdot$s^{-1}	μ	μ
	z.m.	h.p.	pd		h.p.		pd
YB$_6$	—	0.96	1.00(2)	—	11.4	—	—
LaB$_6$	0.93	0.90	0.86(3)	1.4	33.1	32	51.0
CeB$_6$	—	1.06	1.00(5)	1.5	14.2	15	14.2
PrB$_6$	0.85	1.02	0.82(8)	1.3	22.2	22	20.9
NdB$_6$	0.88	1.00	1.01(5)	0.9	22.0	25	25.0
SmB$_6$	1.82	2.86	—	—	0.74	—	—
EuB$_6$	—	0.09	<0.05	—	59.3	—	—
GdB$_6$	—	0.94	1.00(11)	—	9.8	—	9.8
TbB$_6$	—	0.94	—	—	12.2	—	—
YbB$_6$	—	0.05	<0.05	0.09	179.4	76	—
Ref.	[1]	[2]	[3]	[4]	[2]	[4]	[5]

Reflectivity data of single crystals (c), polycrystalline melted samples (p), powders (pd), and polycrystalline films (f) yielded the following values for the optical conduction electron relaxation time τ_{opt}, for the carrier concentration to effective mass ratio $n/(m^*/m_0)$, for m^*/m_0 (calculated using literature n data), and for μ_{opt}, which is proportional to τ/m^* (the values for $n/(m^*/m_0)$ given in parentheses were recalculated from m^*/m_0 and n):

sample	τ_{opt} in 10^{-15} s	$n/(m^*/m_0)$ in 10^{22} cm^{-3}	n in 10^{22} cm^{-3}	m^*/m_0	μ_{opt} in cm$^2 \cdot$V$^{-1} \cdot$s^{-1}	Ref.
LaB$_6$ p	4.85 ± 0.05	4.39 ± 0.02	0.90	0.21	40.5	[6]
p	2.66	(3.9)	1.28	0.33	12.3	[7]
pd	—	4.42	1.4	0.32	23.5	[8]
pd	—	(4.6)	1.30	0.28	28	[5]
f	1.47	(2.5)	1.39	0.56	2.0	[7]
CeB$_6$ c	1.63 ± 0.05	2.04 ± 0.02	—	—	—	[9]
pd	—	(4.4)	1.54	0.35	20	[5]
PrB$_6$ p	2.30 ± 0.05	3.96 ± 0.02	1.02	0.26	15.6	[6]
p	2.66	(3.0)	1.02	0.34	13.8	[7]
pd	6.7*)	4.30	1.3	0.30	13.6	[8]
pd	—	(4.4)	1.20	0.27	30	[5]
f	1.47	(3.1)	1.35	0.43	6.0	[7]
NdB$_6$ p	1.80 ± 0.05	2.90 ± 0.02	1.00	0.35	9.1	[6]
pd	—	(4.6)	1.25	0.27	28	[5]

sample	τ_{opt} in 10^{-15} s	$n/(m^*/m_0)$ in 10^{22} cm^{-3}	n in 10^{22} cm^{-3}	m^*/m_0	μ_{opt} in cm$^2 \cdot$V$^{-1} \cdot$s^{-1}	Ref.
SmB$_6$ p	1.10 ± 0.05	2.09 ± 0.02	2.86	1.42	1.4	[6]
p	2.22	(1.1)	2.86	2.53	1.48	[7]
f	0.44	(1.6)	1.04	0.64	1.21	[7]
GdB$_6$ pd	—	(4.4)	1.32	0.30	29	[5]

*) Plasma relaxation time.

For m^*/m_0 and μ_{opt} for films of LaB$_6$, CeB$_6$, and SmB$_6$ see also Bessaraba, Bychkov [10], Ivanchenko et al. [11], and for LaB$_6$ and PrB$_6$, also Ivanchenko et al. [8].

Relative changes $\Delta\tau/\tau$ and $\Delta(n/m^*)/(n/m^*)$ in single crystals under the action of uniaxial stress referred to the unstrained crystals (data from [6]) were derived from reflectivity (R) data (see p. 83), for the electric field intensity vector $\vec{E}$ of the incident electromagnetic wave parallel and perpendicular to the stress $\vec{p}$. Under the conditions of measurements (r.p. = reflection plane), values for $\Delta(n/m^*)/(n/m^*)$, designated in the following as $\Delta r/r$, and $\Delta\tau/\tau$ amounted to:

sample	conditions		$10^5 \Delta\tau/\tau$	$10^5 \Delta r/r$
LaB$_6$	strain 1.35×10^{-4}	$\vec{E}\|\vec{p}$	5.5 ± 0.5	1.15 ± 0.1
	r.p. = (100); $\sphericalangle(\vec{p}, [110]) \approx 20°$	$\vec{E} \perp \vec{p}$	-5.5 ± 1.3	-3.7 ± 0.9
LaB$_6$	strain 1.7×10^{-4}	$\vec{E}\|\vec{p}$	-8.3 ± 1.0	-1.9 ± 0.2
	$\sphericalangle(r.p., (110)) \approx 15°$; $\sphericalangle(\vec{p}, [100]) \approx 9°$	$\vec{E} \perp \vec{p}$	0.9 ± 0.5	1.9 ± 1.0
CeB$_6$	strain 1.3×10^{-4}	$\vec{E}\|\vec{p}$	18.0 ± 3.0	1.9 ± 0.4
	$\sphericalangle(r.p., (310)) \lesssim 10°$, $\sphericalangle(\vec{p}, [100]) \lesssim 20°$	$\vec{E} \perp \vec{p}$	-3.9 ± 0.7	-4.95 ± 0.8
PrB$_6$	strain 2.5×10^{-4}	$\vec{E}\|\vec{p}$	-0.9 ± 0.2	2.6 ± 0.4
	setting as for CeB$_6$	$\vec{E} \perp \vec{p}$	$\Delta R/R = 0$	
NdB$_6$	strain 1.5×10^{-4}; setting as for CeB$_6$	$\vec{E}\|\vec{p}$	-4.55 ± 0.2	-3.72 ± 0.2

For the resulting shift of the plasma edge and change of the reflectivity see p. 83, Iller [9].

References:

[1] Paderno, Yu. B.; Garf, E. S.; Niemyskii, T.; Pračka, I. (Poroshkovaya Metal. **1969** No. 10, pp. 55/8; Soviet Powder Met. Metal Ceram. **1969** 821/3).

[2] Paderno, Yu. B.; Samsonov, G. V. (Dokl. Akad. Nauk SSSR **137** [1961] 646/7; Proc. Acad. Sci. USSR Phys. Chem. Sect. **136/141** [1961] 293/4).

[3] Grushko, Yu. S.; Paderno, Yu. B.; Mishin, K. Ya.; Molkanov, L. I.; Shadrina, G. A.; Konovalova, E. S.; Dudnik, E. M. (Phys. Status Solidi B **128** [1985] 591/7).

[4] Philips', N. V. Gloeilampenfabrieken (Fr. 1 439 722 [1964/66]; C.A. **65** [1966] 17887).

[5] Ivanchenko, L. A.; Bessaraba, V. I.; Paderno, Yu. B.; Vereshchak, V. M. (Elektron. Str. Fiz. Khim. Svoistva Splavov Soedin. Osn. Perekhodnykh Met. Dokl. 8th Simp., Kiev 1974 [1976], pp. 111/4; C.A. **87** [1977] No. 125852).

[6] Kierzek-Pecold, E. (Phys. Status Solidi **33** [1969] 523/31).

[7] Bessaraba, V. I.; Ivanchenko, L. A.; Paderno, Yu. B. (J. Less-Common Metals **67** [1979] 505/9).

[8] Ivanchenko, L. A.; Paderno, Yu. B.; Pilyankevich, A. N. (Poroshkovaya Metal. **1978** No. 8, pp. 38/48; Soviet Powder Met. Metal Ceram. **17** [1978] 602/9).

[9] Iller, A. (Phys. Status Solidi B **63** [1974] 69/75).

[10] Bessaraba, V. I.; Bychkov, V. P. (Mater. Izdeliya Poluch. Metodom Poroshk. Metall. Dokl. 6th, 7th Nauchn. Konf. Aspir. Molodykh Issled. Inst. Probl. Materialoved. Akad. Nauk Ukr. SSR, Kiev 1972, 1973 [1975], pp. 169/73; C.A. **87** [1977] No. 105729).

[11] Ivanchenko, L. A.; Bessaraba, V. I.; Paderno, Yu. B. (Khim. Svyaz Kristallakh Ikh Fiz. Svoistva **1976** No. 2, pp. 48/51; C.A. **86** [1977] No. 10452).

32.1.5.11.5 Thermionic Emission and Related Properties

General Remarks

Among the binary and ternary rare earth hexaborides, LaB$_6$ appears to be the best thermionic emission material in having both high-current density and excellent stability in the surface composition at high temperatures, Futamoto et al. [1, 2]. Since the study of Lafferty [3] in 1951 on LaB$_6$ and CeB$_6$, the thermionic properties of MB$_6$ were studied extensively. However, widely divergent results were obtained, see the compilations by Fomenko [4], Fomenko, Podchernyaeva [5]. In early studies the sintered and hot-pressed samples used were more or less contaminated, particularly in the case of incongruently vaporizing MB$_6$ (M = Y, Gd, etc.), sometimes even by MB$_4$, which has a lower work function than MB$_6$, see pp. 20 and 150. Contaminating C raises the work function as does adsorbed oxygen, which leads to the formation of a superficial layer of metal oxide plus boron oxide at elevated temperatures, Berrada et al. [6]. Cleaning of the MB$_6$ cathode surface by ion bombardment in an atmosphere of Ar or of a Kr/Xe mixture was recommended by Trigubenko, Tsarev [7]. However, even a 2 h-ion bombardment leaves behind some carbon and oxygen and, moreover, depletes the surface of M (= La, Ce), which raises the work function towards that of boron (~4.4 eV), Berrada et al. [6]. Surface carbon can be removed by heating to ca. 1400 K in an O$_2$ pressure of ca. 10^{-4}Pa, Swanson, McNeely [8]. Oxides evaporate on short heat treatment at 1773 K in ultra high vacuum [6]. However, detailed studies on the congruently vaporizing MB$_6$ (M = La to Sm) revealed, in addition, the necessity of activation in order to get a steady-state surface composition (see p. 90) and thus consistent and reproducible thermionic emission data, see [8] for M = La, Ce, and Sm. For example, the effective work function of polycrystalline samples (M = La to Nd) at 1800 K in a vacuum of 10^{-7} Pa decreased to a minimum during decontamination and activation as a function of time, increased slightly afterwards, and reached a constant value only after 6 h, Berrada et al. [9]. Single crystals, cleaned by slight etching in a dilute HNO$_3$ aqueous solution, were cleaned further successfully and activated at 1923 K, <6×10^{-6} Pa for 1 h. During operation of the cathodes, residual gas poisoning was not observed above 1523 K, if the vacuum was better than 6×10^{-6} Pa [1], and only above 1723 K in a lower vacuum of ≤ 1.33×10^{-4} Pa, Schmidt et al. [10].

The single crystals used in the fundamental studies described in this chapter were all Al-flux grown. According to Davis et al. [11] flat or pointed cathodes can be fabricated from arc float-zone refined MB$_6$ (M = La, Ce, Pr) single-crystal rods by grinding, using a 3000 mesh diamond wheel and final polishing, using a 14000 mesh aluminium oxide-diamond material. In the overall mounting configuration, the MB$_6$ crystal is brazed into a Re cup with a TaC braze, the cup, in turn, being mounted on resistively heatable W filament supports. LaB$_6$ cathodes mounted this way have operated successfully at 1800 K for over 1000 h. This operating temperature can be achieved with ~2.5 W of electrical power. In a second type of mounting (so-called Vogel mount), the cathode is clamped between pyrolytic graphite blocks which are, in turn, held between heavy Mo and W supports. The ohmically heated graphite heats the cathodes by thermal conduction. As there is no possibility of Mo or W evaporating onto the cathode surface, this design is particularly suited to longlife applications [11].

Binary Hexaborides MB_6

Zero-field values for the current density J_0 and the effective work function Φ_{eff} at high temperatures, derived from thermionic measurements are assembled below, together with values for the work function Φ_f at 300 K as derived by the field emission retarding potential (FERP) method. All data refer to (100) single crystals, either bonded to a glassy carbon filament and activated at 1923 K, $< 6 \times 10^{-6}$ Pa for 1 h [1] or pressed onto Ta substrate using an aqueous slurry powdered TaC as binder and activated above 1800 K in a vacuum not defined [8] (J_0 in A/cm^2, Φ_{eff} and Φ_f in eV):

MB_6	1873 K [1]		1600 K [8]		300 K [8]
	J_0	Φ_{eff}	J_0	Φ_{eff}	Φ_f
LaB_6	8.5	2.86(4)	1.0	2.70(5)	2.60(5)
CeB_6	3.1	3.02(3)	1.7	2.62(5)	2.50(5)
PrB_6	1.3	3.16(9)	—	—	—
NdB_6	1.1	3.19(5)	—	—	—
SmB_6	1.6×10^{-3}	4.24(7)	1.4×10^{-4}	3.92(5)	4.30(5)
EuB_6	1.0×10^{-3}	4.32(7)	—	—	—

Current densities of the (100) single crystals activated at 1923 K are shown in **Fig. 29**, p. 74, for the range 1523 to 1923 K [1]. At 1800 K, J_0 for both LaB_6 and CeB_6 exceeded 10 A/cm^2 [8]. For earlier studies of field emission patterns and emission characteristics after flashing treatment at ca. 1673 K, see Okano et al. [12]. Axial output currents >1373 K of fully activated (110) LaB_6, (100) MB_6 (M = Y, La, Ce), and sintered MB_6 (M = La, Pr, Nd) are shown in a figure by Schmidt et al. [10]. At 1673 K, the axial output current of (100) LaB_6 was by four orders larger than that of unstable (100) YB_6, however, it was about half that of (110) LaB_6 [10]. Plots of Φ_{eff} vs. T in the range 1220 to 1800 K (see **Fig. 30**, p. 74) exhibited, for (100) LaB_6 and (100) SmB_6, a break at ca. 1600 and ca. 1450 K, respectively [8]. Auger electron spectroscopy (AES) indicated that high J_0 values and low Φ_{eff} values correlate with low B/M ratios in the surface layer of MB_6 (M = La to Eu), see [2, 8]. However, the energy dependence of the electron reflection coefficient maxima suggested that a conduction band edge shift relative to the Fermi level among the various (001) MB_6 crystals (M = La, Ce, Sm) largely accounts for the work function variation [8]. A similar interpretation was also given by Storms [13] in order to explain the change of Φ_{eff} as a function of room temperature lattice parameter for MB_6; see **Fig. 31**, p. 75, which gives Φ_{eff} at 1500 K for polycrystalline pure YB_6 (see p.150), LaB_6, NdB_6 (both from [13]), CeB_6, PrB_6 (both from [6]), GdB_6, and various compositions of $Nd_{1-x}La_xB_6$ (from [13]). Also shown in this figure are the ranges of Φ_{eff} observed for LaB_6 and NdB_6 of various B/M ratios at the surface. Whereas Φ_{eff} calculated from the measured thermionic current does not differ strongly from that Φ_{eff} derived from the thermionic current corrected for composition B/M, the Richardson work function Φ_0 depends sensitively on the composition effect. Since this is generally not known, the reported Richardson work functions have very little theoretical meaning [13]. The ambiguousness of Φ_0 is also reflected in the nonlinear $\Phi_{eff} = f(T)$ relationships of Fig. 30 [8], p. 74. Richardson plots in the range 1000 to 1500 K [6, 9] and 1373 to 1773 K [10] lead to the following values for Φ_0 (in eV) and emission constant A (in A·cm^{-2}·K^{-2}) for polycrystalline MB_6 after activation at 1800 K for 8 h at 10^{-7} Pa [6, 9] or at (presumably) 1373 K for 0.3 h at ca. 7×10^{-5} Pa [10] (the theoretical value of A is 120.4; the value A = 29 was observed for sintered LaB_6):

| MB$_6$ | sintered sample | | melted sample | polycryst. sample[1] | |
	Φ_0 (for A = 120.4)	Φ_0 (for A = 29)	Φ_0 (for A = 29)	Φ_0	A
YB$_6$	—	—	—	~1	—
LaB$_6$	2.84 to 2.95	2.72 to 2.79	2.67 to 2.83	4.5	100
CeB$_6$	3.06 to 3.14	2.94 to 3.05	2.98 to 3.08	3	1
PrB$_6$	3.08 to 3.21	2.97 to 3.09	—	2.3	—
NdB$_6$	3.15 to 3.27	3.05 to 3.15	3.88 to 4.46	1.6	0.1
EuB$_6$[2]	4.03	3.95		—	—
Ref.	[6, 9]	[6, 9]	[6, 9]	[10]	

[1] The YB$_6$ was very unstable, the PrB$_6$ unstable, and CeB$_6$ was a (100) crystal. – [2] Boron-rich surface layer due to activation.

$\Phi_0 = 2.37 \pm 0.05$ for A = 120.4 for the (100) CeB$_6$ was derived from Φ_{eff} shown in Fig. 30 [8] and contact potential measurements resulted in $\Phi_0 \approx 2.8$ (LaB$_6$), 3.0 (CeB$_6$), 2.5 (PrB$_6$), 3.0 (EuB$_6$), and ~2.6 (YbB$_6$), see a figure in Yamamoto et al. [14]. The ranges of Φ_0 observed by means of an emission microscope (see table above) were interpreted in terms of different preferred crystallite orientations existing at the various points at the surface [9]. The sequence of increasing work functions for LaB$_6$ single-crystal faces was observed to be (346) < (100) < (110) < (111), Swanson et al. [15], see also p. 238. This trend in which the metal-rich crystallographic plane exhibits a lower work function holds also for MB$_6$ with M = Ce to Eu [2], which is in contrast to calculations by [9].

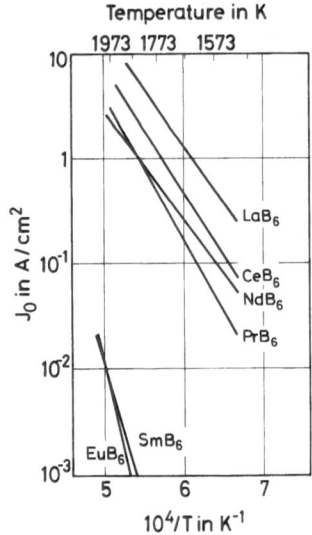

Fig. 29. Current density J$_0$ of activated (100) MB$_6$ single crystals as a function of the heating temperature.

Fig. 30. Effective work function Φ_{eff} vs. temperature for the (100) face of SmB$_6$, LaB$_6$, and CeB$_6$ single crystals (full circles are from earlier measurements or from different crystals; the same scale holds for LaB$_6$ and CeB$_6$).

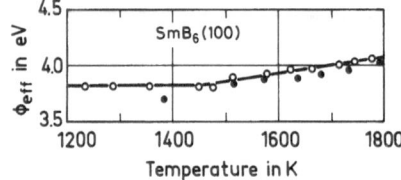

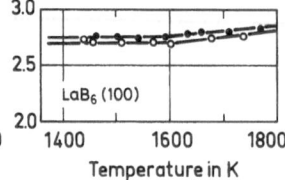

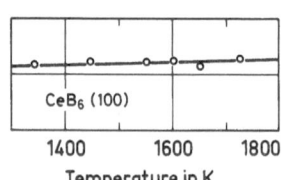

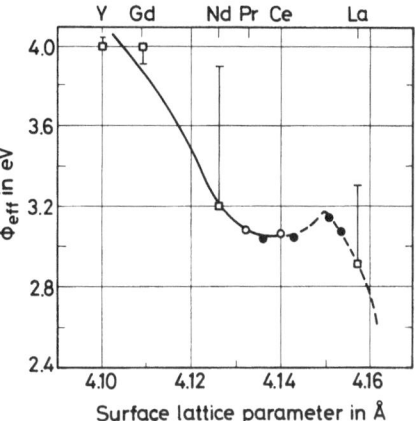

Fig. 31. Relationship between the effective work function Φ_{eff} at 1500 K and the lattice parameter at room temperature of polycrystalline stoichiometric MB_6 (open squares and open circles are from measurements of different authors), of nonstoichiometric LaB_6 and NdB_6 (vertical bars), and of various compositions in the $Nd_{1-x}La_xB_6$ system (full circles).

Data are given in the following for polycrystalline samples which were contaminated in the poor vacuum (see p. 72). In addition, note that MB_6 with M = Sc, Er, Tm, and Lu is unstable (see p. 21) and designated as "MB_6": The effective work function of MB_6 (M = Ce, Pr, Nd) on Ta substrate between 1200 and 1500 K at 10^{-3} Pa was studied as a function of preparation temperature (1773 to 2073 K) in the borothermal reduction by Peshev, Surnev [16]. Effective work functions of MB_6 deposited on various cathode bases were measured between 1000 and 2000 K at $\leqq 3.7 \times 10^{-4}$ Pa and their temperature dependence $\Phi_{eff} = \Phi_0 + (d\Phi/dT)T$, valid in a limited temperature range, was determined (Φ_{eff}, Φ_0 in eV; $d\Phi/dT$ in eV/K, and T in K):

MB_6	base	Φ_{eff} (1500 K)	Φ_{eff} (1800 K)	Φ_0	$d\Phi/dT$	T range	Ref.
SmB_6	W	3.13	—	2.86	1.75×10^{-4}	1100 to 1500	[17]
SmB_6	Ir	3.07	3.11	2.73	2.25×10^{-4}	1100 to 1500	[17]
EuB_6	W	3.17	3.23	2.73	2.9×10^{-4}	1100 to 1500	[17]
EuB_6	Ir	3.56	3.61	3.32	1.56×10^{-4}	1200 to 2000	[17]
GdB_6	W or Ir	2.91	3.07	2.59	2.15×10^{-4}	1000 to 1500	[17]
"TmB_6"	W	3.15	3.27	2.75*)	3.3×10^{-4}*)	1100 to 1800	[18]
"TmB_6"	TaC	3.29	3.40				[18]
"TmB_6"	MB-50 alloy	3.24	3.35				[18]

*) Average for the three bases.

With SmB_6 and EuB_6 on a W base above 1700 K and with "TmB_6" on all three bases tested on aging at $\geqq 800$ K, the work function was found to increase markedly and a metallic deposit (for example, tungsten boride) appeared on the tube walls. However, on an Ir base, the SmB_6 and EuB_6 could be operated up to 1900 K for several hours without giving rise to a metallic deposit; $J_{eff} = 2.2$ A/cm² for SmB_6 and 0.12 A/cm² for EuB_6 at 1900 K. The strong increase of Φ_{eff} observed with GdB_6 on both W and Ir base above 1600 K was ascribed to a space charge limiting mode of operation at the applied anode voltage of 200 V; see Ermakov, Tsarev [17] for MB_6 (M = Sm, Eu, Gd) and Ermakov [18] for "TmB_6". Values for Φ_0 and A derived from Richardson plots for YB_6, LaB_6, and $GdB_6 + GdB_4$ between 1100 and 2000 K, see Kudintseva et al. [19], for MB_6 (M = La to Nd), Kudintseva et al. [20], and for all MB_6 (except M = Sm, Eu, Tb, Tm) deposited on Ta wire cathodes initially covered with sintered Ta powder, see Kudintseva, Tsarev [21]; vacuum and temperature range not defined. Large ranges for both Φ_0 and A for each MB$_6$ (M = Y, La to Tb, Er) were derived from studies at 10^{-4} to 10^{-5} Pa between 1403 and 1573 K after activating the cathodes (same material as in [21]) for 5 min to 120 h. In view of the requirements for reliable data (pure single-phase material, vacuum of $<10^{-6}$ Pa) these determinations and those from [19 to 21] were regarded as unreliable, Trigubenko, Tsarev [7]. Therefore, values for Φ_0 and A and for Φ_{eff} at 1700 K calculated therefrom (except for TbB$_6$) are given below only for the otherwise less-studied compounds MB$_6$ with M = Tb, Dy, Ho, and Yb and the instable compounds "MB$_6$" with M = Sc, Er, Tm, Lu, which presumably are mixtures of several borides (Φ_0 and Φ_{eff} in eV, A in A·cm^{-2}·K^{-2}):

MB$_6$	Φ_0	A	Ref.	Φ_{eff}	Ref.
"ScB$_6$"[1)]	2.96	4.6	[22]	3.46	[24]
TbB$_6$	3.1	120.4	[23]	3.26[2)]	[7]
DyB$_6$	3.53	25.1	[21]	3.76	[24]
HoB$_6$	3.42	13.9	[21]	3.74	[24]
"ErB$_6$"	3.37	9.9	[21]	3.74	[24]
"TmB$_6$"	—	—	—	3.2	[24]
YbB$_6$	3.13	2.5	[21]	3.70	[24]
"LuB$_6$"	3.00	0.36	[21]	3.85	[24]

[1)] According to [19]: ScB$_2$. — [2)] At 1300 K.

The effective work function of a mixture of "TmB$_6$" + TmB$_4$ increased from ca. 2.6 eV at 1000 K to ca. 3.9 eV slightly above 1600 K and decreased at higher temperatures; Φ_{eff} = 3.5 eV and J = 0.22 A/cm^2 at 1900 K, Samsonov et al. [25]. Based on the early assumption that Φ_0 along the MB$_6$ series is least for LaB$_6$, GdB$_6$ (but see Fig. 31, p. 75), and LuB$_6$ and is a maximum for EuB$_6$ and DyB$_6$, the change of Φ_0 with atomic number M was discussed in terms of the number of f electrons and probable f→d transitions. The low Φ_0 of LaB$_6$, GdB$_6$, and LuB$_6$ was attributed to the transition of some B valence electrons to s(M) positions, Samsonov, Shlyuko [26]; for still earlier discussions, see for example, Samsonov [27], Samsonov, Neshpor [28], Samsonov et al. [29], Samsonov, Paderno [30]. For an early discussion of the emission constant A in terms of the electronic structure, see Kmetko [31]. However, A changes sensitively with the surface conditions during measurement, see for example, Trigubenko, Tsarev [7], Zaima et al. [32].

Materials for high-brightness cathodes, among them MB$_6$, were evaluated by defining a figure of merit which is based on Φ and the melting point, see Zaima et al. [32]. Models for predicting work functions of binary compounds were based on a) the properties of M having a low work function, perturbed by the element B of higher work function through its binding with M, Yamamoto et al. [33], or b) the idea of gas adsorption on metal surfaces. In this case the work function of the compound is determined by the higher work-function element (Φ_B = 4.5 eV) which is perturbed by adsorption of lower work-function element M. Both models do not predict the minimum work function observed for each of the compounds, see a figure in the paper, Yamamoto et al. [14].

Ternary Hexaborides M$_{1-x}$La$_x$B$_6$

As the surface layer of mixed hexaborides M$_{1-x}$La$_x$B$_6$ approaches that of LaB$_6$ at high temperatures (see p. 92), it does not seem to be effective to prepare a mixed hexaboride in order to improve the thermionic properties of LaB$_6$, Futamoto et al. [2]. The zero-field current density J$_0$ of various mixed hexaboride single crystals, after cleaning and activation at 1923 K, $<6\times10^{-6}$ Pa for 1 h, ranged between that of CeB$_6$ at low temperatures (1523 K) and that of LaB$_6$ at the highest temperatures studied (1923 K), see **Fig. 32**. Indicated compositions are bulk compositions, also in the following. Values for J_0 and the effective work function Φ_{eff} at 1873 K are:

$M_{1-x}La_xB_6$	J_0 in A/cm^2	Φ_{eff} in eV
$Ce_{0.46}La_{0.54}B_6$	5.2	2.94 ± 0.05
$Pr_{0.91}La_{0.09}B_6$	5.1	2.94 ± 0.05
$Pr_{0.45}La_{0.55}B_6$	5.0	2.95 ± 0.05
$Pr_{0.12}La_{0.88}B_6$	7.0	2.89 ± 0.05
$Sm_{0.42}La_{0.58}B_6$	4.5	2.96 ± 0.06
$Dy_{0.03}La_{0.97}B_6$	5.0	2.94 ± 0.05
LaB_6	8.5	2.86 ± 0.04

Futamoto et al. [1]. Studies on $M_{0.7}La_{0.3}B_6$ with M = Pr and Nd, resulting in lower work functions than for LaB_6, Schmidt, Joy [34], were later not confirmed for the system NdB_6–LaB_6, Storms [35], see Fig. 31, p. 75, from Storms [13]. Polycrystalline cleaned $Y_{0.2}La_{0.8}B_6$, see Schmidt et al. [10] and $Eu_{1-x}La_xB_6$ with x = 0.3 to 0.7, too, exhibited less favorable emission properties than LaB_6, due to their instability, Berrada et al. [6, 9].

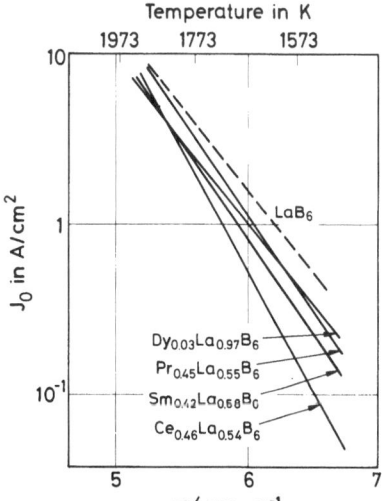

Fig. 32. Current density J_0 of activated $M_{1-x}La_xB_6$ with M = Ce, Pr, Sm, and Dy (see text).

Secondary Emission

Maximum values of the secondary emission factors at room temperature amounted to 0.95 (LaB_6), 0.68 (CeB_6), 0.7 ($GdB_6 + GdB_4$), 0.8 (DyB_6), 0.7 (HoB_6), and to 0.8 for LuB_6 (which does not exist as a pure phase, see p. 21). These values were observed for energies of the primary electrons somewhere between 400 and 600 V, Kudintseva, Tsarev [21]; see also Kudintseva et al. [20] for MB_6 (M = La to Nd) and Kudintseva et al. [19] for M = Y, La, and Gd.

Reflection of Electrons

Reflection coefficients of the specular elastic (R_e) and inelastic reflected (R_{in}) electrons of (100) MB_6 crystals at room temperature as a function of primary electron beam voltage up to ~11 eV were derived from field emission retarding potential (FERP) measurements; for (100) LaB_6, see a figure in the paper. While R_e for (100) SmB_6 was small at threshold, $R_e \approx 0.5$ was observed for both (100) LaB_6 and (100) CeB_6. Such a large value of R_e near the threshold voltage implies a Bragg reflection or a low normal density of electronic states at the vacuum level. For (100) CeB_6 a large threshold value of R_e is also evidenced by the reflection coefficient $R = 0.85 \pm 0.15$ derived from thermionic data (see Fig. 30, p. 74) by means of the Richardson equation. The positions of the four observed R_e maxima in terms of energy relative to the work function barrier Φ_f are almost identical for (100) LaB_6 and (100) CeB_6, see **Fig. 33**. The (100) SmB_6 exhibits only two such maxima and the largest maximum occurs at $\Phi_f + 2.4$ eV rather than $\Phi_f + 2.0$ eV as observed for LaB_6 and CeB_6. Nevertheless, the systematic shift in the R_e maxima with Φ_f was regarded as convincing evidence that the variation in work function for the series MB_6 (M = La, Ce, Sm) was largely due to shifts in the position of the conduction band edge relative to the Fermi level rather than to surface contributions, Swanson, McNeely [8].

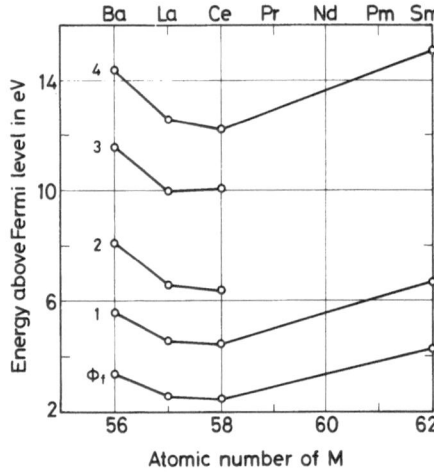

Fig. 33.   (001) FERP work function Φ_f and the succeeding four electron reflection maxima above Φ_f as a function of M in MB_6.

References:

[1] Futamoto, M.; Nakazawa, M.; Kawabe, U. (Surf. Sci. **100** [1980] 470/80).

[2] Futamoto, M.; Nakazawa, M.; Kawabe, U. (Vacuum **33** [1983] 727/32; C.A. **100** [1984] No. 43989).

[3] Lafferty, J. M. (J. Appl. Phys. **22** [1951] 299/309).

[4] Fomenko, V. S. (Emissionnye Svoistva Khimicheskikh Elementov i Ikh Soedinenii, Naukova Dumka, Kiev 1964; Handbook of Thermionic Properties, Plenum Press Data Division, New York 1966, pp. 88/91).

[5] Fomenko, V. S.; Podchernyaeva, I. A. (Emissionnye i Adsorbtsionnye Svoistva Veshchestv i Materialov, Atomizdat, Moscow 1975, pp. 166/75).

[6] Berrada, A.; Mercurio, J.-P.; Etourneau, J.; Hagenmuller, P. (Rev. Intern. Hautes Temp. Refract. **15** [1978] 115/28).

[7] Trigubenko, V. A.; Tsarev, B. M. (Radiotekhn. Elektron. **6** [1961] 1900/5; Radio Eng. Electron. Phys. [USSR] **6** [1961] 1694/8).

[8] Swanson, L. W.; McNeely, D. R. (Surf. Sci. **83** [1979] 11/28).

[9] Berrada, A.; Mercurio, J.-P.; Etourneau, J.; Hagenmuller, P. (J. Less-Common Metals **59** [1978] 7/25).

[10] Schmidt, P. H.; Longinotti, L. D.; Joy, D. C.; Ferris, S. D.; Leamy, H. J.; Fisk, Z. (J. Vac. Sci. Technol. **15** [1978] 1554/60).

[11] Davis, P. R.; Swanson, L. W.; Hutta, J. J.; Jones, D. L. (J. Mater. Sci. 21 [1986] 825/36).
[12] Okano, H.; Futamoto, M.; Hosoki, S.; Kawabe, U. (Shinku 20 No. 4 [1977] 127/35; C.A. 87 [1977] No. 94150).
[13] Storms, E. K. (J. Appl. Phys. 52 [1981] 2961/5).
[14] Yamamoto, S.; Susa, K.; Kawabe, U.; Okano, H. (Japan. J. Appl. Phys. 13 Suppl. 2, Pt. 2 [1974] 209/12).
[15] Swanson, L. W.; Gesley, M. A.; Davis, P. R. (Surf. Sci. 107 [1981] 263/89).
[16] Peshev, P.; Surnev, L. (Dokl. Bolg. Akad. Nauk 19 [1966] 515/8; C.A. 65 [1966] 12880).
[17] Ermakov, S. V.; Tsarev, B. M. (Radiotekhn. Elektron. 10 [1965] 972/5; Radio Eng. Electron. Phys. [USSR] 10 [1965] 833/5; C.A. 63 [1965] 3732).
[18] Ermakov, S. V. (Radiotekhn. Elektron. 9 [1964] 180/1; Radio Eng. Electron. Phys. [USSR] 9 [1964] 142/3; C.A. 60 [1964] 12750).
[19] Kudintseva, G. A.; Neshpor, V. S.; Samsonov, G. V.; Tsarev, B. M.; Paderno, Yu. B. (Vysokotemp. Metallokeram. Materialy 1962 109/12; C.A. 58 [1963] 7468).
[20] Kudintseva, G. A.; Epel'baum, V. A.; Tsarev, B. M. (Bor Tr. Konf. Khim. Bora Ego Soedin., Moscow 1955 [1956], pp. 112/9; C.A. 1960 12780).

[21] Kudintseva, G. A.; Tsarev, B. M. (Radiotekhn. Elektron. 3 [1958] 428/9; Radio Eng. Electron. Phys. [USSR] 3 [1958] 182/5; C.A. 1958 15254).
[22] Samsonov, G. V. (Usp. Khim. 28 [1959] 189/217, 197).
[23] Trigubenko, V. A.; Tsarev, B. M. (Kristallografiya 4 [1959] 542/4; Soviet Phys.-Cryst. 4 [1959] 510/2).
[24] Bondarenko, B. V.; Tsarev, B. M. (Vopr. Teorii Primen. Redkozem. Met. Mater. 1964 86/91; Problems of the Theory and Use of Rare Earth Metals, JPRS-28849 [1965] 108/16; C.A. 62 [1965] 3497).
[25] Samsonov, G. V.; Fomenko, V. S.; Paderno, Yu. B. (Ukr. Fiz. Zh. [Ukr. Ed.] 8 [1963] 700/2; C.A. 59 [1963] 8228).
[26] Samsonov, G. V.; Shlyuko, V. Ya. (Ukr. Fiz. Zh. [Ukr. Ed.] 11 [1966] 437/8; C.A. 65 [1966] 1531).
[27] Samsonov, G. V. (Tugoplavkie Soedineniya Redkozemel'nykh Metallov s Nemetallami, Metallurgiya, Moscow 1964; High-Temperature Compounds of Rare Earth Metals with Nonmetals, Consultants Bureau, New York 1965, pp. 1/280, 21/6; N.S.A. 19 [1965] No. 41073).
[28] Samsonov, G. V.; Neshpor, V. S. (Dokl. Akad. Nauk SSSR 122 [1958] 1021/3; Soviet Phys. Dokl. 3 [1958] 1029/31).
[29] Samsonov, G. V.; Neshpor, V. S.; Paderno, Yu. B. (Ukr. Fiz. Zh. [Ukr. Ed.] 4 [1959] 508/18; N.S.A. 14 [1960] No. 7804).
[30] Samsonov, G. V.; Paderno, Yu. B. (Dopov. Akad Nauk Ukr. RSR 1959 1215/8; C.A. 1961 6329).

[31] Kmetko, E. A. (Phys. Rev. [2] 116 [1959] 895/6).
[32] Zaima, S.; Adachi, H.; Shibata, Y. (J. Vac. Sci. Technol. [2] B 2 [1984] 73/8).
[33] Yamamoto, S.; Susa, K.; Kawabe, U. (J. Chem. Phys. 60 [1974] 4076/80).
[34] Schmidt, P. H.; Joy, D. C. (J. Vac. Sci. Technol. 15 [1978] 1809/10).
[35] Storms, E. K. (J. Appl. Phys. 54 [1983] 1076/81).

32.1.5.12 Optical and Dielectric Properties

Color

Dense MB_6 samples with trivalent M (all except Eu and Yb), hot-pressed or melted, have metallic luster with dark steel blue hue, often with iridescent tarnish, Binder [1] and for M = La to Nd and Er, von Stackelberg, Neumann [2] (for ErB_6, see remarks given in the table below). The colors mentioned by Andrieux [3] for more or less contaminated crystals with M = Y, La, Ce, Nd, Gd, and Er, obtained by electrolysis, agree with those listed by [1] for stoichiometric MB_6. These are given below, together with the change of color observed for increasingly metal deficient samples:

MB_6	stoichiometric	increasing metal deficiency →	
YB_6	blue	blue (gray tinge)	blue-gray
LaB_6	red-violet	violet	blue
CeB_6	blue-violet	blue	blue (gray tinge)
PrB_6	blue	blue	blue-gray
NdB_6	blue (violet tinge)	blue	blue-gray
SmB_6	blue	blue-gray	blue-gray
EuB_6	blue (gray tinge)	gray-blue	gray (blue tinge)
GdB_6	blue	blue	blue-gray
TbB_6	blue	blue-gray	gray
DyB_6	blue	blue (gray tinge)	blue gray
HoB_6	blue	gray-blue	—
ErB_6*)	blue	—	—
YbB_6	blue-gray (black [3])	gray (blue tinge)	gray

*) Does not exist as pure phase, see p. 21.

Binder [1]. The LaB_6, when wetted with aqueous [2] and organic fluids or colorless varnishes, becomes intensively deep red to purple-violet. Self-hardening varnishes stabilize this state [1].

Optical Constants

The refractive index n (see **Fig. 34**) and extinction coefficient k (see **Fig. 35**) for zone-melted MB_6 with M = La, Pr, Nd, Sm in the range 1.2 to 6 eV were derived by Kramers-Kronig analyses. Structures in k at the higher energies, and thus in the absorption coefficient $\alpha = 4\,\pi k/\lambda$ (see a figure in the paper) are probably connected with the interband absorption, Kierzek-Pecold [4].

The spectral emissivity ε_λ at $\lambda = 650\ \mu m$ for powders was measured by Serebryakova et al. [5] at various temperatures between 1123 and 1923 K, using relative small Ta cylinders in the nearly absolute black body arrangement. Values for three selected temperatures are compared with those obtained earlier by Kudintseva et al. [6, 7] for 1823 K:

	YB_6	LaB_6	CeB_6	PrB_6	NdB_6	SmB_6	GdB_6	Ref.
1123 K	0.63	0.68	—	—	0.56	0.71	0.61	[5]
1523 K	0.66	0.69	—	—	0.58	0.69	0.62	[5]
1923 K	0.68	0.71	—	—	0.59	0.67	0.64	[5]
1823 K	0.7[1]	0.7	0.68	0.67	0.64	—	0.65[2]	[6, 7]

[1] From Kudintseva et al. [8]. – [2] Same value in [8] for $GdB_6 + GdB_4$.

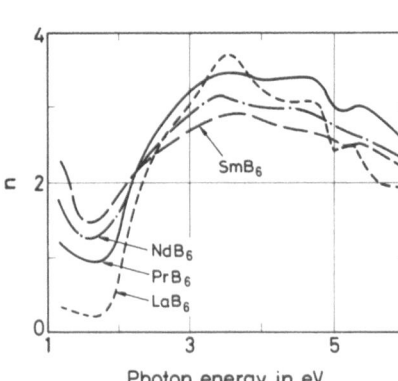

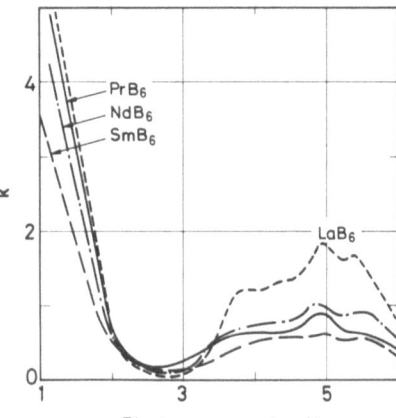

Fig. 34. Refractive index n of zone-melted rare earth hexaborides vs. photon energy.

Fig. 35. Extinction coefficient k of zone-melted rare earth hexaborides vs. photon energy.

For MB_6 (M = Dy, Ho, Yb), $\varepsilon_\lambda \approx 0.7$ [7] (temperature not mentioned). The same value was given for compounds with M = Er, Lu [7], and Tm, Samsonov et al. [9], however, these do not exist as a pure phase (see p. 21).

Transmission. Reflection

For transmission spectra of imperfect films of MB_6 with M = La and Sm in the range 100 to 3000 cm^{-1} in comparison to their reflectivity spectra, see Ivanchenko et al. [10], of MB_6 with M = La, Ce, Sm, Eu, and Yb between 12000 and 30000 cm^{-1}, see Ivanchenko et al. [11], and with M = La, Ce, Sm, and Yb between 12000 and 28000 cm^{-1}, see Bessaraba, Bychkov [12].

The reflectivity spectra of MB_6 with trivalent M = La to Nd and intermediate valent Sm, unlike EuB_6 with divalent Eu, are characterized by a high reflectivity R at low energies and a pronounced reflectivity minimum (plasma edge) at $\omega_{min} \approx 2$ eV. See **Fig. 36**, p. 82, for mechanically polished zone-melted MB_6 with M = La, Pr, Nd, Sm at nearly normal incidence of light in the region 1.2 to 6 eV [4] and **Fig. 37**, p. 82, for Al-flux grown single crystals with M = La, Sm, and Eu between 0.5 and 40 eV. The reflectivity of EuB_6 is rather low at low energies, with a not very marked minimum at $\omega_{min} = 0.4$ eV. All spectra (M = La, Sm, Eu) show a broad maximum with a well-defined structure between 2 and 8 eV, a series of maxima between 10 and 20 eV, and a general intensity decrease at higher energies, Shelykh et al. [13]. The reflectivity of polycrystalline annealed films of LaB_6, PrB_6, and SmB_6 on glass, silicon, quartz, or NaCl, obtained by evaporation and condensation onto heated substrates (500 to 900 K), showed all the main features as did melted samples in the range 100 to 50000 cm^{-1} ($\triangleq 0.012$ to 6.2 eV), however, with less-defined structure above the edge, Bessaraba et al. [14]. A Drude-like behavior was also observed for powders of MB_6 with M = La to Sm, and Gd, whereas the divalent MB_6 exhibited a nearly constant reflectivity (R $\approx 10\%$ for EuB_6, R $\approx 5\%$ for YbB_6) in the whole range studied, 1.2 to 3 eV [11]. Analogous observations were made with imperfect films between 14000 and 30000 cm^{-1} ($\triangleq 1.74$ to 3.72 eV), for M = La, Ce, and Sm [11], Paderno et al. [15], and in addition M = Yb [12]. In contrast, early studies in the range 0.4 to ca. 3.5 μm ($\triangleq 3.1$ to 0.35 eV), on samples not defined, revealed a Drude-like behavior also for YbB_6; $\omega_{min} = 0.65$ μm ($\triangleq 1.91$ eV) for M = La to Nd and 1.5 μm ($\triangleq 0.83$ eV) for M = Yb, Philips' Gloeilampenfabrieken [16].

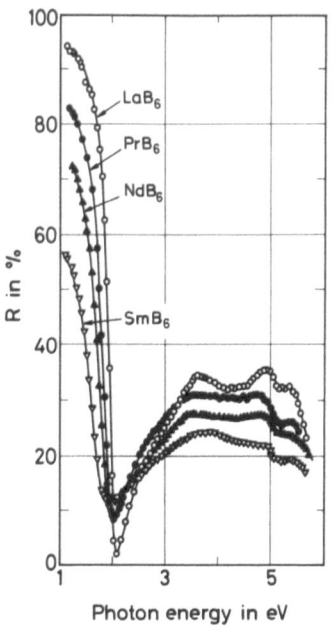

Fig. 36. Reflectivity R of zone-melted crystalline MB$_6$ in the region 1.2 to 6 eV.

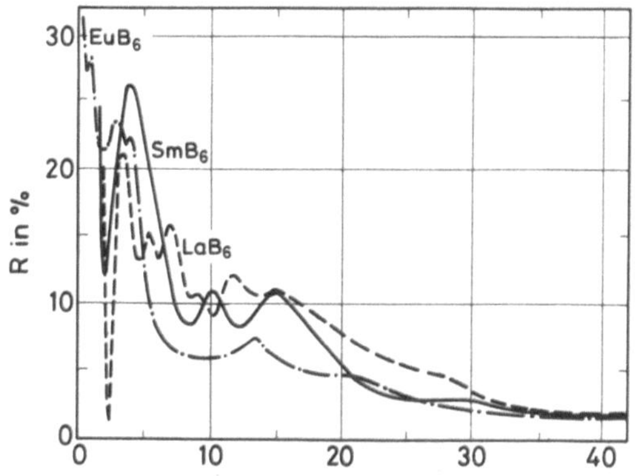

Fig. 37. Reflectivity R of MB$_6$ single crystals grown from an Al flux between 0.5 and 40 eV (R is nearly 100% at low energies for LaB$_6$ and SmB$_6$; not shown in the figure).

The dispersion of the real (ε') and imaginary (ε'') parts of the dielectric constant derived from Kramers-Kronig analysis and the contributions to ε' from free electrons (ε'_f) and bound electrons (ε'_b), see the individual sections, lead to the following picture for the single crystals: In LaB$_6$ and SmB$_6$ the onset of interband transitions, characterized by the minimum of ε'', occurs at $\omega_i = \omega_{min}$, i.e., at 2.1 eV ($\triangleq$16938 cm^{-1}) for LaB$_6$ and 1.8 eV ($\triangleq$14518 cm^{-1}) for SmB$_6$. The plasmon energy of the conduction electrons $\omega_p(\varepsilon'_f = 0) = 4.2$ eV is shifted by ca. 2.2 eV, to $\omega_p^* \approx 2$ eV, owing to the interband transitions. The corresponding sharp peak (P$_{max}$) in the energy loss function $\varepsilon''/[(\varepsilon')^2 + (\varepsilon'')^2]$ is observed at ~2 eV for both compounds. The permeability $1 + \varepsilon'_b$, at $\omega \to 0$, amounts to 4.6 for LaB$_6$, 4.8 for SmB$_6$, and to 4.7 for EuB$_6$ [13]. The following table lists the spectral positions of ω_{min} or ν_{min}, of ω_i, ν_p^*, and P$_{max}$ and the high-frequency permeability ε'_∞ derived from the plot ε' versus square wavelength at $\lambda^2 \to 0$ for polycrystalline melted samples (p), powders (pd), and films (f) with calculated values in parentheses:

sample		ν_{min} in cm^{-1}	ω_{min} in eV	$\omega_i(\varepsilon''_{min})$ in eV	$\nu_p^*(\varepsilon'=0)$ in cm^{-1}	P$_{max}$ in cm^{-1}	ε'_∞	Ref.
LaB$_6$	p	17000[1)]	—	—	—	—	15.15(2)	[4]
	p	17000	—	—	—	15600	13.5	[14]
	p	17000	—	—	15400	16000	12.0	[17]
	pd	16800	(2.08)[2)]	—	16200	—	—	[17]
	f	16000	(1.98)	2.1	—	13600	12.0	[14]
	f	15600	—	—	13600	14000	12.0	[17]

sample		ν_{min} in cm^{-1}	ω_{min} in eV	$\omega_i(\varepsilon''_{min})$ in eV	$\nu_p^*(\varepsilon'=0)$ in cm^{-1}	P_{max} in cm^{-1}	ε'_∞	Ref.
CeB_6	pd	(16534)	2.05	—	—	—	8.26(2)[4]	[11]
	f[3]	18500	—	—	—	—	14.00	[10, 11]
PrB_6	p	16600[1]	—	—	—	—	14.90(2)	[4]
	p	16200	—	—	—	15600	12.0	[14]
	pd	16400	(2.03)[2]	—	15900	—	—	[17]
	f	16000	(1.98)	2.0	—	11200	11.1	[14]
NdB_6	pd	(16373)	2.03	—	—	—	11.60(2)[5]	[11]
SmB_6	p	16200[1]	—	—	—	—	9.38(2)	[4]
	p	15600	—	—	—	13200	10.0	[14]
	f	16000	(1.98)	2.2	—	13600	6.2	[14]
GdB_6	pd	(16212)	2.01	—	—	—	—	[11]

[1] Determined by [14] based on experimental data in [4]. – [2] Determined by [11]. – [3] Imperfect film. – [4] Value from Iller [18] for an imperfect single crystal. – [5] Value from [4] for a polycrystalline sample.

The energy loss function spectra of the MB_6 single crystals with M = La, Sm, Eu (see a figure in the paper) exhibit, in addition to the peak near 2 eV (only for LaB_6, SmB_6), a characteristic peak at 7 to 8 eV due to interband transitions, a peak at ca. 19 eV (only for M = La, Sm) assigned to surface plasmons, and a broad maximum at an energy ranging from 27 to 30 eV due to plasma vibrations of the valence electrons. Similar peaks were observed for β-rhombo-hedral boron, at 8 and 25 eV, i.e., close to the first and second ionization potentials of boron [13]. In the case of polycrystalline films, two maxima observed in the $\varepsilon''(\nu)$ and energy loss functions at 3.6 to 3.9 eV and 4.9 to 5.1 eV were assigned to transitions from p(B) states in the valence band to the 5d(M) conduction band states which are split by the lattice crystal field [14].

Piezoreflectivity

Uniaxial stress-induced shift and deformation of the plasma edge was observed in polar-ized light for two single crystals of LaB_6 of good quality and one each of CeB_6, PrB_6, and NdB_6 of rather poor quality due to a symmetry lowering of the crystals, which makes them anisotropic. The relative change $\Delta R/R$ of the reflectivity resulting from stress between ca. 1.2 and 2.8 eV is due to the change of the free electron concentration, effective mass, relaxation time (see p. 70), and ε'_∞. Figures in the paper show $\Delta R/R$ for the electric field intensity vector $\vec{E}$ of the incident electromagnetic wave parallel and perpendicular to the stress $\vec{p}$. Under the conditions of measurements, the energy shift $\Delta(\hbar\omega)$ of the plasma edge and the relative change $\Delta\varepsilon/\varepsilon$ (referred to data from [4] for p = 0) were as follows (r.p. = reflecting plane):

sample	conditions			$10^5(\Delta\varepsilon/\varepsilon)$	$10^5\Delta(\hbar\omega)$
LaB_6	strain 1.35×10^{-4}	}	$\vec{E}\parallel\vec{p}$	1.9 ± 0.2	0.93 ± 0.15
	r.p. = (100); $\sphericalangle(\vec{p},[110])\approx20°$		$\vec{E}\perp\vec{p}$	-4.2 ± 1.0	-0.7 ± 0.2
LaB_6	strain 1.7×10^{-4}	}	$\vec{E}\parallel\vec{p}$	-3.6 ± 0.5	-2.0 ± 0.35
	$\sphericalangle(r.p.,(110))\approx15°$; $\sphericalangle(\vec{p},[100])\approx9°$		$\vec{E}\perp\vec{p}$	1.6 ± 0.9	$-0.25+0.6$
CeB_6	strain 1.3×10^{-4}	}	$\vec{E}\parallel\vec{p}$	4.6 ± 0.9	2.4 ± 0.8
	$\sphericalangle(r.p.,(310))\leq10°$; $\sphericalangle(\vec{p},[100])\leq20°$		$\vec{E}\perp\vec{p}$	-6.1 ± 1.0	-1.4 ± 0.45

sample	conditions			$10^5(\Delta\varepsilon/\varepsilon)$	$10^5\Delta(\hbar\omega)$
PrB$_6$	strain 2.5×10^{-4}	}	$\vec{E}\parallel\vec{p}$	1.9 ± 0.3	-0.8 ± 0.2
	setting as for CeB$_6$	}	$\vec{E}\perp\vec{p}$	$(\Delta R/R)=0$	
NdB$_6$	strain 1.5×10^{-4}; setting as for CeB$_6$		$\vec{E}\parallel\vec{p}$	-4.15 ± 0.2	-0.6 ± 0.15

The shift of the plasma edge, which is different and even in some cases opposite for $\vec{E}\parallel\vec{p}$ and $\vec{E}\perp\vec{p}$, indicates a splitting of the plasma edge due to the uniaxial stress, Iller [18].

References:

[1] Binder, F. (Radex Rundschau **1977** 52/71, 57; C.A. **87** [1977] No. 27597).

[2] von Stackelberg, M.; Neumann, F. (Z. Physik. Chem. B **19** [1932] 314/20).

[3] Andrieux, L. (Ann. Chim. [Paris] [10] **12** [1929] 423/507, 476).

[4] Kierzek-Pecold, E. (Phys. Status Solidi **33** [1969] 523/31).

[5] Serebryakova, T. I.; Paderno, Yu. B.; Samsonov, G. V. (Opt. Spektroskopiya **8** [1960] 410/2; Opt. Spektrosc. [USSR] **8** [1960] 212/3).

[6] Kudintseva, G. A.; Epel'baum, V. A.; Tsarev, B. M. (Bor Tr. Konf. Khim. Bora Ego Soedin., Moscow 1955 [1956], pp. 112/9; C.A. **1960** 12780).

[7] Kudintseva, G. A; Tsarev, B. M. (Radiotekhn. Elektron. **3** [1958] 428/9; Radio Eng. Electron. Phys. [USSR] **3** [1958] 182/5).

[8] Kudintseva, G. A.; Neshpor, V. S.; Samsonov, G. V. (Vysokotemp. Metallokeram. Mater. **1962** 109/12; C.A. **58** [1963] 7468).

[9] Samsonov, G. V.; Fomenko, V. S.; Paderno, Yu. B. (Ukr. Fiz. Zh. [Ukr. Ed.] **8** [1963] 700/2; C.A. **59** [1963] 8228).

[10] Ivanchenko, L. A.; Bessaraba, V. I.; Paderno, Yu. B. (Khim. Svyaz Kristallakh Ikh Fiz. Svoistva **1976** No. 2, pp. 48/51; C.A. **86** [1977] No. 10452).

[11] Ivanchenko, L. A.; Bessaraba, V. I.; Paderno, Yu. B.; Vereshchak, V. M. (Elektron. Str. Fiz. Khim. Svoistva Splavov Soedin. Osn. Perekhodnykh Metall. Dokl. 8th Simp., Kiev 1974 [1976], pp. 111/4; C.A. **87** [1977] No. 125852).

[12] Bessaraba, V. I.; Bychkov, V. P. (Mater. Izdeliya Poluch. Metodom Poroshk. Metall. Dokl. 6th, 7th Nauchn. Konf. Aspir. Molodykh Issled. Inst. Probl. Materialoved. Akad. Nauk Ukr. SSR, Kiev 1972, 1973 [1975], pp. 169/73; C.A. **87** [1977] No. 105729).

[13] Shelykh, A. I.; Sidorin, K. K.; Karin, M. G.; Bobrikov, V. N.; Korsukova, M. M.; Gurin, V. N.; Smirnov, I. A. (J. Less-Common Metals **82** [1981] 291/6).

[14] Bessaraba, V. I.; Ivanchenko, L. A.; Paderno, Yu. B. (J. Less-Common Metals **67** [1979] 505/9).

[15] Paderno, Yu. B.; Bessaraba, V. I.; Dudnik, E. M.; Ivanchenko, L. A. (Tezisy Dokl. 2nd Vses. Konf. Kristallokhim. Intermetal. Soedin., Lvov, USSR, 1974, pp. 147/8; C.A. **86** [1977] No. 22102).

[16] Philips', N. V. Gloeilampenfabrieken (Fr. 1439722 [1966]; C.A. **65** [1966] 17887).

[17] Ivanchenko, L. A.; Paderno, Yu. B.; Pilyankevich, A. N. (Poroshkovaya Metal. **1978** No. 8, pp. 38/48; Soviet Powder Met. Metal Ceram. **17** [1978] 602/9).

[18] Iller, A. (Phys. Status Solidi B **63** [1974] 69/75).

32.1.5.13 Photoemission and X-Ray Spectra

The X-ray photoelectron spectrum (XPS) of MB_6 with M = La, Nd, Sm, Gd, Tb, and Ho was studied at binding energies E_B up to 16 eV below the Fermi edge E_F. The valence band of LaB_6 having no 4f electrons exhibits two broad weak peaks at $E_B \approx 10$ and 5 eV, Chazalviel et al. [1, 2], which from band structure calculations, consist mainly of B 2s and 2p states, see p. 57. The XPS at lower $|E_B|$ up to E_F is attributed to La 5d states. The valence band region of the other MB_6, after subtracting the valence band spectrum of LaB_6, reveals the multiplet structure of incomplete 4f levels [2]. The corresponding peaks occur at ca. $E_B = 7$ eV for NdB_6; ca. 0, 1, and 3 eV for Sm^{2+} and 7 and 9 eV for Sm^{3+} in intermediate valent SmB_6 with an average Sm valence of 2.7; ca. 10 eV for GdB_6, ca. 4, 9, and 11 eV for TbB_6, and ca. 6 to 7, and 9 eV for HoB_6, see figures in [1]. The 4f lifetime widths (half-width at half-height for a Lorentzian curve) for MB_6, in the case of SmB_6 only the trivalent contribution, were found to obey the relationship $\hbar/\tau = (15-n) \times 0.125$ eV, where n is the number of electrons in the initial 4f configuration. This observation confirms the theoretical prediction from Hirst [3] that the recombination rate should be proportional to the number of 4f holes in the final state. The magnitudes of these line widths can be accounted for in terms of an interatomic Auger process, namely a $5d \rightarrow 4f$ recombination with simultaneous excitation of an electron-hole pair from B 2sp states [2]; see also Chazalviel et al. [4] and for a linear plot of $\hbar/\tau$ vs. $(15-n)$, including (the instable compounds) ErB_6, TmB_6, and LuB_6 also [1].

The photoemission spectrum (PES) of CeB_6, PrB_6, and NdB_6 was calculated and was compared to the respective calculated mixing matrix which considers s–f and p–f mixing, the dynamical d–f Coulomb screening effect as well as the contraction of the 4f wave function on going from Ce to Nd. Figures in the paper down to $E_B = 17$ eV below E_F (and up to 3 eV above E_F) are based on the mixing parameters $(pf\sigma) = -0.231$ eV and $(sf\sigma) = -0.6$ (pfσ) for CeB_6 and $(pf\sigma) = -0.162$ eV for PrB_6. It turns out that the strong peak near $E_B = 2.5$ eV below E_F in the PES of CeB_6 does not correspond to the 4f level which, well-screened by the d–f intraatomic Coulomb interaction, is situated at 4.5 eV. However, in PrB_6 and similarly in NdB_6 the PES peak corresponds to the 4f levels, Ikeda et al. [5]. A more recent calculation of the 4f PES in CeB_6, based on the impurity Anderson model with the only parameter for the occupied 4f level position leads to similar results; calculated peaks are at ~ 2.5 eV (main peak) and ~ 6.25 eV (small peak) below E_F, whereas the energy dependence of the hybridization has a peak at ca. 4.5 eV, see a figure in the paper. The double peak structure is not observed because the valence band is wide, Kasuya et al. [6].

The B K emission spectra near 188 eV ($\triangleq E_B$ of B 1s) of polycrystalline LaB_6, SmB_6, and EuB_6 exhibit a very weak peak (A) near 172 eV, a weak peak (B) near 177 eV, and a main maximum (C) near 182 eV for all, followed by a weak shoulder-like peak (D) in the case of LaB_6 and SmB_6, Okusawa et al. [7]. The main peak C and peak D (at ca. 187 eV), which is characteristic of trivalent M, were also marked in the spectra of MB_6 with M = La, Nd, Sm, (impure) Eu, Gd, and (only peak C) Yb by Okusawa et al. [8]. The spectra reflect the density of states (DOS) of the B 2p orbital of the valence band optically coupled to the B 1s electron at 188 eV. A comparison with the DOS curve for LaB_6, calculated by Hasegawa, Yanase [9] (see p. 216), suggests that peak A, B, and C correspond to the DOS peaks around 15, 10, and 5 eV below E_F, respectively, which all consist mainly of B 2s and 2p contributions. Peak D may correspond to the DOS part just below E_F, which appears as a result of B sp- and La d-orbitals hybridization [7].

Studies of the B K absorption spectrum of LaB_6, SmB_6, and EuB_6 films (in situ evaporation at $< 10^{-4}$ Pa) show almost the same locations of absorption thresholds and no noticeable energy gap between the emission band and the absorption threshold. In addition, the amount of the absorption due to the B 1s excitation is almost the same. This suggests that the amount of the B 2p orbital in the conduction band ist about the same for these hexaborides [7].

Photoelectric yield spectra of MB$_6$ specimens exposed to air do not reflect the bulk states of the materials, see [7] for M = Sm and [8] for M = Y, La, Nd to Gd, and Yb.

Experimental shifts ΔE (in meV) of the K$_{\alpha 1}$ and K$_{\beta 1}$ emission lines of M = Y, La, Ce to Sm, and Gd in pressed MB$_6$ powder relative to MF$_3$, of M = Eu and Yb relative to EuSO$_4$ and YbF$_2$, respectively, are:

MB$_6$	YB$_6$	LaB$_6$	CeB$_6$	PrB$_6$	NdB$_6$	SmB$_6$	EuB$_6$	GdB$_6$	YbB$_6$
$-\Delta E(K_{\alpha 1})$	197 ± 3	84 ± 3	85 ± 4	64 ± 6	73 ± 4	308 ± 4	3 ± 5	54 ± 6	26 ± 11
$-\Delta E(K_{\beta 1})$	138 ± 7	92 ± 7	97 ± 7	75 ± 7	98 ± 8	667 ± 9	26 ± 11	41 ± 14	—

Values were compared with those obtained in relativistic self-consistent field calculations (Dirac-Fock-Slater-Latter) in the effective free ion approximation. The K$_{\alpha 1}$ data show, that M = Y, La, Ce, Pr, and Nd in MB$_6$ have 4f^{n-1} 5d configuration (4d for Y) with a 5d (or 4d) occupancy of ca. 1 (see p. 70) for metallic conduction, that Eu and Yb have 4f^n configuration, and that Sm in SmB$_6$ is intermediate valent with configuration 4f^{6-x}5d^x with x = 0.615 ± 0.007. In contrast, the K$_{\beta 1}$ line shifts for MB$_6$ with metallic conduction could not be explained, neither within the free ion approximation nor by means of calculations involving Wigner-Seitz type boundary conditions for the metal sublattice in MB$_6$, Grushko et al. [10].

Early studies of the X-ray L$_{III}$ absorption spectra of M in MB$_6$ compared to that of M in the oxide (see figures in paper) already indicated that in MB$_6$ the M = La, Ce, Pr, Nd, and Gd are trivalent, Eu and Yb are divalent, whereas Sm, revealing split structures separated by ca. 7 eV, is intermediate valent, containing about 35 to 40% Sm^{2+}, Vainshtein et al. [11]; for M = Sm, Eu, and Yb, Vainshtein et al. [12]; for La and Ce, Vainshtein et al. [13]; for La, Ce, Pr, Gd, and Yb, Troneva [14], for M = La, Gd, Sm, Beaurepaire et al. [15]; for L$_{II}$ and L$_{III}$ spectra of M = La, Nd, Sm, Eu, Gd, and Yb, Okusawa et al. [8]. The features of the L$_{II}$ spectra, and analogously of the L$_{III}$ spectra of La and Ce in MB$_6$ up to ca. 30 eV above the so-called white line were interpreted to result from superpositions of a group of selective lines arising mainly from 2p → nd transitions in the metal. In addition, the weak absorption maximum on the long-wave flank of the white line of Ce in CeB$_6$ was attributed to 2p → 4f quadrupole transitions in the metal. The 4f states (in LaB$_6$ absent) do not take part in the formation of chemical bond [13], see also Samsonov et al. [16]. The line width of the L$_{III}$ white line and the behavior of hyperfine structure at higher energies in various MB$_6$ indicate the predominance of metallic bonds in these compounds [13, 14]. For data on the hyperfine structure in the L$_{III}$ absorption spectra of M in MB$_6$, see [11] (M = Sm, Eu); [14] (M = La, Ce, Pr, Gd, Yb); Troneva [17] (M = La, Yb), and the review in "Seltenerdelemente" B4, 1976, pp. 406/9.

References:

[1] Chazalviel, J.-N.; Campagna, M.; Wertheim, G. K.; Schmidt, P. H.; Longinotti, L. D. (Proc. 12th Rare Earth Res. Conf., Vail, Colo., 1976, pp. 542/51).

[2] Chazalviel, J.-N.; Campagna, M.; Wertheim, G. K.; Schmidt, P. H.; Yafet, Y. (Phys. Rev. Letters **37** [1976] 919/22).

[3] Hirst, L. L. (Phys. Rev. Letters **35** [1975] 1394/6).

[4] Chazalviel, J.-N.; Campagna, M.; Wertheim, G. K.; Schmidt, P. H.; Longinotti, L. D. (Physica B **86/88** [1977] 237/8).

[5] Ikeda, M.; Aoki, Y.; Kasuya, T. (J. Magn. Magn. Mater. **52** [1985] 264/6).

[6] Kasuya, T.; Sakai, O.; Harima, H.; Ikeda, M. (J. Magn. Magn. Mater. **76/77** [1988] 46/52).

[7] Okusawa, M.; Ichikawa, K.; Matsumoto, T.; Tsutsumi, K. (J. Phys. Soc. Japan **51** [1982] 1921/6).

[8] Okusawa, M.; Iwasaki, Y.; Tsutsumi, K.; Aono, M.; Kawai, S. (Japan. J. Appl. Phys. II **17** [1978] 161/3).

[9] Hasegawa, A.; Yanase, A. (J. Phys. F **7** [1977] 1245/60).

[10] Grushko, Yu. S.; Paderno, Yu. B.; Mishin, K. Ya.; Molkanov, L. I.; Shadrina, G. A.; Konovalova, E. S.; Dudnik, E. M. (Phys. Status Solidi B **128** [1985] 591/7).

[11] Vainshtein, E. E.; Blokhin, S. M.; Bril', M. N.; Staryi, I. B.; Paderno, Yu. B. (Zh. Neorgan. Khim. **10** [1965] 121/6; Russ. J. Inorg. Chem. **10** [1965] 64/7).

[12] Vainshtein, E. E.; Blokhin, S. M.; Paderno, Yu. B. (Fiz. Tverd. Tela [Leningrad] **6** [1964] 2909/12; Soviet Phys.-Solid State **6** [1964] 2318/20).

[13] Vainshtein, E. E.; Staryi, I. B.; Blokhin, S. M.; Paderno, Yu. B. (Zh. Strukt. Khim. **3** [1962] 200/7; J. Struct. Chem. [USSR] **3** [1962] 185/91).

[14] Troneva, N. V. (Izv. Akad. Nauk SSSR Ser. Fiz. **28** [1964] 809/10; Bull. Acad. Sci. USSR Phys. Ser. **28** [1964] 717/8).

[15] Beaurepaire, E.; Kappler, J. P.; Krill, G. (Solid State Commun. **57** [1986] 145/9).

[16] Samsonov, G. V.; Vainshtein, E. E.; Paderno, Yu. B. (Fiz. Metal. Metalloved. **13** [1962] 744/9; Phys. Metals Metallog. [USSR] **13** No. 5 [1962] 100/4).

[17] Troneva, N. V. (Izv. Akad. Nauk SSSR Ser. Fiz. **27** [1963] 403/8; Bull. Acad. Sci. USSR Phys. Ser. **27** [1963] 410/4).

32.1.5.14 Electrochemical Behavior

Rare earth hexaboride powders (grain size of 5 to 20 µm), incorporated in carbon paste electrodes, exhibit a resistance to anodic oxidation in aqueous 1 M $NaClO_4$ at pH 1.0, which is between that of corresponding MB_4 and MB_{12} carbon paste electrodes. Since the current-voltage curves measured vs. an AgCl reference electrode display a definite current maximum I_{max} directly proportional to the MB_x concentration in the paste, this anodic oxidation offers a possibility for quantitative phase analysis of boride powders. At a scanning rate 5×10^{-4} V/s, the current maximum was observed at the following value of φ_{max} for pastes with 1% MB_6:

MB_6	LaB_6	CeB_6	SmB_6	EuB_6	GdB_6	DyB_6	YbB_6
φ_{max} in V ..	0.30	0.31	0.29	0.14	0.29	0.33	0.18

Tkach, Paderno [1]. For the change of φ_{max} in the system $Eu_xSm_{1-x}B_6$ with x = 0, 0.3, 0.7, and 1, see the current-voltage curves shown in Tkach, Masyuk [2].

References:

[1] Tkach, A. V.; Paderno, Yu. B. (Dopov. Akad. Nauk Ukr. RSR, Ser. B: Geol., Khim. Biol. Nauki **1984** No. 9, pp. 52/4; C.A. **102** [1985] No. 35077).

[2] Tkach, A. V.; Masyuk, T. V. (Nov. Porosh. i Kompozits. Neorgan. Materialy, Kiev **1983** 129/32; C.A. **101** [1984] No. 236290).

32.1.5.15 Chemical Behavior

32.1.5.15.1 Reactions on Irradiation with Neutrons

Irradiating MB_6 with thermal neutrons causes $^{10}B(n, \alpha)^7Li$ reaction. The MB_6 releases appreciably more He than MB_4. A white deposit, probably Li_2O, formed on the sample surface. Starting MB_6 powders obtained borothermally were contaminated by other borides (M = Y, Dy) or a borate (M = Sm) and, if prepared in a H_2 atmosphere, also by rare earth hydrides. In this case, H_2 was also released during irradiation despite foregoing heat treatment in a vacuum at $\geqq 1973$ K, Hoyt, Zimmerman [1, pp. 2/4], see also Ballowe et al. [2, pp. 4/9].

Dry compacted powders and vacuum hot-pressed dense solids (10^{-4} Pa, 2173 to 2573 K), each encapsulated in gas-tight aluminium capsules, were exposed for 137 to 226 d to reactor core irradiation. The fast (>0.18 MeV) to thermal neutron flux ratio amounted to ca. 1.3. The following table gives, for compact powders (p) and dense solids (s) of theoretical density D', the % ^{10}B burnup produced by a fission density FD, the fractional He release and H_2 release of the sample during irradiation and (in parentheses) the additional release during post-irradiation heating to 1273 K, and the volume increase ΔV. Also given are the estimated maximum operation temperatures op.T produced by the fissions [1]:

sample	D' in %	% ^{10}B burnup	FD in 10^{21} fiss./cm^3	% He release	H_2 release in cm^3/g	ΔV in %	op.T in K
YB_6(p)	67	91	10.6	47	—	6[4]	773
YB_6(s)	92	42	6.7	14(86)	—(6.24)	12	473
SmB_6(p)[1]	60	98.2	10.0	38.2	283	4	—
SmB_6(p)	60	63[2]	6.4	60	210	6	623
SmB_6(s)	86	44	6.4	18(82)	—(3.06)	15	523
EuB_6(p)	61	68	6.8	61	—	5[4]	823
EuB_6(p)	53	65[2]	5.7	64	—	6[4]	773
EuB_6(s)	87	37	5.7	19(51.3)	4.3(8.42)	7	523
DyB_6(s)	87	43[2]	6.3	82	—	[3]	573
DyB_6(s)[1]	87	37	5.7	94.0	—	[3]	—

[1] Data from Hoyt [3]. – [2] Calculated value. – [3] Sample fragmented. – [4] Only 4% according to [3].

Essentially complete He release occurred on post-irradiation heating up to 1273 K in the case of YB_6 and SmB_6 [1]. For earlier data on gas release and swelling of MB_6 with M = Y, Sm, and Eu, see Holden et al. [4] and for cold-pressed compacts also [2, p. 23]. Due to the different thermal neutron cross sections of Y and Eu, the irradiation-induced activities of the EuB_6 and YB_6 samples amounted to 180 rad/h and 100 to 200 mrad/h, respectively, at 25.4 cm. All the dense solid and powder compact MB_6 samples suffered such structure damage during irradiation that the structures were approaching an amorphous state. Studies on a hot-pressed YB_6 specimen enriched with 92% ^{10}B of the total boron to 40% ^{10}B burnup showed that the burnup evidently took place only near the surface and, due to gross self-shielding from thermal neutrons, not in the center. The contribution from fast neutrons to the total structure damage was small, the lattice contracted by 0.2%. For a discussion of the degree of damage in MB_6 compared to other borides, see the paper, Cummings, Clark [5]. The loose friable powders changed on irradiation to firm solids ("radiation" or "fission" sintering) which showed no evidence of cracking or fragmentation [1]; for MB_6 with M = Y, Sm, and Eu, see also Hoyt, Zimmerman [6]. Hot-pressed YB_6 and SmB_6 underwent pronounced softening during irradiation, for changes in the Vickers hardness and the density, see the paper [1].

References:

[1] Hoyt, E. W.; Zimmerman, D. L. (GEAP-3743-Pt. 1 [1962] 1/41; C.A. **59** [1963] 12331).
[2] Ballowe, W. C.; Holden, A. N.; Ozeroff, W. J.; Williamson, H. E.; Weidenbaum, B. (GEAP-3355-Pt. 2 [1960] 1/132, 1/31; N.S.A. **15** [1961] No. 26541).
[3] Hoyt, E. W. (Proc. 2nd Conf. Rare Earth Res., Glenwood Springs, Colo., 1961 [1962], pp. 287/99; N.S.A. **16** [1962] No. 32072).

[4] Holden, A. N.; Hoyt, E. W.; Cummings, W. V.; Zimmerman, D. L. (Prop. Reactor Mater. Eff. Radiat. Damage Proc. Intern. Conf., Berkely Castle, Engl., 1961 [1962], pp. 457/74; C. A. **58** [1963] 2092).

[5] Cummings, W. V.; Clark, W. I. (GEAP-3743-Pt. 3 [1962] 1/41; C. A. **59** [1963] 14666).

[6] Hoyt, E. W.; Zimmerman, D. L. (GEAP-3743-Pt. 2 [1962] 1/18; N.S.A. **17** [1963] No. 20512; C. A. **58** [1963] 8596).

32.1.5.15.2 Thermal Stability

Vaporization. Dissociation

In view of the thermal behavior of polycrystalline MB_6 in a vacuum of $\lesssim 1 \times 10^{-4}$ Pa, three main classes are distinguished, see the review by Etourneau et al. [1]. Discrepancies exist in the classification of EuB_6.

(1) MB_6 with M = La, Ce, Pr, Nd, Sm evaporate congruently, Smith [2], up to their melting points, Etourneau et al. [3 to 5], Hagenmuller et al. [6]; i.e.,

$$MB_6(s) \rightarrow M(g) + 6\ B(g)$$

from mass spectrometric studies [2]. Remarkable sublimation takes place above 2073 K; for weight changes observed after a 1 h-treatment, see **Fig. 38** [3]. The composition of impure SmB_6 was found to change in the course of congruent evaporation up to melting, Dudnik et al. [7]. According to these authors [7] and Ames, McGrath [8] EuB_6 also belongs to this class. However, EuB_6 is difficult to obtain free of Eu metal, and reproducible ion intensities of Eu(g) and B(g) could not be measured on sixteen samples [8]. For additional data on SmB_6 and EuB_6, see under (3).

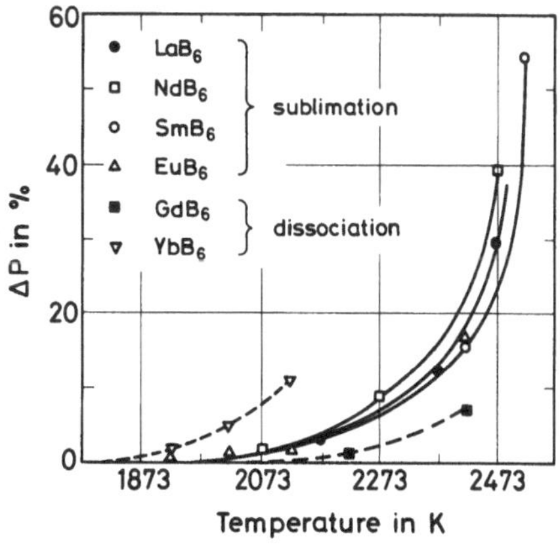

Fig. 38.   Weight change ΔP on sublimation and dissociation of rare earth hexaborides during 1 h-treatments in a vacuum of $\sim 133\ \mu Pa$.

(2) MB_6 with M = Y, Gd, Tb, Dy lose boron to give tetraboride [2]:

$$MB_6(s) \rightarrow MB_4(s) + 2\ B(g)$$

Dissociation takes place at 1823 K [3 to 6] and is associated with a change in color of the sample surface from blue to yellowish within 1 h. The weight change is rather small up to 2273 K, see Fig. 38 for GdB_6 [3]. This dissociation path was confirmed for GdB_6 and TbB_6, and was also observed for HoB_6, confirming suggestions by [2], and for ErB_6 contaminated with ErB_{12} (but see p. 21). Two different DyB_6 samples changed color on heating to 2288 K from greyish blue to bright blue, however, they did not decompose even on repeated heating [8].

Calculations of the congruently vaporizing composition (CVC) in the system Dy–B indicated that DyB$_6$ belongs either to this group (2) or to group (3), see below, Storms, Mueller [9].

(3) This group comprises the MB$_6$, which dissociate with preferential loss of M to give a compound richer in boron or solid boron: The EuB$_6$ dissociated to give rhombohedral β B at the melting point near 2673 K [4 to 6]:

$$EuB_6(s) \rightarrow Eu(g) + 6 \ B(s)$$

For the observation of congruent vaporization of EuB$_6$ [7, 8] see path (1). However, the recently observed quasi-CVC of a single-crystal surface (see the table on p. 91), Futamoto et al. [10], and the calculated CVC [9], see table below, indicate that EuB$_6$, loses metal preferentially. With YbB$_6$ the process is

$$2 \ YbB_6(s) \rightarrow YbB_{12}(s) + Yb(g)$$

at 1923 K [3 to 6], but above 2023 K a higher boride, so-called YbB$_{100}$, Smith, Gilles [11], was formed owing to the decomposition of YbB$_{12}$ [3]. For the dissociation of YbB$_6$ to so-called YbB$_{100}$, studied $\geqq$1973 K, see [8]. The evaporation of YbB$_6$ on heating to melting followed the above equation in the initial stage [7]. The calculated CVC for YbB$_6$ was boron, as it was for EuB$_6$, and may be for SmB$_6$ which is in contrast to all observations, see path (1). Similarly, for DyB$_6$ (see path (2)), a CVC of DyB$_{12}$ was proposed [9].

The above experimental results are based on mass spectrometric studies during free evaporation and Knudsen effusion of MB$_6$ from ZrB$_2$ crucibles, followed by X-ray analysis of the residue and condensate [2]; on weight change studies during free evaporation from BN supports and X-ray, metallographic, and chemical analyses of condensate and residue [3 to 6]; free evaporation from a water-cooled Cu crucible by electron beam heating and X-ray and chemical analyses of fused residues and of condensates [7]; mass spectrometric studies during Knudsen effusion from ZrB$_2$-lined W crucibles and X-ray analysis of the residue [8]. Mass spectrometric studies of the evaporation from (reactive) W crucibles indicated dissociation via MB$_6$(s) $\rightarrow$ M(g) + 6 B(s) for studied M = Ce, Sm, Gd, and Tb between 1900 and 2300 K (but see p. 96). At higher temperatures, boron also evaporated, Gordienko et al. [12].

Values for the congruent vaporizing compositions (CVC) of several rare earth boride systems (and thus also of MB$_6$) were estimated by assuming stabilities for the MB$_6$ and MB$_4$ phases and by using the known vapor pressure of the elements. They are assembled below together with those CVC measured for M = La, Nd, and Gd (see the corresponding individual sections):

M in MB$_x$	Y	La	Ce	Pr	Nd	Sm
CVC	YB$_{4.02}$	LaB$_{6.04}$	CeB$_{6.04}$	PrB$_{6.05}$	NdB$_{6.06}$	B
M in MB$_x$	Eu	Gd	Tb	Dy	Er*)	Yb
CVC	B	GdB$_{4.03}$	TbB$_{4.05}$	DyB$_{12}$	ErB$_{4.1}$	B

*) ErB$_6$ does not exist, see p. 21.

Storms, Mueller [9]. For the dependence of the CVC on temperature, purity, and bulk composition of LaB$_6$ see p. 205.

Thermodynamics of Evaporation of MB$_6$ and Solid Solutions (M, La)B$_6$

Mass spectrometric studies of the free evaporation from the thoroughly cleaned (100) surface of single crystals and Auger electron spectroscopy (AES) of the surface led to the following conclusions: The vaporization product from cleaned MB$_6$ surfaces (M = La to Eu) consists only of atomic species M and B. In the initial stage, vaporization proceeds noncongruently: For the activation energies of evaporation, $E_B \gg E_M$ was found and/or B/M ratios in the vapor flux of $\ll 6$, see Goldstein, Szostak [14] for MB_6 with M = La, Ce, Eu at $\leqq 1973$ K, Futamoto et al. [10] for CeB_6 between ~ 1923 and 2273 K, and [13] for M = La, Ce, Sm between 1600 and 1800 K. In addition, E_B and E_M, and the B/M ratio in the vapor flux were initially irreproducible. After prolonged heating above 1800 K, B/M ≈ 6 in the vapor and reproducible $E_B = E_M$ were observed (no values given); i.e., an apparent congruently vaporizing composition (CVC) was obtained, which also resulted in the lowest vaporization rate [13]. Surface studies in a vacuum $\leqq 6 \times 10^{-6}$ Pa indicated that MB_6 with M = La to Eu required 1 h at $\gtrsim 1923$ K to achieve a steady-state surface composition (quasi CVC). The approximate B/M ratio in the surface layer of so-heated MB_6 remained nearly constant afterwards between 1673 and 1973 K and increased (E_M decreased) on going from LaB_6 to EuB_6, whereby the surface of LaB_6 was most enriched with metal atoms, that of EuB_6 was enriched with boron atoms (see the table below). As the diffusion of elements from inside the bulk is negligible, the surface composition is determined predominantly by evaporation behavior of component species, at least in the temperature region 1673 to 2273 K where hexaborides are generally used as thermionic emitters. The evaporation rate is of the order of ten (100) atomic layers per min at 1973 K for LaB_6 which shows the minimum evaporation rate. As a general rule, the higher the pure metal (La to Eu) vapor pressure, the lower the evaporation activation energy of metal from the hexaboride (see a figure in the paper), Futamoto et al. [10]. The following assemblage gives values for E_M and E_B, and for the approximate B/M ratio of the steady-state surface composition at 1923 K from [10] and, for comparison, E_M and E_B (in parentheses) measured by [14] under noncongruent vaporization conditions:

MB_6	E_M in eV	E_B in eV	B/M[*]
LaB_6	5.81 ± 0.10 (5.5)	5.74 ± 0.08 (5.5)	4.7 ± 0.6
CeB_6	5.50 ± 0.05 (4.5)	5.52 ± 0.05 (5.6)	5.4 ± 0.2
PrB_6	5.41 ± 0.15	5.40 ± 0.09	5.7 ± 0.5
NdB_6	5.40 ± 0.10	5.41 ± 0.10	5.6 ± 0.5
SmB_6	4.83 ± 0.02	5.04 ± 0.05	6.0 ± 0.4
EuB_6	4.86 ± 0.09 (3.6)	4.88 ± 0.08 (5.6)	12.4 ± 0.7

[*] Determined by AES based on elemental sensitivities of B(KLL) and M(MNN) transitions without further corrections [10].

By comparison, reported literature data for the activation energy of boron from pure crystalline boron range from 5.3 to 5.5 eV [10].

Values of the preferred evaporation of metal below ca. 2300 K and of the evaporation of boron above this temperature, observed on polycrystalline samples of MB_6 with M = Ce and Sm and also Gd and Tb (but see p. 89 and below), see Gordienko et al. [12]. Knudsen data for polycrystalline samples in the steady state were presented as total free enthalpy of vaporization for $MB_6(s) \rightarrow M(g) + 6 B(g)$ with M = La, Ce, Pr, Sm; see figures in the paper. With EuB_6, the measured values of $-\ln P_{Eu} P_B^6$ ranged from 69.61 to 56.18 over the limited temperature range 2214 to 2314 K for the four best (yet irreproducible) runs. The 2nd and 3rd law entropy and enthalpy changes at the temperature T of congruent vaporization and average value of ΔH_{298}° derived from the 2nd and 3rd law values corrected to 298 K are as follows (1 J $\widehat{=}$ 0.239 cal):

MB_6	T in K	ΔS_T° (2nd) in cal·mol^{-1}·K^{-1}	ΔS_T° (3rd) in cal·mol^{-1}·K^{-1}	ΔH_T° (2nd) in kcal/mol	ΔH_{298}° (av.) in kcal/mol
LaB_6	2275	232.7 ± 20.2	239.1	834.4	854.6 ± 37.0
CeB_6	2275	212.9 ± 14.7	228.0	795.9	846.6 ± 37.4

MB$_6$	T in K	ΔS_T° (2nd) in cal·mol^{-1}·K^{-1}	ΔS_T° (3rd) in cal·mol^{-1}·K^{-1}	ΔH_T° (2nd) in kcal/mol	ΔH_{298}° (av.) in kcal/mol
PrB$_6$	2285	221.6 ± 27.5	223.4	833.3	872.9 ± 51.8
SmB$_6$	2215	253.6 ± 21.7	222.9	834.3	832.6 ± 49.8
EuB$_6$	2255	—	212.7	—	812.4 ± 45.0

Values for the decomposition $MB_6(s) \rightarrow MB_4(s) + 2B(g)$ for M = Gd, Tb, Ho, and Er are:

MB$_6$	T in K	ΔS_T° (2nd) in cal·mol^{-1}·K^{-1}	ΔS_T° (3rd) in cal·mol^{-1}·K^{-1}	ΔH_T° (2nd) in kcal/mol	ΔH_{298}° (av.) in kcal/mol
GdB$_6$	2270	39.9 ± 3.5	65.1	170.1	202.3 ± 26
TbB$_6$	2170	52.8 ± 2.8	68.2	196.0	214.9 ± 19
HoB$_6$	2185	63.1 ± 3.5	69.2	200.5	210.4 ± 15
ErB$_6$	2200	74.1 ± 3.5	69.6	229.2	227.6 ± 15

The TbB$_6$ and ErB$_6$ (but see p. 21) were contaminated with TbB$_{12}$ and ErB$_{12}$, respectively, Ames, McGrath [8].

The different vaporization paths were discussed in terms of a relationship between the stabilities of the hexaborides and the size of the metal atom [8]. The evaporation properties of MB$_6$ with M = La to Sm under equilibrium vaporization conditions appear to be controlled by the breaking of the M–B bond. This would explain the high metal concentration in the surface layer for the MB$_6$ in which M–B bond energy and E_M are large. For EuB$_6$, the breaking of B–B bond is considered to control the evaporation behavior [10].

When a mixed hexaboride $M_{1-x}La_xB_6$ with M = Ce, Pr, or Sm was isothermally heated at a high temperature ($\gtrsim$ 1873 K), the M metal which has the lower value of activation energy for evaporation from the binary hexaboride evaporated from the (110) surface preferentially with respect to La in the initial stage. After some time, dependent on temperature, the M/La ratio in the vapor approached a constant value; see a figure in the paper for $Sm_{0.42}La_{0.58}B_6$ between 1953 and 2143 K as a function of heating time ($\leq$ 1 h). This behavior indicates that the surface layer composition approaches a steady state, which is more enriched with La metal than the bulk composition [10]. In the following, activation energies of evaporation from the (100) plane of several mixed hexaborides under steady-state conditions from [10] are given together with the surface composition La/(La + M) after heating for 1 h at 1873 to 1923 K to reach the steady-state surface compositions, as determined by AES by Futamoto et al. [15]:

$M_{1-x}La_xB_6$	E_{La} in eV	E_M in eV	E_B in eV	La/(La + M)
Ce$_{0.46}$La$_{0.54}$B$_6$	5.89 ± 0.05	5.87 ± 0.05	5.79 ± 0.07	—
Pr$_{0.45}$La$_{0.55}$B$_6$	5.82 ± 0.12	5.81 ± 0.10	5.80 ± 0.10	0.73 ± 0.06
Sm$_{0.42}$La$_{0.58}$B$_6$	5.63 ± 0.07	5.44 ± 0.10	5.51 ± 0.09	0.76 ± 0.03

So-heat-treated $Pr_{0.12}La_{0.88}B_6$ and $Pr_{0.91}La_{0.09}B_6$ had values of La/(La + M) of 0.90 ± 0.02 and 0.38 ± 0.05, respectively. The surface layer was 2 to 3 nm thick. The thickness did not depend on the heating time once a steady-state surface composition was obtained [15]. Even after a steady-state evaporation is attained, the M atom will preferentially evaporate from the surface since the M–B bond energy is lower than that of La–B [10]. The enrichment of the surface layer with La accounts for the fact that mixed hexaborides $M_{1-x}La_xB_6$ show thermionic emission properties similar to LaB$_6$ [15], Nakazawa et al. [16]. When the temperature was raised in steps from room temperature to 1973 K, the surface composition exhibited three stages (see figures in the paper): Below 1273 K, the surface was similar to the bulk composition, i.e., no atomic diffusion or evaporation occurred. Between 1273 and 1673 K, the surface changed gradually to a La-enriched state. Above that temperature, the surface composition was kept nearly constant even with long heating. The ratio $La/(La + M)$ of $M_{1-x}La_xB_6$ at 1873 K amounted to:

$Ce_{0.5}La_{0.5}B_6$	$Pr_{0.45}La_{0.55}B_6$	$Pr_{0.9}La_{0.1}B_6$	$Sm_{0.4}La_{0.6}B_6$	$Dy_{0.1}La_{0.9}B_6$
0.85	0.86	0.57	0.76	0.9

Surface and bulk compositions of sputter-cleaned samples are identical at room temperature [16].

References:

[1] Etourneau, J.; Mercurio, J.-P.; Hagenmuller, P. (in: Matkovich, V. I., Boron and Refractory Borides, Springer, Berlin – Heidelberg – New York 1977, pp. 115/38).

[2] Smith, P. K. (COO-1140-103 [1964] 1/485, 127/38, 177/9; N.S.A. **18** [1964] No. 25319; Diss. Univ. Kansas 1964, pp. 1/482; Diss. Abstr. **25** [1964/65] No. 5591).

[3] Etourneau, J.; Mercurio, J.-P.; Naslain, R.; Hagenmuller, P. (Colloq. Intern. Centre Natl. Rech. Sci. [Paris] No. 205 [1972] 429/38).

[4] Etourneau, J.; Mercurio, J.-P.; Naslain, R. (Compt. Rend. C **275** [1972] 273/6).

[5] Etourneau, J.; Mercurio, J.-P.; Naslain, R.; Hagenmuller, P. (Bor. Poluch. Strukt. Svoistva Mater. 4th Mezhdunar. Simp. Boru, Tbilisi 1972 [1974], pp. 228/32; C.A. **83** [1975] No. 151148).

[6] Hagenmuller, P.; Etourneau, J.; Mercurio, J.-P.; Naslain, R. N. (Redkozemel. Metal. Splavy Soedin. Mater. 7th Soveshch., Moscow 1972 [1973], pp. 229/36; C.A. **80** [1974] No. 152305).

[7] Dudnik, E. M.; Bessaraba, V. I.; Paderno, Yu. B. (Poroshkovaya Metal. **1976** No. 12, pp. 60/2; Soviet Powder Met. Metal Ceram. **15** [1976] 945/7).

[8] Ames, L. L.; McGrath, L. (High Temp. Sci. **7** [1975] 44/54).

[9] Storms, E. K.; Mueller, B. A. (J. Appl. Phys. **52** [1981] 2966/70).

[10] Futamoto, M.; Nakazawa, M.; Kawabe, U. (Vacuum **33** [1983] 727/32).

[11] Smith, P. K.; Gilles, P. W. (J. Inorg. Nucl. Chem. **26** [1964] 1465/7).

[12] Gordienko, S. P.; Fesenko, V. V.; Fenochka, B. V. (Zh. Fiz. Khim. **40** [1966] 3092/4; Russ. J. Phys. Chem. **40** [1966] 1659/60).

[13] Swanson, L. W.; McNeely, D. R. (Surf. Sci. **83** [1979] 11/28).

[14] Goldstein, B.; Szostak, D. J. (Surf. Sci. **74** [1978] 461/78, 466/7, 477).

[15] Futamoto, M.; Nakazawa, M.; Kawabe, U. (Surf. Sci. **100** [1980] 470/80).

[16] Nakazawa, M.; Futamoto, M.; Usami, K.; Kawabe, U. (J. Appl. Phys. **52** [1981] 6917/20).

32.1.5.15.3 Reactions with Gases

Reactions at Various Temperatures

The catalytic activity of hot-pressed MB_6 samples during atomic H recombination on their surfaces was characterized by the following values of the recombination coefficient γ of atomic H: γ (in 10^{-2}) = 4.51 (LaB_6), 4.16 (YbB_6), 3.69 (EuB_6), 3.60 (SmB_6), and $\lesssim 3.10$ (MB_6 with M = Ce, Nd, Gd, Tb). The phenomenological coefficients are $L_{11}^* = 1.79 \times 10^{-2} J \cdot cm^{-2} \cdot s^{-1} \cdot K^{-1}$ and $L_{12}^* \approx 2.5 \times 10^{-5}$ g-atom $\cdot cm^{-2} \cdot s^{-1}$ for all the compounds, Lavrenko et al. [1].

The rare earth hexaborides do not react with H$_2$ and rare gases, see "Single-Crystal Growth", p. 28, and also not with CO, whereas with N$_2$ and NH$_3$, nitrides are formed at high temperatures, presumably above ca. 1223 K, see the review by Binder [2].

MB$_6$, as was shown for LaB$_6$, reacts with oxygen to form a mixed oxide layer (20 to 30 Å) of La$_2$O$_3$ and a boron oxide, the proportions of which depend on temperature and O$_2$ partial pressure. Photoelectron spectra (ESCA-XPS) reveal that a melted, polished LaB$_6$ sample oxidizes when kept for a prolonged time in air at room temperature. No oxidation by residual gases was observed between 298 and 873 K in a vacuum of 10^{-6} to 10^{-7} Pa. However, oxidation does take place above 873 K in a vacuum of 10^{-5} to 10^{-6} Pa; oxides are removed during a short heat treatment ≧1773 K, Berrada et al. [3, 12]. Powders (grain size of ≦10 μm) of MB$_6$ with M = La, Nd, Sm, and Dy on an Al$_2$O$_3$ support, kept in a furnace without circulation of air, were not oxidized appreciably at temperatures up to 923 K and retained their color. After 50 h at 923 K, the weight increase was 5%. Ignition took place at 1022 K for LaB$_6$, 983 K for NdB$_6$, 966 K for SmB$_6$, and at 1006 K for DyB$_6$, associated with a change in color to black and a weight increase of ca. 10%. Studies at 1023, 1123, and 1223 K indicated rapid oxidation during the first ca. 5 h (see figures in the paper), which became progressively more active at the higher temperatures. On prolonged heating at 1023 K, oxidation slowed down due to the formation of a protective stable dark grey surface film which, after 80 h, had the overall composition "MB$_4$O$_9$" (M = Nd, Dy) or "MB$_5$O$_4$" (M = La, Sm). In contrast, the protective film formed in the first period at 1123 and 1223 K decomposed with time (still after 175 h) to products having the color of the corresponding rare earth oxides after 80 h. At 1223 K, the oxidation product had the overall composition "M$_2$B$_5$O$_{10}$" (M = La, Nd, Sm, Dy) which melted between 1423 and 1523 K. At higher temperatures (≧1273 K), the MB$_6$ powder reacted with the supporting Al$_2$O$_3$ (see p. 99), Timofeeva, Timofeeva [4]. For weight gains of MB$_6$ (M = La, Sm, Eu, Gd, Dy, Yb) at 1073 K over periods up to 200 h, see Timofeeva, Timofeeva [5]. Recent differential thermal analyses of powders (≦ 40 μm) during heating at 12 K/min led to the following results by means of additional thermogravimetric, X-ray, and petrographic studies: Heating of MB$_6$ (M = La to Eu) contaminated by up to 0.6 wt% B$_2$O$_3$ and of Eu$_{0.51}$Pr$_{0.51}$B$_6$ in air up to 1273 K is accompanied by the formation of surface oxide films. The films contain crystalline and vitreous boron oxide and borates. The boron oxide appears above a temperature between 1013 and 1043 K in the case of MB$_6$, M = La to Sm, and borates of the general formula M(BO$_2$)$_3$ begin to form at some higher temperature. The films protect these MB$_6$ against further oxidation. With Eu$_{0.99}$B$_6$ and Eu$_{0.51}$Pr$_{0.51}$B$_6$, boron oxide formation begins as low as 948 and 963 K, respectively, and some borate formation occurs at 1013 and 1023 K, respectively. The borate formed with Eu$_{0.99}$B$_6$ at 1173 K is EuB$_4$O$_7$. The film melts at 1253 K, forms a vitreous surface film on the boride grains, and the rate of oxidation of the Eu$_{0.99}$B$_6$ increases once again. In the temperature range 1063 to 1143 K, in which resistive thick-film pastes in the manufacture of composite resistors are usually calcined, the changes in weight of MB$_6$ (M = La to Sm) amount to 11 to 17%, whereas for Eu$_{0.99}$B$_6$ and Eu$_{0.51}$Pr$_{0.51}$B$_6$ the changes in weight are 37 to 40%, Islamgaliev et al. [6].

In moist air of 100% relative humidity, powders (grain size of 56 μm) ignited at 923 K (SmB$_6$) and 853 K (EuB$_6$) compared to 938 and 858 K, respectively, in dry air, Manuev et al. [7]. In a steam autoclave at ≧ 573 K, MB$_6$ oxidized to MBO$_3$. After 100 h at 573 K and 8.9 MPa, specimens of MB$_6$ with M = La, Sm, Eu, Gd, Dy, and Yb fractured, and the chunks were covered with a white film [5]. More or less complete conversion into MBO$_3$ (M = Y, Sm, Eu, and Dy) was observed after 500 h at 673 K, 10.3 MPa, Hoyt [8]; for M = Y, Sm, and Eu, also Ballowe et al. [9].

Gases such as NO$_2$, SO$_2$, SO$_3$, and CO$_2$ oxidize MB$_6$ at about the same temperatures as does air. Halogens decompose MB$_6$ above ca. 523 K, dry HCl does so above ca. 923 K, see the review by Binder [2].

Field-Induced Reactions

He, Ne, and H_2 were tested as an imaging gas in field-ion microscopy (FIM). The gas of (5 to 8) $\times 10^{-3}$ Pa was introduced into the equipment having a vacuum of better than 10^{-6} Pa. The FIM images at ca. 70 K of MB_6 (M = La to Eu) and of $Sm_{0.42}La_{0.58}B_6$ single-crystal tips in He and of LaB_6 and $Sm_{0.42}La_{0.58}B_6$, the only ones studied in Ne, all corresponded to the tip orientation. This was interpreted in terms of field evaporation of both B and M, as is observed in vacuum. However, in H_2 atmosphere, B atoms were preferably etched away by chemical reaction between B atoms and H_2 at the imaging conditions, thus leading to the appearance of additional planes in the FIM image, Futamoto, Kawabe [10].

The efficiency of making atomic iodine negative ions at an uncleaned hot MB_6 (M = Y, La, Gd) surface was investigated, using the mechanism of surface ionization. Molecular iodine vapor of ca. 1 to 10^{-6} Pa was allowed to impinge on a MB_6 coated W filament of a Hull (d.c.) magnetron. In the case of GdB_6 the maximum atomic iodine negative ion current density, which was space-charge limited, was 15 mA/cm^2. It occurred at <1873 K for iodine vapor pressures of 10^{-3} Pa and higher; for details, see the paper, Dong et al. [11].

References:

[1] Lavrenko, V. A.; Vasil'ev, A. V.; Garf, E. S.; Paderno, Yu. B. (Redkozemel. Metal. Ikh Soedin. **1970** 114/8; C.A. **76** [1972] No. 28163).

[2] Binder, F. (Radex Rundschau **1977** 52/71, 63; C.A. **87** [1977] No. 27597).

[3] Berrada, A.; Mercurio, J.-P.; Etourneau, J.; Hagenmuller, P. (Rev. Intern. Hautes Temp. Refract. **15** [1978] 115/28).

[4] Timofeeva, E. N.; Timofeeva, N. I. (Zh. Prikl. Khim. **44** [1971] 1400/2; J. Appl. Chem. [USSR] **44** [1971] 1414/6).

[5] Timofeeva, N. I.; Timofeeva, E. N. (Izv. Akad. Nauk SSSR Neorgan. Materialy **4** [1968] 1789/91; Inorg. Materials [USSR] **4** [1968] 1559/61).

[6] Islamgaliev, R. K.; Zyrin, A. V.; Semenov-Kobzar', A. A.; Shulishova, O. I.; Shcherbak, I. A. (Poroshkovaya Metal. **1987** No. 12, pp. 36/9; Soviet Powder Met. Metal Ceram. **26** [1987] 980/3).

[7] Manuev, N. V.; Popov, E. I.; Poyarkov, V. G.; Finaev, Yu. A. (Vestsi Akad. Navuk Belarusk. SSR Ser. Fiz. Energ. Navuk No. 3 [1973] 132/5; C.A. **80** [1974] No. 40195).

[8] Hoyt, E. W. (Proc. 2nd Rare Earth Research Conf., Glenwood Springs, Colo., 1961 [1962], pp. 287/99; N.S.A. **16** [1962] No. 32072).

[9] Ballowe, W. C.; Holden, A. N.; Ozeroff, W. J.; Williamson, H. E.; Weidenbaum, B. (GEAP-3355-Pt. 2 [1960], pp. 1/31; N.S.A. **15** [1961] No. 26541).

[10] Futamoto, M.; Kawabe, U. (Surf. Sci. **93** [1980] L117/L123).

[11] Dong, W. D.; Kilpatrick, W. D.; Teem, J. M.; Zuccaro, D. E. (Progr. Astronaut. Aeron. **9** [1963] 269/89; C.A. **62** [1965] 6332).

[12] Berrada, A.; Mercurio, J.-P.; Etourneau, J.; Alexandre, F.; Theeten, J. B.; Duc, T. M. (Surf. Sci. **72** [1978] 177/88).

32.1.5.15.4 Reactions with Elements

The MB_6 cathodes on heating are very reactive with most refractory elements, Davis et al. [1]. The reaction of carbon with MB_6 to B_4C and lanthanide carbides took place on hot-pressing MB_6 in graphite dies in a vacuum at 2500 K, Hoyt, Chorné [2]. Arc melting of mixtures of C and MB_6 in an oxygen-purged Ar atmosphere led to the following conclusions: There exists a solid solubility of C in studied MB$_6$ with M = La, Sm, and Gd (unchanged X-ray pattern of MB$_6$), which amounted to ca. 15 at% C in GdB$_6$. Larger amounts of C (ca. 10 wt%) reacted with GdB$_6$ (unlike LaB$_6$ and SmB$_6$), DyB$_6$, and mixtures of YbB$_6$ + YbB$_4$ or of HoB$_4$ + HoB$_6$ + HoB$_{12}$ to give MB$_2$C$_2$ (M = Gd, Dy, Ho, Yb) plus MB$_6$ (M = Gd, Yb) or MB$_4$ (M = Dy), Smith [3, pp. 15/23, 37]. During Knudsen evaporation of NdB$_6$ from a C crucible, the hexaboride was not attacked [3, p. 328]. Aluminium does not react with MB$_6$ (M = Y, Sm, Eu) in a vacuum during 2 h, either at 823 K or above its melting point at 1273 K (see "Single-Crystal Growth", p. 28). The metals Ti, Zr, Ta, W, Ni, Fe, and 304 stainless steel form borides: The incomplete 2 h-reaction with a Ti matrix (1273 K, vacuum) yields TiB and a limited solid solution (Ti, M) with M = Y, Sm, Eu. The more reactive Zr matrix forms at that temperature within 1 h both ZrB and ZrB$_2$ in the case of YB$_6$ and SmB$_6$, only ZrB$_2$ with EuB$_6$, and (Zr, M) for all, Antony, Cummings [4]. Arc melting of GdB$_6$ with Zr in an Ar atmosphere purged of O$_2$ gives GdB$_4$ plus ZrB$_2$. In the presence of free (added) boron at start, ZrB$_{12}$ is formed additionally. The Ta reduces LaB$_6$ on arc melting to metallic La with formation of TaB, Ta$_{1.6}$B, or Ta$_{2.4}$B depending on temperature. Mixtures of Ta, GdB$_6$, and (added) B give GdB$_6$, GdB$_4$, and TaB$_2$ on arc melting. During Knudsen vaporization of MB$_6$ in W crucibles, MB$_4$ (M = La, Nd, Tb, Dy, Yb) and tungsten borides were formed [3, pp. 325/46]. Under the conditions of thermionic emission, MB$_6$ (M = Sm, Eu) react with W above 1700 K to form tungsten boride, probably W$_2$B$_5$. No such reaction was observed with GdB$_6$, Ermakov, Tsarev [5]. Reaction of MB$_6$ (M = Y, Sm, Eu) with a Ni matrix (1273 K, 2 h, vacuum) gave Ni$_2$B and another nickel boride (Ni$_4$B$_3$, Weidenbaum et al. [6]), and with an Fe matrix, Fe$_2$B and FeB besides traces of unreacted components. Products of the reaction with 304 stainless steel were Fe$_2$B, FeB, and (for SmB$_6$, EuB$_6$) also Ni$_2$B. No reaction took place with Cu at 1273 K and with Ag at 1173 K [4].

References:

[1]   Davis, P. R.; Swanson, L. W.; Hutta, J. J.; Jones, D. L. (J. Mater. Sci. **21** [1986] 825/36, 834).
[2]   Hoyt, E. W.; Chorné, J. (GEAP-3332 [1960] 1/15, 5; C.A. **1961** 1338).
[3]   Smith, P. K. (COO-1140-103 [1964] 1/485; N.S.A. **18** [1964] No. 25319; Diss. Univ. Kansas 1964, pp. 1/482; Diss. Abstr. **25** [1964/65] No. 5591).
[4]   Antony, K. C.; Cummings, W. V. (GEAP-3530 [1960] 1/25; C.A. **55** [1961] 8232).
[5]   Ermakov, S. V.; Tsarev, B. M. (Radiotekhn. Elektron. **10** [1965] 972/5; Radio Eng. Electron. Phys. [USSR] **10** [1965] 833/5; C.A. **63** [1965] 3732).
[6]   Weidenbaum, B.; Hoyt, E. W.; Zimmerman, D. L.; Cummings, W. V.; Antony, K. C. (Mater. Fuels High-Temp. Nucl. Energy Appl. Proc. Natl. Top. Meeting Am. Soc., San Diego, Calif., 1962 [1964], pp. 314/44, 333; N.S.A. **18** [1964] No. 22519).

32.1.5.15.5   Reactions with Compounds

Reactions with Water

The hexaborides resist attack by water, see Kosolapova, Domasevich [1] for fine powders of MB$_6$ with M = La, Ce, Pr, and Nd. The slight solubility in boiling water after 500 h, characterized by the rare earth content of the water (200 cm^3), related to its content in the 10g-powder material investigated, amounted to a value in the range 22 to 82 ppm for M = La, Sm, Eu, Gd, Dy, and Yb, Timofeeva, Timofeeva [2]. The quantity of boron dissolved from more or less impure YB$_6$ and EuB$_6$ after 500 h was ca. 60 ppm and 340 ppm, respectively, still increasing with time in the case of EuB$_6$, see a figure in the paper, Ballowe et al. [3].

Reactions with Acids

General Remarks

In general, the MB_6 are readily decomposed in oxidizing media, whereas they exhibit (fairly) high resistance to attack by nonoxidizing media, see for example, Andrieux [4] and the review by Binder [5]. Some of the solubility studies were performed with the intention of finding a solution suitable for a) the determination of free boron contaminating the MB_6, see Kugai, Nazarchuk [6, 7], or b) the purification of MB_6, especially LaB_6, from oxygen-containing impurities, see Nazarchuk et al. [8]. The latter affect the stability of MB_6 severely by causing hydrolysis in moist air. Suitable selective solutions appear to be H_2SO_4 (3:1 and 2:1) for a) [7] and 6 N HCl for LaB_6 in case b) [8].

A suitable etchant for MB_6 is HNO_3 (1:2), Binder [5]; for YB_6 and EuB_6 a mixture of $20 H_2O + 15 HF + 10 HNO_3$ was used by Grossman et al. [9].

HNO_3

Hexaborides MB_6 (M = Y, La, Ce, Nd, Gd, Er, Yb) are decomposed rapidly by dilute nitric acid solutions. Violent reaction takes place with concentrated HNO_3, Andrieux [4], for CeB_6 also Andrieux [10], the heat evolved amounted to (in kcal/mol): -777.3 ± 0.8 (LaB_6), -835.9 ± 3.0 (CeB_6), -785.5 ± 2.0 (PrB_6), and -764.7 ± 3.5 (NdB_6) or -3.25, -3.50, -3.29, and -3.20 MJ/mol, respectively, Kosolapova, Domasevich [1]. Dissolution of powders in concentrated HNO_3 and HNO_3 (1:1) proceeds within one minute, if the acid is in excess, Binder [5]. For complete dissolution of MB_6 (M = Y, La to Sm, Gd) in HNO_3 (1:1) upon weak heating for 5 min, see Kugai, Nazarchuk [6], and for dissolution of MB_6 (M = La to Nd) treated for 1 h in a 100-fold excess of HNO_3 (D = 1.34 g/cm^3 and dilutions 1:1 and 1:5) at temperatures of 376 to 393 K, see [1] which also gives the amount of B_2H_6 and B_4H_{10} formed during the decomposition in HNO_3 (1:20). The solubility of EuB_6 and DyB_6 in 0.1 to 6 N HNO_3 at 298 K (3d-studies) increased with the acid concentration; for erroneous values, see the paper, Ryabchikov et al. [11].

HF and HCl

The MB_6 (M = Y, La, Ce, Nd, Gd, Er, Yb) resisted attack by concentrated HF and HCl in the cold, Andrieux [4]; for CeB_6 also Andrieux [10]. According to Binder [5], this holds especially for bulk material. With concentrated HF, slow reaction takes place on boiling in the presence of air [5]. Fairly high resistance to attack by hydrochloric acid of any concentration, even in the presence of oxidizing agents was also observed for MB_6 (M = La to Nd) of particle size 3.8 to 4.7 μm, kept for 1 h in a 100-fold excess of reagent at high temperatures (reflux condenser). The table below compares the amount of insoluble residues in these solutions with those observed in sulfuric acid solutions and in aqueous alkali solutions:

medium	T in K	insoluble residue in %			
		LaB_6	CeB_6	PrB_6	NdB_6
HCl (D = 1.19 g/cm^3)	388	96.2	95.0	95.6	95.3
HCl (1:1)	383	98.0	95.2	90.8	94.6
HCl (1:5)	378	98.6	98.5	98.2	98.9
HCl (D = 1.19) + 50% citric acid	383	99.6	98.1	99.2	98.9
HCl (D = 1.19) + 0.2 N Trilon B	383	98.7	98.0	98.9	99.2
H_2SO_4(D = 1.84 g/cm^3)	553	0.0	0.0	0.0	0.0
H_2SO_4 (1:1)	523	88.7	89.6	89.4	96.0

medium	T in K	insoluble residue in %			
		LaB$_6$	CeB$_6$	PrB$_6$	NdB$_6$
H$_2$SO$_4$ (1:5)	393	98.7	95.5	98.2	98.6
H$_2$SO$_4$ (1:4) + 50% citric acid	398	98.6	98.0	98.7	99.0
H$_2$SO$_4$ (1:4) + 25% oxalic acid	483	99.0	98.2	99.6	99.5
H$_2$SO$_4$ (1:4) + 0.2 N Trilon B	401	98.9	97.5	98.1	99.3
NaOH (10%)	375	98.5	98.9	99.3	99.2
NaOH (40%)	378	99.7	99.6	99.9	98.9

For ratios of M/B of insoluble residues and solutions, see the paper, Kosolapova, Domasevich [1]. According to earlier 2 h-studies the stability of MB$_6$ in HCl (1:1) decreased in the order LaB$_6$ > PrB$_6$ > GdB$_6$ > NdB$_6$ > CeB$_6$ > SmB$_6$ > YB$_6$, for details see the papers, Kugai, Nazarchuk [6], Kugai [12]. Powders of La$_{1-x}$B$_6$ (x = 0, 0.06, 0.14) and Sm$_{1-x}$B$_6$ (x = 0, 0.1, 0.19, 0.27), both containing various amounts of oxygen, were boiled repeatedly each time for 1 h in 6 N or 12 N solutions of HCl (or H$_2$SO$_4$). Analyses of the cooled solutions revealed that the slight dissolution of M$_{1-x}$B$_6$ occurred without a change in their composition, except for Sm$_{0.73}$B$_6$. In this case the undissolved residue after 4 h in 6 N HCl (and 6 N H$_2$SO$_4$) was depleted to Sm$_{0.65}$B$_6$. The solubility of La$_{1-x}$B$_6$ in these acid solutions increases with increasing x, whereas that of Sm$_{1-x}$B$_6$ decreases with increasing x (see figures in the paper). Studies with La$_{1.0}$B$_6$ in 6 N HCl showed that in 1 to 2 h primarily the oxide containing boride phase dissolved, whereas the characteristic solubility of La$_{1.0}$B$_6$ is insignificant, Nazarchuk et al. [8]. The solubility of EuB$_6$ and DyB$_6$ in 3 N to 10 N HCl at 298 K (3d-studies) increased with the acid concentration; for erroneous values, see the paper, Ryabchikov et al. [11].

H$_2$SO$_4$

Powders of MB$_6$ (M = La to Nd) kept for 1 h in a 100-fold excess of concentrated H$_2$SO$_4$ at high temperatures decompose completely (see table above). Free sulfur, sulfur oxides, B$_2$H$_6$, B$_4$H$_{10}$, and H$_2$ are formed, Kosolapova, Domasevich [1]. Complete decomposition takes also place in H$_2$SO$_4$ (3:1) at 483 K (2 h) and in H$_2$SO$_4$ (2:1) at 463 K (3 to 4 h), on use of 0.5 g MB$_6$ (M = La, Ce, Nd) and 50 mL of each reagent. More dilute sulfuric acid solutions (1:1) attack MB$_6$ much less, Kugai, Nazarchuk [7]. For the amount of insoluble residue in various solutions (100-fold excess of reagent, 1 h), see the table above from [1]. According to earlier 2 h-studies, the stability of MB$_6$ in H$_2$SO$_4$ (1:1) decreased in the order LaB$_6$ > GdB$_6$ > CeB$_6$ > NdB$_6$ > SmB$_6$ > YB$_6$ > PrB$_6$; for details see the papers, Kugai, Nazarchuk [6], Kugai [12]. The solubility of La$_{1-x}$B$_6$ (x = 0, 0.06, 0.14) and Sm$_{1-x}$B$_6$ (x = 0, 0.1, 0.19, 0.27) in 6 N and 12 N H$_2$SO$_4$ depends on x in the same way as it does in 6 N and 12 N HCl, Nazarchuk et al. [8]; see above. The solubility of EuB$_6$ in 0.1 to 3 N H$_2$SO$_4$ at 298 K (3d-studies) was higher than that of DyB$_6$. On a further increase in the acid concentration to 14 N H$_2$SO$_4$, the solubility of EuB$_6$ decreased, unlike that of DyB$_6$; for erroneous values, see the paper, Ryabchikov et al. [11].

Reactions with Mixtures of Acids and Oxidizing or Complex-Forming Agents

Powdered MB$_6$ (M = Y, La to Sm, Gd) were decomposed completely by mixtures of HNO$_3$ and H$_2$O$_2$ and by aqua regia in the cold. The compounds react with a sulfuric acid solution of KIO$_3$ or K$_5$IO$_6$ to liberate iodine and are also unstable in a sulfuric acid solution of Ce(SO$_4$)$_2$, Kugai, Nazarchuk [6]. For the solubility of MB$_6$ (M = La to Nd) in acids containing complex-forming additions such as oxalic acid, citric acid, and ethylenediaminetetraacetic acid (Trilon B), see the table on pp. 97/8.

Reactions with Alkali Solutions

MB_6 (M = Y, La, Ce, Nd, Gd, Er, Yb) resist attack by alkali solutions in the cold, Andrieux [4], for CeB_6 also Andrieux [10]. Powdered MB_6 (M = Y, La to Sm, Gd) hardly dissolved at all in aqueous alkali solutions upon heating, Kugai, Nazarchuk [6]. The table on p. 98 gives the amount of insoluble residue observed after keeping MB_6 (M = La to Nd) with a particle size of 3.8 to 4.7 μm for 1 h in a 100-fold excess of a NaOH solution at a high temperature, Kosolapova, Domasevich [1]. Mixtures of solutions of NaOH and H_2O_2 also do not act on MB_6 (M = Y, La to Sm, Gd) upon heating, with the exception of CeB_6 which is partially decomposed by such a mixture [6].

Reactions with Melts

In general, alkaline melts decompose the hexaborides rapidly, whereas neutral or acid, nonoxidizing salt melts attack these compounds rather weakly, especially in the absence of air, see the review by Binder [5]. The MB_6 (M = Y, La, Ce, Nd, Gd, Er, Yb) decompose when fused with alkalis and with the nitrates, carbonates, and bisulfates of the alkali metals, Andrieux [4], for CeB_6 also Andrieux [10].

Reactions with Solid Compounds

Studied MB_6 (M = Y, La, Ce, Nd, Gd, Er, Yb) reacted violently when heated with oxidants such as Na_2O_2 and PbO_2, Andrieux [4]; for CeB_6 also Andrieux [10]. Fine powders of MB_6 (M = La, Nd, Sm, and Dy) kept in a furnace without circulation of air reacted with the supporting Al_2O_3 above ca. 1273 K. The oxidation products underwent vitrification at 1323 to 1373 K, Timofeeva, Timofeeva [13]. Dense samples of MB_6 (M = Y, Sm, Eu, and Dy) reacted with Al_2O_3 as crucible material on heating at 1500 and 2200 K to give metal borate plus boron oxide, Hoyt [14].

The MB_6 do not react with BN. However, due to its high volatility at high temperatures, BN is unsuitable as crucible material in the hot-pressing process or vaporization of MB_6. Studies in the system Gd–B–C reveal that B_4C is also unsuitable as crucible material. Mixtures of ZrB_2 and MB_6 (M = Nd, Gd, Dy) contaminated by MB_4 do not react on arc melting to give ZrB_{12}. The solid solubility of GdB_6 in ZrB_2 at high temperatures was estimated to be < 5 mol%, that of ZrB_2 in GdB_6 to be ≤ 5 mol%. Before use of ZrB_2 crucibles in MB_6 vaporization studies, the crucible requires outgassing for 20 h at 2523 K in a vacuum and an additional heat treatment to remove elemental boron totally. From thermodynamic estimations, also HfB_2 should not react with MB_6 and thus likewise be suitable as crucible material, Smith [15].

References:

[1] Kosolapova, T. Ya.; Domasevich, L. T. (Poroshkovaya Metal. **1970** No. 5, pp. 1/5; Soviet Powder Met. Metal Ceram. **1970** 353/6).

[2] Timofeeva, N. I.; Timofeeva, E. N. (Izv. Akad. Nauk SSSR Neorgan. Materialy **4** [1968] 1789/91; Inorg. Materials [USSR] **4** [1968] 1559/61).

[3] Ballowe, W. C.; Holden, A. N.; Ozeroff, W. J.; Williamson, H. E.; Weidenbaum, B. (GEAP-3355-Pt. 2 [1960] pp. 1/31, 21; N.S.A. **15** [1961] No. 26541).

[4] Andrieux, L. (Ann. Chim. [Paris] [10] **12** [1929] 423/507, 476).

[5] Binder, F. (Radex Rundschau **1977** 52/71, 63; C.A. **87** [1977] No. 27597).

[6] Kugai, L. N.; Nazarchuk, T. N. (Zh. Analit. Khim. **16** [1961] 205/8; J. Anal. Chem. [USSR] **16** [1961] 213/6).

[7] Kugai, L. N.; Nazarchuk, T. N. (Poroshkovaya Metal. **1975** No. 6, pp. 99/101; Soviet Powder Met. Metal Ceram. **14** [1975] 509/11).

[8]   Nazarchuk, T. N.; Dudnik, E. M.; Paderno, Yu. B.; Yukhimenko, E. V. (Poroshkovaya Metal. **1984** No. 3, pp. 84/8; Soviet Powder Met. Metal Ceram. **23** [1984] 242/5).

[9]   Grossman, L. N.; Hoyt, E. W.; Ingold, I. H.; Kaznoff, A. I.; Sanderson, M. J. (GEST-2009 [1962] pp. 7-1/7-26, 7-6; N.S.A. **17** [1963] No. 14819).

[10] Andrieux, L. (Compt. Rend. **186** [1928] 1736/8).

[11] Ryabchikov, D. I.; Vagina, N. S.; Kozlov, V. G. (Zh. Neorgan. Khim. **13** [1968] 588/9; Russ. J. Inorg. Chem. **13** [1968] 304/5).

[12] Kugai, L. N. (Byul. Inst. Metallokeram. Spets. Splavov Akad. Nauk Ukr. SSR No. 6 [1960/61] 45/51; C.A. **56** [1962] 12300).

[13] Timofeeva, E. N.; Timofeeva, N. I. (Zh. Prikl. Khim. **44** [1971] 1400/2; J. Appl. Chem. [USSR] **44** [1971] 1414/6).

[14] Hoyt, E. W. (Proc. 2nd Conf. Rare Earth Res., Glennwood Springs, Colo., 1961 [1962], pp. 287/99; N.S.A. **16** [1962] No. 32072).

[15] Smith, P. K. (COO-1140-103 [1964] 1/485, 167/76; N.S.A. **18** [1964] No. 25319; Diss. Univ. Kansas 1964, pp. 1/482; Diss. Abstr. **25** [1964/65] No. 5591).

32.1.6 Comparative Data for MB$_{12}$

Only Sc, Y, and the smaller M = Tb to Lu form MB$_{12}$. Under high pressure GdB$_{12}$ is also obtained. The compounds have a very narrow range of homogeneity.

32.1.6.1 Preparation

The dodecaborides can be prepared from the elements, or more usually by reduction of M$_2$O$_3$ with amorphous or crystalline boron in vacuum.

From Oxides and Boron

MB$_{12}$ with M = Tm, Yb, and Lu were obtained in two steps. At first, powders were prepared by borothermal reduction under vacuum at 2473 K according to:

$$M_2O_3 + 27\ B \rightarrow 2\ MB_{12} + 3\ BO$$

Then fused samples were prepared by melting the powder in an arc furnace or by floating-zone method. These products contained a considerable amount of MB$_4$ and MB$_6$. Therefore they were crushed into fine powder and the impurities dissolved away with dilute HNO$_3$ before remelting until the products were found to be single phase, Iga et al. [1], see also Iga et al. [2], Kasaya, Iga [3].

The synthesis of single-phase MB$_{12}$ with M = Y, Tb to Lu by borothermal reduction of M$_2$O$_3$ in vacuum is hindered by the very narrow range of homogeneity and the different evaporation rates of the components. So a 3 to 7% B excess has to be used for the formation of predominantly MB$_{12}$ in ZrB$_2$ crucibles in the first step. In a second step, the samples were held 1 to 1.5 h in the vacuum at a temperature close to the melting point of the MB$_6$ impurities, Moiseenko et al. [4]. The samples were made from thoroughly mixed M$_2$O$_3$ and B powders, pressed at (1 to 2)$\times$10^3 kg/cm^2 ($\sim$100 to 200 MPa) and heated 1 to 1.5 h at 1$\times$10^{-5} Torr ($\sim$1.3 mPa) at temperature T$_1$ and T$_2$, for the first and the second step, respectively, and a B excess as follows:

M in MB_{12}	Y	Tb	Dy	Ho	Er	Tm	Yb	Lu
T_1 in K	1953	1873	1723	1723	1723	1723	1873	1723
T_2 in K	1993	2033	2023	2093	2093	2113	2013	2113
B excess in wt%	7	6	4	4	5	5	6	5

For M = Tb and Yb the holding time in the second step was 5 h, for the others 1 to 1.5 h. The stoichiometry was proved by X-ray, metallographic, and chemical analyses, Moiseenko, Odintsov [5], Moiseenko et al. [26], see also Moiseenko, Odintsov [6]. Slightly M-deficient samples were obtained under similar conditions at a compacting pressure of 300 kg/cm² (~ 30 MPa), and heating only at T_1 in vacuum of < 10⁻⁴ Torr (≲13 mPa) by Paderno, Odintsov [24], see also Odintsov [25].

The applied pressure influences the degree of reaction completeness. At low pressure the samples are porous. Diffusion, contact phenomena, and related effects influence the process, and the reaction time increases by 20 to 30%. At high pressure the samples are dense, however, the reaction time is increased 1.5 to 2 times owing to the difficulty of removal of gaseous reaction products. The optimal pressure is 90 to 120 kg/cm² (~9 to 12 MPa), Moiseenko, Odintsov [7], Odintsov, Moiseenko [8]. Earlier, the same group of authors obtained MB_{12} (M = Y, Tb to Lu) by sintering briquets of M_2O_3 + B that had previously been compacted at 8 tons/cm² (~800 MPa) under Ar at 2073 to 2173 K for 2 h, Paderno et al. [9], see also Odintsov, Yakovenko [10]. MB_{12} with M = Y, Dy, Ho, Er, Tm, and Lu were prepared by the reaction of M_2O_3 with excess of amorphous B at 1673 to 1873 K in 1 h in a He atmosphere, Binder et al. [11], La Placa et al. [12]. For the preparation of ErB_{12}, TmB_{12}, and LuB_{12} in vacuum at 1373 to 1873 K in ZrB_2 or BN crucibles see also Sturgeon, Eick [13, 14]. MB_{12} obtained from M_2O_3 and B in vacuum were hot-pressed at 150 to 200 kg/cm² (~15 to 20 MPa) and 2000 to 2500 K for 5 to 40 min in a graphite die under vacuum to yield high-density pellets, Hoyt [21].

Single crystals of ScB_{12} and YB_{12} were obtained by the reaction of cold-pressed M_2O_3 and B (B:M = 7:1) in an induction coupled Ar plasma torch apparatus on a BN plate. The samples were heated 2 min to 2273 to 2773 K, quenched in water and washed with concentrated HCl. Crystals could be easily hand-picked under the microscope, Matkovich et al. [15, 16].

From the Elements

The light blue compounds MB_{12} with M = Y, Dy, Er, and Yb were prepared from the pure metals and boron which were hot-pressed or arc-melted under purified Ar, Schwetz et al. [17]. MB_{12} with M = Sc, Y, Ho, Er, Tm, and Lu were also prepared by arc melting from the elements by Fisk [18]. TbB_{12} and YbB_{12} (which were not obtained from the oxides) were also obtained by this method from the metals and crystalline boron under 1 atm Ar, La Placa et al. [19, 20].

High-pressure high-temperature techniques have been used to synthesize GdB_{12} in addition to the known TbB_{12}. However, NdB_{12} and SmB_{12} were not obtained. The stoichiometric mixtures of the elements were exposed to >60 kbar (>6 GPa) at 2373 K (M = Gd) or 65 kbar (6.5 GPa) at 1923 K (M = Tb) in a BN crucible surrounded by a graphite heater, Cannon et al. [22, 23].

References:

[1] Iga, F.; Takakuwa, Y.; Takahashi, T.; Kasaya, M.; Kasuya, T.; Sagawa, T. (Solid State Commun. **50** [1984] 903/5).

[2] Iga, F.; Kasaya, M.; Kasuya, T. (J. Magn. Magn. Mater. **52** [1985] 279/82).

[3] Kasaya, M.; Iga, F. (AIP [Am. Inst. Phys.] Conf. Proc. No. 140 [1986] 11/8).

[4] Moiseenko, L. L.; Odintsov, V. V.; Ivashina, G. A. (Izv. Vysshikh Uchebn. Zavedenii Fiz. **20** No. 5 [1977] 159; Soviet Phys. J. **20** [1977] 697/8).

[5] Moiseenko, L. L.; Odintsov, V. V. (Izv. Sibirsk. Otd. Akad. Nauk SSSR Ser. Khim. Nauk **1978** No. 5, pp. 48/50; C.A. **89** [1978] No. 225253).

[6] Moiseenko, L. L.; Odintsov, V. V. (J. Less-Common Metals **67** [1979] 237/43).

[7] Moiseenko, L. L.; Odintsov, V. V. (Tugoplavkie Soedin. Redkozemel. Met. Mater. Vses. 3rd Semin., Novosibirsk 1977 [1979], pp. 34/6; C.A. **93** [1980] No. 10321).

[8] Odintsov, V. V.; Moiseenko, L. L. (Poroshkovaya Metal. **1977** No. 10, pp. 33/4; Soviet Powder Met. Metal Ceram. **16** [1977] 766/7).

[9] Paderno, Yu. B.; Odintsov, V. V.; Timofeeva, I. I.; Klochkov, L. A. (Teplofiz. Vys. Temp. **9** [1971] 200/2; High Temp. [USSR] **9** [1971] 175/7).

[10] Odintsov, V. V.; Yakovenko, S. N. (Poroshkovaya Metal. **1984** No. 1, pp. 71/3; Soviet Powder Met. Metal Ceram. **23** [1984] 68/9).

[11] Binder, I.; La Placa, S.; Post, B. (Boron Synth. Struct. Prop. Proc. Conf., Asbury Park, N. J., 1959 [1960], pp. 86/93, 87; C.A. **1961** 10175).

[12] La Placa, S.; Binder, I.; Post, B. (J. Inorg. Nucl. Chem. **18** [1961] 113/7).

[13] Sturgeon, G. D.; Eick, H. A. (Proc. 3rd Conf. Rare Earth Res., Clearwater, Fla., 1963 [1964], pp. 87/97).

[14] Sturgeon, G. D.; Eick, H. A. (TID-17968 [1961] 1/10; N.S.A. **17** [1963] No. 14216).

[15] Matkovich, V. I.; Economy, J.; Giese, R. F., Jr.; Barrett, R. (Acta Cryst. **19** [1965] 1056/8).

[16] Matkovich, V. I.; Economy, J.; Giese, R. F., Jr.; Barrett, R. B. (AD-608444 [1964] 1/16; C.A. **63** [1964] 1283).

[17] Schwetz, K.; Ettmayer, P.; Kieffer, R.; Lipp, A. (Radex Rundschau **1972** 257/65).

[18] Fisk, Z. (Diss. Univ. California 1969, pp. 1/96, 17; Diss. Abstr. Intern. B **30** [1970] 4333).

[19] La Placa, S.; Noonan, D.; Post, B. (Acta Cryst. **16** [1963] 1182).

[20] La Placa, S.; Noonan, D.; Post, B. (BNL-6741 [1963] 1/8; N.S.A. **17** [1963] No. 39020).

[21] Hoyt, E. W. (Proc. 2nd Conf. Rare Earth Res., Glenwood Springs, Colo., 1961 [1962], pp. 287/99, 291).

[22] Cannon, J. F.; Cannon, D. M.; Hall, H. T. (J. Less-Common Metals **56** [1977] 83/90).

[23] Cannon, J. F.; Cannon, D. M.; Hall, H. T. (High-Pressure Sci. Technol. 6th AIRAPT Conf., Boulder, Colo., 1977 [1979], Vol. 1, pp. 1000/6).

[24] Paderno, Yu. B.; Odintsov, V. V. (Metalloterm. Protsessy Khim. Metall. Mater. Konf., Novosibirsk 1971, pp. 39/43; C.A. **80** [1974] No. 66230).

[25] Odintsov, V. V. (Tekhnol. Poluch. Nov. Mater. **1972** 85/7; C.A. **79** [1973] No. 147784).

[26] Moiseenko, L. L.; Odintsov, V. V.; Ivashina, G. A. (Deposited Doc. VINITI-859-77 [1977] 1/10; C.A. **90** [1979] No. 80107).

32.1.6.2 Crystallographic Properties. Density

All of the rare earth dodecaborides except possibly ScB$_{12}$ (see p. 127) crystallize in the face-centered cubic UB$_{12}$ structure, space group Fm3m-O$_h^5$ (No. 225), Z = 4. The structure is related to the NaCl type in which M occupy the Na sites (0, 0, 0) and the B$_{12}$ cubooctahedra occupy the Cl sites (B in ½, x, x with x = 0.1706 in the case of YB$_{12}$, see p. 156). However, the B$_{12}$ groups form a rigid three-dimensional framework of B atoms (see **Fig. 39**) with inter B–B bond lengths of ~1.68 Å and intra B–B bond lengths of ~1.80 Å. The UB$_{12}$ type can also be described as a face-centered cubic array of M atoms, with each M located in the center of a B$_{24}$ cubooctahedron, see the review by Spear [1]. The following lattice parameters a (in Å) and densities D (in g/cm^3) are given:

MB$_{12}$	YB$_{12}$	TbB$_{12}$	DyB$_{12}$	HoB$_{12}$	ErB$_{12}$	TmB$_{12}$	YbB$_{12}$	LuB$_{12}$
a [2]	7.503	7.505	7.498	7.495	7.486	7.476	7.471	7.463
a [11]	7.500	7.504	7.501	7.492	7.484	7.476	7.469	7.464
a [1]	7.5008	7.504	7.4992	7.492	7.4832	7.476	7.4690	7.464
D [11]	3.444	4.538	4.600	4.655	4.706	4.756	4.824	4.868

Values a = 7.508 Å for TbB$_{12}$ to 7.464 Å for LuB$_{12}$ with YB$_{12}$ between TbB$_{12}$ and DyB$_{12}$ and D = 3.442 g/cm^3 for YB$_{12}$ and D = 4.530 g/cm^3 for TbB$_{12}$ to 4.867 g/cm^3 for LuB$_{12}$ are given by Odintsov et al. [6]. Nearly the same values at 293 K are published by Paderno et al. [3], Paderno, Odintsov [4], see also Yukhimenko et al. [5] and the not-strictly-linear graphs of the lattice constants as a function of the atomic number given by Kasaya, Iga [16] and Moiseenko et al. [18]. For additional references see [1]. At 78 K a = 7.504 Å for TbB$_{12}$ to 7.461 Å for LuB$_{12}$ [3]. A critical survey on crystal chemistry and physical properties is given by Etourneau [7]. The lattice constants of GdB$_{12}$ and TbB$_{12}$ obtained under high pressure are 7.524(1) and 7.509(1) Å. Interplanar spacings and intensities of the powder diagram of GdB$_{12}$ are listed, Cannon et al. [8 to 10].

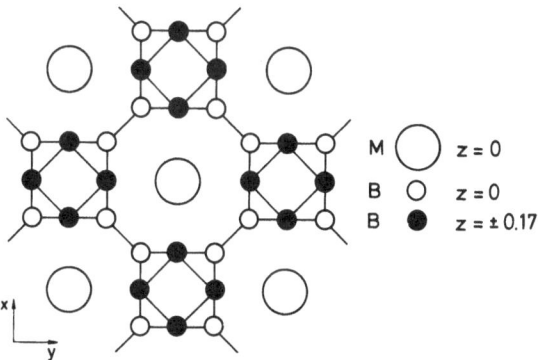

M ◯ z = 0
B ○ z = 0
B ● z = ± 0.17

Fig. 39. Projection of the face-centered cubic MB$_{12}$ structure along the z axis.

Interatomic distances listed by [11] refer to the early assumed parameter x = ⅙ for B; distances M–B ranged from 2.797 Å for TbB$_{12}$ to 2.782 Å for LuB$_{12}$ and B–B from 1.769 Å for TbB$_{12}$ to 1.759 Å for LuB$_{12}$, see also La Placa et al. [12], Binder et al. [13]. As a typical powder pattern, that of TmB$_{12}$ is given in [13]. Structural determinants in higher borides such as MB$_{12}$ are discussed by Matkovich et al. [14, 15]. From the linear change of the lattice constants with the M^{3+} radii, trivalent M is implicated and the slope indicates the rigid three-dimensional B framework [1]. For lattice constants vs. atomic radius see also Schwetz et al. [17].

References:

[1] Spear, K. E. (in: Alper, A. M., Refractory Materials, Phase Diagrams, Vol. 4, Academic, New York 1976, pp. 91/159, 112/4).

[2] Moiseenko, L. L.; Paderno, Yu. B. (Elektron. Str. Fiz. Khim. Svoistva Tugoplavkikh Soedin. Splavov Dokl. Vses. 9th Simp., Ivano-Frankovsk, Ukr. SSR, 1979 [1980], pp. 128/31; C.A. **94** [1981] No. 218173).

[3] Paderno, Yu. B.; Odintsov, V. V.; Timofeeva, I. I.; Klochkov, L. A. (Teplofiz. Vys. Temp. **9** [1971] 200/2; High Temp. [USSR] **9** [1971] 175/7).

[4] Paderno, Yu. B.; Odintsov, V. V. (Metalloterm. Protsessy Khim. Metall. Mater. Konf., Novosibirsk 1971, pp. 39/43; C.A. **80** [1974] No. 66230).

[5] Yukhimenko, E. V.; Odintsov, V. V.; Kotlyar, E. E.; Paderno, Yu. B. (Poroshkovaya Metal. **1971** No. 11, pp. 52/5; Soviet Powder Met. Metal Ceram. **10** [1971] 895/7).

[6] Odintsov, V. V.; Paderno, Yu. B.; Goryachev, Yu. M. (Khim. Svyaz Krist. Poluprovodn. Polumet. **1973** 177/83; C.A. **81** [1974] No. 144812).

[7] Etourneau, J. (J. Less-Common Metals **110** [1985] 267/81).

[8] Cannon, J. F.; Cannon, D. M.; Hall, H. T. (J. Less-Common Metals **56** [1977] 83/90, 86).

[9] Cannon, J. F.; Hall, H. T. (Rare Earths Mod. Sci. Technol. [1977/78] 219/24).

[10] Cannon, J. F.; Cannon, D. M.; Hall, H. T. (High-Pressure Sci. Technol. 6th AIRAPT Conf., Boulder, Colo., 1977 [1979], Vol. 1, pp. 1000/6).

[11] La Placa, S.; Noonan, D.; Post, B. (BNL-6741 [1963] 1/8; N.S.A. **17** [1963] No. 39020).

[12] La Placa, S.; Binder, I.; Post, B. (J. Inorg. Nucl. Chem. **18** [1961] 113/7).

[13] Binder, I.; La Placa, S.; Post, B. (Boron Synth. Struct. Prop. Proc. Conf., Asbury Park, N. J., 1959 [1960], pp. 86/93, 87; C.A. **1961** 10175).

[14] Matkovich, V. I.; Giese, R. F., Jr.; Economy, J. (Z. Krist. **122** [1965] 116/30).

[15] Matkovich, V. I.; Economy, J. (in: Matkovich, V. I., Boron and Refractory Borides, Springer, Berlin – New York 1977, pp. 78/95, 82, 84).

[16] Kasaya, M.; Iga, F. (AIP [Am. Inst. Phys.] Conf. Proc. No. 140 [1986] 11/8).

[17] Schwetz, K.; Ettmayer, P.; Kieffer, R.; Lipp. A. (Radex Rundschau **1972** 257/65, 259).

[18] Moiseenko, L. L.; Odintsov, V. V.; Grushko, Yu. S. (Izv. Akad. Nauk SSSR Neorgan. Materialy **15** [1979] 695/7; Inorg. Materials [USSR] **15** [1979] 542/3).

32.1.6.3 Microhardness

The microhardness H_μ in kg/mm^2 ($\triangleq 0.98 \times 10^7$ Pa) at 30 to 200 g load depends markedly on the load below ~ 100 g, becomes independent of it above ~ 100 g, and takes the following values at 100 g and 293 K (10 s loading time):

MB$_{12}$	YB$_{12}$	TbB$_{12}$	DyB$_{12}$	HoB$_{12}$	ErB$_{12}$	TmB$_{12}$	YbB$_{12}$	LuB$_{12}$
H_μ	3200	2600	2400	2700	2800	3000	~ 1900	2900

The hardness of the dodecaborides is higher than that of the components and is due to the strengthening of the interatomic bonds and to the presence of predominant covalent B–B bonds, Odintsov [1], see also Goryachev et al. [2]. However, the Knoop microhardness at 100 g load is given as 2330, 2310, 2390, and 2400 kg/mm^2 (± 50) for YB$_{12}$, DyB$_{12}$, ErB$_{12}$, and YbB$_{12}$ by Schwetz et al. [3].

References:

[1] Odintsov, V. V. (Izv. Akad. Nauk SSSR Neorgan. Materialy **10** [1974] 366/7; Inorg. Materials [USSR] **10** [1974] 317/8).

[2] Goryachev, Yu. M.; Odintsov, V. V.; Paderno, Yu. B. (Metallofizika No. 37 [1971] 29/36; C.A. **77** [1972] No. 81514).

[3] Schwetz, K.; Ettmayer, P.; Kieffer, R.; Lipp, A. (Radex Rundschau **1972** 257/65).

32.1.6.4 Thermal Properties

The thermal expansion coefficient α (in K^{-1}) was measured by X-ray methods in the range 1) at 78 to 300 K and dilatometrically in the temperature range 2) at 300 to 1300 K ($\pm 2\%$):

MB_{12}	YB_{12}	TbB_{12}	DyB_{12}	HoB_{12}	ErB_{12}	TmB_{12}	YbB_{12}	LuB_{12}
$10^6\alpha$ (1)	2.000	2.300	2.670	2.003	2.004	2.008	2.009	2.010
$10^6\alpha$ (2)	5.2	5.0	5.5	5.2	5.2	5.2	5.7	5.0
$10^6\alpha$ (300 K)	3.6	3.2	4.2	3.6	3.7	3.85	3.75	3.4

The temperature dependence of α is given in a figure. From α at 300 K, the root mean square amplitudes of the thermal vibrations were calculated. The values are between 0.041 and 0.047 Å, nearly the same for all MB_{12} whereas they differ considerably in the pure metals, Paderno et al. [1], see also Goryachev et al. [2], Odintsov et al. [3]. From X-ray diffraction measurements $\alpha = (3.0 \pm 0.2) \times 10^{-6}\ K^{-1}$ for both HoB_{12} (293 to 623 K) and ErB_{12} (113 to 693 K). This extremely small value reflects the rigid inflexible B framework, La Placa et al. [9].

Whereas TmB_{12} vaporizes congruently with formation of gaseous Tm and B, MB_{12} with M = Tb, Ho, Er form solid MB_6 besides gaseous B as shown by mass spectroscopic investigations of the evaporization reactions at 2173 K using Knudsen cell effusion method, McGrath [4].

Incongruent melting temperature T_f in K and Debye temperature Θ_D in K are as follows:

MB_{12}	YB_{12}	TbB_{12}	DyB_{12}	HoB_{12}	ErB_{12}	TmB_{12}	YbB_{12}	LuB_{12}
T_f [3]	2923	2700*)	2523	2723	2573	2673	—	2623
T_f [10]	2473	2473	2373	2373	2353	2453	2473	2443
Θ_D [1, 3]	1500	1245	1095	1234	1216	1190	1187	1250
Θ_D [5]	1064	834	871	886	858	886	—	856

*) Read from a figure in [8].

Melting temperatures from ~ 2900 K for GdB_{12} to ~ 2500 K for ErB_{12} may be read from a figure given by Goryachev, Kovenskaya [8].

The standard entropy S°_{298} in $J \cdot mol^{-1} \cdot K^{-1}$ of dodecaborides was evaluated by the method of additive comparative calculation (± 13):

M in MB_{12}	Sc	Y	Tb	Dy	Ho	Er	Tm	Yb	Lu
S°_{298}	96	109	130	138	138	134	138	121	113

Allowance was made for magnetic contributions. The additive method of calculation is applicable due to the rigid B network, Borovikova, Fesenko [6].

The thermal conductivities λ of MB_{12} with M = Y, Dy, Ho, Er, and Tm are higher than those of the pure metals and vary inversely with temperature at ~ 300 to ~ 1000 K, adopting a constant value at high temperature. Due to the fairly rigid lattice, the fraction of the phonon component in λ is enhanced. The narrowing of the distance between conduction electrons causes an enhancement of the electronic fraction. A figure $\lambda = f(T)$ is given. The values are in the range ~ 0.4 to ~ 0.2 $W \cdot cm^{-1} \cdot K^{-1}$. The Wiedemann-Franz law holds but the Lorenz number departs from theory, Odintsov et al. [7]. Values of λ at 293 K (in $cal \cdot cm^{-1} \cdot K^{-1} \cdot s^{-1}$):

MB_{12}	YB_{12}	DyB_{12}	HoB_{12}	ErB_{12}	TmB_{12}	LuB_{12}
λ	0.065	0.178	0.550	0.288	0.140	0.720

Compared with ZrB_{12} the electronic part in λ is lower for YB_{12} and higher for the other investigated MB_{12} [2]. For Nernst-Ettingshausen coefficients see p. 108.

References:

[1]   Paderno, Yu. B.; Odintsov, V. V.; Timofeeva, I. I.; Klochkov, L. A. (Teplofiz. Vys. Temp. **9**
      [1971] 200/2; High Temp. [USSR] **9** [1971] 175/7).

[2]   Goryachev, Yu. M.; Odintsov, V. V.; Paderno, Yu. B. (Metallofizika No. 37 [1971] 29/36;
      C.A. **77** [1972] No. 81514).

[3]   Odintsov, V. V.; Paderno, Yu. B.; Goryachev, Yu. M. (Khim. Svyaz Krist. Poluprovodn.
      Polumet. **1973** 177/83; C.A. **81** [1974] No. 144812).

[4]   McGrath, L. C. (Diss. New Mexico State Univ. 1972, pp. 1/236, 161/2; Diss. Abstr. Intern. B
      33 [1972] 2546/7).

[5]   Dutchak, Ya. I.; Fedyshin, Ya. I.; Paderno, Yu. B.; Vadets, D. I.; Odintsov, V. V. (Tezisy
      Dokl. 2nd Vses. Konf. Kristallokhim. Intermetal. Soedin., Lvov, USSR, 1974, p. 149; C.A. **86**
      [1977] No. 10825).

[6]   Borovikova, M. S.; Fesenko, V. V. (J. Less-Common Metals **117** [1986] 287/91).

[7]   Odintsov, V. V.; Lesnaya, M. I.; L'vov, S. N. (At. Energiya SSSR **35** No. 3 [1973] 194; At.
      Energy [USSR] **35** [1973] 834/5; C.A. **80** [1974] No. 7732).

[8]   Goryachev, Yu. M.; Kovenskaya, B. A. (J. Less-Common Metals **67** [1979] 273/9).

[9]   La Placa, S.; Binder, I.; Post, B. (J. Inorg. Nucl. Chem. **18** [1961] 113/7).

[10] Spear, K. E. (in: Alper, A. M., Refractroy Materials, Phase Diagrams, Vol. 4, Academic, New
      York 1976, pp. 91/159, 134).

32.1.6.5 Magnetic Properties

The temperature dependence of the magnetic susceptibility χ of MB_{12} (M = Y, Tb to Lu) was
measured over the range 90 to 1200 K on polycrystalline samples. For M = Tb to Yb the inverse
susceptibility $1/\chi$ vs. T plot is linear thus obeying the Curie-Weiss law. The paramagnetic
properties are mainly due to 4f electrons. The measured χ_{mol} in 10^{-3} cm^3/mol at 300 K, the Curie
constant C in $cm^3 \cdot K \cdot mol^{-1}$, the effective magnetic moment μ_{eff} in μ_B of MB_{12} and M^{3+}, and the
paramagnetic Curie temperature Θ_p in K are summarized in the following table:

MB_{12}	χ_{mol}	C	$\mu_{eff}(MB_{12})$	$\mu_{eff}(M^{3+})$	Θ_p
TbB_{12}	33.92	11.58	9.62 ± 0.03	9.72	-41.4 ± 1.0
DyB_{12}	41.70	13.92	10.55 ± 0.04	10.64	-33.8 ± 0.5
HoB_{12}	41.72	13.83	10.52 ± 0.05	10.60	-31.5 ± 0.5
ErB_{12}	34.66	11.42	9.57 ± 0.05	9.58	-29.5 ± 0.5
TmB_{12}	21.41	7.03	7.50 ± 0.06	7.56	-28.3 ± 1.0
YbB_{12}	5.55	2.29	4.28 ± 0.03	4.50	-112.2 ± 1.0

The magnetic moments of MB_{12}, calculated according to Van Vleck's theory, are close to,
however somewhat smaller than that of M^{3+}. This means trivalent M in MB_{12}. The negative sign
of Θ_p indicates antiferromagnetic behavior. Ruderman-Kittel-Kasuya-Yosida exchange interactions are assumed. For YB_{12} and LuB_{12} at 300 K, specific susceptibilities $\chi = 0.3 \times 10^{-6}$ and
-0.32×10^{-6} cm^3/g are measured and $\chi = 0.41 \times 10^{-6}$ and -0.24×10^{-6} cm^3/g are calculated.
For YB_{12} the temperature dependence in the low-temperature region can be related to the
predominance of the temperature-dependent paramagnetism of the ion core. With increasing
temperature, χ decreases and becomes negative near 500 K. Possibly the χ vs. T curve of LuB_{12}
behaves analogously below 90 K, Moiseenko, Odintsov [1]; for MB_{12} with M = Tb to Tm see also
Moiseenko et al. [2, 3]. The samples used in the earlier investigations (at 100 to 300 K, Odintsov
et al. [4], up to 1000 K, Odintsov, Moiseenko [5]) contained ferromagnetic (and other) impuri-

ties. Therefore the results of [4] and [5] deviate considerably from that of [1 to 3]. From susceptibility measurements, the Néel temperature $T_N = 6.5$ K was obtained for HoB_{12} and ErB_{12}, from resistivity data $T_N = 4.2$ K for TmB_{12}, Matthias et al. [6], Fisk [7]. $\Theta_p = -23$ K and $T_N < 10$ K for HoB_{12} and $\Theta_p = -20$ K and $T_N < 10$ K for ErB_{12} is given in Etourneau [8], Etourneau, Hagenmuller [9].

References:

[1] Moiseenko, L. L.; Odintsov, V. V. (J. Less-Common Metals **67** [1979] 237/43).

[2] Moiseenko, L. L.; Odintsov, V. V.; Ishchenko, G. N. (Izv. Vysshikh Uchebn. Zavedenii Fiz. **20** No. 12 [1977] 96/101; Soviet Phys. J. **20** [1977] 1613/6).

[3] Moiseenko, L. L.; Odintsov, V. V.; Ishchenko, G. N. (Tugoplavkie Soedin. Redkozemel. Met. Mater. Vses. 3rd Semin., Novosibirsk 1977 [1979], pp. 55/9; C.A. **93** [1980] No. 59818).

[4] Odintsov, V. V.; Kostetskii, I. I.; L'vov, S. N. (Izv. Akad. Nauk SSSR Neorgan. Materialy **9** [1973] 944/7; Inorg. Materials [USSR] **9** [1973] 844/7).

[5] Odintsov, V. V.; Moiseenko, L. L. (Elektron. Str. Fiz. Khim. Svoistva Splavov Soedin. Osn. Perekhodnykh Met. Dokl. 8th Simp., Kiev 1974 [1976], pp. 107/11; C.A. **87** [1977] No. 160803).

[6] Matthias, B. T.; Geballe, T. H.; Andres, K.; Corenzwit, E.; Hull, G. W.; Maita, J. P. (Science **159** [1968] 530).

[7] Fisk, Z. (Diss. Univ. California 1969, pp. 1/96, 4; Diss. Abstr. Intern. B **30** [1970] 4333).

[8] Etourneau, J. (J. Less-Common Metals **110** [1985] 267/81, 273).

[9] Etourneau, J.; Hagenmuller, P. (Phil. Mag. [8] B **52** [1985] 589/610, 599).

32.1.6.6 Electrical Properties

32.1.6.6.1 Electronic Structure

From XPS measurements it is concluded that Tm and Lu are trivalent in MB_{12}, however, YbB_{12} is a valence fluctuating compound with the valency $+2.9$ at room temperature. The intraatomic Coulomb correlation energy between Yb^{2+} and Yb^{3+} ions is about 6.0 eV; for details see "Rare Earth Elements" C11b, Section 32.1.8.16. The binding energy of the 4f level in TmB_{12} and LuB_{12} is 5.0 and 7.7 eV, respectively, Iga et al. [1]. The energy band structures of YB_{12}, YbB_{12}, and LuB_{12} were calculated by MO-LCAO methods. The results (see the figure in the paper) are used to explain the metallic character of YB_{12} and LuB_{12} which is determined by partial electron filling of the $(2t_{2g})^2$ and $(1t_{1g})^4$ levels, respectively. The semiconduction of YbB_{12} should be associated with the filling of the $(3e_g)^4$ level and its partial overlap with the adjacent $1a_{2u}$ and $2t_{2g}$ levels, Odintsov et al. [2]; they are at a distance of 0.120 and 0.121 eV from $3e_g$, Odintsov et al. [3]; the energy spectra are tabulated for the center of the first Brillouin zone by Goryachev et al. [4]. From a comparison of the fine structure of the B K_α emission band in B and MB_{12} (M = Dy to Lu), it is concluded that the 4f electrons do not participate in the formation of MB_{12}, Odintsov, Moiseenko [5]. The spectra indicate the absence of any donor-acceptor M–B interaction (M = Y, Dy to Lu) and the presence of two valence bands: one of M and one of B which interact through an almost empty conduction band, Odintsov et al. [6]. Electron configuration and bonding are considered in connection with thermionic emission studies by Samsonov et al. [7, 8].

References:

[1] Iga, F.; Takakuwa, Y.; Takahashi, T.; Kasaya, M.; Kasuya, T.; Sagawa, T. (Solid State Commun. **50** [1984] 903/5).

[2] Odintsov, V. V.; Paderno, Yu. B.; Goryachev, Yu. M. (Zh. Strukt. Khim. **12** [1971] 344/6; J. Struct. Chem. [USSR] **12** [1971] 323/5).

[3] Odintsov, V. V.; Paderno, Yu. B.; Goryachev, Yu. M. (Khim. Svyaz Krist. Poluprovodn. Polumet. **1973** 177/83; C.A. **81** [1974] No. 144812).

[4] Goryachev, Yu. M.; Odintsov, V. V.; Paderno, Yu. B. (Metallofizika No. 37 [1971] 29/36; C.A. **77** [1972] No. 81514).

[5] Odintsov, V. V.; Moiseenko, L. L. (Elektron. Str. Fiz. Khim. Svoistva Splavov Soedin. Osn. Perekhodnykh Met. Dokl. 8th Simp., Kiev 1974 [1976], pp. 107/11; C.A. **87** [1977] No. 160803).

[6] Odintsov, V. V.; Zhurakovskii, E. A.; Paderno, Yu. B.; Vasilenko, N. N.; Goryachev, Yu. M. (Konfiguratsionnye Predstavleniya Elektron. Str. Fiz. Materialoved. Mater. 2nd Nauchn. Semin. Konfiguratsionnoi Modeli Kondens. Sostoyaaniya Veshchestva, Lvov 1976 [1977], pp. 76/82; C.A. **89** [1978] No. 169399).

[7] Samsonov, G. V.; Okhremchuk, L. N.; Podchernyaeva, I. A.; Fomenko, V. S.; Odintsov, V. V. (Bor: Poluch. Strukt. Svoistva Mater. 4th Mezhdunar. Simp. Boru, Tbilisi 1972 [1974], pp. 245/7; C.A. **83** [1975] No. 140574).

[8] Samsonov, G. V.; Okhremchuk, L. N.; Podchernyaeva, I. A.; Fomenko, V. S.; Odintsov, V. V. (Izv. Akad. Nauk SSSR Neorgan. Materialy **10** [1974] 270/2; Inorg. Materials [USSR] **10** [1974] 231/3).

32.1.6.6.2 Resistivity. Magnetoelectric and Thermoelectric Properties

The electrical resistivity ϱ in $\mu\Omega \cdot cm$, the thermoelectric power S in $\mu V/K$, the Hall coefficient R_H in 10^{-4} cm^3/C, and the Nernst-Ettingshausen coefficient Q in 10^{-4} cm$^2 \cdot s^{-1} \cdot K^{-1}$ at 300 K of samples with 14 to 20% porositiy were measured and electron concentration n in 10^{22} cm^{-3}, mobility μ in cm$^2 \cdot V^{-1} \cdot s^{-1}$, and effective mass m* in m$_0$ are calculated:

MB$_{12}$	ϱ	$-S$	$-R_H$	Q	n	μ	m*
YB$_{12}$	11.56	3.2	4.82	2.17	1.29	41.7	1.54
TbB$_{12}$	16.65	2.8	4.73	2.09	1.32	28.4	1.88
DyB$_{12}$	13.32	2.9	4.78	1.88	1.31	35.9	1.50
HoB$_{12}$	13.85	3.3	4.74	1.73	1.32	34.2	1.54
ErB$_{12}$	14.18	1.6	4.71	1.74	1.32	33.2	1.27
TmB$_{12}$	15.95	1.9	4.76	1.82	1.31	29.8	1.47
YbB$_{12}$	184.4	3.9	9.91	2.30	0.63	5.2	9.37
LuB$_{12}$	12.93	2.8	4.79	2.08	1.30	37.0	1.54

Electron conductivity prevails, however, YbB$_{12}$ deviates from the other investigated MB$_{12}$. It is analogous to CeB$_6$ and a one-band concept seems not justified, Moiseenko, Paderno [1]. For similar values of ϱ, S, R$_H$, n, and μ obtained on samples with 15 to 20% porosity see Odintsov et al. [2], Goryachev et al. [3], of ϱ and S see Moiseenko [4] and of R$_H$ and Q see Moiseenko, Paderno [5]. Higher n and lower μ are tabulated by Nemchenko [6]. The temperature dependence of ϱ below room temperature is shown in **Fig. 40**, that of S in **Fig. 41**. ϱ is smaller than for M and MB$_6$ owing to more covalent B–B bonds. It changes only slightly from YB$_{12}$ to LuB$_{12}$ but some increase with increasing 4f electrons can be seen. The conductivity has metallic character (except YbB$_{12}$) and at low temperature, scattering occurs on both impurities and localized magnetic 4f states. The temperature behavior of S indicates that at least two types of current carriers are present in the conduction band [4]. From R$_H$ between 77 and 300 K it is also assumed that not only electrons are the current carriers [5]. Above room temperature the ϱ vs. T curves increase linearly except for YbB_{12}, see the figure for M = Y, Dy to Tm up to ~1000 K in Odintsov et al. [7], for M = Y, Tb to Lu up to 1173 K in [2], see also [3]. The course of S is monotonously more negative with increasing temperature in the same range [2, 3]. The resistivity of YbB_{12} has a maximum near 1073 K [2] and S has a minimum near 773 K [2, 3].

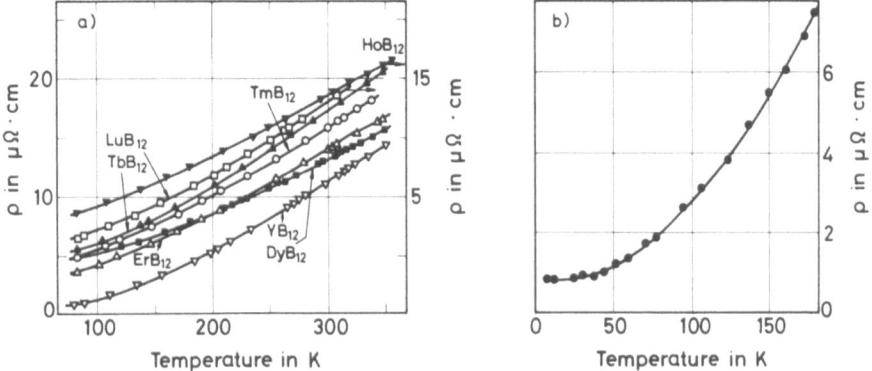

Fig. 40.   Temperature dependence of the electrical resistivity ϱ of vacuum-sintered MB_{12} samples with M = Y, Dy, Er, Tm, Tb (Fig. 40a), left hand scale), and Ho or Lu (Fig. 40a), right hand scale). The temperature dependence of ϱ for ErB_{12} in the low-temperature region is shown in Fig. 40b).

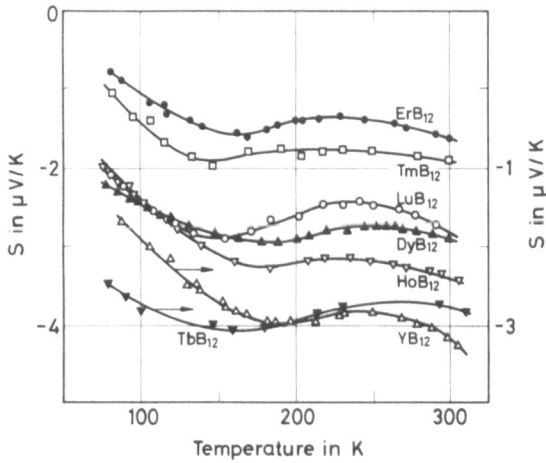

Fig. 41.   Temperature dependence of the thermoelectric power S of vacuum-sintered MB_{12} samples with M = Er, Tm, Lu, Dy, Ho (left hand scale) and Tb, Y (right hand scale).

Supraconducting transition temperatures $T_c = 0.39$, 4.7, and 0.48 K for ScB_{12}, YB_{12}, and LuB_{12} are reported. YB_{12} shows incomplete superconductive transition, Matthias et al. [8], Fisk [9].

References:

[1] Moiseenko, L. L.; Paderno, Yu. B. (Elektron. Str. Fiz. Khim. Svoistva Tugoplavkikh Soedin. Splavov Dokl. Vses. 9th Simp., Ivano-Frankovsk, Ukr. SSR, 1979 [1980], pp. 128/31; C.A. **94** [1981] No. 218173).

[2] Odintsov, V. V.; Paderno, Yu. B.; Goryachev, Yu. M. (Khim. Svyaz Krist. Poluprovodn. Polumet. **1973** 177/83; C.A. **81** [1974] No. 144812).

[3] Goryachev, Yu. M.; Odintsov, V. V.; Paderno, Yu. B. (Metallofizika No. 37 [1971] 29/36; C.A. **77** [1972] No. 81514).

[4] Moiseenko, L. L. (Ukr. Fiz. Zh. **27** [1982] 1340/3; C.A. **97** [1982] No. 206260).

[5] Moiseenko, L. L.; Paderno, Yu. B. (Izv. Vysshikh Uchebn. Zavedenii Fiz. **25** No. 10 [1982] 119/20; C.A. **97** [1982] No. 228212).

[6] Nemchenko, V. F. (Elektron. Str. Fiz. Khim. Svoistva Tugoplavkikh Soedin. Splavov Dokl. Vses. 9th Simp., Ivano-Frankovsk, Ukr. SSR, 1979 [1980], pp. 147/51; C.A. **95** [1981] No. 33856).

[7] Odintsov, V. V.; Lesnaya, M. I.; L'vov, S. N. (At. Energiya SSSR **35** No. 3 [1973] 194; At. Energy [USSR] **35** [1973] 834/5; C.A. **80** [1974] No. 7732).

[8] Matthias, B. T.; Geballe, T. H.; Andres, K.; Corenzwit, E.; Hull, G. W.; Maita, J. P. (Science **159** [1968] 530).

[9] Fisk, Z. (Diss. Univ. California 1969, pp. 1/96, 4; Diss. Abstr. Intern. B **30** [1970] 4333).

32.1.6.6.3 Thermionic Emission

The thermionic emission of compact MB$_{12}$ was measured. The temperature dependence of the work function Φ in eV is tabulated:

MB$_{12}$	$\Phi_0 + d\Phi/dT \cdot T$	T in K	Φ(1900 K)
YB$_{12}$	$3.78 + 4 \times 10^{-4}$ T	1650 to 2000	4.53
TbB$_{12}$	$d\Phi/dT \approx 1 \times 10^{-4}$ T	1770 to 1990	4.50
DyB$_{12}$	$3.50 + 5 \times 10^{-4}$ T	1650 to 1960	4.45
HoB$_{12}$	$3.67 + 4 \times 10^{-4}$ T	1750 to 2000	4.43
ErB$_{12}$	$3.95 + 2.5 \times 10^{-4}$ T	1770 to 2000	4.42
TmB$_{12}$	$3.12 + 6 \times 10^{-4}$ T	1680 to 1930	4.35
YbB$_{12}$	$3.84 + 2 \times 10^{-4}$ T	1700 to 1960	4.20
LuB$_{12}$	$3.77 + 2.8 \times 10^{-4}$ T	1550 to 1950	4.30

For powders, nonreproducible results are found. Φ is lowest for YbB$_{12}$ with the smallest number of s + d electrons.

Reference:

Samsonov, G. V.; Okhremchuk, L. N.; Podchernyaeva, I. A.; Fomenko, V. S.; Odintsov, V. V. (Izv. Akad. Nauk SSR Neorgan. Materialy **10** [1974] 270/2; Inorg. Materials [USSR] **10** [1974] 231/3, Bor: Poluch. Strukt. Svoistva Mater. 4th Mezhdunar. Simp. Boru, Tbilisi 1972 [1974], pp. 245/7; C.A. **83** [1975] No. 140574).

32.1.6.7 Electrochemical Behavior

The anodic oxidation polarization curves of paste electrodes consisting of 1% MB_{12} powder (5 to 20 µm) in C powder and Si organic binder were measured in aqueous 1 m $NaClO_4$ at pH 1.0 vs. an AgCl reference electrode. The current voltage curves display a definite maximum at $\varphi_{max} = 0.77$, 0.80, 0.78, 0.82, and 0.80 V for M = Dy, Ho, Er, Tm, and Lu, respectively. Thus MB_{12} show increased resistance to anodic oxidation compared to MB_4 ($\varphi_{max} \approx 0.1$ V) and MB_6 ($\varphi_{max} \approx 0.3$ V). The basic process is the ionization of the metal. The abrupt decrease of the current after reaching the maximum is attributed to diffusion limitations with depletion of the cathode surface from M.

Reference:

Tkach, A. V.; Paderno, Yu. B. (Dopov. Akad. Nauk Ukr. RSR Ser. B: Geol. Khim. Biol. Nauki **1984** No. 9, pp. 52/4; C.A. **102** [1985] No. 35077).

32.1.6.8 Chemical Behavior

Due to the rigid covalent B–B bonds, MB_{12} with M = Y, Ho, Er, Tm, and Lu show high oxidation resistance. The oxidation of porous samples in air begins at 823 to 873 K. At first, a sharp mass increase occurs, followed by stabilization after coating of the pore openings by oxide. The scale is dense, strongly adherent, and does not flake off, Odintsov, Yakovenko [1]. YB_{12}, "LaB_{12}", and ErB_{12} were heated in Al_2O_3 crucibles at 1500 to 2200 K in air. They show less oxidation resistance than the transition metal diborides but are more resistant than B, MB_4, and MB_6. In addition, all the compounds reacted with the crucible material, Hoyt [4].

MB_{12} with M = Y, Dy to Lu are resistant to HCl and H_2SO_4 and are decomposed by oxidizing acids. The insoluble residues (in %) are given in the following table after a) 24 h at room temperature and b) 1 h at the boiling point ($= 373$ to 383 K in dilute acids and 523 to 533 K in concentrated H_2SO_4):

MB_{12}	HCl(1:1)		HNO_3(1:1)		H_2SO_4(1:1)		aqua regia		H_2SO_4 conc.		H_2SO_4 (1:1) + HNO_3	
	a	b	a	b	a	b	a	b	a	b	a	b
YB_{12}	100	99.5	12	11.5	100	99.3	11	10	100	99.5	11	0
DyB_{12}	100	99.5	0	2 to 3	99.8	99.8	0	1.2	—	—	11	0
HoB_{12}	100	98.0	0	0	100	99.0	0	0	—	—	0	0
ErB_{12}	99.8	99.3	0	0	100	99.1	0	0	—	—	2.0	0
TmB_{12}	99.5	99.5	0	0	100	99.3	0	0	100	95.9	0	0
YbB_{12}	100	99.4	17	6 to 7	100	40.0*)	17	1 to 2	—	—	15	0
LuB_{12}	100	99.0	0	0	—	99.5	0	0	—	—	2.0	0

*) Obviously a misprint in [2].

The insoluble residue was invariably identical with the original boride. YB_{12} and YbB_{12} exhibit higher resistance than the other studied MB_{12}, for instance MB_{12} with M = Ho, Er, Tm, and Lu decompose in HNO_3 and aqua regia in 7 to 8 h at room temperature and in 45 to 60 min at the boiling point. In a mixture of concentrated H_2SO_4 and HNO_3 at the boiling point, YB_{12} is decomposed completely within 1 h, TmB_{12} only to 19%, Yukhimenko et al. [2].

From X-ray investigations in the systems ErB$_{12}$–DyB$_{12}$ and YbB$_{12}$–DyB$_{12}$, it is concluded that all isotypic MB$_{12}$ show complete solid solubility. Solid solubility of hypothetic LaB$_{12}$, SmB$_{12}$, GdB$_{12}$ in stable MB$_{12}$ is very limited, Schwetz et al. [3].

References:

[1] Odintsov, V. V.; Yakovenko, S. N. (Poroshkovaya Metal. **1984** No. 1, pp. 71/3; Soviet Powder Met. Metal Ceram. **23** [1984] 68/9).
[2] Yukhimenko, E. V.; Odintsov, V. V.; Kotlyar, E. E.; Paderno, Yu. B. (Poroshkovaya Metal. **1971** No. 11, pp. 52/5; Soviet Powder Met. Metal Ceram. **1971** 895/7).
[3] Schwetz, K.; Ettmayer, P.; Kieffer, R.; Lipp, A. (Radex Rundschau **1972** 257/65, 259).
[4] Hoyt, E. W. (Proc. 2nd Conf. Rare Earth Res., Glenwood Springs, Colo., 1961 [1962], pp. 287/99, 292).

32.1.7 Comparative Data for MB$_{\sim 66}$

32.1.7.1 General

The general formula MB$_{66}$, commonly used for the most B-rich compound in the M–B systems, corresponds to the random occupancy of 50% of the metal sites in the structure, see f.i. YB$_{66}$, the most intensely studied compound of this row. Sometimes the name hectoboride is used. A discussion on the composition of the first discovered "YB$_{70}$" (1960), see p. 158, up to ErB$_{66}$ (1973), see "Rare Earth Elements" C 11b, Section 32.1.8.14, is given in Spear [1].

During X-ray and metallographic investigations of quasi-binary MB$_{12}$–M′B$_{12}$ systems, the existence of mixed hectoboride phases is assumed for DyB$_{12}$–ErB$_{12}$ and DyB$_{12}$–YbB$_{12}$ as well as for DyB$_{12}$ with hypothetical LaB$_{12}$ and SmB$_{12}$ and for YB$_{12}$ with hypothetical GdB$_{12}$, Schwetz et al. [2].

References:

[1] Spear, K. E. (in: Alper, A. M., Refractory Materials, Phase Diagrams, Vol. 4, Academic, New York 1976, pp. 91/159, 102/3).
[2] Schwetz, K.; Ettmayer, P.; Kieffer, R.; Lipp, A. (Radex Rundschau **1972** 257/65).

32.1.7.2 Preparation

Compounds MB$_{66}$ with M = Sm, Gd, and Yb are synthesized in 3 steps: Powder preparation, cold pressing, and zone melting in He or vacuum, Golikova, Tadzhiev [1]. Arc melting of the mixed elements in purified Ar on a H$_2$O cooled Cu hearth was used for M = Y, La, Ce, Pr, Nd, Sm, Gd, Dy, Er, and Yb. For M = La, Ce, and Pr the MB$_{66}$ is not obtained. It is expected, however, that all M from Nd to Lu form MB$_{66}$, Spear, Solovyev [2]. In the same way, so-called hectoborides with M = Y, Sm, Gd to Lu were obtained, those with M = La, Ce, Pr, Nd, and Eu with an atomic radius >1.822 Å were not obtained, Schwetz et al. [3]. For the composition of these phases see Spear [4]. Crystals of GdB$_{66}$ and DyB$_{66}$ were prepared by the floating-zone method in a high-frequency furnace, Golikova et al. [5].

References:

[1] Golikova, O. A.; Tadzhiev, A. (J. Less-Common Metals **82** [1981] 169/71).
[2] Spear, K. E.; Solovyev, G. I. (NBS-SP-364 [1972] 597/604; C.A. **77** [1972] No. 157317).
[3] Schwetz, K.; Ettmayer, P.; Kieffer, R.; Lipp, A. (J. Less-Common Metals **26** [1972] 99/104).
[4] Spear, K. E. (in: Alper, A. M., Refractory Materials, Phase Diagrams, Vol. 4, Academic, New York 1976, pp. 91/159, 102).
[5] Golikova, O. A.; Domashevskaya, E. P.; Tadzhiev, A.; Terekhov, V. A. (Fiz. Tverd. Tela [Leningrad] **30** [1988] 899/901; Soviet Phys.-Solid State **30** [1988] 521/2).

32.1.7.3 Crystallographic Properties

The high boron compounds $MB_{\sim 66}$ crystallize in the fcc YB_{66} structure, see p. 159, space group $Fm\overline{3}c\text{-}O_h^6$ (No. 226), Z = 24. The basic unit of the structure is a 156-atom supericosahedron of 13 B_{12} icosahedra (one B_{12} icosahedron in the center of twelve B_{12} icosahedra). The crystal framework consists of two interpenetrating fcc supericosahedra, differing by 90° rotation of both supericosahedra. There are eight of these supericosahedra and thus 1248 icosahedra B atoms in the cell. The M positions and the locations of the remaining B atoms can be most easily pictured by referring to an octant of the cell shown in **Fig. 42**, p. 114. Six metal atom sites exist which are statistically half-filled. The center of the octant is occupied by a 36- or a 48-B atom group in a random fashion, yielding the average formula $M_3[B_{12}(B_{12})_{12}]_1$-$(B_{36})_{1/2}(B_{48})_{1/2}$ for one octant. The supericosahedron units are not isolated but tightly bonded by B–B bond lengths of the same order as those internal to each B_{12} icosahedron. The following lattice parameters a in Å are summarized by Spear [1]:

M in MB_{66}	M-rich	Ref.	single phase	Ref.	B-rich	Ref.
Y	23.451	[2]			23.446	[3]
					23.440	[4]
Nd	23.508	[2]				
Pm			23.495	[1]*)		
Sm			23.474	[2]	23.487	[3]
Gd	23.476	[2]			23.474	[3]
Tb					23.457	[3]
Dy	23.466	[1]	23.441	[3]	23.419	[1]
			23.422	[2]		
Ho					23.441	[3]
Er	23.428	[3]	23.438	[2]		
			23.440	[5]		
Tm					23.433	[3]
Yb			23.415	[2]	23.422	[3]
Lu					23.412	[3]

*) Estimated by [1].

A summary of the lattice structure of MB_{66} is also given by Golikova [6]. From the linear dependence of the lattice constants on the M^{3+} radii (see **Fig. 43**, p. 114), trivalent M is indicated and from the slope a more rigid B framework than in other M borides. Much B–B bonding is similar to that in pure B [1]. Observed and calculated intensities of the powder diagram of "DyB$_{66}$" and HoB$_{66}$ are given in [3].

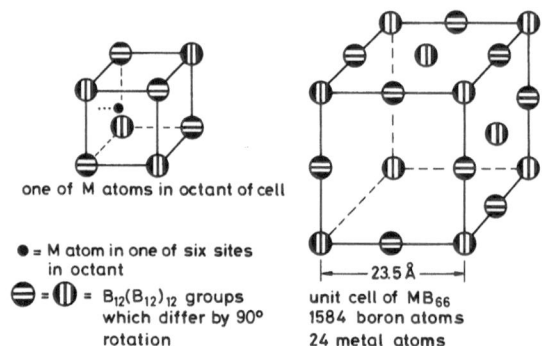

one of M atoms in octant of cell

● = M atom in one of six sites in octant

⊜ = ⓪ = B$_{12}$(B$_{12}$)$_{12}$ groups which differ by 90° rotation

unit cell of MB$_{66}$
1584 boron atoms
24 metal atoms

Fig. 42.   Schematic drawing of the face-centered cubic MB$_{66}$ structure.

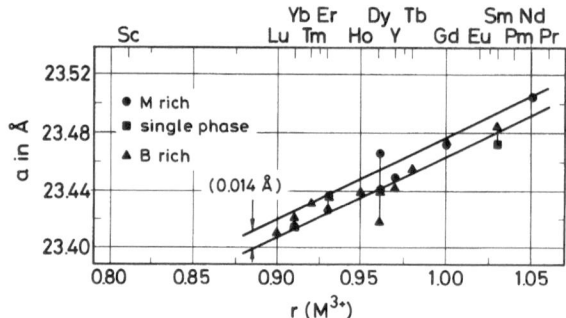

Fig. 43.   Lattice constants of MB$_{66}$ versus M^{3+} ionic radii.

The structure is, to some extent, similar to the amorphous structure of a continuous disordered network with hundreds of atoms and changes of the short-range order in the unit cell, Golikova [7]. The amorphous concept was considered earlier by Golikova [8], Golikova, Tadzhiev [9]. The variations of the short-range order are represented by the B–B bond lengths increasing from the center to the outer icosahedra in the range 1.72 to 2.17 Å, Golikova [7], Golikova et al. [10].

References:

[1] Spear, K. E. (in: Alper, A. M., Refractory Materials, Phase Diagrams, Vol. 4, Academic, New York 1976, pp. 91/159, 114/6, 120/6).
[2] Spear, K. E.; Solovyev, G. I. (NBS-SP-364 [1972] 597/604; C. A. **77** [1972] No. 157317).
[3] Schwetz, K.; Ettmayer, P.; Kieffer, R.; Lipp, A. (J. Less-Common Metals **26** [1972] 99/104).
[4] Richards, S. M.; Kasper, J. S. (Acta Cryst. B **25** [1969] 237/51).
[5] Nichols, M. C.; Mar, R. W. (Inorg. Chem. **12** [1973] 1710/1).
[6] Golikova, O. A. (Phys. Status Solidi A **101** [1987] 277/314).

[7] Golikova, O. A. (Fiz. Tverd. Tela [Leningrad] **29** [1987] 2869/72; Soviet Phys.-Solid State **29** [1987] 1652/4).

[8] Golikova, O. A. (Phys. Status Solidi A **51** [1979] 11/40, 15, 31).

[9] Golikova, O. A.; Tadzhiev, A. (J. Non-Cryst. Solids **87** [1986] 64/9).

[10] Golikova, O. A.; Domashevskaya, E. P.; Tadzhiev, A.; Terekhov, V. A. (Fiz. Tverd. Tela [Leningrad] **30** [1988] 899/901; Soviet Phys.-Solid State **30** [1988] 521/2).

32.1.7.4 Hardness. Thermal Properties

MB_{66} are extremely hard. The Vickers microhardness at 200 g load was determined by Spear, Solovyev [1], the Knoop microhardness at 100 g load by Schwetz et al. [2]. H_μ in kp/mm^2 ($\hat{=}\,10^7$ Pa) of polycrystalline samples is as follows:

MB_{66}	YB_{66}	NdB_{66}	SmB_{66}*)	GdB_{66}	TbB_{66}	DyB_{66}	HoB_{66}	ErB_{66}	TmB_{66}	YbB_{66}	LuB_{66}
$H_\mu \pm 200$ [1] ...	3860	3910	3610	3660	—	3840	—	3635	—	3960	—
$H_\mu \pm 50$ [2]	2620	—	2510	2460	2340	2540	2340	2440	2300	2410	2460

*) Sample annealed 2 h at 2273 K; before annealing $H_\mu = 3710$ kp/mm^2 ($\hat{=}\,37$ GPa).

The melting temperature is 2423 ± 100 K for MB_{66} with M = Y, Nd, Sm, Gd, Dy, Er, and Yb [1], 2298 to 2423 K for all MB_{66}, Spear [3].

The thermal conductivity λ vs. temperature between ~300 and 900 K of GdB_{66} is shown in **Fig. 44** as a typical example for the group of MB_{66}. It is constant over a wide temperature range and thus comparable to that of amorphous B, Golikova, Tadzhiev [4], see also Golikova [5, 6]. The behavior of λ in the low-temperature region down to ~0.1 K is shown in **Fig. 45**, p. 116, for YB_{66} and compared with literature data on YB_n (n = 61.7 and 66), amorphous B and amorphous SiO_2. The behavior of YB_n resembles greatly that of glasses with λ approximately proportional to T^2 at low temperature. Very strong phonon scattering leads to nearly temperature-independent λ above 10 K, Türkes et al. [7], see also Golikova [8].

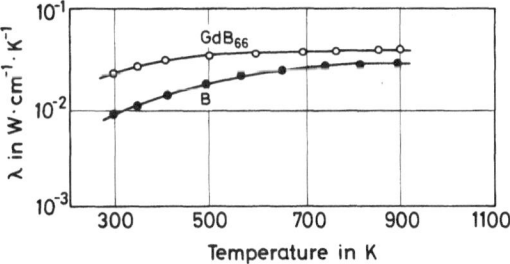

Fig. 44.   Temperature dependence of the thermal conductivity λ of zone-melted GdB_{66} and amorphous boron.

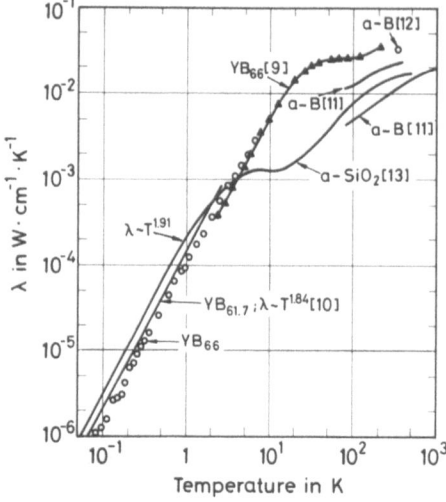

Fig. 45. Temperature dependence of the thermal conductivity λ of slightly nonstoichiometric YB$_{66}$ single crystals and of amorphous boron and SiO$_2$ (see text).

References:

[1] Spear, K. E.; Solovyev, G. I. (NBS-SP-364 [1972] 597/604; C.A. **77** [1972] No. 157317).

[2] Schwetz, K.; Ettmayer, P.; Kieffer, R.; Lipp, A. (J. Less-Common Metals **26** [1972] 99/104).

[3] Spear, K. E. (in: Alper, A. M., Refractory Materials, Phase Diagrams, Vol. 4, Academic, New York 1976, pp. 91/159, 134).

[4] Golikova, O. A.; Tadzhiev, A. (J. Non-Cryst. Solids **87** [1986] 64/9).

[5] Golikova, O. A. (Phys. Status Solidi A **51** [1979] 11/40, 31).

[6] Golikova, O. A. (Tr. Vses. Konf. Fiz. Poluprovodn., Baku 1982, Vol. 2, pp. 140/1; C.A. **99** [1983] No. 114239).

[7] Türkes, P. R. H.; Swartz, E. T.; Pohl, R. O. (AIP [Am. Inst. Phys.] Conf. Proc. No. 140 [1986] 346/61, 350/2).

[8] Golikova, O. A. (Phys. Status Solidi A **101** [1987] 277/314, 301/4).

[9] Slack, G. A.; Oliver, D. W.; Horn, F. H. (Phys. Rev. [3] B **4** [1971] 1714/20).

[10] Raychaudhuri, A. K.; Peech, J. M.; Pohl, R. O. (Phonon Scattering Condens. Matter Proc. 3rd Intern. Conf., Brown Univ., Providence 1979 [1980], p. 45/8).

[11] Golikova, O. A.; Zaitsev, V. K.; Orlov, V. M.; Petrov, A. V.; Stilbans, L. S.; Tkalenko, E. N. (Phys. Status Solidi A **21** [1974] 405/12).

[12] Talley, C. P.; Line, L. E.; Overman, Q. D. (Boron Synth. Struct. Prop. Proc. Conf., Asbury Park, N. J., 1959 [1960], p. 94 from [7]).

[13] Raychaudhuri, A. K. (Diss. Cornell Univ. 1980 from [7]).

32.1.7.5 Magnetic, Electrical, and Optical Properties

The semiconducting compounds HoB$_{66}$, ErB$_{66}$, and YbB$_{66}$ have negative paramagnetic Curie temperatures $\Theta_p = -5$, -5, and -30 K with effective moments very close to that of the free M^{3+}, Etourneau [1], Etourneau, Hagenmuller [2].

The density of states for MB$_{66}$ is proposed to be similar to that of amorphous B. The Fermi level E$_F$ is situated at the tail of the density of states, which is higher when E$_F$ is closer to the mobility edge of the valence band E_v (for $\Delta E = E_F - E_v$, see table below, together with T_0, which was derived from the temperature dependence of the conductivity and is about the reciprocal density of localized states at E_F). According to this model the tail for all MB_{66} is the same, but the position of E_F is different. YbB_{66} has lowest ΔE and T_0 values, GdB_{66} highest, and SmB_{66} takes an intermediate position. Calculations of the density of states effective mass for 300 K gives the very low value $m^* \approx 0.02\ m_0$, Golikova et al. [3], or $0.01\ m_0$, Golikova, Tadzhiev [4]. The density of states model is also discussed by Golikova, Tadzhiev [5]. The temperature coefficient of ΔE is $\gamma \approx 10^{-4}$ eV/K [5], see also [6]. The top of the valence band is at 180.0 and 178.6 eV relative to the core 1s level of B for GdB_{66} and DyB_{66}, Golikova et al. [7].

The electrical conductivity σ and the thermoelectric power S were measured at room temperature to ~ 1500 K for SmB_{66}, GdB_{66}, and YbB_{66}, Golikova et al. [8], σ at 100 to 300 K and S at 300 K, Golikova et al. [3 to 5]. In addition, σ and S up to ~ 1000 K are given for SmB_{66} by [4] and for DyB_{66} by [5]. The Hall coefficient was measured for SmB_{66}, GdB_{66}, and YbB_{66} at 100 to 300 K [3]. The results (of presumably zone-melted samples) at 300 K are summarized in the following table together with the energy gap E_g from σ or S, the Hall mobility μ, the carrier concentration n, the reciprocal density of states at the Fermi level T_0 (from the proportionality of σ and $\exp[(T_0/T)^{1/4}]$), and the activation energy ΔE at $T = 0$ K:

MB_{66}	YB_{66}	SmB_{66}	GdB_{66}	DyB_{66}	YbB_{66}
σ in $\Omega^{-1} \cdot cm^{-1}$ ...	3×10^{-3}	10×10^{-3}	2×10^{-3}	7.5×10^{-3}	6×10^{-3}
S in $\mu V/K$	340	100	390	140	270
$E_g(\sigma)$ in eV	1.00	0.80	0.87	0.72	1.27
$E_g(S)$ in eV	0.88	0.63	0.68	0.70	1.10
μ in $cm^2 \cdot V^{-1} \cdot s^{-1}$.	10	15	15	10	5
n in cm^{-3}	—	2.7×10^{16}	0.15×10^{16}	—	1.2×10^{16}
T_0 in K	—	2×10^7	4×10^7	—	0.12×10^7
ΔE in eV	—	0.15	0.20	—	0.10

The temperature dependence of σ and n is shown in **Fig. 46** [3, 4], see also [5]. As a typical example for the high-temperature behavior σ and S of DyB_{66} are shown in **Fig. 47**, p. 118, [5], see also Golikova et al. [9]. These semiconductors cannot be considered as heavily doped β-rhombohedral B as concluded for YB_{66}, see p. 161. At low temperature the activation energy is variable and a hopping type conductivity is assumed similar to that in amorphous semiconductors with

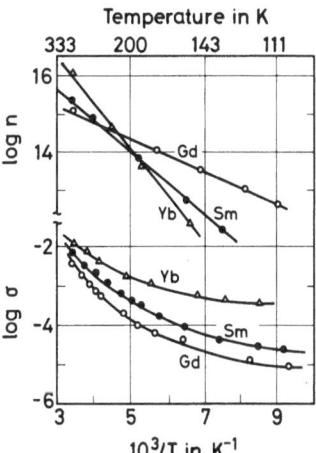

Fig. 46. Temperature dependence of the electrical conductivity σ (in $\Omega^{-1} \cdot cm^{-1}$) and of the carrier concentration n (in cm^{-3}) of MB_{66} with M = Sm, Gd, and Yb.

transitions from E_v to E_F (Mott's law is obeyed). In the higher temperature range transition to band conduction mechanism takes place with transitions from E_v to the conduction band E_c (intrinsic conduction), see the review by Golikova [6, p. 301], [3 to 5]. The calculated mobility of current carriers at E_v is 1 to 10 cm$^2 \cdot$ V$^{-1} \cdot$ s^{-1} at 300 K, close to the Hall mobility. This shows that the main contribution to conduction at room temperature comes from current carriers moving on nonlocalized states of the valence band [5, 6]. The E_g value obtained from optical studies on GdB$_{66}$ at 300 K (see **Fig. 48**) agrees with those given in the table above, Golikova et al. [9].

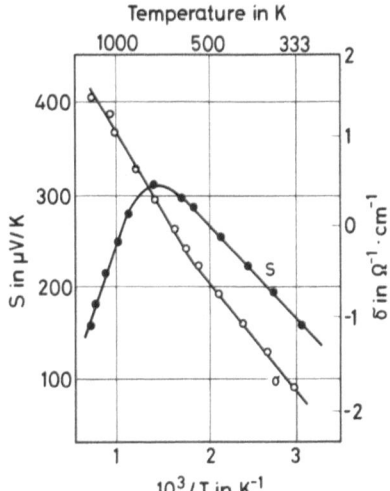

Fig. 47.   Temperature dependence of the electrical conductivity σ and of the thermoelectric power S of DyB$_{66}$.

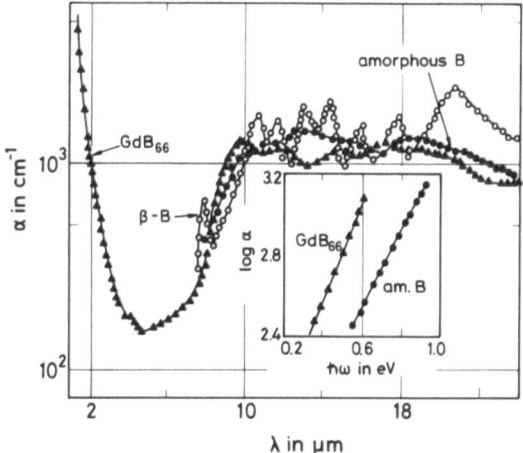

Fig. 48.   Optical absorption coefficient α of GdB$_{66}$, β B, and amorphous B at 300 K. The inset shows the long-wavelength tails.

The IR absorption spectra of MB$_{66}$ are compared with that of amorphous B and of β-rhombohedral B (with M = Gd as a typical example of the row), see Fig. 48. Urbach's tail is observed and there is no distinct structure similar to the spectra of amorphous solids [5, 6, 9].

The spectra of α- and β-rhombohedral B, GdB_{66} and amorphous B illustrate the increasing complexity of the crystal structure in this direction, Golikova [10].

References:

[1] Etourneau, J. (J. Less-Common Metals **110** [1985] 267/81, 273).
[2] Etourneau, J.; Hagenmuller, P. (Phil. Mag. [8] B **52** [1985] 589/610, 599).
[3] Golikova, O. A.; Orlov, V. M.; Panteleeva, G. V.; Tadzhiev, A. A.; Dzhafarov, E. O. (Fiz. Tekhn. Poluprov. **14** [1980] 1405/8; Soviet Phys. Semicond. **14** [1980] 833/4).
[4] Golikova, O. A.; Tadzhiev, A. (J. Less-Common Metals **82** [1981] 169/71).
[5] Golikova, O. A.; Tadzhiev, A. (J. Non-Cryst. Solids **87** [1986] 64/9).
[6] Golikova, O. A. (Phys. Status Solidi A **101** [1987] 277/314, 293/4, 300/1, 305).
[7] Golikova, O. A.; Domashevskaya, E. P.; Tadzhiev, A.; Terekhov, V. A. (Fiz. Tverd. Tela [Leningrad] **30** [1988] 899/901; Soviet Phys.-Solid State **30** [1988] 521/2).
[8] Golikova, O. A.; Orlov, V. M.; Panteleeva, G. V.; Tadzhiev, A. A. (Izv. Akad. Nauk SSSR Neorgan. Materialy **16** [1980] 1118/9; C.A. **93** [1980] No. 86318).
[9] Golikova, O. A.; Orlov, V. M.; Tadzhiev, A. A. (Sint. Svoistva Soedin. Redkozem. Elem. **1982** 95/102; C.A. **99** [1983] No. 204131).
[10] Golikova, O. A. (Fiz. Tverd. Tela [Leningrad] **29** [1987] 2869/72; Soviet Phys.-Solid State **29** [1987] 1652/4).

32.1.8 Data for Individual Borides

32.1.8.1 Scandium Borides

Additional data for ScB_2 are found in Section 32.1.1, on p. 2.

General Reference:

Gschneidner, K. A., Jr. in: Horovitz, C. T., Scandium, Its Occurrence, Chemistry, Physics, Metallurgy, Biology, and Technology, Academic, London – New York 1975, pp. 152/251, 156, 159/62.

32.1.8.1.1 The Sc–B System

The assessed provisional phase diagram is given in **Fig. 49**, p. 120, from Massalski [1]. Earlier phase diagrams are given by Spear [2 to 4]. According to the figures in [2, 3], vaporization of the Sc-rich compositions sets in above ~ 2800°C. The system Sc–B differs from other lanthanide boron systems, which often exhibit MB_4, MB_6, and "MB_{66}". Earlier, the existence of ScB_4 and ScB_6 was reported, see for instance Samsonov et al. [5], Samsonov [6]. At ambient pressure ScB_2 and ScB_{12} melt congruently at 2250 and 2040°C, respectively, according to the figure. In vacuum ScB_2 begins to sublime without decomposition at 1800°C, Etourneau et al. [7]. A high-boron phase, i.e., a solid solution of Sc in β B with the approximate composition ScB_{28} is reported by Callmer [8], see also Carlsson, Lundström [9]. The maximum solubility of Sc in β B is 3.5 at% [8] according to [1]. The solid solution of Sc in rhombohedral β B appears at $B:Sc \geqq 20$ and is the only phase above $B:Sc = 100$ [7], Peshev et al. [10].

The heat of formation for hypothetical ScB was calculated with the help of electronegativities and density parameters of the components as $\Delta H = -80$ kJ/mol, Miedema et al. [11]. The values $\Delta H = -54$, -77, and -76 kJ/mol were calculated by Niessen, de Boer [12] according to the semiempirical model proposed by Miedema et al. [13] for Sc_2B, ScB, and ScB_2.

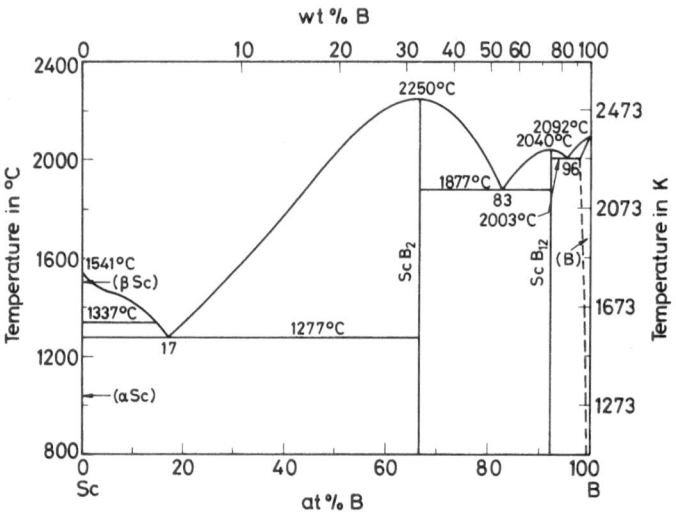

Fig. 49.   The Sc–B phase diagram.

References:

[1] Massalski, T. B. (Binary Alloy Phase Diagrams ASM 1986, Vol. 1, Metals Park, Ohio, pp. 381/2).

[2] Spear, K. E. (in: Alper, A. M., Refractory Materials, Phase Diagrams, Vol. 4, Academic, New York 1976, pp. 91/159, 153/4).

[3] Spear, K. E. (in: Matkovich, V. I., Boron and Refractory Borides, Springer, Berlin – New York 1977, pp. 439/56, 454).

[4] Spear, K. E. (NBS-SP-496-Vol. 2 [1978] 744/62, 758; C.A. **89** [1978] No. 31531).

[5] Samsonov, G. V.; Verkhoglyadova, T. S.; Vdovenko, S. A. (Vopr. Teor. Primen. Redkozem. Metal. Akad. Nauk SSSR **1964** 163/5; JPRS-28849 [1965] 213/6; N.S.A. **19** [1965] No. 23014).

[6] Samsonov, G. V. (Tugoplavkie Soedineniya Redkozemel'nykh Metallov s Nemetallami, Metallurgiya, Moscow 1964; High-Temperature Compounds of Rare Earth Metals with Nonmetals, Consultants Bureau, New York 1965, pp. 1/280, 1/100, 60/4).

[7] Etourneau, J.; Mercurio, J.-P.; Naslain, R.; Hagenmuller, P. (Colloq. Intern. Centre Natl. Rech. Sci. [Paris] No. 205 [1972] 429/38).

[8] Callmer, B. (J. Solid State Chem. **23** [1978] 391/8).

[9] Carlsson, J.-O.; Lundström, T. (J. Less-Common Metals **22** [1970] 317/20).

[10] Peshev, P.; Etourneau, J.; Naslain, R. (Mater. Res. Bull. **5** [1970] 319/27).

[11] Miedema, A. R.; de Boer, F. R.; Boom, R. (Proc. 12th Rare Earth Res. Conf., Vail, Colo., 1976, Vol. 2, pp. 838/46).

[12] Niessen, A. K.; de Boer, F. R. (J. Less-Common Metals **82** [1981] 75/80).

[13] Miedema, A. R.; de Châtel, P. F.; de Boer, F. R. (Physica B+C **100** [1980] 1/28).

32.1.8.1.2 ScB$_2$

Formation and Preparation

The compound was prepared by a borothermal reaction of Sc$_2$O$_3$ in vacuum at 1473 to 2473 K according to $2Sc_2O_3 + 14B \rightarrow 4ScB_2 + 3B_2O_2$. A very pure product was obtained in 1 h at 2073 K, Etourneau et al. [1]. A vacuum of 10^{-6} Torr (~133 µPa) and a Ta crucible were used by Peshev et al. [2]. The borothermal reduction of Sc$_2$O$_3$ was also used by Przybylska et al. [3], Samsonov [4]; the reaction with B$_4$C according to $Sc_2O_3 + B_4C + 2C \rightarrow 2ScB_2 + 3CO$ has a certain advantage over the above reaction with B only at >2173 K according to theoretical considerations. In practice, 2 h at 2123 K were employed, Samsonov et al. [5] or 1.5 h at 2173 K, Samsonov [7, p. 60]. Direct synthesis with simultaneous hot pressing was used by Shulishova, Shcherbak [6]. Sc and B powders were mixed and pelletized. The pellets were placed in ZrO$_2$ or recrystallized Al$_2$O$_3$ crucibles and heated under Ar in an induction furnace. After crushing and grinding, the powder was hot pressed in graphite molds under Ar at >1673 K and >2000 psi ($\triangleq$13.8 MPa). Surface contaminants from the graphite mold were removed by grinding, McAlister et al. [20].

Theoretical considerations indicated that the formation of ScB$_2$ according to reactions (1) to (4):

$$(1) \quad 0.5\,Sc_2O_3 + B_2O_3 + 4.5\,C \rightarrow ScB_2 + 4.5\,CO$$
$$(2) \quad 0.5\,Sc_2O_3 + 0.5\,B_4C + C \rightarrow ScB_2 + 1.5\,CO$$
$$(3) \quad 0.5\,Sc_2O_3 + 2\,B + 1.5\,C \rightarrow ScB_2 + 1.5\,CO$$
$$(4) \quad 0.5\,Sc_2O_3 + 3.5\,B \rightarrow ScB_2 + 1.5\,BO$$

is possible at >1591, 1587, 1495, and 2137 K, Peshev [14].

Crystallographic and Mechanical Properties

According to X-ray powder investigations, ScB$_2$ is hexagonal, AlB$_2$ type, space group P6/mmm-D$_{6h}^1$ (No. 191), a = 3.148, c = 3.516 Å, Etourneau et al. [1], see also Peshev et al. [2]. a = 3.146, c = 3.517 Å, c/a = 1.12; Z = 1 is given in a summary by Samsonov [7, p. 38] according to Zhuravlev, Stepanova [8], see also Zhurakovskii, Dzeganovskii [9], Shulishova, Shcherbak [6]. The atomic positions are: Sc in 1(a) (0,0,0) and B in 2(d) (1/3,2/3,1/2) and (2/3,1/3,1/2). Interplanar spacings and intensities of the powder diagram are tabulated [8]. Structural data of ScB$_2$ in comparison to MB$_2$ with M = Ti, V, Cr, and Mn indicate that the relatively large Sc atoms cause a significant stretching of the B–B bonds with a corresponding compression of Sc–Sc, Topor, Kleppa [13, pp. 1012/4].

The density is $D_{exp} = 3.642 \pm 0.005$ g/cm^3 and $D_{calc} = 3.665$ g/cm^3, Etourneau et al. [1], Peshev et al. [2]. The measured density of hot-pressed samples was given earlier as 3.56 g/cm^3, Samsonov [4], 3.65 g/cm^3, Samsonov [7, p. 60], Kudintseva et al. [11] and $D_{calc} = 3.67$ g/cm^3, Zhuravlev, Stepanova [8].

The microhardness at a 50 g load is 1742 ± 337 kg/mm^2 (~17.1 GPa) for sintered ScB$_2$ [4], 1330 ± 80 kg/mm^2 (~13.0 GPa) [5], [7, p. 44]; 2630 kg/mm^2 (~25.8 GPa) at a 200 g load, Samsonov, Paderno [12].

Thermal Properties

The mean thermal expansion coefficient determined by X-rays at 293 to 873 K is $\alpha = (6.8 \pm 0.5) \times 10^{-6}$ K^{-1} along a and $\alpha = (7.6 \pm 0.5) \times 10^{-6}$ K^{-1} along c, Zhuravlev, Stepanova [15]. The mean value $\bar{\alpha} = 9.48 \times 10^{-6}$ K^{-1} at 293 to 1073 K was given earlier by Samsonov [4]; $\bar{\alpha} = (7$ to $8) \times 10^{-6}$ K^{-1} at 343 to 1273 K, Samsonov et al. [5].

In vacuum of 10^{-6} Torr ($\triangleq 133$ µPa) ScB_2 begins to sublime without decomposition at 2073 K and at 2473 K the weight loss is $\sim 50\%$ after 30 min (a weight loss vs. temperature curve is given) [1, 2].

The standard enthalpy of formation $\Delta H° = -307.0 \pm 14.8$ kJ/mol ($\triangleq 73.4 \pm 3.5$ kcal/mol) was obtained for $Sc(s) + 2B(s) \rightarrow ScB_2(s)$ from the calorimetrically determined melting and solution enthalpies in liquid Pt–Sc–B alloy at 1400 K. The systematic variation of $\Delta H°$ in the first-row transition metal diborides is discussed in relation to the structural parameters and the exchange in M–M, B–B, and M–B bonding, Topor, Kleppa [13]. The thermodynamic data $\Delta H^°_{298} = -264.8$, $\Delta G^°_{298} = -279.5$, both in kJ/mol, and $\Delta S^°_{298} = 49.4$ J·mol^{-1}·K^{-1} were estimated from heats of reaction, Peshev [14].

ScB_2 melts at 2523 K [4]; see also Fig. 49, p. 120. It melts above 2500 K as taken from a figure given by Topor, Kleppa [13]. The low-temperature heat capacity is $C_p = \gamma T + \alpha T^3$ with $\gamma = 2.2$ mJ·K^{-2}·mol^{-1} and $\alpha = 1.175 \times 10^{-2}$ mJ·K^{-4}·mol^{-1} at 1.3 to 10 K; a figure C_p/T vs. T^2 is given. The Debye temperature is $\Theta_D = 550$ K, Castaing et al. [16], however, Θ_D truely is 790 K, Gschneidner [17].

Magnetic and Electrical Properties

The magnetic susceptibility χ was measured at 77 to 625 K. It decreases only very slightly and is about $\chi = 100 \times 10^{-6}$ cm^3/mol in the whole temperature range. Therefore ScB_2 seems to be a Pauli paramagnet, Peshev et al. [2].

The isotropic Knight shifts K, electrical quadrupole coupling constants e^2qQ/h, and the line widths ΔH derived from the central resonance of ^{45}Sc and ^{11}B are essentially temperature independent between 4 and 300 K: With respect to a concentrated solution of $Sc(NO_3)_3$ at 16 MHz K(^{45}Sc) = 0.07%, e^2qQ/h = 6.1(2) to 6.2(3) MHz, ΔH = 38(3) to 40(3) Oe. With respect to triethyl borate K(^{11}B) = 0.00(1) to $-0.006(8)\%$, e^2qQ/h = 0.50(6) to 0.53(7) MHz, ΔH = 12 Oe field independent (frequency range from 4 to 21.5 MHz). K(Sc) is lower than for metallic Sc due to reduced orbital and Pauli contributions. From comparison within the 3d series in ScB_2 the d-electron effects are negligible, Carter, Swartz [18]; K_{ScB_2} to K_{Sc} = 0.23, Barnes et al. [19]. The difference between calculated and measured shifts of B in ScB_2, TiB_2, VB_2, and CrB_2 vs. the calculated M d-orbital densities at the Fermi level is linear. This behavior suggests that the negative B Knight shifts result from interatomic conduction electron polarization, McAlister et al. [20].

The electronic structure of ScB_2 was determined by an LCAO-MO approach and compared with that of the first-row transition metal diborides. The rigid band concept (see McAlister et al. [20]) is certainly justified. The lowest occupied bands are very broad and associated with B–B and Sc–B bonding. These merge in the region of the Fermi level into a series of flat bands which are localized on the Sc 3d orbital. The higher vacant orbitals have a wide energy range owing to Sc 4s and 4p orbital interactions with the B valence orbital. The bands are especially wide along the Γ-K, Γ-M, A-L, and A-H directions of the Brillouin zone. The broadness is indicative of the delocalized nature of the bonding and arises from interactions along the hexagonal and trigonal planes of B and Sc, respectively. The calculated Fermi level is -5.6 eV. In contrast to other studied MB_2 in ScB_2, the B 2p orbitals contribute $\sim 25\%$ to the Fermi level states, the main contribution is from the Sc 3d orbital. From the density of states the charge of Sc = $+2.28$ is obtained. A large electron drift towards B gives rise to a substantial ionic contribution to the bonding. The donated electrons populate the B 2p orbitals. The bonding within the studied diborides is explained with the help of solid-state calculations at a special point and by quasi-molecular cluster (e.g. $Sc_{18}B_{36}$) calculations, Armstrong [23]; see also [24]. Computationally, a model band calculation using the APW method was carried out by

McAlister et al. [20], which predicts qualitatively both the experimentally observed B soft X-ray K emission spectrum and other physical properties. The total and partial densities of states are shown in a figure [20]. Electronic specific heat data show that the density of states at the Fermi level is high (ca. 10 states·Ry^{-1}·cell^{-1}, see a figure in [20]) and typical for dominantly d-character band [16]. X-ray spectra indicate that the Sc 4s electrons participate essentially in the Sc–Sc bond. The Sc–B bond is probably formed by Sc 4s and 3d electrons and contributions of B p states. The covalent component is considerably polarized [21]. The mixing of B and Sc wave functions is discussed in [22]. Relative to the bonding energy of the C 1s electron (of 285.0 eV), the bonding energy of B 1s electrons in ScB$_2$ is 187.0 eV (pure B: 187.8 eV) from photoelectron spectra indicating a high donor ability of Sc. Conclusions are drawn on the type and strength of the chemical bond in borides, Aleshin et al. [25]. The electron transfer from M to B is assumed and discussed for the first-row transition metal diborides by [13].

ScB$_2$ is a typical metallic conductor. Between $\sim$293 and 1073 K the electrical resistivity rises linearly from $\varrho \approx 55$ to $140\,\mu\Omega\cdot$cm (taken from a figure), Samsonov et al. [5]; $\varrho \approx 180\,\mu\Omega\cdot$cm at 1073 K is given in a figure by Samsonov [7, p. 62]. The thermoelectric power S is only weakly dependent on temperature up to 773 K and is near $-7\,\mu$V/K [7, pp. 62/3]; $|S| = 7.7\,\mu$V/K [4].

The thermionic properties of ScB$_2$ on Ta, W, or Mo cathodes were originally studied at 1100 to 1600 K. The work function is $\Phi = 2.29 \pm 0.08$ eV, the Richardson constant $A \approx 10^{-5}\,$A$\cdot$cm$^{-2}\cdot$K^{-2}. The density of the emission current is $j = 5 \times 10^{-6}\,$A/cm^2 (presumably at 1400 K), Fomenko [28]. Later $\Phi = 2.80 \pm 0.05$ eV, $A = 4.6 \pm 2.1\,$A$\cdot$cm$^{-2}\cdot$K^{-2} was obtained for the stabilized emission. The previously observed regularity, relative to some physical properties in MB$_2$ (M = Sc, Ti, V, Cr), is verified by the thermoelectronic emission, Kudintseva et al. [11]. Studies at 1080 to 1600 K give the effective work function $\Phi_T = 3.76$ eV and $j = 4.3 \times 10^{-4}\,$A/cm^2 at 1600 K, Samsonov et al. [30]. For relations between thermoelectric emission and physical properties see also Samsonov, Zhuravlev [31].

Spectral Emission and High-Energy Spectra

The spectral emittance is $\varepsilon = 0.89$ at 1073 to 2073 K at 650 nm, Samsonov et al. [32], see also Samsonov [4].

The X-ray K absorption spectrum of Sc in ScB$_2$ was measured and compared with that of Sc in metallic Sc. Whereas the position of the absorption edge and of the main absorption maximum are shifted to shorter wavelengths relative to metallic Sc the longwave maximum remains constant. Pecularities in the K absorption edge are due to greater absorption in the 4s region, Zhurakovskii, Dzeganovskii [9]. The fine structure of the Kβ_5 band and the Kβ'' satellite observed in the X-ray fluorescence spectrum of Sc in metallic Sc, Sc$_2$O$_3$, ScN, ScB$_2$, and ScC by Zhurakovskii et al. [10] was already described in "Seltenerdelemente" B 4, 1976, pp. 33/4. The X-ray M$_{II,III}$ emission was studied and compared to the relative narrow one of metallic Sc by Shulakov et al. [29].

The B soft X-ray K emission spectrum of ScB$_2$ was measured and a systematic variation with metal atomic number was observed in the series ScB$_2$, TiB$_2$, VB$_2$, and CrB$_2$ as in other physical properties. From the steepness and location of the K emission profiles they are identified as emission edges associated with the Fermi cutoff. The intensity yields a measure of the change in p-orbital density at B sites at the Fermi level along the series, McAlister et al. [20]. The B Kα band was measured by ultra-longwave X-ray spectrometry in the same series of diborides. The band shape is identical within the series. In comparison with B, the asymmetry index, position, and width of the band are changed, Zhurakovskii, Vasilenko [21]; the longwave shift is 0.3 eV, revealing a negative charge on B, Lyakhovskaya et al. [22].

Chemical Behavior

ScB$_2$ sublimes in vacuum of 133 µPa at $\geqq$ 2073 K without decomposition, see p. 122.

The resistance of ScB$_2$ powder to oxidation in air was studied from 773 to 1173 K in 60 min. No decomposition was observed up to 873 K. At 973, 1073, and 1173 K after 60 min 9.5, 20.8, and 24.7% of the initial ScB$_2$ was decomposed, respectively. In addition to B$_2$O$_3$ and Sc$_2$O$_3$ the formation of scandium oxide boride is assumed, Lyutaya, Akinina [26]. Weight gain isotherms at 773, 873, and 973 K up to 240 min are given by Samsonov et al. [5], [7, pp. 62/3]. ScB$_2$ does not decompose in water. In the following table, the decomposition rate r in various acids and the percentage c of the initial compound remaining undecomposed after 1 h, at temperatures T are given.

reagent	T in K	r in g·h^{-1}·(g ScB$_2$)$^{-1}$	c in %
HCl (D = 1.19 g/cm^3)	353	0.75	25.0
H$_2$SO$_4$ (D = 1.84 g/cm^3)	573	0.67	32.8
HNO$_3$ (D = 1.4 g/cm^3)	393	0.88	22.5
HCl (1:1)	373	0.78	21.4
H$_2$SO$_4$ (1:1)	423	0.65	34.5
HNO$_3$ (1:1)	383	0.84	15.8

The Sc:B ratio is virtually unchanged in the residue and the solution. The resistance is lower than for ScB$_{12}$ owing to structural reasons [26], see also Lyutaya, Akinina [27].

References:

[1] Etourneau, J.; Mercurio, J.-P.; Naslain, R.; Hagenmuller, P. (Colloq. Intern. Centre Natl. Rech. Sci. [Paris] No. 205 [1972] 429/38).

[2] Peshev, P.; Etourneau, J.; Naslain, R. (Mater. Res. Bull. 5 [1970] 319/27).

[3] Przybylska, M.; Reddoch, A. H.; Ritter, G. J. (J. Am. Chem. Soc. 85 [1963] 407/11).

[4] Samsonov, G. V. (Dokl. Akad. Nauk SSSR 133 [1960] 1344/6; Proc. Acad. Sci. USSR Chem. Sect. 130/135 [1960] 969/71).

[5] Samsonov, G. V.; Verkhoglyadova, T. S.; Vdovenko, S. A. (Vopr. Teor. Primen. Redkozem. Metal. Akad. Nauk SSSR 1964 163/5; JPRS-28849 [1965] 213/6; N.S.A. 19 [1965] No. 23014).

[6] Shulishova, O. I.; Shcherbak, I. A. (Izv. Akad. Nauk SSSR Neorgan. Materialy 3 [1967] 1495/7; Inorg. Materials [USSR] 3 [1967] 1304/6).

[7] Samsonov, G. V. (Tugoplavkie Soedineniya Redkozemel'nykh Metallov s Nemetallami, Metallurgiya, Moscow, 1964; High-Temperature Compounds of Rare Earth Metals with Nonmetals, Consultants Bureau, New York 1965, pp. 1/280, 1/100).

[8] Zhuravlev, N. N.; Stepanova, A. A. (Kristallografiya 3 [1958] 83/5; Soviet Phys.-Cryst. 3 [1958] 76/7).

[9] Zhurakovskii, E. A.; Dzeganovskii, V. P. (Dokl. Akad. Nauk SSSR 150 [1963] 1260/2; Soviet Phys.-Dokl. 8 [1963] 594/6).

[10] Zhurakovskii, E. A.; Vladimirova, A. A.; Dzeganovskii, V. P. (Dokl. Akad. Nauk SSSR 170 [1966] 548/51; Soviet Phys.-Dokl. 11 [1966] 814/7).

[11] Kudintseva, G. A.; Neshpor, V. S.; Samsonov, G. V.; Tsarev, B. M.; Paderno, Yu. B. (Vysokotemp. Metallokeram. Mater. 1962 109/12; C.A. 58 [1963] 7468).

[12] Samsonov, G. V.; Paderno, Yu. B. (Boridy Redkozemel'nykh Metallov 1961 1/94; AEC-tr-5264 [1961] 1/104, 59).

[13] Topor, L.; Kleppa, O. J. (J. Chem. Thermodyn. 17 [1985] 1003/16).

[14] Peshev, P. (Rev. Intern. Hautes Temp. Refract. **4** [1967] 289/96, 291).

[15] Zhuravlev, N. N.; Stepanova, A. A. (Poroshkovaya Metal. **1964** No. 6, pp. 83/4; Soviet Powder Met. Metal Ceram. **1964** 510).

[16] Castaing, J.; Caudron, R.; Toupance, G.; Costa, P. (Solid State Commun. **7** [1969] 1453/6).

[17] Gschneidner, K. A., Jr. (in: Horovitz, C. T., Scandium, Its Occurrence, Chemistry, Physics, Metallurgy, Biology, and Technology, Academic, London–New York 1975, pp. 152/251, 161).

[18] Carter, G. C.; Swartz, J. C. (J. Phys. Chem. Solids **32** [1971] 2415/21).

[19] Barnes, R. G.; Creel, R. B.; Torgeson, D. R. (J. Chem. Phys. **53** [1970] 3762/3).

[20] McAlister, A. J.; Cuthill, J. R.; Williams, M. L.; Dobbyn, R. C. (Proc. Intern. Symp. X-Ray Spectra Electron. Struct. Matter. **1972/73** 426/48; C.A. **81** [1974] No. 113051).

[21] Zhurakhovskii, E. A.; Vasilenko, N. N. (Izv. Akad. Nauk Kaz. SSR Ser. Fiz. Mat. **9** No. 4 [1971] 50/3; C.A. **75** [1971] No. 135358).

[22] Lyakhovskaya, I. I.; Zimkina, T. M.; Fomichev, V. A. (Izv. Akad. Nauk SSSR Ser. Fiz. **36** [1972] 393/6; Bull. Acad. Sci. USSR Phys. Ser. **36** [1972] 360/3).

[23] Armstrong, D. R. (Theor. Chim. Acta **64** [1983] 137/52).

[24] Armstrong, D. R. (J. Less-Common Metals **82** [1981] 357).

[25] Aleshin, V. G.; Kosolapova, T. Ya.; Nemoshkalenko, V. V.; Serebryakova, T. I.; Chudinov, N. G. (J. Less-Common Metals **67** [1978] 173/7).

[26] Lyutaya, M. D.; Akinina, Z. S. (Izv. Akad. Nauk SSSR Neorgan. Materialy **1** [1965] 1039/43; Inorg. Materials [USSR] **1** [1965] 953/7).

[27] Lyutaya, M. D.; Akinina, Z. S. (Zavodsk. Lab. **31** [1965] 1066/8; Ind. Lab. [USSR] **31** [1965] 1321/2).

[28] Fomenko, V. S. (Radiotekhn. Elektron. **6** [1961] 1406; Radio Eng. Electron. Phys. **6** [1961] 1249/50).

[29] Shulakov, A. S.; Lyakhovskaya, I. I.; Fomichev, V. A. (Fiz. Tverd. Tela [Leningrad] **15** [1973] 2246/8; Soviet Phys.-Solid State **15** [1973] 1503/4).

[30] Samsonov, G. V.; Fomenko, V. S.; Paderno, Yu. B. (Ukr. Fiz. Zh. **8** [1963] 700/2; C.A. **59** [1963] 8228).

[31] Samsonov, G. V.; Zhuravlev, N. N. (Byull. Inst. Metallokeram. Spets. Splavov Akad. Nauk Ukr. SSR No. 5 [1960] 43/51; C.A. **1961** 14003).

[32] Samsonov, G. V.; Fomenko, V. S.; Paderno, Yu. B. (Ogneupory **27** No. 1 [1962] 40/2; C.A. **56** [1962] 13828).

32.1.8.1.3 ScB$_4$ (?)

Recent reviews on binary rare earth borides do not state the existence of ScB$_4$, Etourneau [1], Kost et al. [2]. The failure of synthetic efforts by borothermal reduction of Sc$_2$O$_3$ under vacuum between 1473 and 2473 K and with compositions B/Sc between 2 and 100, reported by Peshev et al. [3], or by Przybylska et al. [4] agree with the Sc–B system shown by Spear [5, p. 154] and results by Etourneau et al. [6], which shows no intermediate phase between ScB$_2$ and ScB$_{12}$, see p. 119. Previous reports [7, 8] on the ScB$_4$ synthesis thus have not been verified. Samsonov [7] has described a phase with a composition close to ScB$_4$ formed on reduction of Sc$_2$O$_3$ by B or B$_4$C. A tetragonal unit cell with a = 7.7 and c = 3.64 Å was given and also the electrical resistance, the thermoelectric power, the thermal expansion coefficient, and the microhardness were reported [7]. A ScB$_4$ sample, though not further characterized, was mentioned by Shulishova, Shcherbak [8].

References:

[1] Etourneau, J. (J. Less-Common Metals **110** [1985] 267/81).
[2] Kost, M. E.; Shilov, A. L.; Mikheeva, V. I.; Uspenskaya, S. I.; et al. (in: Eliseev, A. A., Soedineniya Redkozemel'nykh Elementov. Gidridy, Boridy, Karbidy, Fosfidy, Pniktidy Khal'kogenidy, Psevdogalogenidy, Moscow 1983, pp. 40/68).
[3] Peshev, P.; Etourneau, J.; Naslain, R. (Mater. Res. Bull. **5** [1970] 319/27).
[4] Przybylska, M.; Reddoch, A. H.; Ritter, G. J. (J. Am. Chem. Soc. **85** [1963] 407/11).
[5] Spear, K. E. (in: Alper, A. M., Refractory Materials, Phase Diagrams, Vol. 4, Academic, New York 1976, pp. 91/159).
[6] Etourneau, J.; Mercurio, J.-P.; Naslain, R.; Hagenmuller, P. (Colloq. Intern. Centre Natl. Rech. Sci. [Paris] No. 205 [1972] 429/38).
[7] Samsonov, G. V. (Tugoplavkie Soedineniya Redkozemel'nykh Metallov s Nemetallami, Metallurgiya, Moscow, 1964; High-Temperature Compounds of Rare Earth Metals with Nonmetals, Consultants Bureau, New York 1965, pp. 1/280, 1/100).
[8] Shulishova, O. I.; Shcherbak, I. A. (Izv. Akad. Nauk SSSR Neorgan. Materialy **3** [1967] 1495/7; Inorg. Materials [USSR] **3** [1967] 1304/6).

32.1.8.1.4 ScB_6 (?)

If earlier experimental results, which could not be confirmed, are discounted, the synthesis of ScB_6 has not yet been accomplished, see for example Samsonov et al. [1, 2], see also p. 21.

Theoretical investigations using structural data of group II and III metal hexaborides of the CaB_6 type yielded a high polarizing power of the scandium atom on a boron atom and it resulted the strongest polar metal-boron bond within the hexaboride series. One might expect the scandium atom to exert a strong influence on the boron sublattice, causing an excessive extension of the boron octahedra and the loss of their stability. This was reflected by an exceptionally large difference between intraoctahedral (b_e) and interoctahedral (b_i) boron–boron bond lengths calculated for a hypothetical ScB_6 compound: the ratio was $b_e/b_i = 1.38$ compared to $b_e/b_i = 1.18$ for YB_6, which is regarded as upper limit for a stable CaB_6 structure. Therefore, the small ionic radius of Sc, normally thought in the literature to be too small to stabilize the CaB_6 structure, was assumed not to be the actual reason for the instability of ScB_6. However, the existence of ScB_6 could not be ruled out at high pressure, Barantseva, Paderno [3, 4]. The possibility of preparing a metastable ScB_6 phase under high pressure was discussed. Calculated bond lengths between boron atoms for the metastable phase were close to those of yttrium hexaboride, see a table in the paper [4].

Theoretical lattice constants: $a = 4.06$ Å from extrapolated MB_6 data, Tsolovski, Peshev [5] and $a = 4.08$ Å calculated for a hypothetical ScB_6 compound [3, 4]; $a = 3.74$ Å was calculated for a metastable hypothetical ScB_6 phase (see above) [4].

References:

[1] Samsonov, G. V.; Kondrashov, A. I.; Okhremchuk, L. N.; Podchernyaeva, I. A.; Siman, N. I.; Fomenko, V. S. (Poroshkovaya Metal. **1977** No. 1, pp. 21/8; Soviet Powder Met. Metal Ceram. **16** [1977] 16/22).
[2] Samsonov, G. V.; Kondrashov, A. I.; Okhremchuk, L. N.; Podchernyaeva, I. A.; Siman, N. I.; Fomenko, V. S. (J. Less-Common Metals **67** [1979] 415/8).
[3] Barantseva, I. G.; Paderno, Yu. B. (Poroshkovaya Metal. **1981** No. 9, pp. 56/60; Soviet Powder Met. Metal Ceram. **20** [1981] 635/8).

[4] Barantseva, I. G.; Paderno, Yu. B. (Fiz. Tekh. Vys. Davlenii **8** [1982] 50/4; C.A. **98** [1983] No. 190527).

[5] Tsolovski, I. A.; Peshev, P. D. (Dokl. Bolg. Akad. Nauk **25** [1972] 209/12; C.A. **77** [1972] No. 25711).

32.1.8.1.5 ScB$_{12}$

ScB$_{12}$ was obtained like the diboride by heating Sc$_2$O$_3$ and B. A product designed to yield an atomic ratio B:Sc = 6 was obtained in 10 min at 1598 K and contained ScB$_2$ and some Sc$_2$O$_3$. By heating for an additional 8 h at 1823 K the diboride lost Sc yielding the dodecaboride, Przybylska et al. [1]. Borothermal reduction of Sc$_2$O$_3$ at 1873 to 2073 K was used by Peshev et al. [2], Etourneau et al. [3], see also Lyutaya, Akinina [4].

X-ray powder investigations indicated ScB$_{12}$ to be face-centered cubic, UB$_{12}$ type, with a = 7.422 ± 0.002 Å (interplanar spacings and intensities are tabulated) [1], but later single-crystal data indicated it to be tetragonal, with a = 5.22, c = 7.35 Å. Space group I4/mmm-D$_{4h}^{17}$ (No. 139). The crystals were twinned on the (111) plane, Matkovich et al. [7, 10]. It is suggested that ScB$_{12}$ may transform to a cubic cell (corresponding to the cell of MB$_{12}$) at elevated temperatures. Relations of the size of M in MB$_{12}$ are discussed and Sc is found too small for the usual cubic structure [10]. From powder data a = 5.2347(2), c = 7.3583(4) Å is given, Callmer [8]. However, in a recent study on ScB$_{12}$ single crystals, obtained from metallic Sc and fine crystalline B by fusion under Ar, the face-centered cubic structure was confirmed with a = 7.402(3) Å. This indicates that ScB$_{12}$ belongs to the UB$_{12}$ type with space group Fm$\overline{3}$m-O$_h^5$ (No. 225). The coordinates of the atoms were recalculated by the least-squares method to R = 0.033 and are as follows: 4 Sc at 4 (a) (0,0,0), 48 B at 48 (i) (1/2, 0.1685(3), 0.1685(3)) with anisotropic temperature parameters. The Sc is located in the center of the B$_{24}$ cubooctahedron with the Sc–B distance 2.752(1) Å being greater than the sum of the atomic radii. The coordination polyhedron around B is a square pyramid with two additional Sc opposite the lateral faces. The B–B distance from the center to the vertex is 1.705(2) Å, to the base of the pyramid 1.764(2) Å, Bruskov et al. [9]. The calculated density is 2.890 g/cm^3 for the tetragonal cell. The experimental density is 2.883 ± 0.005 g/cm^3.

ScB$_{12}$ melts congruently at 2315 ± 50 K [2, 3], see also Samsonov et al. [5] and p. 119. ScB$_{12}$ is diamagnetic with χ = − 65 × 10^{-6} cm^3/mol [2].

The oxidation resistance of ScB$_{12}$ in air was studied between 773 and 1173 K. No decomposition was observed up to 873 K after 60 min. At 973, 1073, and 1173 K after 60 min 2.8, 25.3, and 26.6% of the initial ScB$_{12}$ were decomposed. ScB$_{12}$ is not decomposed in water. In the following table, the decomposition rate r in various acids and the percentage c of the initial compound remaining undecomposed after the reaction time t at temperatures T are given:

reagent	t in h	T in K	r in g·h^{-1}·(g ScB$_{12}$)$^{-1}$	c in %
HCl (D = 1.19 g/cm^3)	5	353	0.002	99.2
H$_2$SO$_4$ (D = 1.84 g/cm^3)	4	573	0.03	86.4
HNO$_3$ (D = 1.4 g/cm^3)	1	393	0.81	19.2
HCl (1:1)	5	373	0.002	98.7
H$_2$SO$_4$ (1:1)	4	423	0.027	89.4
HNO$_3$ (1:1)	1	383	0.82	17.8

The resistance is higher than for ScB_2 on structural reasons. So it is possible to remove ScB_2 impurities from ScB_{12} with dilute HCl. The ratio Sc:B in residues and solutions is virtually unchanged [4], see also Lyutaya, Akinina [6].

References:

[1] Przybylska, M.; Reddoch, A. H.; Ritter, G. J. (J. Am. Chem. Soc. **85** [1963] 407/11).
[2] Peshev, P.; Etourneau, J.; Naslain, R. (Mater. Res. Bull. **5** [1970] 319/27).
[3] Etourneau, J.; Mercurio, J.-P.; Naslain, R.; Hagenmuller, P. (Colloq. Intern. Centre Natl. Rech. Sci. [Paris] No. 205 [1972] 429/38).
[4] Lyutaya, M. D.; Akinina, Z. S. (Izv. Akad. Nauk SSSR Neorgan. Materialy **1** [1965] 1039/43; Inorg. Materials [USSR] **1** [1965] 953/7).
[5] Samsonov, G. V.; Goryachev, Yu. M.; Kovenskaya, B. A.; Arabei, B. G. (Bor: Poluch. Strukt. Svoistva Mater. 4th Mezhdunar. Simp. Boru, Tbilisi 1972 [1974], pp. 171/4).
[6] Lyutaya, M. D.; Akinina, Z. S. (Zavodsk. Lab. **31** [1965] 1066/8; Ind. Lab. [USSR] **31** [1965] 1321/2).
[7] Matkovich, V. I.; Economy, J.; Giese, R. F., Jr.; Barrett, R. (Acta Cryst. **19** [1965] 1056/8).
[8] Callmer, B. (J. Solid State Chem. **23** [1978] 391/8).
[9] Bruskov, V. A.; Zavalii, L. V.; Kuz'ma, Yu. B. (Izv. Akad. Nauk SSSR Neorgan. Materialy **24** [1988] 506/7; Inorg. Materials [USSR] **24** [1988] 420/1).
[10] Matkovich, V. I.; Economy, J.; Giese, R. F., Jr.; Barrett, R. B. (AD-608 444 [1964] 1/16; C.A. **63** [1964] 1283).

32.1.8.1.6 Solid Solutions of Sc in β Boron ($ScB_{\sim28}$)

Samples with the initial composition ScB_{20} were prepared by arc-melting crystalline B and Sc on a water-cooled Cu hearth under purified Ar. Large single crystals (composition not given) are obtained by adding Sc to a B melt, Carlsson, Lundström [1]. The rapidly cooled nominal ScB_{20} sample consisted of ScB_{12} and $ScB_{\sim28}$. According to microprobe, the latter contained 3.2 at% ($\cong 12\pm2$ wt%) Sc. Single-crystal structure refinement gives 3.5 at% ($\cong 13.1$ wt%) Sc corresponding to the solubility limit, Callmer [2]. Single-crystal diffractometric investigations show $ScB_{\sim28}$ to be hexagonal with a = 10.9658(4), c = 24.0875(13) Å and space group $R\bar{3}m$-D_{3d}^5 (No. 166). For the rhombohedral cell a = 10.225 Å, $\alpha = 64.85°$ is given [2]. From powder photographs a = 10.9620(7), c = 24.0752(27) Å are obtained [1]. The three-dimensional β-B network is only slightly changed. Two Sc atoms are inserted interstitially, one Sc atom replaces two B atoms at an intermediate position. The atomic positions are as follows:

atom	position	x	y	z	partial occupancy(%)
B(1)	36(i)	0.17428(22)	0.17305(23)	0.17433(8)	
B(2)	36(i)	0.31847(24)	0.29386(23)	0.12889(8)	
B(3)	36(i)	0.26165(24)	0.21742(24)	0.41839(9)	
B(4)	36(i)	0.23783(26)	0.25242(26)	0.34743(9)	92.5(10)
B(5)	18(h)	0.05411(17)	0.10822(17)	0.94334(12)	
B(6)	18(h)	0.08460(17)	0.16920(17)	0.01299(12)	
B(7)	18(h)	0.10827(17)	0.21655(17)	0.88852(12)	
B(8)	18(h)	0.16859(18)	0.33719(18)	0.02772(12)	
B(9)	18(h)	0.12986(17)	0.25973(17)	0.76774(12)	
B(10)	18(h)	0.10172(17)	0.20345(17)	0.69826(12)	

atom	position	x	y	z	partial occupancy(%)
B(11)	18(h)	0.05810(17)	0.11620(17)	0.32820(13)	
B(12)	18(h)	0.09033(17)	0.18067(17)	0.39764(12)	
B(13)	18(h)	0.05585(30)	0.11170(30)	0.55603(22)	61.3(14)
B(14)	6(c)	0	0	0.38524(22)	
B(15)	3(b)	0	0	1/2	
Sc(2)	18(h)	0.20454(10)	0.40908(10)	0.17601(7)	31.4(3)
Sc(3)	6(c)	0	0	0.23539(5)	72.7(5)
Sc(4)	18(f)	0.3713(10)	0	0	5.7(3)

The numbering of the atoms follows that used for solid solutions of M = Cr, Mn, Fe, and Cu in β B, which have a M(1) position. For structural details and comparison with pure β B and these solid solutions in β B see the original paper. The Sc–Sc and Sc–B distances <4.0 and <3.0 Å are (in Å):

Sc(2)–B(13)	2.097(5)	Sc(3)–6B(1)	2.405(2)	Sc(4)–2B(4)	0.860(3)
–B(12)	2.356(3)	–3B(9)	2.467(3)	–2B(8)	2.166(7)
–2B(3)	2.387(3)	–3B(11)	2.492(3)	–2B(3)	2.340(2)
–2B(13)	2.403(3)	–3B(10)	2.506(3)	–2B(10)	2.341(2)
–2B(2)	2.427(2)			–2B(11)	2.501(7)
–2B(1)	2.439(3)			–2B(12)	2.547(7)
–2B(3)	2.446(3)			–2B(4)	2.765(11)
–2B(2)	2.453(2)			–2B(6)	2.814(11)
–B(15)	2.456(2)			–Sc(4)	2.822(18)
–2Sc(2)	2.486(2)				

Callmer [2]. The microhardness at a 50 g load is 4210 kp/mm² (~41.3 GPa) compared to 3420 kp/mm² (~33.5 GPa) for pure β B [1].

References:

[1] Carlsson, J.-O.; Lundström, T. (J. Less-Common Metals 22 [1970] 317/20).
[2] Callmer, B. (J. Solid State Chem. 23 [1978] 391/8).

32.1.8.2 Yttrium Borides

Additional data for YB_2, YB_4, YB_6, YB_{12}, and $YB_{\sim 66}$ are found in Section 32.1.1, 32.1.4, 32.1.5, 32.1.6, and 32.1.7, on pp. 2, 6, 21, 100, and 112, respectively.

32.1.8.2.1 The Y–B System

Data up to 1961 are critically considered by Elliott [10].

The phase diagram of the system Y–B shows the existence of five compounds: YB_2, YB_4, YB_6, YB_{12}, and $YB_{\sim 66}$, see **Fig. 50**, p. 130, from Spear [1]. It was derived from metallographic and X-ray diffraction studies of Lundin, Klodt [2] revised for the B-rich region of the system by

Oliver, Brower [3]. The liquidus temperature decreases from the melting point of β Y at 1552°C
to the eutectic point S → α Y + YB$_2$ at 1290°C and 25.5 at% B. The α Y apparently transforms to
β Y by a peritectic reaction. The boron solubility in the metal is lower than 1 at%. The YB$_2$ and
YB$_4$ melt congruently at 2100 and 2800°C, respectively. YB$_6$ and YB$_{12}$ form peritectically at 2600
and 2200°C, respectively, the peritectic melts contain 89.5 and 95 at% B, respectively [2]. In
addition, a high-boron compound YB$_{70}$ is assumed [2], which was later reported as compound
YB$_n$ melting at ~2100°C congruently and forming a eutectic with YB$_{12}$ at 2050°C [3] in addition
to that with β B found by Seybolt [4]. The composition varies from 100 > n > 20 and is
commonly considered to be n = 66. Precise measurements of density and lattice parameter
indicate the congruent melting point at n = 61.75 and the stoichiometric composition at n = 68,
Slack et al. [5]. A high-B compound YB$_{\sim 50}$ with ~1700 atoms in the cubic cell and a
complicated structure was mentioned by Seybolt [4].

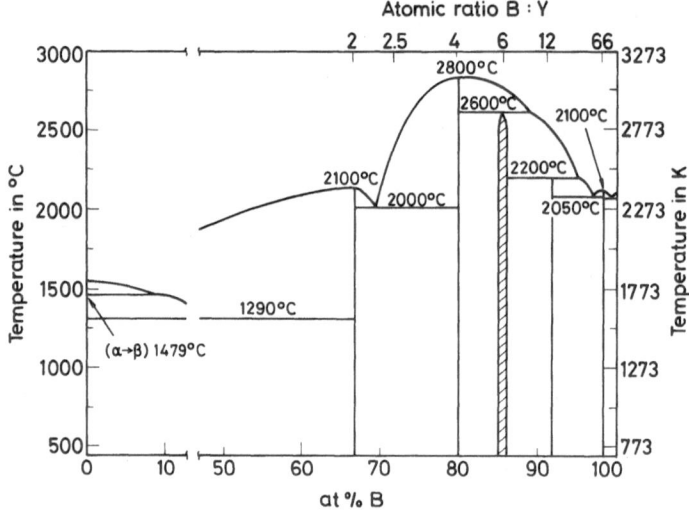

Fig. 50.   The Y–B phase diagram.

The heats of formation ΔH were predicted by Miedema et al. [9] for some compositions Y : B
within the system with the help of electronegativity and density parameters of the components:

Y : B	5:1	2:1	1:1	1:2	1:5
− ΔH in kJ/mol	26	52	75	84	56

Samples with the compositions YB$_{2.5}$, YB$_5$, YB$_{6.4}$, and YB$_{14}$ in the two-phase regions YB$_2$ + YB$_4$,
YB$_4$ + YB$_6$, YB$_6$ + YB$_{12}$, and YB$_{12}$ + YB$_{66}$ were prepared and their thermionic properties were
studied by Jaskie [6, 7], Jaskie, Jacobson [8].

References:

[1] Spear, K. E. (in: Alper, A. M., Refractory Materials, Phase Diagrams, Vol. 4, Academic, New
     York 1976, pp. 91/159, 148/9).
[2] Lundin, C.; Klodt, D.; quoted by Lundin, C. E. (in: Spedding, F. H.; Daane, A. H., The Rare
     Earths, Wiley, New York 1961, pp. 224/385, 247/8).
[3] Oliver, D. W.; Brower, G. D. (J. Cryst. Growth **11** [1971] 185/90).
[4] Seybolt, A. V. (Trans. Am. Soc. Metals **52** [1960] 971/89).

[5] Slack, G. A.; Oliver, D. W.; Brower, G. D.; Young, J. D. (J. Phys. Chem. Solids **38** [1977] 45/9).

[6] Jaskie, J. E. (Diss. Arizona State Univ. 1981, pp. 1/123; Diss. Abstr. Intern. B **42** [1982] 4098).

[7] Jaskie, J. E. (DOE-ET-15419-T1 [1981] 1/121; C.A. **99** [1983] No. 80946).

[8] Jaskie, J. E.; Jacobson, D. (Proc. Intersoc. Energy Convers. Eng. Conf. **15** No. 3 [1980] 2331/3; C.A. **94** [1981] No. 142615).

[9] Miedema, A. R.; de Boer, F. R.; Boom, R. (Proc. 12th Rare Earth Res. Conf., Vail, Colo., 1976, Vol. 2, pp. 838/46).

[10] Elliot, R. P. (Constitution of Binary Alloys, 1st Suppl., McGraw-Hill, New York 1965, pp. 143/5).

32.1.8.2.2   YB$_2$

Preparation

Stoichiometric amounts of Y and B were mixed and heated under Ar in a silit furnace at 1573 K. The product contained 4.4% YB$_4$, Markovskii et al. [1]. Direct synthesis from high-purity elements in vacuum at 1320 to 1520 K for 15 min results in stoichiometric YB$_2$ only when a 100 to 200% Y excess is used owing to the high vapor pressure of Y compared with B, Manelis et al. [2]. By floating-zone techniques, small elongated single crystals were obtained. They could easily be cleaved along (001) planes of the hexagonal lattice into flat elongated plates, Johnson [3], Johnson, Daane [4, 5].

Physical Properties

YB$_2$ is hexagonal, AlB$_2$ type, space group P6/mmm-D$_{6h}^1$ (No. 191) with a = 3.300 ± 0.004, c = 3.838 ± 0.005 Å, Zhuravlev et al. [6]; a = 3.290, c = 3.835 Å [2], a = 3.298 ± 0.002, c = 3.843 ± 0.004 Å, c/a = 1.165, Lundin, Klodt [7], a = 3.3036 ± 0.0005, c = 3.8427 ± 0.0005 Å, and c/a = 1.1632. Interplanar spacings and line intensities of the powder diagram are given. The measured density is D$_{exp}$ = 4.86 g/cm^3, the calculated density is D$_{calc}$ = 5.07 g/cm^3 [1].

The calculated mean thermal expansion coefficients between ∼293 and 873 K are α = (9.4 ± 1.0) × 10^{-6} K^{-1} and (8.5 ± 0.9) × 10^{-6} K^{-1} in the a and c directions, respectively [6]. YB$_2$ melts at 2220 ± 95 K [2], see, however, the phase diagram [7], where 2373 K (2100°C) is given, see also Kober [8], Samsonov et al. [13]. The magnetic susceptibility was measured between 70 and 570 K and shows YB$_2$ to be a typical diamagnet, apparently because of the ionic skeleton of the crystal [2].

The NMR isotropic Knight shift K, quadrupole coupling constant e^{2}qQ/h, and line width ΔH in YB$_2$ were determined at 300 K as follows: K(^{89}Y) is 0.20(3)% at 2.7258 MHz (compared with the ^{14}N resonance of HNO$_3$ at 4.0115 MHz using a ratio ν(^{14}N):ν(^{89}Y) = 1.47462); ΔH = 12(2) or 20(3) Oe in the absorption or dispersion mode, respectively, at 2.7 MHz. For the ^{11}B resonance K = −0.004(10)% with respect to triethyl borate (?) is obtained, e^{2}qQ/h = 0.34(2) MHz, and ΔH = 8(1) Oe at 4 to 16 MHz. Comparison is made within the 4d diboride series and with Y metal. K is lower than for Y metal (0.37%) primarily owing to reduced orbital and Pauli contributions. The ^{89}Y lineshape shows some asymmetry probably owing to a second phase present in the sample, Carter, Swartz [9]. K(^{89}Y) = 0.16(1)%, K(^{11}B) = −0.005(1)%, and e^{2}qQ/h(^{11}B) = 0.398(10) MHz at 300 K. The results at 77 K differ negligibly, Barnes et al. [10]. A schematic diagram of the electronic structure is shown in Fig. 52, p. 137, deduced from the data for LaB$_6$ and fourth-group transition-metal diborides, Tanaka, Ishizawa [14].

The electrical resistivity of a polycrystalline sample is $16 \pm 0.6 \, \mu\Omega \cdot cm$ [3]; single crystals in the basal plane have $39 \pm 2.6 \, \mu\Omega \cdot cm$. The measured Hall coefficient (c axis parallel to the magnetic field) $R_H = -(20.5 \pm 2.5) \times 10^{-11} \, m^3/C$ is in agreement with the calculated value $-22.65 \times 10^{-11} \, m^3/C$ for one electron per Y [3 to 5].

The emission properties of YB_2 were determined at 1500 to 2000 K. The values of the work function Φ are lower than those for YB_4, YB_6, and YB_{12}; $\Phi_T = 3.23$ to 3.46 eV and emission current density $j_{max} = 3.7 \times 10^{-3}$ to 0.9 A/cm². Figures for Φ_T vs. temperature and current-voltage characteristics are presented in the paper. It is confirmed that the basic contribution to the thermoemission comes from the electrons of Y and not from those of B as had been assumed earlier [2]. That B accepts electrons for filling the 2p orbitals is stated in [8].

Chemical Behavior

YB_2 is stable up to the melting point [2]. It is decomposed by dilute HCl (1:10) even at room temperature with evolution of boranes. The overall hydrolysis is simplified by the reaction: $YB_2 + 4.8 H_2O \rightarrow Y(OH)_3 + 0.9 B_2O_2 + B_{0.2}H_{0.3} + 3.15 H_2$; the B_2O_2 forms a hydrate $B_2O_2 \cdot x H_2O$ [1]. YB_2 reacts with B on heating in a violent exothermic reaction to give YB_4, Oliver, Brower [11]. There is no noticeable solubility of Cr in YB_2, Kuz'ma et al. [12].

References:

[1] Markovskii, L. Ya.; Vekshina, N. V.; Kondrashev, Yu. D.; Voevodskaya, T. K.; Pitirimov, B. Z. (Zh. Prikl. Khim. **42** [1969] 2690/51; J. Appl. Chem. [USSR] **42** [1969] 2549/53).

[2] Manelis, R. M.; Telyukova, T. M.; Grishina, L. P. (Izv. Akad. Nauk SSSR Neorgan. Materialy **6** [1970] 1184/5; Inorg. Materials [USSR] **6** [1970] 1035/6).

[3] Johnson, R. W. (Diss. Iowa State Univ. 1962, pp. 1/68, 21, 34, 43; Diss. Abstr. **23** [1962/63] 850).

[4] Johnson, R. W.; Daane, A. H. (J. Chem. Phys. **38** [1963] 425/32).

[5] Johnson, R. W.; Daane, A. H. (IS-473 [1962] 1/30, 5; N.S.A. **17** [1963] No. 11306).

[6] Zhuravlev, N. N.; Belousova, I. A.; Manelis, R. M.; Belousova, N. A. (Kristallografiya **15** [1970] 836/8; Soviet Phys.-Cryst. **15** [1970] 723/4).

[7] Lundin, C.; Klodt, D.; quoted by Lundin, C. E. (in: Spedding, F. H.; Daane, A. H., The Rare Earths, Wiley, New York 1961, pp. 224/385, 248).

[8] Kober, V. I. (Elektron. Str. Fiz. Khim. Svoistva Tugoplavkikh Soedin. Splavov Dokl. 9th Vses. Simp., Ivano-Frankovsk, USSR, 1979 [1980], pp. 253/8; C.A. **95** [1981] No. 121237).

[9] Carter, G. C.; Swartz, J. C. (J. Phys. Chem. Solids **32** [1971] 2415/21).

[10] Barnes, R. G.; Creel, R. B.; Torgeson, D. R. (J. Chem. Phys. **53** [1970] 3762/3).

[11] Oliver, D. W.; Brower, G. D. (J. Cryst. Growth **11** [1971] 185/90).

[12] Kuz'ma, Yu. B.; Sobolev, A. S.; Furtak, M. P. (Izv. Akad. Nauk SSSR Neorgan. Materialy **6** [1970] 2205/6; Inorg. Materials [USSR] **6** [1970] 1936/7).

[13] Samsonov, G. V.; Goryachev, Yu. M.; Kovenskaya, B. A.; Arabei, B. G. (Bor: Poluch. Strukt. Svoistva Mater. 4th Mezhdunar. Simp. Boru, Tbilisi 1972 [1974], pp. 171/4).

[14] Tanaka, T.; Ishizawa, Y. (J. Phys. C **18** [1985] 4933/40, 4939).

32.1.8.2.3 "YB$_3$"

A phase YB$_3$, tetragonal with a = 3.78, c = 3.55 Å, was reported by Binder [1], see also Lundin [2]. The semiconducting material YB$_3$ with high electrical resistance and polymeric structure prepared by [3], and used in current-controlling devices for electrical circuits, is a mixture of YB$_2$ and YB$_4$.

References:

[1] Binder, I. (Powder Met. Bull. **7** [1956] 74/5).
[2] Lundin, C. E. (in: Spedding, F. H.; Daane, A. H., The Rare Earths, Wiley, New York 1961, pp. 224/385, 248).
[3] Energy Conversion Devices, Inc. (Brit. 1260191 [1969/72]; C.A. **76** [1972] No. 65359).

32.1.8.2.4 YB$_4$

YB$_4$ is a congruently at 3073 K (or lower) melting phase in the Y–B system, see p. 130, Lundin [1], cf. Manelis et al. [2].

Formation. Preparation

Polycrystalline Samples

Polycrystalline samples were prepared from Y$_2$O$_3$ with reduction by either C or B at temperatures between 1473 and 2073 K, Binder [3]. Dense pellets with 96% of the X-ray densitiy were obtained by vacuum hot pressing: The samples were heated in a vacuum up to 2200 K for 50 min. A pressure up to 211 kg/cm^2 ($\triangleq$ 20.7 MPa) was applied gradually as the material was heated, Hoyt, Chorné [4]. The borothermal reduction of briquetted B/Y$_2$O$_3$ intimate mixtures in a tantalum crucible under vacuum gave single-phase YB$_4$ after 60 min at 2073 to 2273 K and at a molar proportion of B/Y$_2$O$_3$ = 11 in the charge. The surface layer (~0.5 mm) was discarded from the completely reduced products. Pressing and vacuum sintering of the powdered YB$_4$ yielded samples with a porosity of 22 to 26%. YB$_4$ was also obtained from charges richer in boron by heating to above the decompositon temperatures of the higher borides, Manelis et al. [2], similarly described in [5]. The reaction products from Y$_2$O$_3$ + 11B at 2173 K for 1 h under vacuum were closer to the desired composition than those at lower temperature, but they contained small amounts of YB$_6$, Kudintseva et al. [6]. Pellets of B/Y$_2$O$_3$ mixtures reacted to form YB$_4$ within 2 h at 1873 K under vacuum, Tanaka et al. [7]. YB$_4$ formed during thermal decomposition of the hexaboride in a solar vacuum furnace, Peshev [8].

Single-phase YB$_4$ was obtained from the elements in the stoichiometric ratio by heating in a BN crucible under Ar at 1873 to 1973 K for 15 to 30 min, followed by treatment with hydrochloric acid and separation of the crystals by a heavy liquid (classification). Chemical analyses indicated compositions from YB$_{3.7}$ to YB$_{3.9}$ and in one case, YB$_{4.03}$, Giese et al. [9]. Arc melting of mixtures of Y and B under He gave single-phase YB$_4$, Buschow, Creyghton [10]. Rods of similarly prepared YB$_4$ were used to grow single crystals of YB$_4$ by the floating-zone method with heating by an induction coil, Johnson, Daane [11, 12].

Single Crystals

High-purity single crystals of YB$_4$, up to 80 mm long and 8 mm in diameter, were prepared by multiple zone refining using the floating-zone method. YB$_4$ powder from a borothermal reduction of a stoichiometric charge was ground to an average particle diameter of about 5 μm, compressed to a rectangular rod of $10 \times 10 \times 100$ mm^3 after adding an ethanol-camphor solution, re-pressed hydrostatically in order to achieve a higher and more uniform density, reshaped to a cylinder of 10 mm diameter, and sintered at 2173 K for 30 min under vacuum in a BN crucible. Triple zone refining of this rod under a helium pressure of 700 kPa at a growth rate of 5 to 15 mm/h yielded crystals with residual resistance ratio $\varrho_{300}/\varrho_{4.2} = 54$; no metal impurities were detected by fluorescent X-ray analysis. With an initial boron deficit or excess, the zone refining was either impossible or it yielded inferior purity crystals, respectively. A growth rate of 5 mm/h was most suitable. Higher rates gave rise to a polycrystalline rim around the central crystal. The feed rate was reduced by 10% compared to the pulling rate. The molten-zone temperature was held at $T_m = 2883 \pm 10$ K. The growth direction was always near [001], Tanaka et al. [7]. Crystal growth by the Czochralski method gave growth directions also along [001] and 10° from [011] in two experiments with growth rates of 1 and 2 mm/min, respectively. The residual resistance ratio, 2.8, of the as-grown-crystals increased to 11.5 after annealing for 1 d at 2073 K, Bressel et al. [13].

Attempts at crystal growth in an aluminium melt gave Al-contaminated tetraboride [9]. Whereas more recently, Guette et al. [14] grew crystals by this method without mentioning any contamination.

Atomic ratios of between 3.81 and 4.25 for B:Y and between 65 and 94.85 for Al:Y, and equilibration for longer than 10 h at above 1673 K, gave golden metallic single crystals of YB$_4$ with polyhedral shape and diameters up to 4 mm, Okada, Atoda [15].

YB$_4$ single crystals are lavender in color with a metallic luster, Tanaka et al. [7]. Other samples were described as grayish brown, Samsonov [16].

Crystallographic Properties

YB$_4$ is isomorphous with the rare earth tetraborides: tetragonal UB$_4$ type which is described on p. 8. The following values, in Å, were reported for the lattice parameters a and c, respectively, at room temperature: 7.098 ± 0.003, 4.018 ± 0.001, Bressel et al. [13], 7.110 ± 0.004, 4.033 ± 0.005, Zhuravlev et al. [17], 7.105, 4.020, Buschow, Creyghton [10], 7.12 ± 0.05, 4.04 ± 0.05, Manelis et al. [2], 7.111 ± 0.001, 4.012 ± 0.005, Lundin [1], 7.09, 4.01, Binder [3]; from single crystals the values 7.0993 ± 0.0007, 4.0179 ± 0.0007, Okada, Atoda [15], 7.111, 4.017 ± 0.002, Guette et al. [14], 7.11 ± 0.01, 4.02 ± 0.01, Giese et al. [9, 18], 7.08, 4.00, Kudintseva et al. [6], 7.086, 4.012, Rudman et al. [19] were reported. The temperature dependence of a and c is shown in Fig. 5, p. 11, with a = 7.085 and c = 4.013 at presumably 4.2 K, Berrada et al. [20].

An X-ray single-crystal structure determination with R = 0.041 (for further data see p. 8) showed the metal atom surrounded by 16 boron atoms at distances from 2.729 to 2.856 Å and with two additional at 3.069 Å. The shortest interboron distance is 1.629 Å between the octahedra, whereas all the other B–B distances range from 1.721 to 1.809 Å [14]. A fractional occupation of any boron positions (see below) could not be confirmed [14]. Previously a single-crystal structure determination with inferior reliability (R = 0.113) indicated only an 80% occupation of the equatorial position (B3) of the boron octahedron, which in fact, would reduce most of these entities to square pyramids, Giese et al. [9, 18]. The positional parameters of [9] are consistent with those of the later paper [14].

YbB$_4$ single crystals cleave along (001) planes, Johnson, Daane [11].

Mechanical Properties

Density

The experimental density D of crystals grown in an Al flux is 4.29 ± 0.02 g/cm^3, Guette et al. [14], D = 4.26 g/cm^3, Okada, Atoda [15]. The X-ray density of 4.32 g/cm^3 was never reached by the samples prepared from the elements; densities were between 3.98 and 4.20 g/cm^3, Giese et al. [9]. The vacuum hot-pressed YB$_4$ samples of Hoyt, Chorné [4] are described to have 96% of the X-ray density, which is given as 4.47 g/cm^3 [4]. (This value is considerably higher than that (4.36 g/cm^3 [2, 5, 16]) which corresponds to the smallest of the puplished unit cells.)

Hardness

The Vickers hardness by 100 g load of vacuum hot-pressed YB$_4$ is 1200 kg/mm^2 ($\cong$ 12 GPa), Hoyt, Chorné [4]. The microhardness is 2850 ± 100 kg/mm^2 ($\cong$ 28 GPa) and the bending strength $\sigma = 290$ kg/cm^2 ($\cong$ 28.4 MPa) for samples with a porosity of 22 to 26%, grain size 10 to 20 μm, Meerson et al. [5], cf. Manelis et al. [2]. The Knoop microhardness of a single crystal was 1580 to 1740 kg/mm^2 ($\cong$ 15.5 to 17.1 GPa) on the (001) surface and 1190 to 2600 kg/mm^2 ($\cong$ 11.7 to 25.5 GPa) on (211), both at 100 g load [15].

Thermal Properties

Thermal Expansion

The mean thermal expansion coefficient α of YB$_4$ perpendicular and parallel to the tetragonal axis is $(7.6 \pm 0.5) \times 10^{-6}$ K^{-1} and $(6.4 \pm 0.6) \times 10^{-6}$ K^{-1}, respectively, as derived from X-ray data for temperatures between 298 and 773 to 973 K, Zhuravlev et al. [17]. For α of a sintered sample see p. 13.

Melting Point. Vaporization

A melting point of 2883 ± 10 K was deduced from zone-refining experiments, Tanaka et al. [7]. Congruent melting at 3073 K is implied in the Y–B-phase diagram by Lundin [1]. A melting or decomposition temperature of 2273 K is given by Hoyt, Chorné [4], whereas a stability at least up to 2750 K is stated by Meerson et al. [5]. An estimate for the melting point, $T_m = 3020$ K, was made from the thermal expansion coefficient, Severyanina et al. [21].

The congruently vaporizing composition of the Y–B system was estimated from the known vapor pressures of the elements as YB$_{4.02}$, Storms, Mueller [22].

Enthalpy of Formation

The standard molar enthalpy of formation of YB$_4$ was determined calorimetrically from the reaction of YB$_4$ or Y + 4 B with solid Pt near 1400 K which yields a liquid alloy. This gave $\Delta H^\circ_{f,298} = -(261.7 \pm 8.9)$ kJ/mol. The value is consistent with that for LaB$_6$: 1/5 ΔH°_f(YB$_4$) $\approx$ 1/7 ΔH°_f(LaB$_6$), see p. 209, Topor, Kleppa [23].

For standard entropy, thermal conductivity, and Debye temperature, see pp. 13/4.

Magnetic Properties

YB$_4$ is a diamagnetic compound with the susceptibility $\chi_g = -24 \times 10^{-8}$ cm^3/g at 300 K that is only slightly temperature dependent in the range 77 to 300 K, Buschow, Creyghton [10]. A similar behavior above 77 K was observed by Bressel et al. [13]. Below 77 K the susceptibility increases and becomes positive below 28 K, indicating the presence of paramagnetic impurities [13]. A positive susceptibility $\chi_g = 2.33 \times 10^{-6}$ cm³/g was found at room temperature, but no temperature dependence was measured, Paderno et al. [24].

Electrical Properties

Electronic Structure. Fermi Surface. de Haas-van Alphen Effect

For the general bonding model of the UB_4-type rare earth tetraborides, developed by Lipscomb, Britton [25], see p. 17. Attempts to verify the model by Hall-effect measurements on YB_4 showed that the simple relation $R_H = -1/(ne)$ between Hall constant and carrier concentration n does not hold, as the Fermi sphere corresponding to one electron overlapped into four zones, Johnson, Daane [11]. From the interpretation of the de Haas-van Alphen effect, Tanaka, Ishizawa [26], the tetraboride structure is expected to need four more valence electrons from Y per unit cell to fill the bonding orbitals than predicted from the model of [25].

The de Haas-van Alphen effect (dHvA) was measured on a high-purity single crystal bounded by {100} and {001} planes. Magnetic fields B up to 6 T and a temperature of about 1.5 K were used. Many frequency branches were observed in the frequency range 2×10^2 to 14.5×10^2 T and the dHvA amplitudes were measured for cyclotron mass determinations. For a figure of the frequency dependence on the magnetic field direction with respect to the crystal axes, see the paper; rotation was within the three principal planes (110), (001), and (100). Some of the branches could be traced only partially. The α frequencies were detected through all the angular regions measured. The γ frequencies have the largest amplitude but show a drastic angular dependence and could not be detected in some regions. The observed dHvA frequencies F and cyclotron masses m^*/m_0 for the various branches for high-symmetry directions are:

B parallel to	[001]	[001]	[010]	[010]	[010]	[010]	[010]
branch	α	γ	α	β	γ	δ	ϵ
F in 10^2 T	2.0	14.4	4.95	4.25	6.30	2.85	8.05
m^*/m_0	0.13	—	0.23	0.2	0.30	—	0.4
B parallel to	[110]	[110]	[110]	[110]	[110]	[110]	[110]
branch	α	β	γ	δ	ν	ω	λ
F in 10^2 T	4.90	4.48	12.65	3.65	10.05	6.05	7.67
m^*/m_0	0.24	0.22	—	0.2	—	—	—

The models shown in **Fig. 51** for the α and γ Fermi surface were deduced from the dHvA frequencies, even though it is difficult to construct a whole Fermi surface from the existing limited data. The β and δ Fermi surfaces are also expected to be closed. The volume contained in the α Fermi surface corresponds to 0.008 carriers per unit cell. The same order is expected also for the β surface, whereas estimates for δ and γ Fermi surfaces are 0.005 and 0.03, respectively. It seems that the carriers contained in the γ Fermi surface are compensated by some of the α, β, and δ Fermi surfaces. From consideration of the small carrier concentration and the nearly compensated carriers, YB_4 is proposed to be a semimetal with a carrier concentration of about 0.03 per unit cell. This is also supported by a discussion of the rather large Hall coefficient, see p. 138, Tanaka, Ishizawa [26].

The dHvA results, together with crystal structural features, lead to the interpretation of the electronic structure as a superposition of those of YB_6 and YB_2 as illustrated in **Fig. 52**. This model predicts YB_4 to be a semimetal with a low density of states at the Fermi level, consistent with the experimental facts, Tanaka, Ishizawa [26]. For the B Kα emission band see p. 18.

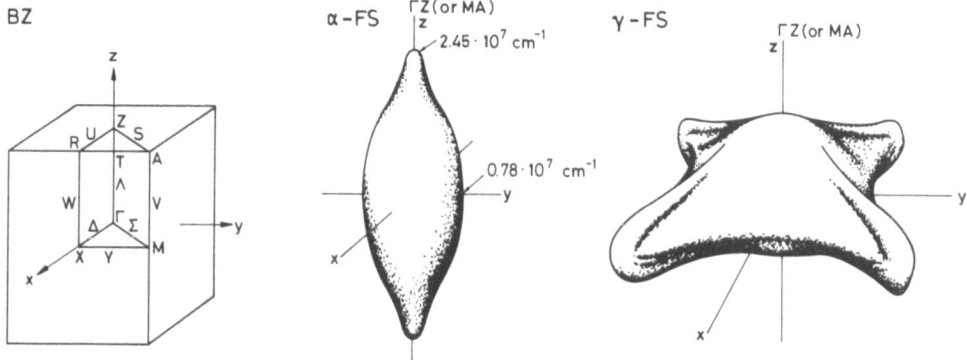

Fig. 51. Brillouin zone (BZ) and proposed Fermi surfaces (FS) of YB$_4$.

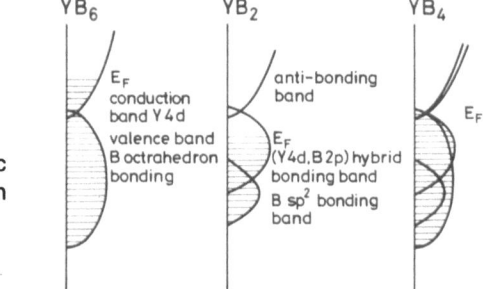

Fig. 52. Schematic diagram of the electronic structure of YB$_4$, interpreted as superposition of those of YB$_6$ and YB$_2$.

Electrical Conductivity

The electrical resistivity of a high-purity single crystal at room temperature $\varrho(300)$ is 26.2 μΩ·cm for I∥[100], Tanaka, Ishizawa [26]. As-grown single crystals at room temperature showed $\varrho = 31.3 \pm 1.0$ μΩ·cm and $\varrho = 28.5 \pm 1.3$ μΩ·cm parallel and perpendicular to c, respectively [11], and $\varrho = 27$ and 23 μΩ·cm parallel to c before and after annealing for 1 d at 2073 K, respectively, Bressel et al. [13]. For a sintered sample $\varrho = 35.3$ μΩ·cm is given by Paderno et al. [27], $\varrho = 34.85$ μΩ·cm by Severyanina et al. [28]; see p. 18.

Above 70 K up to room temperature, log ϱ for a Czochralski grown single crystal is nearly proportional to log T, whereas below ~25 K it is nearly temperature-independent according to a figure in [13]. The resistivity at 4.2 K was $\varrho = 9.8$ and 2 μΩ·cm before and after annealing, respectively. Thus the residual resistance ratio RRR = $\varrho(300)/\varrho(4.2)$ increased considerably from 2.8 to 11.5, which is close to that observed in an unannealed Al-flux grown crystal, Bressel et al. [13]. The highest RRR = 54 was observed in a crystal purified by multiple zone refining [26]. The temperature coefficient of ϱ between 300 and 1025 K is 5.4×10^{-3} K^{-1} [28].

The electrical resistance of YB$_4$ on Si prepared by co-deposition of Y and B on the cold substrate was higher than that of bulk YB$_4$. It could be decreased by annealing under nitrogen, Ryan, Roberts [29].

Hall Effect. Carrier Concentration. Mobility

The Hall coefficient is $R_H = -(213 \pm 9) \times 10^{-11}$ m^3/C for a single crystal with H∥c, indicating negative charge carriers, Johnson, Daane [11]. $R_H = -70.2 \times 10^{-11}$ m^3/C is given by Paderno et al. [27] for a sintered sample. For derived carrier concentrations, see p. 19.

Interpretation of the experimental Hall coefficient $R_H = -213 \times 10^{-11}$ m^3/C, within a one-band model yields a carrier concentration n of 0.6 per unit cell. The result, using a two-band model, should be smaller. A more realistic carrier concentration of 0.1 per unit cell is obtained by making some assumptions on the relative mobilities of electrons and holes, based on the observed effective masses. This value is more consistent with the result from the dHvA effect of 0.03 per unit cell, Tanaka, Ishizawa [26].

The failure of the simple relation to correlate the observed Hall coefficient with the predicted carrier concentration in M^{3+} tetraborides had already been stated by Johnson, Daane [11] and similarly by Auskern, Aronson [30]. The Hall measurements of [11] were interpreted to correspond to 0.2 conduction electrons per metal atom which requires an apparent oxidation state close to II, similar to that of Sm in SmB$_6$, Etourneau et al. [31, p.134]. The carrier concentration in YB$_4$ (and MB$_4$, M = Gd to Tm) was concluded to be lower than in Y, Gd to Tm metals from a simple relation between the Debye temperature, the molecular weight, and the electrical resistivity [28].

Thermionic Properties

The thermionic properties were studied with the YB$_4$ powder dispersed in an organic binder and placed on a Ta substrate in a thickness of 70 to 80 µm. After evacuation to $(2$ to $5) \times 10^{-7}$ Torr ($\triangleq 27$ to 67 µPa), conditioning of the YB$_4$ cathode at 1530 to 1890 K lowered the effective work function Φ_{eff} from 3.65 to 3.45 eV. The Richardson plot gave $\Phi_R = 3.2$ eV and the Richardson constant $A_R = 1.58 \times 10^3$ A·cm^{-2}·K^{-2}. Compared to the other Y borides, YB$_4$ has the highest emission current density, I = 0.284 A/cm^2 at 1890 K; but the emission properties are considerably inferior to those of LaB$_6$, Meerson et al. [5], see also Manelis et al. [32]. The temperature dependence was determined as $\Phi_{eff} = 3.2 + 2.5 \times 10^{-4} \cdot T$, in eV, between 1530 and 1890 K [32]. Low values $\Phi_R = 2.08$ eV and $A_R = 4.47 \times 10^{-2}$ A·cm^{-2}·K^{-2} before and $\Phi_R = 2.38$ eV and $A_R = 0.1$ A·cm^{-2}·K^{-2} after activation at 1550 K were derived from the Richardson plot; $\Phi_{eff} = 2.18 + 7.5 \times 10^{-4} \cdot T$ between 1473 and 1873 K. The current density was 0.6 A/cm^2 at 1873 K and a field strength of 5500 V/cm, Kudintseva et al. [6]. These values could not be confirmed [33].

The thermionic properties of yttrium boride samples with bulk compositions "YB$_3$" and "YB$_{5.1}$" were studied at 1190 to 1850 K, giving work functions $\Phi_R = 3.28$ and 3.95 eV, respectively, from Richardson plots; the experimental emission currents were extrapolated to zero field by Schottky plots. During operation, the surface composition and work function of any yttrium boride cathode will move to that of YB$_4$, which is the congruently vaporizing phase, Jaskie, Jacobson [33], see also [34].

Chemical Behavior. Radiation Effects. Use

Oxidation of YB$_4$ single crystals in air began above 1033 K, giving YBO$_3$ and B$_2$O$_3$. The temperature dependence of the oxidation rate indicated an activation energy of 95.6 kcal/mol ($\triangleq 400$ kJ/mol) [15].

YB$_4$ was found to be stable against H$_2$O steam under 6.9×10^6 Pa ($\triangleq 1000$ psi) at 773 K for 140 h; under the same conditions, other borides (YB$_6$, ErB$_4$, DyB$_4$) were completely converted to borates, Hoyt [35], cf. [36, 37]. YB$_4$ films on Si substrates had an increased resistance to steam oxidation (for 30 min at 1273 K) when Si had been added during deposition of the film, Ryan, Roberts [29].

Yttrium tetraboride powder did not dissolve on boiling for 30 min in 1:1 diluted HCl or 15% NaOH solutions, but was completely dissolved within 5 min in cold 1:1 HNO$_3$ or within 10 min in boiling 1:1 H$_2$SO$_4$ [2, 5].

The retention of helium formed from ^{10}B in the n, α reaction appears to be somewhat better than that for the hexaborides of Y, Sm, and Eu. Loosely packed powders welded together under irradiation, whereas the volume of solidified pellets increased considerably, e.g. by 19% at a 70% ^{10}B burnup, Holden et al. [38]; slightly different figures were given in another report on this study [35].

A layer of yttrium boride on a semiconductor substrate deposited from electron-beam-evaporated YB₄ was used as boron source for doping. There is a good ohmic contact between boride and substrate, Ishaq et al. [39].

References:

[1] Lundin, C. E. (in: Spedding, F. H.; Daane, A. H., The Rare Earths, Wiley, New York 1961, pp. 224/385, 247/8).
[2] Manelis, R. M.; Meerson, G. A.; Zhuravlev, N. N.; Telyukova, T.; Stepanova, A. A.; Gramm, N. V. (Poroshkovaya Metal. **1966** No. 11, pp. 77/84; Soviet Powder Met. Metal Ceram. **1966** 904/9).
[3] Binder, I. (Powder Met. Bull. **7** [1956] 74/5).
[4] Hoyt, E. W.; Chorné, J. (GEAP-3332 [1960] 1/15; C.A. **1961** 1338).
[5] Meerson, G. A.; Zhuravlev, N. N.; Manelis, R. M.; Runov, A. D.; Stepanova, A. A.; Grishina, L. P.; Gramm, N. V. (Izv. Akad. Nauk SSSR Neorgan. Materialy **2** [1966] 608/16; Inorg. Materials [USSR] **2** [1966] 527/33).
[6] Kudintseva, G. A.; Kuznetsova, G. M.; Bondarenko, V. P.; Selinova, N. F.; Shlyuko, V. Ya. (Poroshkovaya Metal. **1968** No. 2, pp. 45/53; Soviet Powder Met. Metal Ceram. **1968** 115/20).
[7] Tanaka, T.; Otani, S.; Ishizawa, Y. (J. Less-Common Metals **102** [1984] 281/7).
[8] Peshev, P. (Izv. Otd. Khim. Nauki Bulg. Akad. Nauk **4** [1971] 267/81; C.A. **76** [1972] No. 67516).
[9] Giese, R. F., Jr.; Matkovich, V. I.; Economy, J. (Z. Krist. **122** [1965] 423/32).
[10] Buschow, K. H. J.; Creyghton, J. H. N. (J. Chem. Phys. **57** [1972] 3910/4).

[11] Johnson, R. W.; Daane, A. H. (J. Chem. Phys. **38** [1963] 425/32).
[12] Johnson, R. W.; Daane, A. H. (IS-473 [1962]; N.S.A. **17** [1963] No. 11306).
[13] Bressel, B.; Chevalier, B.; Etourneau, J.; Hagenmuller, P. (J. Cryst. Growth **47** [1979] 429/33).
[14] Guette, A.; Vlasse, M.; Etourneau, J.; Naslain, R. (Compt. Rend. C **291** [1980] 145/8).
[15] Okada, S.; Atoda, T. (Yogyo Kyokaishi **89** [1981] 339/45; C.A. **95** [1981] No. 89085).
[16] Samsonov, G. V. (Tugoplavkie Soedineniya Redkozemel'nykh Metallov s Nemetallami, Metallurgiya, Moscow 1964; High-Temperature Compounds of Rare Earth Metals with Nonmetals, Consultants Bureau, New York 1965, pp. 1/280).
[17] Zhuravlev, N. N.; Belousova, I. A.; Manelis, R. M.; Belousova, N. A. (Kristallografiya **15** [1970] 836/8; Soviet Phys.-Cryst. **15** [1970/71] 723/4).
[18] Giese, R. F., Jr.; Matkovich, V. I.; Economy, J. (AD-613474 [1965] 1/18; C.A. **63** [1965] 7720).
[19] Rudman, R.; La Placa, S.; Post, B. (Acta Cryst. **16** [1963] A29).
[20] Berrada, A.; Mercurio, J.-P.; Chevalier, B.; Etourneau, J.; Hagenmuller, P.; Lalanne, M.; Gianduzzo, J. C.; Georges, R. (Mater. Res. Bull. **11** [1976] 1519/26).

[21] Severyanina, E. N.; Dudnik, E. M.; Paderno, Yu. B. (Poroshkovaya Metal. **1973** No. 12, pp. 72/4; Soviet Powder Met. Metal Ceram. **12** [1973] 1001/2).
[22] Storms, E. K.; Mueller, B. A. (J. Appl. Phys. **52** [1981] 2966/70).
[23] Topor, L.; Kleppa, O. J. (High Temp. Sci. **22** [1986] 139/44).

[24] Paderno, Yu. B.; Severyanina, E. N.; Fedchenko, R. G. (Izv. Akad. Nauk SSSR Neorgan. Materialy 10 [1974] 1900; Inorg. Materials [USSR] 10 [1974] 1633/4).

[25] Lipscomb, W. N.; Britton, D. (J. Chem. Phys. 33 [1960] 275/80).

[26] Tanaka, T.; Ishizawa, Y. (J. Phys. C 18 [1985] 4933/40).

[27] Paderno, Yu. B.; Severyanina, E. N.; Dudnik, E. M.; Lazorenko, V. (V Sb. Splavy Redk. Met. Osob. Fiz. Khim. Svoistv. 1975 118/21; C.A. 85 [1976] No. 12590).

[28] Severyanina, E. N.; Dudnik, E. M.; Paderno, Yu. B. (Poroshkovaya Metal. 1974 No. 10, pp. 83/5; Soviet Powder Met. Metal Ceram. 13 [1974] 843/5).

[29] Ryan, J. G.; Roberts, S. (Thin Solid Films 135 [1986] 9/19).

[30] Auskern, A. B.; Aronson, S. (J. Chem. Phys. 49 [1968] 172).

[31] Etourneau, J.; Mercurio, J.-P.; Hagenmuller, P. (in: Matkovich, V. I., Boron and Refractory Borides, Springer, Berlin – New York 1977, pp. 115/38).

[32] Manelis, R. M.; Telyukova, L. P.; Grishina, L. P. (Izv. Akad. Nauk SSSR Neorgan. Materialy 6 [1970] 1184/5; Inorg. Materials [USSR] 6 [1970] 1035/6).

[33] Jaskie, J.; Jacobson, D. (Proc. Intersoc. Energy Convers. Eng. Conf. 15 [1980] 2331/3; C.A. 94 [1981] No. 142615).

[34] Jaskie, J. E. (Diss. Arizona State Univ. 1981, pp. 1/123; Diss. Abstr. Intern. B 42 [1982] 4098).

[35] Hoyt, E. W. (Proc. 2nd Conf. Rare Earth Res., Glenwood Springs, Colo., 1961 [1962], pp. 287/99; N.S.A. 16 [1962] No. 32072).

[36] Hoyt, E. W.; Chorné, J.; Cummings, W. V. (GEAP-3548 [1960] 1/17; C.A. 1961 16369).

[37] Hoyt, E. W.; Zimmerman, D. L. (GEAP-3743-Pt. II [1962] 1/18; N.S.A. 17 [1963] No. 20512).

[38] Holden, A. N.; Hoyt, E. W.; Cummings, W. V.; Zimmerman, D. L. (Prop. Reactor Mater. Eff. Radiat. Damage Proc. Intern. Conf., Berkeley Castle, Gloucestershire, Engl., 1961 [1962], pp. 457/74).

[39] Ishaq, M. H.; Roberts, S.; Ryan, J. G. (U.S. 4490193 [1983] 1/6; C.A. 102 [1985] No. 104610).

32.1.8.2.5 YB_6

32.1.8.2.5.1 Preparation and Stoichiometry

The existence of a homogeneity range for YB_6 is indicated by the spread of lattice parameters, electrical resistance ratios, and superconducting transition temperatures, Fisk et al. [1, 2]. For additional data, see p. 22.

Preparation of Polycrystalline Samples

In general, it is difficult to synthesize blue YB_6 free of YB_4. The reported formation temperatures are nearly the same as the reported decomposition temperatures (see p. 153). Thus, at a given temperature, the reaction time can be regarded as a dominant parameter. Sometimes it is possible to remove a surface layer of YB_4 from a YB_6 core (see "Densification", p. 142). Formation of this YB_4 layer is frequently observed in YB_6 synthesis because of decomposition on the surface of the YB_6 already formed. Boron used in the syntheses was usually contaminated by carbon.

Synthesis directly from the elements was carried out in an arc melter under Ar. The product was remelted into rods. During the final zone melting slightly below 2873 K under Ar, large grains formed from the boron-rich zone created with the zone-leveling technique, Johnson [3].

The borothermal method, $Y_2O_3 + 15B \rightarrow 2YB_6 + 3/2B_2O_2\uparrow$, is carried out under vacuum with amorphous boron. Crystalline boron leads to a higher carbon content in the product. Vapor-pressure measurements show that the formation of YB$_6$ begins to take place vigorously at 1793 K. A sharp pressure rise was observed at 1793 to 1813 K. Theoretically, the formation of YB$_6$ becomes possible above about 1900 K (see a figure in the paper), Bondarenko et al. [4]. Optimum conditions for YB$_6$ preparation consist of mixing the oxide with an excess of 5% of amorphous boron for 8 h in alcohol in a rubber-lined steel ball mill containing rubber-coated steel balls. The charge was dried at 343 to 353 K, rubbed through a 42 µm kapron sieve, compacted to briquets, and heated in a vacuum furnace with a tungsten heater, 2073 K, 10^{-6} Torr ($\triangleq 133$ µPa), 1 h. The product obtained was stoichiometric, Meerson, Mamedov [5], and the content of Fe, Si, Al, Mg, Ca, Pb, Cu, and Zn was less than 10^{-3}%, Mamedov et al. [6]. For systematic investigations of impurities resulting from various mill, mortar, and sieve materials, etc., see the paper [5]. In the same laboratory single-phase YB$_6$ containing less than 0.01% of impurities was obtained by mixing stoichiometric amounts of starting materials in an agate mortar, with subsequent sieving, briquetting at a pressure of 600 daN/cm^2 ($\triangleq 60$ MPa), and reducing in a tantalum crucible (vacuum furnace with tungsten heater, initial vacuum (2 to 5)$\times 10^{-5}$ Torr) at 1973 K, 1 h. Different amounts of YB$_4$ were found in products obtained below 1973 K, as well as above 1973 K (studies from 1673 to 2273 K), Manelis et al. [7]. Within an investigated temperature range of 1673 to 1973 K, a single-phase product was obtained between 1773 and 1823 K under 10^{-2} to 10^{-3} Torr ($\triangleq 1.33$ to 0.133 Pa) for 1 h, using a BN crucible, Kawabe et al. [8]. The starting materials with an additional boron excess of 15% were heated in a ZrB$_2$ crucible at 10^{-4} Torr ($\triangleq 13.3$ mPa) in two steps: about 30 min at 1273 K and then about 1 h at 1673 K. The cooled samples were reground and the second step was repeated, Sobczak, Sienko [9]. The same laboratory published a synthesis using stoichiometric amounts of Y$_2$O$_3$ and B at about 1973 K under 10^{-5} Torr ($\triangleq 1.33$ mPa). The samples were reground and reheated, Hiebl, Sienko [10]. Variation of the reaction temperature (1173 to 2173 K) and boron excess (0 to 5%), as well as experiments investigating the addition of YB$_6$ to a starting charge, were examined by Kudintseva et al. [11]. Reaction temperatures were also investigated by Bondarenko et al. [12]. These experiments as well as the resulting optimum preparation conditions led to a product contaminated with YB$_4$. The optimum conditions were given as follows: starting materials are pressed into compacts at a pressure of 0.6 to 0.8 t/cm^2 ($\triangleq 58.8$ to 78.5 MPa), heated in a vacuum furnace equipped with a boronized graphite heater, initial vacuum 10^{-2} to 10^{-3} Torr ($\triangleq 1.33$ to 0.133 Pa), at 1973 K for 1 h, sieved, pressed again, and reheated at 1873 K for 40 min or at 1973 K for 30 min. The product contained 57.8% Y, 41.4% B, and 0.1% carbon [11]. A single-phase product was obtained by preparing YB$_6$ and subsequently melting it in an argon-arc furnace in order to remove small amounts of oxides. Final crushing to a powder was carried out in cyclohexane and the sample was dried under 10^{-7} Torr ($\triangleq 13.3$ µPa), Aono et al. [13]. For the preparation of highly impure YB$_6$ under semilaboratory conditions, electrovacuum furnace, 1923 K, 90 min, see Polishchuk [14]. YB$_6$ with the intended by-products YB$_4$ and YB$_4$ plus YB$_{12}$ were prepared under vacuum at 1873 K from charges of molar ratios B:Y$_2$O$_3$=14, 20, and 26. Samples so prepared had different transition temperatures into the superconducting state (see p. 149), indicating a variation in the stoichiometry of YB$_6$ [2].

The YB$_6$ formation via $Y_2O_3 + 3B_4C \rightarrow 2YB_6 + 3CO\uparrow$ (boron carbide method) in vacuum appears to begin at 1243 K. The compound was prepared at 2073 to 2173 K with a yield of 92 to 93%, Kudintseva et al. [15]. Reaction at 1673, 1823, and 1973 K gave YB$_6$ admixed with other borides. Y$_2$O$_3$ and B$_2$O$_3$ react in the presence of carbon in a similar manner [8].

Densification of Polycrystalline Samples

Relatively dense samples (porosity ca. 29%) were prepared by pressing of YB_6 powder mixed with a 10% solution of polyvinyl alcohol at a pressure of 4 to 5 t/cm^2 ($\triangleq$ 392 to 490 MPa) and subsequently sintering at 2650 K for 30 min at $\leqq (1 \text{ to } 3) \times 10^{-5}$ Torr ($\triangleq$1.33 to 4 mPa). Studies of samples sintered at a temperature of ca. 1900 to 2700 K show that the higher the sintering temperature, the higher the relative density and linear shrinkage of the samples, see figures in the paper. Increasing the sintering temperature to above ca. 2600 K leads to a drop in the relative density as a result of extensive evaporation (formation of a dense gray surface layer of YB_4). Evaporation during sintering can be decreased by covering the sample with YB_6 powder. YB_6 sintered at 2650 K has a microstructure consisting of a grain size of 20 to 30 μm predominantly, Meerson et al. [16]. Porosity values of 22 to 26% are published in [7].

YB_6 powder was readily hot pressed, but a gray-brown-colored YB_4 phase formed on the surface, the thickness of which depended on pressing conditions and increased with increasing temperature and holding time. Optimum conditions are 220 kg/cm^2 ($\triangleq$ 21.6 MPa) pressure, 2273 ± 20 K, 20 min reaction time [11]. YB_6 with a density of 92 to 96% was prepared by hot pressing YB_6 powder (obtained from $Y_2O_3 + 18$ B) between 1773 and 1973 K and a pressure of 4000 to 6000 psi (27.5 to 41.4 MPa) in graphite dies under vacuum with a pressing time of 10 to 15 min, Kaznoff et al. [17]. Hot-pressed YB_6 obtained at 1773, 2073, and 2573 K with density values of 90, 99, and 99%, respectively, contained some YB_4 in the bulk, Kaznoff, Wilson [18, 19]. After heating this material in a tantalum resistance furnace at 2273 K and 10^{-6} Torr ($\triangleq$133 μPa) for up to 6 h the surface was coated with a YB_4 layer, whereas the interior of the sample was free of YB_4. Large porosity increases were noted. The grain size increased from 11 to 14.4 μm. After 6 h, cracks developed in the grains. No grain growth took place at 2173 K within 10 to 20 h [18].

Single-Crystal Growth

Single crystals of YB_6, all presumably nonstoichiometric (B-rich), were prepared from arc-melted buttons using an Al–Ga flux growth technique (95 at% Ga/5 at% Al) executed in an Al_2O_3 crucible under Ar at 1823 K. The stoichiometry of the crystals was varied by varying the starting charge nominal composition from YB_6 to YB_9. Charges more Y-rich than YB_6 produced YB_4 crystals along with YB_6 [1].

Preparation of Films

Thin films of YB_6 were prepared using a dual-electron-beam co-evaporation method. The boron to yttrium ratio on a single-crystal silicon substrate was 5.9:1, if the evaporation rate for boron was 0.31 nm/s and that for yttrium was 0.22 nm/s. It is necessary to deposit boron at low rates in order to avoid source material explosions caused by localized overheating, Ryan, Roberts [20]. Films of 300 to 1200 nm thickness were prepared by magnetic field-assisted d.c. sputtering on hot (1273 K) single-crystal sapphire substrates. The sputtering target (20 mm diameter) was produced by co-melting of high-purity yttrium and boron in an arc furnace with a ratio of B/Y $\approx$6.0, Schneider et al. [21].

References:

[1] Fisk, Z.; Schmidt, P. H.; Longinotti, L. D. (Mater. Res. Bull. **11** [1976] 1019/22).

[2] Fisk, Z.; Lawson, A. C.; Fitzgerald, R. W. (Mater. Res. Bull. **9** [1974] 633/6).

[3] Johnson, R. W. (J. Appl. Phys. **34** [1963] 352/5).

[4] Bondarenko, V. P.; Bilyk, I. I.; Shlyuko, V. Ya. (Poroshkovaya Metal. **1966** No. 9, pp. 43/9; Soviet Powder Met. Metal Ceram. **1966** 713/7).

[5] Meerson, G. A.; Mamedov, F. G. (Izv. Akad. Nauk SSSR Neorgan. Materialy **3** [1967] 802/7; Inorg. Materials [USSR] **3** [1967] 717/21).

[6] Mamedov, F. G.; Meerson, G. A.; Zhuravlev, N. N.; Umanskii, Ya. S. (Izv. Akad. Nauk SSSR Neorgan. Materialy **3** [1967] 950/6; Inorg. Materials [USSR] **3** [1967] 851/6).

[7] Manelis, R. M.; Meerson, G. A.; Zhuravlev, N. N.; Telyukova, T. M.; Stepanova, A. A.; Gramm, N. V. (Poroshkovaya Metal. **1966** No. 11, pp. 77/84; Soviet Powder Met. Metal Ceram. **1966** 904/9).

[8] Kawabe, U. T.; Yamamoto, S.; Tokorozawa, S.; Aita, T. K.; Honda, Y. H. (Ger. Offen. 2442537 [1975]; C.A. **83** [1975] No. 124943).

[9] Sobczak, R. J.; Sienko, M. J. (J. Less-Common Metals **67** [1979] 167/71).

[10] Hiebl, K.; Sienko, M. J. (Inorg. Chem. **19** [1980] 2179/80).

[11] Kudintseva, G. A.; Kuznetsova, G. M.; Bondarenko, V. P.; Selivanova, N. F.; Shlyuko, V. Ya. (Poroshkovaya Metal. **1968** No. 2, pp. 45/53; Soviet Powder Met. Metal Ceram. **1968** 115/20).

[12] Bondarenko, V. P.; Selivanova, N. F.; Shlyuko, V. Ya. (Visn. Kiivs'k. Politekhn. Inst. Ser. Mashinobuduv. **3** [1967] 174/9; C.A. **68** [1968] No. 118872).

[13] Aono, M.; Kawai, S.; Kono, S.; Okusawa, M.; Sagawa, T.; Takehana, Y. (J. Phys. Chem. Solids **37** [1976] 215/9).

[14] Polishchuk, V. S. (Poroshkovaya Metal. **1965** No. 7, pp. 100/7; Soviet Powder Met. Metal Ceram. **1965** 596/601).

[15] Kudintseva, G. A.; Poliakova, M. D.; Samsonov, G. V.; Tsarev, B. M. (Fiz. Metal. Metalloved. **6** [1958] 272/5; Phys. Metals Metallog. [USSR] **6** No. 2 [1958] 83/6).

[16] Meerson, G. A.; Manelis, R. M.; Telyukova, T. M. (Izv. Akad. Nauk SSSR Neorgan. Materialy **2** [1966] 291/8; Inorg. Materials [USSR] **2** [1966] 250/5).

[17] Kaznoff, A. I.; Hoyt, E. W.; Grossman, L. N. (Advan. Energy Convers. **3** [1963] 167/73; C.A. **61** [1964] 3782).

[18] Kaznoff, A. I.; Wilson, J. G. (GEST-2009 [1962] 6-1/6-57, 6-7, 6-17/6-18; N.S.A. **17** [1963] No. 14819).

[19] Kaznoff, A. I.; Wilson, J. G. (GEST-2009 [1962] 7-1/7-26, 7-4/7-5; N.S.A. **17** [1963] No. 14819).

[20] Ryan, J. G.; Roberts, S. (Thin Solid Films **135** [1986] 9/19).

[21] Schneider, R.; Geerk, J.; Rietschel, H. (Europhys. Letters **4** [1987] 845/9).

32.1.8.2.5.2 Crystallographic Properties. Density

YB$_6$ has the cubic CaB$_6$ structure. The lattice constant a (in Å) of nonstoichiometric (B-rich) single crystals, grown from an Al–Ga flux, change from a = 4.1021 to 4.1027, if the starting charge nominal composition was varied from YB$_6$ to YB$_9$, Fisk et al. [1]. Lattice constants of polycrystalline samples are given in the order of preparation methods. From elements: a = 4.103 ± 0.001, Johnson, Daane [2, 3], a = 4.08, Kiessling [4], and a = 4.0997 ± 0.0005 Å for YB$_6$ at the Y-rich boundary, Jaskie, Jacobson [5]. From the borothermal method: a = 4.0994 ± 0.001, Hiebl, Sienko [6]; a = 4.10, Sobzcak, Sienko [7]; a = 4.102 with error limits up to 1×10^{-3}, Meerson et al. [8, 9], Meerson, Mamedov [10], Manelis et al. [11], Mamedov et al. [12], Zhuravlev et al. [13]. The corresponding calculated density is D$_{calc}$ = 3.68 g/cm^3 for stoichiometric YB$_6$ [12], and is D$_{calc}$ = 3.70 g/cm^3 for a composition Y:B = 1:5.9 [8, 9, 11]. Preparation from fused salt electrolysis: a = 4.07, Allard [14], and a = 4.113, Blum, Bertaut [15]. D$_{exp}$ = 3.72 g/cm^3 was measured at 288 K by Andrieux [16]. From boron carbide method:

a = 4.126, D_{calc} = 3.63 g/cm^3 compared to D_{exp} = 4.1 g/cm^3 determined pycnometrically in ethylbenzene, Tvorogov [17]; a = 4.128, D_{calc} = 3.633 g/cm^3, D_{exp} = 3.64 ± 0.04 g/cm^3, Kudintseva et al. [18].

References:

[1] Fisk, Z.; Schmidt, P. H.; Longinotti, L. D. (Mater. Res. Bull. **11** [1976] 1019/22).

[2] Johnson, R. W.; Daane, A. H. (J. Chem. Phys. **38** [1963] 425/32).

[3] Johnson, R. W.; Daane, A. H. (IS-473 [1962] 1/30, 17; N.S.A. **17** [1963] No. 11306).

[4] Kiessling, R. (Acta Chem. Scand. **4** [1950] 209/27, 210, 219).

[5] Jaskie, J.; Jacobson, D. (Proc. Intersoc. Energy Convers. Eng. Conf. **15** [1980] 2331/3; C. A. **94** [1981] No. 142615).

[6] Hiebl, K.; Sienko, M. J. (Inorg. Chem. **19** [1980] 2179/80).

[7] Sobczak, R. J.; Sienko, M. J. (J. Less-Common Metals **67** [1979] 167/71).

[8] Meerson, G. A.; Zhuravlev, N. N.; Manelis, R. M.; Runov, A. D.; Stepanova, A. A.; Grishina, L. P.; Gramm, N. V. (Izv. Akad. Nauk SSSR Neorgan. Materialy **2** [1966] 608/16; Inorg. Materials [USSR] **2** [1966] 527/33).

[9] Meerson, G. A.; Manelis, R. M.; Telyukova, T. M. (Izv. Akad. Nauk SSSR Neorgan. Materialy **2** [1966] 291/8; Inorg. Materials [USSR] **2** [1966] 250/5).

[10] Meerson, G. A.; Mamedov, F. G. (Izv. Akad. Nauk SSSR Neorgan. Materialy **3** [1967] 802/7; Inorg. Materials [USSR] **3** [1967] 717/21).

[11] Manelis, R. M.; Meerson, G. A.; Zhuravlev, N. N.; Telyukova, T. M.; Stepanova, A. A.; Gramm, N. V. (Poroshkovaya Metal. **1966** No. 11, pp. 77/84; Soviet Powder Met. Metal Ceram. **1966** 904/9).

[12] Mamedov, F. G.; Meerson, G. A.; Zhuravlev, N. N.; Umanskii, Ya. S. (Izv. Akad. Nauk SSSR Neorgan. Materialy **3** [1967] 950/6; Inorg. Materials [USSR] **3** [1967] 851/6).

[13] Zhuravlev, N. N.; Belousova, I. A.; Manelis, R. M.; Belousova, N. A. (Kristallografiya **15** [1970] 836/8; Soviet Phys.-Cryst. **15** [1970] 723/4).

[14] Allard, G. (Bull. Soc. Chim. France [4] **51** [1932] 1213/5).

[15] Blum, P.; Bertaut, F. (Acta Cryst. **7** [1954] 81/6).

[16] Andrieux, M. L. (Ann. Phys. [Paris] [10] **12** [1929] 423/507, 476).

[17] Tvorogov, N. N. (Zh. Neorgan. Khim. **4** [1959] 1961/6; Russ. J. Inorg. Chem. **4** [1959] 890/3).

[18] Kudintseva, G. A.; Poliakova, M. D.; Samsonov, G. V.; Tsarev, B. M. (Fiz. Metal. Metalloved. **6** [1958] 272/5; Phys. Metals Metallog. [USSR] **6** No. 2 [1958] 83/6).

32.1.8.2.5.3 Lattice Vibrations. Electron-Phonon Coupling

Inelastic neutron scattering experiments on polycrystalline YB$_6$ (energy-loss mode, excitation energy E_0 = 64.3 meV) exhibit the lowest frequency at 10 meV (see a figure in the paper), which (like that of LaB$_6$ and CeB$_6$) was attributed to the acoustic translational F_{1u} mode, Schell et al. [1]. For features in the acoustic phonon part derived from electron tunneling studies, see p. 145. Raman-scattering data of other trivalent hexaborides, extrapolated to YB$_6$ (a = 4.113 Å used as lattice constant), lead to the following modes: 86.8 meV (F_{2g}), 149.4 meV (E_g), and 163.7 meV (A_{1g}), Schell [2]. A phonon density of states model which considers central forces between nearest-neighbor and next-nearest-neighbor boron atoms, together with a metal nearest-neighbor boron-atom central force and a Keating-type angular force constant was fitted to these four frequencies (for details, see [2]) to give the phonon density of states curve shown in **Fig. 53**. The frequency of the torsional F_{1g} mode is predicted to lie at 30 meV.

The low-frequency vibrations involving the undeformed B$_6$ octahedron of YB$_6$ were all below those of LaB$_6$ [1]. For the phonon dispersion of YB$_6$ see a figure in the paper [2].

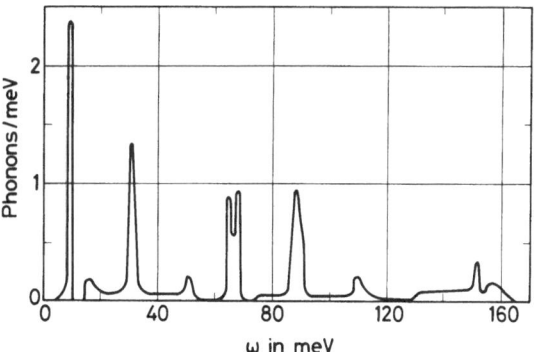

Fig. 53. Phonon density of states for YB$_6$.

The relatively small frequencies of the low-frequency vibrations in YB$_6$ compared to LaB$_6$, in spite of the lighter mass of Y, indicate a stronger electron-phonon coupling in YB$_6$ [1], Kunii et al. [3]. The evaluated electronic structure (see p. 147) and the phonon modes were used to calculate the Eliashberg functions within the rigid-muffin-tin approximation for electron-phonon coupling. It turned out that nonlocal corrections to the electron-phonon coupling were essential for those modes where the B$_6$ octahedra move as a whole (for details and a figure, see the paper). Omission of these corrections led to a large overestimate of the McMillan parameter λ and the superconducting transition temperature T$_c$. The hypothesis that the boron sublattice is mainly responsible for superconductivity is discussed [1]. However, electron tunneling measurements into superconducting YB$_6$ (see p. 149) indicate strong electron-phonon coupling of conduction electrons with acoustic phonons, and the high T$_c$ was well described by the Eliashberg equations in this low-energy region without proximity effect corrections. The Eliashberg function reveals a shoulder at 4.9 meV (origin not clear) and a wide peak around 8.5 meV in the case of a YB$_6$ film at 1.2 K, without any significant contribution above ca. 16 meV, Schneider et al. [4]. A broad main peak at 11 meV (Einstein-like mode of Y) and a small peak above 20 meV were observed with a single crystal at 1.8 K. The rapid change in the 4d character on the Y site at the Fermi level (see p. 147) as well as the T^3 law in resistivity at low temperatures and the nearly linear behavior above 60 K (see p. 149) are consistent with the strong electron-phonon coupling with Y modes [3].

References:

[1] Schell, G.; Winter, H.; Rietschel, H.; Gompf, F. (Phys. Rev. [3] B **25** [1982] 1589/99).

[2] Schell, G. (KFK-3001 [1980] 1/88; INIS Atomindex **11** [1980] No. 561 104).

[3] Kunii, S.; Kasuya, T.; Kadowaki, K.; Date, M.; Woods, S. B. (Solid State Commun. **52** [1984] 659/61).

[4] Schneider, R.; Geerk, J.; Rietschel, H. (Europhys. Letters **4** [1987] 845/9).

32.1.8.2.5.4 Mechanical and Thermal Properties

For the density see p. 143. The microhardness of dense samples is $3264 \pm 21\,\text{kg/mm}^2$ ($\triangleq 32\,\text{GPa}$) at a load of 50 g, Kudintseva et al. [1], and $2575 \pm 100\,\text{kg/mm}^2$ ($\triangleq 25.3\,\text{GPa}$), Meerson et al. [2, 3], Manelis et al. [4], at a load of 100 g [3]. A microhardness (load 100 g) of $2737 \pm 129\,\text{kg/mm}^2$ ($\triangleq 26.8\,\text{GPa}$) is published by Kondrashev [5]. A value similar to that of [2 to 4] is read from a figure in Kovenskaya et al. [12]. Investigations of the flexural strength σ in dependence of the sintering temperature above ~ 2150 K show a monotonous increase up to 380 to 400 kg/cm^2 ($\triangleq 37.3$ to 39.2 MPa) at 2650 K. Further increase of the sintering temperature leads to a drop in the strength, see a figure in [3]. The value $\sigma = 270\,\text{kg/cm}^2$ ($\triangleq 26.5$ MPa) of a sample sintered at 2450 K [2, 4] agrees with the data of [3]. Hot-pressed YB_6 samples, contaminated with YB_4, are mechanically stable in thermal cycling experiments (inductive heating) between 773 and 1973 K for 50 cycles. Conditions: cycle period of 2 min, 30 s at full power (mean heat-up rate 3500 K/min), and 90 s at zero power; vacuum below 10^{-6} Torr ($\triangleq 133\,\mu\text{Pa}$). At upper temperatures of 2273 K and higher, YB_6 specimens generally break and their grains show many microcracks, Kaznoff, Wilson [6].

The mean thermal expansion coefficient $\alpha = (6.02 \pm 0.6) \times 10^{-6}\,\text{K}^{-1}$ of YB_6 prepared by borothermal method was calculated from the lattice constants between 293 and 873 K, Zhuravlev et al. [7]; $\alpha \approx 6.2 \times 10^{-6}\,\text{K}^{-1}$ at ~ 300 K, from a figure in [12].

YB_6 melts incongruently: The melting point $T_m = 2873$ K, published by Schreyer et al. [8, 9], is in agreement with the phase diagram, see p. 130. Early values are 2780 K [2] and 2573 K [1]. $T_m \approx 2950$ K and the Debye temperature $\Theta_D \approx 770$ K are read from figures in [12].

The heat of formation $\Delta H^\circ_{298} = -24\,\text{kcal/mol}$ ($\triangleq -100$ kJ/mol) was calculated from data of vapor pressure experiments (samples obtained by boron carbide method) using the entropy values 10.58 and 10.52 cal·mol^{-1}·K^{-1} ($\triangleq 44.3$ and 44 J·mol^{-1}·K^{-1}) for YB_6 and Y_2O_3, respectively [1].

The molar heat capacity as a function of temperature follows the equation $C_p = 24.56 + 21.27 \times 10^{-3}\,T$ cal·mol^{-1}·K^{-1} ($\triangleq C_p = 102.8 + 8.9 \times 10^{-2}\,T$ J·mol^{-1}·K^{-1}), based on the Debye temperature and the melting point (no values given), Bondarenko et al. [10]. At low temperatures an anomaly associated with the transition into the superconducting state ($T_c \approx 7$ K) was observed. The extrapolated electronic heat capacity coefficient γ amounts to roughly 1×10^{-4} cal·K^{-2}·(g-atom B)$^{-1}$ ($\triangleq 418\,\mu\text{J}$·K^{-2}·(g-atom B)$^{-1}$), Matthias et al. [11].

References:

[1] Kudintseva, G. A.; Poliakova, M. D.; Samsonov, G. V.; Tsarev, B. M. (Fiz. Metal. Metalloved. **6** [1958] 272/5; Phys. Metals Metallog. [USSR] **6** No. 2 [1958] 83/6).

[2] Meerson, G. A.; Zhuravlev, N. N.; Manelis, R. M.; Runov, A. D.; Stepanova, A. A.; Grishina, L. P.; Gramm, N. V. (Izv. Akad. Nauk SSSR Neorgan. Materialy **2** [1966] 608/16; Inorg. Materials [USSR] **2** [1966] 527/33).

[3] Meerson, G. A.; Manelis, R. M.; Telyukova, T. M. (Izv. Akad. Nauk SSSR Neorgan. Materialy **2** [1966] 291/8; Inorg. Materials [USSR] **2** [1966] 250/5).

[4] Manelis, R. M.; Meerson, G. A.; Zhuravlev, N. N.; Telyukova, T. M.; Stepanova, A. A.; Gramm, N. V. (Poroshkovaya Metal. **1966** No. 11, pp. 77/84; Soviet Powder Met. Metal Ceram. **1966** 904/9).

[5] Kondrashev, A. I. (Poluch. Issled. Svoistv Nov. Mater. Mater. 10th Nauchn. Konf. Aspir. Molodykh Issled. Inst. Probl. Materialoved. Akad. Nauk Ukr. SSR, Kiev 1976 [1978], pp. 48/52; C.A. **91** [1979] No. 111144).

[6] Kaznoff, A. I.; Wilson, J. G. (GEST-2009 [1962] 7-1/7-26; N.S.A. **17** [1963] No. 14819).

[7] Zhuravlev, N. N.; Belousova, I. A.; Manelis, R. M.; Belousova, N. A. (Kristallografiya **15** [1970] 836/8; Soviet Phys.-Cryst. **15** [1970] 723/4).

[8] Schreyer, J. M.; Schmitt, C. R.; Hays, R. A.; Farwell, D. (Sol. Energy **25** [1980] 179/85; C.A. **94** [1981] No. 106502).

[9] Schreyer, J. M.; Schmitt, C. R.; Hays, R. A.; Farwell, D. (AES Coat. Sol. Collect. 2nd Symp. Proc., Winter Park, Fla., 1979, Paper No. 14, pp. 1/10; C.A. **94** [1981] No. 159825).

[10] Bondarenko, V. P.; Bilyk, I. I.; Shlyuko, V. Ya. (Poroshkovaya Metal. **1966** No. 9, pp. 43/9; Soviet Powder Met. Metal Ceram. **1966** 713/7).

[11] Matthias, B. T.; Geballe, T. H.; Andres, K.; Corenzwit, E.; Hull, G. W.; Maita, J. P. (Science **159** [1968] 530).

[12] Kovenskaya, B. A.; Kondrashov, A. I.; Dudnik, E. M.; Kolotun, V. F. (Tugoplavkie Soedin. Redkozemel. Met. Mater. 3rd Vses. Semin., Novosibirsk 1977 [1979], pp. 36/9; C.A. **93** [1980] No. 60231).

32.1.8.2.5.5 Magnetic and Electrical Properties

Magnetization

Magnetization measurements at various low temperatures, plotted in **Fig. 54**, indicate that YB$_6$ behaves as a type II superconductor, Kunii et al. [1].

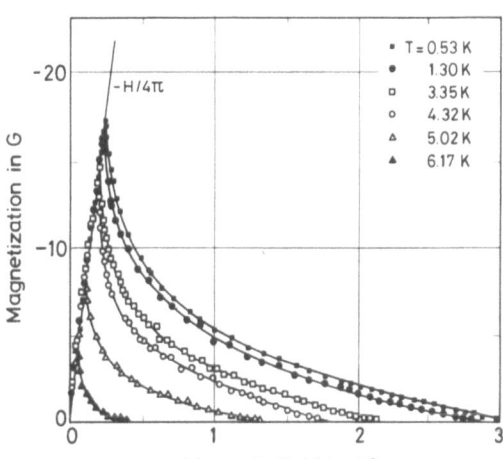

Fig. 54. Diamagnetic magnetization of YB$_6$ at various low temperatures.

Electronic Structure

Fig. 55, p. 148, shows the X-ray photoemission spectrum (XPS) of YB$_6$, related to the Fermi energy $E_F = 0$. Strong peaks at $E_B = -45.6$ and -25.3 eV correspond to the Y 4s and 4p core levels, respectively. The valence band region (shaded area) can be explained by the calculation of Longuet-Higgins, Roberts [2] (see p. 57), i.e., a model involving weak covalent bonding between the boron and metal atoms. In this picture, the peak centered at $E_B \approx -14$ eV corresponds to the A band (T$_{1u}$ states) and the broad peak at ca. -7 eV to the overlapping B and F bands (T$_{2g}$, T$''_{1u}$, A$'_{1g}$ states). The lowest lying E band (A$_{1g}$ states), expected from this calculation as a weak peak, is presumably concealed by the Y 4p peak. The tail of the spectrum which extends to the Fermi level may be attributed to conduction electrons. The conduction electrons are assumed to be in a relatively wide conduction band, as the spectral intensity due to the conduction electrons is weak, Aono et al. [3]. A schematic diagram of the electronic structure is shown in Fig. 52, p. 137, deduced from data for LaB$_6$ and fourth-group transition-metal diborides, Tanaka, Ishizawa [32].

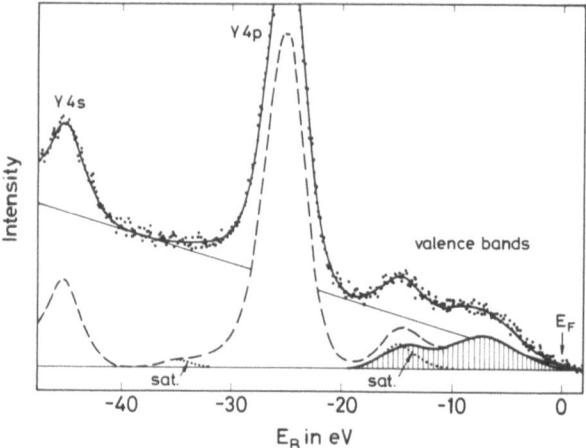

Fig. 55. Photoemission spectrum of YB_6. The dashed
curve was obtained by subtracting the background.
The shaded area corresponds to the valence band region
corrected for the background plus Y 4p satellite peak.

Self-consistent band structure calculations were carried out with the fast-symmetrized-cluster approach, using the Hedin-Lundqvist formula for exchange and correlation. The values 0.5 and 0.2 were chosen for the muffin tin spheres of metal and boron atoms, respectively. The self-consistent potentials were used to obtain the one-particle Green's functions. The boron octahedron states, of the same sequence as found by [2], had the following energies:

states	A_{1g}	T_{1u}	A'_{1g}, T''_{1u}, E'_g	T_{2g}	T_{2u}
energy in Ry	0.01	0.2 to 0.45	0.3 to 0.68	0.5 to 0.9	≥ 0.95

T_{2g} and T_{2u} states are energetically separated, which leads to the range of low total density of states (DOS) with the minimum at 0.9 Ry. The states with significant contribution to the DOS from metal atom regions lie above ~0.85 Ry. The Fermi energy is calculated to $E_F = 1.061$ Ry and $n(E_F) = 8.87$ states·Ry^{-1}·$spin^{-1}$·$cell^{-1}$, Schell et al. [4]. These results represent an improvement of preliminary calculations published by Schell et al. [5] and are compatible with XPS measurements made by Aono et al. [3]. Augmented plane-wave (APW) band calculations reveal that the character of the lowest conduction electron band at the Γ point is Γ_{25}, that is mostly 2p on the boron site. The Fermi level is a little bit below this Γ_{25} point and the 4d character on the Y site shows a steep increase near the Fermi level due to the crossover effect, Aoki [6]. According to early calculations by a discrete variational method in the Hartree-Fock-Slater model, with full crystal potentials generated from neutral Y and B atom configurations and an exchange parameter $\alpha = 1.0$, the Fermi level lies in the middle of a broad conduction band; $n(E_F) = 6.05$ states/Ry. The Fermi energy intersects a single band which has minima at X_1 and Σ_4 (see a figure in the paper). A belly-neck Fermi surface is predicted. From a comparison with LaB_6 it was concluded that the relatively high superconducting transition temperature T_c of YB_6 cannot be explained by a simple density-of-states effect, Walch et al. [7]. Studies on T_c of pure YB_6 and its various solid solutions led to the conclusion that the Fermi energy of YB_6 coincides with a maximum in the density-of-states curve, Hiebl, Sienko [8].

For charge carrier properties derived from the Hall effect, see p. 150.

Electrical Conductivity

The electrical resistivity $\varrho = 40.5 \pm 1.6\ \mu\Omega \cdot cm$ was measured at room temperature on a polycrystalline sample prepared by zone melting, Johnson, Daane [9, 10]; $\varrho = 38.7\ \mu\Omega \cdot cm$, Kovenskaya et al. [11]. A similar value was obtained for a single crystal. Its electrical resistivity varies as T^3 between the superconducting transition temperature T_c (ca. 7 K) and about 20 K, and nearly as T above 60 K up to room temperature, see a figure in the paper [1]. The residual electrical resistance ratio $\varrho(295\ K)/\varrho(ca.\ T_c)$ of flux-grown crystals changed from 5.4 to 2.4, if the starting charge nominal composition was changed from YB$_6$ to YB$_9$, Fisk et al. [12]. Films on sapphire substrates had residual resistance ratios of 1.0 to 1.7, Schneider et al. [31].

The room temperature resistivity of YB$_6$ films deposited either onto single crystal p-Si $\langle 100 \rangle$ substrates or onto thermally oxidized silicon substrates (both finally annealed in N$_2$ for 60 min) decreases with increasing annealing temperature (1073 to 1323 K). For annealing temperatures $\geqq 1173$ K, this decrease is stronger for the p-Si $\langle 100 \rangle$ films due to substrate dependent boron out-diffusion (see p. 154). The room temperature resistivity is 88 $\mu\Omega \cdot cm$ for a film on SiO$_2$ annealed in N$_2$ at 1273 K. Films are no longer electrically conductive ($\varrho \geqslant 10^3\ \mu\Omega \cdot cm$) after 30 min steam oxidation at 1273 K, Ryan, Roberts [13].

YB$_6$ is a type II superconductor (see Fig. 54, p. 147) with a superconducting transition temperature $T_c = 7.5$ K for a single crystal [1] and $T_c = 7.16$ K for a sputtered film on sapphire substrate [31]. Flux-grown single crystals had $T_c = 5.8$ K, independent of the starting charge composition, which varied from YB$_6$ to YB$_9$ [12]. But for polycrystalline YB$_6$ samples prepared by borothermal reduction, T_c values varied depending on the B/Y$_2$O$_3$ ratio of the starting charge as follows: $T_c < 1.5$, 6.8 ± 0.1, and 6.3 ± 0.3 K at B/Y$_2$O$_3$ = 14, 20, and 26, respectively. $T_c = 6.8$ K for arc-melted YB$_6$, Fisk et al. [14]. $T_c = 6.0 \pm 0.1$ K, Sobczak, Sienko [15]; $T_c \approx 5.5$ K (read from a figure) [8]; $T_c = 6.5$ to 7.1 K, Matthias et al. [16]. Replacing small amounts of yttrium by other rare earth elements lowers T_c remarkably, see p. 65. Values for T_c were calculated from the Eliashberg function: Coupling of electrons to acoustic phonons results in $T_c = 7.06$ K [31], whereas coupling of electrons to mainly nonacoustic phonons results in $T_c = 3.85$ or 23 K, considering or disregarding nonlocal corrections, respectively [4]. For the energy gap, see below.

Tunnel Effect and Point Contact Spectroscopy

Electron tunneling was measured with the GaAs Schottky barrier-type tunneling method (BTT) on a cleaved (100) surface at 1.8 K, both in the superconducting and in the normal state, superconductivity of the specimens was quenched with a magnetic field [1]. Recent measurements were performed at 1.2 and 8 K on a sputtered YB$_6$ film using a YB$_6$/YB$_6$–oxide/In tunnel junction [31]. The normalized tunneling conductance versus voltage plot, given in **Fig. 56**, p. 150, shows strong coupling deviations from the BCS (Bardeen, Cooper, Schrieffer) curve and leads to an energy gap $\Delta = 1.22$ meV [1], to be compared to $\Delta = 1.24$ meV for the film [31], and $2\Delta = 2.414$ meV from critical field studies, Kadowaki [17]. Thus, YB$_6$ is a medium to strong-coupling superconductor with $2\Delta/kT_c = 4.02$ [31] or 3.8 [1]. The second derivative of the current-voltage characteristic, $-d^2I/dV^2$ as a function of voltage shows two peaks corresponding to structures (acoustic phonons, see p. 144) in the Eliashberg function at 4.9 and 8.5 eV in the case of the film [31] and at only ~ 11 meV and above 20 meV in the case of the single crystal [1]. Barrierless metal point contacts (MPC) between YB$_6$ needle and YB$_6$ (100)-cleavage surface show, in the superconducting state at 4.2 K, a reverse energy dependence of the differential resistance dV/dI compared with BTT. In the case of MPC at zero or low external magnetic field, the differential resistance dV/dI as a function of bias voltage suddenly rises divergently at $\pm 2\Delta$, whereas in the GaAs Schottky barrier tunneling it is the differential conductance dI/dV that suddenly increases. The reason for a sudden increase of resistance in

MPC spectra was assumed to be that the electron, having the different character and the same phase in the direct coupling system, may be strongly repulsive. The structures at $\pm 2\Delta$ and that of ± 8 mV disappear with increasing magnetic field. At 5.5 kOe (at 4.2 K), the MPC spectrum is reduced to that of the clean normal metal contact, Kunii [18].

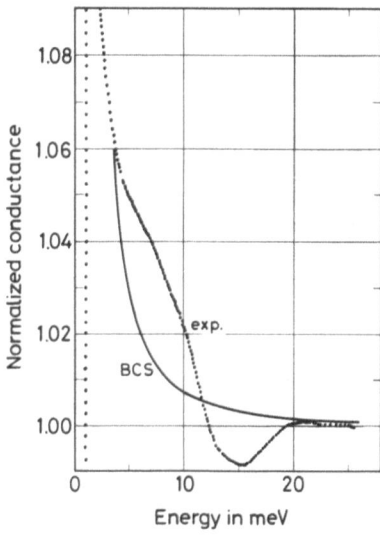

Fig. 56.   Normalized tunneling conductance versus voltage for a superconducting YB_6 single crystal at 1.8 K.

Hall Effect. Thermoelectric Power. Charge Carrier Properties

The Hall coefficient of a nearly stoichiometric sample which was prepared by zone melting was measured twice, with the sample inverted for the second measurement: $R_H = -(44.7 \pm 2.2) \times 10^{-11}$ m³/C and $R_H = -(44.6 \pm 2.4) \times 10^{-11}$ m³/C compared with the calculated value of $R_H = -43.2 \times 10^{-11}$ m³/C assuming one conduction electron per Y ion, Johnson, Daane [9, 10]. Later, $R_H = -55.9 \times 10^{-11}$ m³/C was measured which gives a concentration of charge carriers of $n_H = 1.12 \times 10^{22}$ cm⁻³, corresponding to 0.77 electrons per Y atom; Hall mobility $\mu_H = 14.4 \times 10^4$ cm²·V⁻¹·s⁻¹; effective mass $m^* = 1.03\,m_0$, Kovenskaya et al. [11].

The thermoelectric power is negative: $S = -4.11$ µV/K [11], and $S = -4.6$ µV/K (against copper), Kudintseva et al. [19].

Electron Emission

Under the conditions of prolonged thermionic emission, YB_6 and mixtures of $YB_6 + YB_4$ or $YB_6 + YB_{12}$ undergo vaporization. The measured thermionic work function is assumed to reflect the degree of transition of the surface towards YB_4, which has a lower work function (see p. 138) than YB_6, Jaskie, Jacobson [20]. In view of the fact that it is difficult to prepare YB_6 cathodes not contaminated with YB_4, see for example Kudintseva et al. [21], this may explain in part the diverging data on thermionic emission properties reported for unactivated YB_6 cathodes. Cathodes were produced either by hot pressing [20, 21], Samsonov et al. [22, 23], Wolski [24] or by coating a Ta substrate [21], Trigubenko, Tsarev [25], Meerson et al. [26] or, less preferably, a W ribbon, Ermakov et al. [27]; for details, see [20, 21, 26], Kaznoff et al. [28]. Values for the work function Φ_R and the Richardson constant A (derived by the Richardson straight-line method), the work function Φ_T at $A = 120$ A·cm⁻²·K⁻² (calculated from the Richardson-Dushman equation), and its temperature dependence Φ_T (T), measured in the temperature ranges given, are summarized below:

temperature range in K	$\Phi_T{}^{a)}$ in eV	$\Phi_T(T) = \Phi_0 + (d\Phi/dT)T$ in eV	Φ_R in eV	A in $A \cdot cm^{-2} \cdot K^{-2}$	Ref.
1660 to 1710	4.25 to 4.3	$3.58 + 4.0 \times 10^{-4}$ T	—	—	[29]
1500 to 1890	4.09 to 4.23	$3.57 + 3.5 \times 10^{-4}$ T	—	—	[22]
1583 to 1873	3.38 to 3.60	$2.4 + 7.5 \times 10^{-4}$ T$^{b)}$	3.45	1.77	[21]
1000 to 1800	2.99 to 3.15	$2.78 + 2.075 \times 10^{-4}$ T	—	—	[27]
1660 to 1790	4.2 to 4.3	—	5.31	2.24×10^5	[26]
1173 to 1573$^{c)}$	—	—	2.3 to 3.3	0.06 to ~160	[25]
~1200 to 1500	2.43 to 2.49	$2.22 + 1.79 \times 10^{-4}$ T$^{d)}$	2.22 ± 0.05	15	[19]
1583 to 1873$^{e)}$	3.19 to 3.45	$2.06 + 8.7 \times 10^{-4}$ T$^{b)}$	1.87	1.6×10^{-2}	[21]
1680 to 1950$^{f)}$	4.48 to 4.52	$4.21 + 1.6 \times 10^{-4}$ T	—	—	[22]

a) All values except those by [26], [29] were calculated from Φ_T(T). – b) Formula is valid for Celsius degrees. – c) Measured after processing the cathode for up to 2 h. – d) Published in [27]. – e) After activating the cathode at 1583 K for 15 h. – f) After annealing the cathode for 1.5 h at 1780 K.

$\Phi_R = 3.95$ eV was derived from measurements on a sample with the bulk composition YB$_{5.1}$ in the range from 1350 to 1820 K [20]; and $\Phi = 3.6$ eV was determined on a YB$_6$ film on SiO$_2$ substrate, evaluated from metal/oxide semiconductor capacitance-voltage technique by Ryan, Roberts [13]. A cathode activated at 1583 K for 15 h, unlike the unactivated one, had a work function close to that of a YB$_4$ cathode [21]. The contrary change, an increase of the work function to a value close to that of YB$_{12}$, was observed in [22] on annealing a YB$_6$ cathode for 1.5 h at 1780 K and was explained with a boron enrichment of the surface during decomposition. According to Meerson et al. [26], Φ_T of unactivated cathodes on Ta substrates increases from 4.20 to 4.30 eV (4.25 to 4.32 eV) between 1660 and 1710 K and it decreases at higher temperatures to 4.20 eV (4.12 eV) at 1840 K. Values in parentheses refer to cathodes conditioned for times between 15 and 40 min. After 40 min at 1840 K, partial decomposition was observed. The emission current density at an electric field strength of ca. 90 V/m increases from 5.17×10^{-5} to 9.68×10^{-4} A/cm^2 between 1660 and 1790 K. A figure in the paper shows the variation of emission current density with anode voltage at different temperatures [26]. A value as high as 0.68 A/cm^2 reported for the emission current density of a cathode, used at 1673 K by Lee Shau-Chung et al. [30], appears to originate from YB$_4$, see [21]. For current-voltage characteristics from hot-pressed YB$_6$ cathodes working under conditions of high-pulse voltage at temperatures between 1843 to 2109 K, see [24].

References:

[1] Kunii, S.; Kasuya, T.; Kadowaki, K.; Date, M.; Woods, S. B. (Solid State Commun. **52** [1984] 659/61).

[2] Longuet-Higgins, H. C.; Roberts, M. de V. (Proc. Roy. Soc. [London] A **224** [1954] 336/47).

[3] Aono, M.; Kawai, S.; Kono, S.; Okusawa, M.; Sagawa, T.; Takehana, Y. (J. Phys. Chem. Solids **37** [1976] 215/9).

[4] Schell, G.; Winter, H.; Rietschel, H.; Gompf, F. (Phys. Rev. [3] B **25** [1982] 1589/99).

[5] Schell, G.; Winter, H.; Rietschel, H. (Supercond. d-f-Band Met. Proc. Conf., San Diego, Calif., 1979 [1980], pp. 465/71; C.A. **93** [1980] No. 124363).

[6] Aoki, Y. (Diss. Univ. Tohoku 1982 from [1]).

[7] Walch, P. F.; Ellis, D. E.; Mueller, F. M. (Phys. Rev. [3] B **15** [1977] 1859/66).

[8] Hiebl, K.; Sienko, M. J. (Inorg. Chem. **19** [1980] 2179/80).

[9] Johnson, R. W.; Daane, A. H. (J. Chem. Phys. **38** [1963] 425/32).

[10] Johnson, R. W.; Daane, A. H. (IS-473 [1962] 1/30, 13, 17, 23; N.S.A. **17** [1963] No. 11306).

[11] Kovenskaya, B. A.; Kondrashov, A. I.; Dudnik, E. M.; Kolotun, V. F. (Tugoplavkie Soedin. Redkozemel. Met. Mater. 3rd Vses. Semin., Novosibirsk 1977 [1979], pp. 36/9; C.A. **93** [1980] No. 60231).

[12] Fisk, Z.; Schmidt, P. H.; Longinotti, L. D. (Mater. Res. Bull. **11** [1976] 1019/22).

[13] Ryan, J. G.; Roberts, S. (Thin Solid Films **135** [1986] 9/19).

[14] Fisk, Z.; Lawson, A. C.; Fitzgerald, R. W. (Mater. Res. Bull. **9** [1974] 633/6).

[15] Sobczak, R. J.; Sienko, M. J. (J. Less-Common Metals **67** [1979] 167/71).

[16] Matthias, B. T.; Geballe, T. H.; Andres, K.; Corenzwit, E.; Hull, G. W.; Maita, J. P. (Science **159** [1968] 530).

[17] Kadowaki, K. (from [1]).

[18] Kunii, S. (J. Magn. Magn. Mater. **63/64** [1987] 673/6).

[19] Kudintseva, G. A.; Poliakova, M. D.; Samsonov, G. V.; Tsarev, B. M. (Fiz. Metal. Metalloved. **6** [1958] 272/5; Phys. Metals Metallog. [USSR] **6** No. 2 [1958] 83/6).

[20] Jaskie, J.; Jacobson, D. (Proc. Intersoc. Energy Convers. Eng. Conf. **15** [1980] 2331/3; C.A. **94** [1981] No. 142615).

[21] Kudintseva, G. A.; Kuznetsova, G. M.; Bondarenko, V. P.; Selivanova, N. F.; Shlyuko, V. Ya. (Poroshkovaya Metal. **1968** No. 2, pp. 45/53; Soviet Powder Met. Metal Ceram. **1968** 115/20).

[22] Samsonov, G. V.; Kondrashov, A. I.; Okhremchuk, L. N.; Podchernyaeva, I. A.; Siman, N. I.; Fomenko, V. S. (Poroshkovaya Metal. **1977** No. 1, pp. 21/8; Soviet Powder Met. Metal Ceram. **16** [1977] 16/22).

[23] Samsonov, G. V.; Kondrashov, A. I.; Okhremchuk, L. N.; Podchernyaeva, I. A.; Siman, N. I.; Fomenko, V. S. (J. Less-Common Metals **67** [1979] 415/8).

[24] Wolski, W. (Elektronika **11** No. 4 [1970] 149/55; C.A. **74** [1971] No. 17055).

[25] Trigubenko, V. A.; Tsarev, B. M. (Radiotekhn. Elektron. **6** [1961] 1900/5; Radio Eng. Electron. Phys. [USSR] **6** [1961] 1694/8; C.A. **57** [1962] 187).

[26] Meerson, G. A.; Zhuravlev, N. N.; Manelis, R. M.; Runov, A. D.; Stepanova, A. A.; Grishina, L. P.; Gramm, N. V. (Izv. Akad. Nauk SSSR Neorgan. Materialy **2** [1966] 608/16; Inorg. Materials [USSR] **2** [1966] 527/33).

[27] Ermakov, S. V.; Mamedov, F. G.; Meerson, G. A.; Tsarev, B. M. (Izv. Akad. Nauk SSSR Neorgan. Materialy **3** [1967] 1094/5; Inorg. Materials [USSR] **3** [1967] 975/6).

[28] Kaznoff, A. I.; Hoyt, E. W.; Grossman, L. N. (Advan. Energy Convers. **3** [1963] 167/73; C.A. **61** [1964] 3782).

[29] Manelis, R. M.; Telyukova, T. M.; Grishina, L. P. (Izv. Akad. Nauk SSSR Neorgan. Materialy **6** [1970] 1184/5; Inorg. Materials [USSR] **6** [1970] 1035/6).

[30] Lee Shau-Chung; Liau Shen-Hung; Liu Kwei Tshian (Tien Tzu Hsueh Pao No. 1 [1965] 48/56; C.A. **66** [1967] No. 14890).

[31] Schneider, R.; Geerk, J.; Rietschel, H. (Europhys. Letters **4** [1987] 845/9).

[32] Tanaka, T.; Ishizawa, Y. (J. Phys. C **18** [1985] 4933/40, 4939).

32.1.8.2.5.6 Optical Properties

YB$_6$ is blue, Mamedov et al. [1], blue-black or blue-violet, Kaznoff et al. [2].

The thermal radiation coefficient $\varepsilon_\lambda = 0.7$ is found for $\lambda = 0.655$ μm at 1773 K, Kudintseva et al. [3]. This value is at variance with $\varepsilon_\lambda = 0.80$ for $\lambda = 0.65$ μm found by measurements in the temperature range of 973 to 2173 K, Kaznoff et al. [2, 4].

Investigations of solar energy recovery carried out with plasma-sprayed YB$_6$ coatings deposited onto steel substrates show high absorptivity in the wavelength representing the sun spectrum and low emissivity in the IR region, Schreyer et al. [5, 6].

For the X-ray photoemission spectrum, see "Electronic Structure", p. 147.

References:

[1] Mamedov, F. G.; Meerson, G. A.; Zhuravlev, N. N.; Umanskii, Ya. S. (Izv. Akad. Nauk SSSR Neorgan. Materialy 3 [1967] 950/6; Inorg. Materials [USSR] 3 [1967] 851/6).
[2] Kaznoff, A. I.; Grossman, L. N.; Hoyt, E. W. (in: Grossman, L. N.; Hoyt, E. W.; Ingold, J. H.; Kaznoff, A. I.; Sanderson, M. J., GEST-2009 [1962] 9-1/9-15; N.S.A. 17 [1963] No. 14819).
[3] Kudintseva, G. A.; Poliakova, M. D.; Samsonov, G. V.; Tsarev, B. M. (Fiz. Metal. Metalloved. 6 [1958] 272/5; Phys. Metals Metallog. [USSR] 6 No. 2 [1958] 83/6).
[4] Kaznoff, A. I.; Hoyt, E. W.; Grossman, L. N. (Advan. Energy Convers. 3 [1963] 167/73; C.A. 61 [1964] 3782).
[5] Schreyer, J. M.; Schmitt, C. R.; Hays, R. A.; Farwell, D. (Solar Energy 25 [1980] 179/85; C.A. 94 [1981] No. 106502).
[6] Schreyer, J. M.; Schmitt, C. R.; Hays, R. A.; Farwell, D. (AES Coat. Sol. Collect. 2nd Symp. Proc., Winter Park, Fla., 1979, Paper No. 14, pp. 1/10; C.A. 94 [1981] No. 159825).

32.1.8.2.5.7 Chemical Behavior

On Heating

The thermal decomposition of YB$_6$ under vacuum leads to the formation of YB$_4$ and B which volatilizes. This was confirmed in effusion studies in the range of 1873 to 1973 K. Sublimation studies of the Langmuir type, carried out at 1773 to 2273 K under 10^{-5} to 10^{-6} Torr ($\cong 1.33$ to 0.133 mPa), show that YB$_4$ forms as a porous layer on the surface. The associated color change of the surface from blue-black to reddish gold is visible within about 0.5 h at 1773 K and in a few minutes at 2273 K. The substrate consists of YB$_6$. These phases were confirmed by X-ray studies. Free boron and higher borides were not detected in the solid phase. Sublimation tests around 1873 K show a gradual decrease in the rate of loss of material which settles to a constant rate in a period of several hours. This is interpreted to be the result of buildup of the porous YB$_4$ layer. Higher temperature runs do not exhibit this behavior, as a steady state is achieved. The steady-state weight loss w (in $g \cdot min^{-1} \cdot cm^{-2}$) of 95% dense YB$_6$ in the range of 1873 to 2273 K is given by $w = (8.36 \times 10^5 T^{-1/2}) \cdot exp(-37400/T)$, Kaznoff et al. [1, 2]. Samples of YB$_6$ lost up to 10.5% in weight during a 30-min exposure at 2650 K, Meerson et al. [3]. Formation of a dense gray YB$_4$ layer on YB$_6$ samples was observed at 1800 to 1825 K by Meerson et al. [4]. Similar decomposition temperatures were noted by Samsonov et al. [5, 6] and they agree with most of the observations made during the preparation of YB$_6$ (see p. 140) and electron emission (see p. 150). Decomposition is reported for a YB$_6$ cathode activated at 1583 K for 15 h, Kudintseva et al. [7]. Plasma-sprayed coatings of YB$_6$ on steel substrates withstand heating in a resistance furnace at 1073 K for 20 h without cracking or flaking. The specimens tested in solar flux show no change in composition up to 947 K in 20 min, but at 1338 K YB_6 began to change to YB_4. About 83 to 96% of the incident energy was recovered from solar flux producing coating temperatures of 488 to 589 K, and 75 to 78% recovery was obtained from a solar flux producing coating temperatures of 956 to 1092 K. The difference in the energy recovery efficiency is attributed to the differences in specimen temperature and heat flux, Schreyer et al. [8, 9]. For the boron out-diffusion on heating YB_6 on Si or SiO_2 substrate, see below.

Solubility

YB_6 boiled for 30 min in HCl does not react, whereas it completely dissolves within 5 min in HNO_3 in the cold, and within 10 min in boiling H_2SO_4 (all acids half-concentrated). YB_6 does not dissolve in boiling 15% NaOH [4], Manelis et al. [10].

Other Reactions

YB_6 films deposited onto Si or SiO_2 were not able to withstand a 30-min steam oxidation at 1273 K without total loss of electrical conductivity and film integrity, Ryan, Roberts [11]. YB_6 begins to react at 2373 to 2423 K with graphite to form a second phase which was assumed to be a carbide of Y with a microhardness of 1657 ± 187 kg/mm^2 ($\cong 16.3$ GPa), Kudintseva et al. [12].

Boron out-diffusion effects of YB_6 films deposited onto p-Si $\langle 100 \rangle$ or SiO_2 substrates were investigated by annealing the films for 60 min in N_2. Significant out-diffusion of boron into the silicon substrate occurs at annealing temperatures of 1173 K and above, whereas the SiO_2 substrate behaves as a barrier to boron out-diffusion for short annealing times [11]. The concentration profile of boron in a p-Si $\langle 100 \rangle$ substrate ($\varrho \approx 12\ \Omega \cdot cm$) was determined after heating a YB_6 (and LaB_6) film at 1273 K for 270 min in N_2, for a figure see the paper. The results show that YB_6 can be used for boron doping of p-Si, thereby establishing ohmic contacts between the boride and these doped regions, Ishaq et al. [13].

Hot-pressed YB_6 is not attacked by caesium vapor (nominally 1.82 Torr ($\cong 243$ Pa) at 573 K [1]) in a special test chamber at 1273 K, test duration 350 h. The composite structures of YB_6–Ta and YB_6–Mo, produced by hot pressing, show interfacial failures which could not be attributed to caesium attack, Kaznoff, Hoyt [14], Kaznoff et al. [1, 2].

YB_6 was bonded to tantalum and molybdenum (for the use as emitter material) by hot pressing the powders in graphite dies. The reaction layer between Ta and YB_6 was found to be primarily Ta_2B and TaB. Volatilization of some Y was noted if the hot pressing was done in Mo dies. The bonding of YB_6 to Mo was less successful and the reaction zone between the two materials was less well defined [1, 2].

References:

[1] Kaznoff, A. I.; Hoyt, E. W.; Grossman, L. N. (Advan. Energy Convers. 3 [1963] 167/73; C. A. 61 [1964] 3782).
[2] Kaznoff, A. I.; Grossman, L. N.; Hoyt, E. W. (in: Grossman, L. N.; Hoyt, E. W.; Ingold, J. H.; Kaznoff, A. I.; Sanderson, M. J., GEST-2009 [1962] 9-1/9-15; N.S.A. 17 [1963] No. 14819).
[3] Meerson, G. A.; Manelis, R. M.; Telyukova, T. M. (Izv. Akad. Nauk SSSR Neorgan. Materialy 2 [1966] 291/8; Inorg. Materials [USSR] 2 [1966] 250/5).
[4] Meerson, G. A.; Zhuravlev, N. N.; Manelis, R. M.; Runov, A. D.; Stepanova, A. A.; Grishina, L. P.; Gramm, N. V. (Izv. Akad. Nauk SSSR Neorgan. Materialy 2 [1966] 608/16; Inorg. Materials [USSR] 2 [1966] 527/33).

[5] Samsonov, G. V.; Kondrashov, A. I.; Okhremchuk, L. N.; Podchernyaeva, I. A.; Siman, N. I.; Fomenko, V. S. (Poroshkovaya Metal. **1977** No. 1, pp. 21/8; Soviet Powder Met. Metal Ceram. **16** [1977] 16/22).

[6] Samsonov, G. V.; Kondrashov, A. I.; Okhremchuk, L. N.; Podchernyaeva, I. A.; Siman, N. I.; Fomenko, V. S. (J. Less-Common Metals **67** [1979] 415/8).

[7] Kudintseva, G. A.; Kuznetsova, G. M.; Bondarenko, V. P.; Selivanova, N. F.; Shlyuko, V. Ya. (Poroshkovaya Metal. **1968** No. 2, pp. 45/53; Soviet Powder Met. Metal Ceram. **1968** 115/20).

[8] Schreyer, J. M.; Schmitt, C. R.; Hays, R. A.; Farwell, D. (Solar Energy **25** [1980] 179/85; C. A. **94** [1981] No. 106502).

[9] Schreyer, J. M.; Schmitt, C. R.; Hays, R. A.; Farwell, D. (AES Coat. Sol. Collect. 2nd Symp. Proc., Winter Park, Fla., 1979, Paper No. 14, pp. 1/10; C.A. **94** [1981] No. 159825).

[10] Manelis, R. M.; Meerson, G. A.; Zhuravlev, N. N.; Telyukova, T. M.; Stepanova, A. A.; Gramm, N. V. (Poroshkovaya Metal. **1966** No. 11, pp. 77/84; Soviet Powder Met. Metal Ceram. **1966** 904/9).

[11] Ryan, J. G.; Roberts, S. (Thin Solid Films **135** [1986] 9/19).

[12] Kudintseva, G. A.; Poliakova, M. D.; Samsonov, G. V.; Tsarev, B. M. (Fiz. Metal. Metalloved. **6** [1958] 272/5; Phys. Metals Metallog. [USSR] **6** No. 2 [1958] 83/6).

[13] Ishaq, M. H.; Roberts, S.; Ryan, J. G. (U.S. 4490193 [1984]; C.A. **102** [1985] No. 104610).

[14] Kaznoff, A. I.; Hoyt, E. W. (in: Grossman, L. N.; Hoyt, E. W.; Ingold, J. H.; Kaznoff, A. I.; Sanderson, M. J., GEST-2009 [1962] 5-1/5-34, 5-2, 5-23; N.S.A. **17** [1963] No. 14819).

32.1.8.2.6 $Y_{1-x}Sc_xB_6$ (?)

The preparation of $Y_{0.5}Sc_{0.5}B_6$ was attempted analogously to a method given for $La_{1-x}Sc_{0.5x}Y_{0.5x}B_6$ solid solutions (see p. 272), but the product consisted of a multiphase system containing $(Y,Sc)B_6$, ScB_2, and ScB_{12}. (However, solid solutions of YB_6 with hypothetical ScB_6 are not expected to form, see p. 126.)

Reference:

Kondrashev, A. I. (Poluch. Issled. Svoistv Nov. Mater. Mater. 10th Nauchn. Konf. Aspir. Molodykh Issled. Inst. Probl. Materialoved. Akad. Nauk Ukr. SSR, Kiev 1976 [1978], pp. 48/52; C.A. **91** [1979] No. 111144).

32.1.8.2.7 YB$_{12}$

Preparation

The compound was prepared by a vacuum thermal reduction of Y_2O_3 by amorphous B at 1975 K and $(2 \text{ to } 5) \times 10^{-5}$ Torr ($\triangleq 2.7$ to 6.7 mPa) in a Ta crucible in a cooled quartz cover. The product had 22 to 26% porosity, Meerson et al. [1]. The reduction temperature is limited by the decomposition processes which tend to decrease the B content and give YB_4, Manelis et al. [2]. YB_{12} with 80 to 90% of the theoretical density was obtained by hot pressing at 2250 K and 210 kg/cm^2 ($\triangleq 20.6$ MPa), Holden et al. [3]. A sample prepared by floating-zone melting contained a finely divided phase richer in B, Johnson [4], Johnson, Daane [5, 6].

Crystallographic, Mechanical, and Thermal Properties

YB_{12} is face-centered cubic with a = 7.500 ± 0.001 Å [4 to 6], see also Matkovich et al. [7]. a = 7.506 ± 0.002 Å [1, 2], 7.589 ± 0.002 Å at 298 K, space group $Fm\bar{3}m$-O_h^5 (No. 225), UB_{12} type, Zhuravlev et al. [8]. Observed and calculated structure factors are listed and an electron density projection on the (001) plane is shown. Atomic positions: Y is in (0, 0, 0) and B in (1/2, 0.1706, 0.1706) at R = 0.061. There are two types of B–B separations: Within the cubooctahedron B–B = 1.809 ± 0.024 Å, external B–B = 1.684 ± 0.024 Å. The Y–B distance is 2.783 ± 0.012 Å [7], see also Matkovich et al. [9]. The calculated density is D_{calc} = 3.43 [1, 2] or 3.442 g/cm³, and the measured density is D_{exp} = 3.33 g/cm³ [4 to 6].

The microhardness is 2500 ± 150 kg/mm² ($\triangleq 24.5$ GPa), the bending strength 165 kg/cm² ($\triangleq 16.2$ MPa) [1,2]. The compound melts at 2473 K, Samsonov et al. [19]. The mean thermal expansion coefficient between 298 and 873 K is $\alpha = (6.6 \pm 0.6) \times 10^{-6}$ K^{-1} [8].

Electrical and Optical Properties

The B atoms are acceptors and use the electrons for filling their 2p orbital, Kober [10]. A calculation of the electronic energy band structure of cubic YB_{12} was carried out by a nonrelativistic LAPW (linearized augmented plane wave) method with potentials of muffin tin form. Eigenvalues and major charge components at the Γ point of the Brillouin zone are tabulated for the range -0.275 to $+1.263$ Ry ($\triangleq -3.7$ to 17.2 eV), with the Fermi level E_F at 0.837 Ry ($\triangleq 11.38$ eV). The Y s- and p-like density of states are small, the Y d-like is significant in structure and magnitude between 9 and 10 eV, somewhat below the Γ'_{25} and Γ_{12} levels. This d character does not correspond to the d band but represents hybridization with B states. The B s-like density of states predominates in the lower energy region, the p-like closer to the Fermi energy. For calculations disregarding Y and for comparison with the literature, see the papers, Switendick [15,16]. The energy band structure obtained by the relativistic self-consistent APW method is shown in **Fig. 57** near the Fermi energy. The bands up to the 19th are made mostly by the s, p orbits on B forming the bonding orbitals and are completely occupied. The 20th and the 21st band degenerate on the Δ axes and are composed of t_{2g}(Y) and p (B) states. In contrast to other fcc metals, the degenerate part of t_{2g} becomes the lower band because of the anisotropic p–d mixing character. This is the bonding orbit between the t_{2g}(Y) and the antibonding p (B) orbit. One electron occupies these conduction bands. The specific heat coefficient $\gamma = 2.36$ mJ·mol^{-1}·K^{-2} was calculated and the $\langle 100 \rangle$ plane cyclotron masses in the lower band amount to 0.71 and 0.79, in the upper band to 0.27 and 0.52 of the free electron mass, Harima et al. [17].

As expected from the foregoing calculations, YB_{12} is metallic with trivalent Y. The temperature dependence of the electrical resistivity below room temperature is shown in **Fig. 58** together with the curve for the also metallic LuB_{12}. The residual resistivity is 1.7 $\mu\Omega$·cm, Kasaya et al. [18]. The electrical resistivity of a polycrystalline sample with the measured density of only 3.11 g/cm³ is $\varrho = 94.8$ $\mu\Omega$·cm at room temperature. The Hall constant R_H was measured and corrected either for porosity or a second phase: $R_H = (-103$ and $-89) \times 10^{-11}$ m³/C, respectively. On the assumption of a Y-deficient lattice, the calculated value is $R_H = -86.5 \times 10^{-11}$ m³/C, corresponding to about one free electron per Y atom [4 to 6].

From measurements at 1670 to 1730 K, the work function of YB_{12} on a Ta substrate is $\Phi_T = 4.6$ and the emission current density is $j_{max} = 5.52 \times 10^{-6}$ to 2.01×10^{-5} A/cm². The main contribution to the thermoemission is made by the Y atoms, Manelis et al. [11], Meerson et al. [1]; the thermionic properties are inferior to LaB_6. Compared with YB_4 and YB_6 the Φ_T value is largest, the j_{max} value lowest for YB_{12}, Manelis et al. [12].

In reflected white light the metallic YB_{12} is light blue, Oliver, Brower [13].

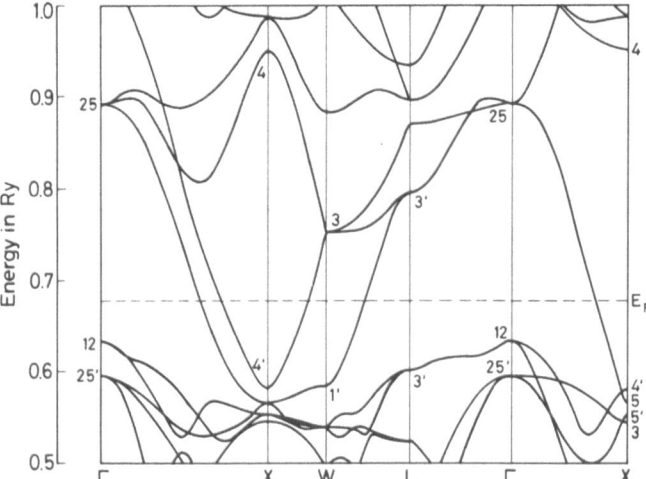

Fig. 57. Self-consistent energy band structure of YB$_{12}$ in the vicinity of E$_F$. The symmetry labels denote rotational operations about a Y site. The energies are relative to the average potential of the interstitial region.

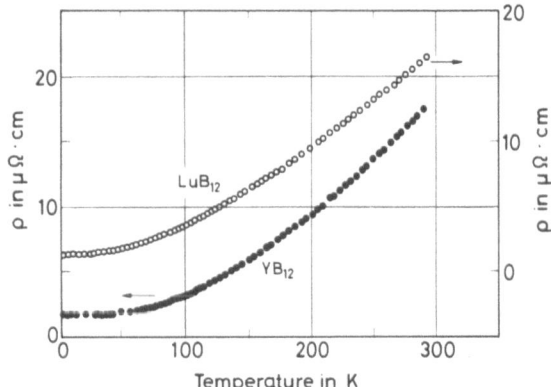

Fig. 58. Temperature dependence of the electrical resistivity ϱ of YB$_{12}$ and LuB$_{12}$.

Chemical Behavior

YB$_{12}$ dissolves in 1:1 HNO$_3$ with a slight deposit and does not dissolve in 1:1 H$_2$SO$_4$ and 1:1 HCl on boiling for 30 min. It does not decompose in 15% NaOH [1], whereas it dissolves in 15% NaOH leaving a slight residue [2]. YB$_{12}$ does not dissolve noticeable amounts of Cr, Kuz'ma et al. [14] or "GdB$_{12}$" (only obtained under high pressure, see p. 100), Schwetz et al. [20].

References:

[1] Meerson, G. A.; Zhuravlev, N. N., Manelis, R. M.; Runov, A. D.; Stepanova, A. A.; Grishina, L. P.; Gramm, N. V. (Izv. Akad. Nauk SSSR Neorgan. Materialy **2** [1966] 608/16; Inorg. Materials [USSR] **2** [1966] 527/33).

 [2] Manelis, R. M.; Meerson, G. A.; Zhuravlev, N. N.; Telyukova, T. M.; Stepanova, A. A.;
     Gramm, N. V. (Poroshkovaya Metal. **1966** No. 11, pp. 77/84; Soviet Powder Met. Metal
     Ceram. **1966** 904/9).
 [3] Holden, A. N.; Hoyt, E. W.; Cummings, W. V.; Antony, K. C.; Zimmerman, D. L. (Plansee
     Proc. 4th Semin., Reutte/Tyrol, Austria, 1961 [1962], pp. 615/44, 626).
 [4] Johnson, R. W. (Diss. Iowa State Univ. 1962, pp. 1/68, 23, 43; Diss. Abstr. **23** [1962] 850).
 [5] Johnson, R. W.; Daane, A. H. (J. Chem. Phys. **38** [1963] 425/32).
 [6] Johnson, R. W.; Daane, A. H. (IS-473 [1962] 1/30; N.S.A. **17** [1963] No. 11306).
 [7] Matkovich, V. I.; Economy, J.; Giese, R. F., Jr.; Barrett, R. (Acta Cryst. **19** [1965] 1056/8).
 [8] Zhuravlev, N. N.; Belousova, I. A.; Manelis, R. M.; Belousova, N. A. (Kristallografiya **15**
     [1970] 836/8; Soviet Phys.-Cryst. **15** [1970] 723/4).
 [9] Matkovich, V. I.; Economy, J.; Giese, R. F., Jr.; Barrett, R. B. (AD-608444 [1964] 1/16; C. A.
     63 [1964] 1283).
[10] Kober, V. I. (Elektron. Str. Fiz. Khim. Svoistva Tugoplavkikh Soedin. Splavov Dokl. 9th
     Vses. Simp., Ivano-Frankovsk, USSR, 1979 [1980], pp. 253/8; C. A. **95** [1981] No. 121237).

[11] Manelis, R. M.; Telyukova, T. M.; Grishina, L. P. (Izv. Akad. Nauk SSSR Neorgan. Materialy
     6 [1970] 1184/5; Inorg. Materials [USSR] **6** [1970] 1035/6).
[12] Manelis, R. M.; Grishina, L. P.; Runov, A. D. (Radiotekhn. Electron. **11** [1966] 2098/100;
     Radio Eng. Electron. Phys. [USSR] **11** [1966] 1855/6; C. A. **66** [1967] No. 23353).
[13] Oliver, D. W.; Brower, G. D. (J. Cryst. Growth **11** [1971] 185/90).
[14] Kuz'ma, Yu. B.; Sobolev, A. S.; Furtak, M. P. (Izv. Akad. Nauk SSSR Neorgan. Materialy **6**
     [1970] 2205/6; Inorg. Materials [USSR] **6** [1970] 1936/7).
[15] Switendick, A. C. (SAND-85-2732-C [1985] 1/14; C. A. **106** [1987] No. 90433).
[16] Switendick, A. C. (AIP [Am. Inst. Phys.] Conf. Proc. No. 140 [1986] 260/73; C. A. **105** [1986]
     No. 30323).
[17] Harima, H.; Yanase, A.; Kasuya T. (J. Magn. Magn. Mater. **47/48** [1985] 567/9).
[18] Kasaya, M.; Iga, F.; Takigawa, M.; Kasuya, T. (J. Magn. Magn. Mater. **47/48** [1985] 429/35).
[19] Samsonov, G. V.; Goryachev, Yu. M.; Kovenskaya, B. A.; Arabei, B. G. (Bor: Poluch.
     Strukt. Svoistva Mater. 4th Mezhdunar. Simp. Boru, Tbilisi 1972 [1974], pp. 171/4).
[20] Schwetz, K.; Ettmayer, P.; Kieffer, R.; Lipp, A. (Radex Rundschau **1972** 257/65).

32.1.8.2.8 YB$_{\sim 66}$

The most B-rich compound in the system Y–B has a composition YB_n with $20 < n < 100$.
Commonly it is considered to be $n \approx 66$, see. p. 129.

Preparation and Use

YB_{66} is a convenient monochromator for soft X-ray synchrotron radiation because of its
large d value of 5.86 Å for the (400) reflection, toughness, and structureless transmission curve
from 1070 to 2080 eV. However, it is necessary to improve the quality and the size of the
subgrain-free region in the single crystals, Tanaka et al. [1,2].

Polycrystalline $YB_{\sim 66}$ was prepared from YB_6 and amorphous B in a ratio $B:Y = 62$, which
corresponds to the congruently melting composition. The constituents were mixed, pressed
into a rod, repressed hydrostatically into a cylinder, and heated at 2173 K under vacuum in a BN
crucible inserted in a graphite susceptor. The obtained feed rod was thinned at one end. For
the preparation of single crystals the floating-zone method is used with indirect heating in a
high-pressure furnace at 2×10^5 Pa He. The radiative heating was done by a W ring which was

inductively heated to 3073 K. The growth direction is controlled to the [100] or [110] direction by a seeding process. The crystals were grown at 10 mm/h and a rotation of 12 rev/min. When the molten zone passed through the thinned end of the feed rod, the speed was gradually increased to 18 mm/h and the W temperature was lowered to 2973 K. Single crystals up to 11 mm in diameter and 50 mm in length were obtained, Tanaka et al. [1]. The W temperature was lowered to 2873 K and a double zone pass was used by Tanaka et al. [2]. Single crystals could not be grown by a sealed Bridgman technique or by a Czochralski technique using a BN crucible because of the incompatibility of the melt with refractories. Therefore crystal boules with [100] and [111] axes were grown by a three step Czochralski pedestal method. In the first step polycrystalline YB$_{66}$ was synthesized in an r.f. heated water-cooled Cu boat. By cooling, it was possible to control the violent exothermic reaction $YB_2 + 2B \rightarrow YB_4$. The brittle material used as a pedestal was pulled from this boat through a side arm in the quartz envelope without fracturing. Crystals of good quality were grown by the pedestal method which avoids contamination from the refractory. Zone refining could be done in the same apparatus, Oliver, Brower [3]. The apparatus and its operation are described in detail by Oliver et al. [4]. Congruently melting YB$_{61.75}$ was grown from a melt with the same composition (1.59 mol% very pure Y). It was very difficult to grow crystals with n different from 61.75. At both sides of this composition, a segregation coefficient k = 0.3 exists, which is the ratio between the deviation of the Y concentration from 1.59 mol% in the solid and in the liquid. A stoichiometric crystal YB$_{68}$ would be grown from a starting melt composition of ~1.2 mol% Y. Good single crystals have only been grown from melts with 1.45 to 1.96 mol% Y, Slack et al. [10].

Crystallographic Properties

According to powder and single-crystal data, YB$_{66}$ is face-centered cubic with a = 23.440 ± 0.006 Å, space group Fm$\overline{3}$c-O$_h^6$ (No. 226), Richards, Kasper [5], see also Richards [13], Slack et al. [9]; a = 23.43 Å for a [110] crystal [1] and a = 23.445 ± 0.005 Å for a congruently melting crystal YB$_{61.75 \pm 0.2}$ (1.59 mol% Y), which was grown from a melt with 1.59 mol% Y of high purity. The lattice parameter increases with increasing Y content from 23.440 ± 0.006 Å to 23.482 ± 0.003 Å in the presence of β B or YB$_{12}$ for crystals grown from melts with less or more Y than the congruent composition, respectively, Slack et al. [10]. A detailed structure analysis was made. The final refinement converged to R = 0.067. The atomic coordinates and occupancies, where relevant, are listed in the following table:

atom	position	fractional occupancy	coordinates		
			x	y	z
1(Y)	48f	0.5	0.05448(5)	1/4	1/4
2(B)	96i	1.0	0	0.03716(24)	0.05923(24)
3(B)	96i	1.0	0	0.07598(24)	0.11659(25)
4(B)	96i	1.0	0	0.03889(24)	0.18145(23)
5(B)	96i	1.0	0	0.14835(23)	0.24188(24)
6(B)	96i	1.0	0	0.18593(23)	0.17186(24)
7(B)	192j	1.0	0.03845(18)	0.14027(16)	0.12163(17)
8(B)	192j	1.0	0.03921(19)	0.08171(17)	0.22939(16)
9(B)	192j	1.0	0.06306(17)	0.07748(17)	0.15860(17)
10(B)	192j	1.0	0.06365(18)	0.14595(17)	0.19491(19)
11(B)	192j	0.71(2)	0.13193(33)	0.17493(35)	0.19719(36)
12(B)	192j	0.65(2)	0.23420(53)	0.15866(50)	0.30080(62)
13(B)	192j	0.28(2)	0.17343(71)	0.12776(64)	0.25728(85)
14(B)	64g	0.279(18)	0.23407(82)	0.23407(82)	0.23407(82)

In the unit cell there are 24 Y and approximately 1584 B atoms, 1248 of which are contained in eight thirteen-icosahedron units of 156 atoms each. Each $(B_{12})_{13}$ unit is a cluster of twelve B_{12} icosahedra grouped icosahedrally around a thirteenth; see **Fig. 59**. The extensive B framework (see Fig. 59) contains channels along the fourfold axes at (1/4, 1/4, x) with the half-filled Y sites and the remaining B atoms in it. These B atoms were assumed to form two cage-like configurations, B_{36} and B_{48}, which statistically occur about (1/4, 1/4, 1/4), Richards, Kasper [5], Richards [13]. For the congruently melting composition $YB_{61.75}$, a more accurate number of B atoms = 1628 ± 4 is computed from the density and lattice constant. The eight units of $(B_{12})_{13}$ icosahedra plus eight units of altogether 48 B atoms in the channels would require 1632 B atoms. The number of Y atoms is 26.4 corresponding to 55.0% occupancy. A 50% occupancy would correspond to the stoichiometric YB_{68} [10]. The Y atoms coordinate with cage B atoms and icosahedral B atoms surrounding them. Owing to the short distances between two adjacent Y sites, only one of them can be occupied and pairs having one vacant and one occupied site occur [5]. The surrounding B atoms hinder the movement of Y to its empty neighbor site. Under stress, however, this might become possible. For a schematic representation of the structure and comparison with a transmission electron microscope picture see [10]. The B bond lengths in the central icosahedron are 1.719 and 1.742 Å, in outer icosahedra 1.752 to 1.863 Å, and intericosahedral 1.624 to 1.823 Å. The Y–B bonds are from 2.691 to 2.768 Å. For details such as the B–B bond lenghts in the B_{36} and B_{48} units and a comparison with β-rhombohedral B see the papers, Richards, Kasper [5], Richards [13]. Comparison with the isotypic ThB_{66} is made by Naslain et al. [6]. Some reflections in the X-ray diffraction pattern not characteristic of the space group $Fm\overline{3}c\text{-}O_h^6$ are explained by ordering of Y atoms, and possibly of the different kinds of nonicosahedral B atoms. Any major departure from the icosahedral B framework is not considered, Kasper [7]. From the temperature dependence of the mean-free-path of phonons, randomness of B and Y atom positions is believed, Slack et al. [9]. For the crystal chemistry of boron-rich phases and of boron see Naslain [8].

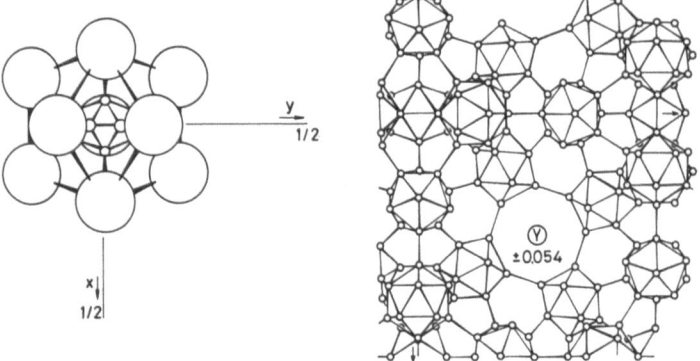

Fig. 59.   Crystal structure of YB_{66}. The thirteen-icosahedron unit is shown left, and the arrangement of the units in the xy0 plane is shown right (see text).

Mechanical and Thermal Properties

The measured density D_{exp} in g/cm³ of $YB_{61.75}$ is $D_{exp} = 2.5678 \pm 0.005$ in good agreement with $D_{calc} = 2.575$ calculated under the assumption of 1632 B atoms in the unit cell [10]. The earlier calculations assuming approximately 1584 B atoms resulted in a lower $D_{calc} = 2.482$ for YB_{66} compared to $D_{exp} = 2.5195 \pm 0.0006$ at 303 K [13], 2.568 ± 0.005 [3, 9], or 2.52 [5]. The range of densities within the single-phase YB$_n$ is D$_{exp}$ = 2.578 to 2.566 at the high and low Y end, respectively [10].

The elastic constants (in GPa) c$_{11}$ = 380 ± 30, c$_{44}$ = 160 ± 8, and c$_{12}$ = 40 ± 50 have been calculated for YB$_n$ from the sound velocities [10], see also [3]. The elastic anisotropy constant is A = 2 c$_{44}$/(c$_{11}$ − c$_{12}$) = 0.96 ± 0.14 [10].

At 300 K the sound velocity in a YB$_{66}$ crystal at 10^7 Hz and for [100] propagation is v = 7.9 × 10^5 cm/s for a transverse wave and v = 12.1 × 10^5 for a longitudinal wave [9]. For a longitudinal wave v = 11.8 × 10^5 cm/s for YB$_{61.75}$ at 6 × 10^7 Hz and for [111] propagation. The acoustic attenuation for transverse and longitudinal waves propagating in [100] direction was measured for YB$_{66}$ and YB$_{61.75}$ at 0.06, 0.19, and 1.15 GHz vs. temperature from ~2 to ~300 K. A figure in the paper shows the results for YB$_{61.7}$ at 0.06 GHz and YB$_{66}$ at 1.15 GHz, both for transverse waves propagating in the [100] direction. At 1.15 GHz the attenuation was too large to measure above 11 K. At 0.06 GHz it is mainly diffraction-limited below 200 K, and above 200 K it begins to rise steeply. The attenuation for longitudinal waves in YB$_{61.7}$ has a similar magnitude and temperature dependence, and is at least a factor of 3 larger at 0.19 GHz. The higher attenuation compared to that of β B is possibly due to the random occupancy of Y sites in the lattice [10]. The energy of the highest acoustic phonons is $\bar{v}$ = 80 cm^{-1} as obtained from the sound velocity and the lattice constant. From the Debye temperature $\bar{v}$ = 904 cm^{-1} is obtained. The ratio 80/904 (~9%) gives the fraction of the total range of phonon energies actually lying within the acoustic branch and taking part in thermal transport. The mean free path of phonons l computed from the heat capacity of all phonon branches decreases between 2 and 300 K from 7 × 10^{-4} to ~4 × 10^{-8} cm, respectively to 1.8 × 10^{-5} cm, when only the actual carriers of the thermal energy, the acoustic phonons, are considered. At Θ$_D$ the mean free path is l = 2 × 10^{-5} cm. The low l values at low temperatures (l at 3 K is much less than the sample diameter) are explained by the random structure frozen-in in the melt-grown crystals. Near Θ$_D$ phonon scattering is dominant [9].

The Debye temperature Θ$_D$ = 1300 ± 50 K of YB$_n$ is calculated from the sound velocity [10], see also [9]. Some preliminary specific heat capacity measurements at low temperature gave a measured Θ$_D$ = 1200 K, Bilir [11].

The temperature dependence of the thermal conductivity λ of YB$_{66}$ resembles that of glass and is as follows:

T in K	3	6	10	20	30	45 to 100
λ in 10^{-3} W·cm^{-1}·K^{-1}	0.62	2.3	5.2	12.5	17	20

The data points show an upward trend at 150 to 300 K caused by uncompensated thermal radiation losses. Correcting for this gives λ = 0.02 W·cm^{-1}·K^{-1} at 300 K [9], see also [3]. Below a few K the law λ ≈ 10^{-4} T$^{<2}$ holds, as shown in Fig. 45, p. 116, measured down to ~0.1 K. Above ~10 K λ becomes nearly independent of temperature owing to strong phonon scattering the cause of which is unknown, Türkes et al. [14].

Electrical and Optical Properties

The electrical resistivity ϱ of YB$_{61.5}$ decreases from ~5 × 10^5 to ~3 × 10^{-2} Ω·cm (read from a given figure) between 140 and 1000 K. Samples from YB$_{61.5}$ to YB$_{66}$ are invariably p type and for all samples ϱ is almost the same: 360 Ω·cm at room temperature [10], ϱ = 620 Ω·cm [1, 2]; earlier the order 10^6 Ω·cm was given for semiconducting YB$_{66}$, as for β B [3]. The conduction is a type of hopping process with very low mobility <10 cm^2·V^{-1}·s^{-1} at room temperature. To a first approximation YB$_{61.5}$ is a B lattice with 2.05 × 10^{21} cm^{-3} Y impurities. Thus the ϱ vs. T curve is comparable to that of heavily doped β B [10]. The conclusion of heavily doped β-rhombo-hedral B was later contradicted for MB$_{66}$ by Golikova, Tadzhiev [12]; see p. 116.

The crystals show a metallic luster [2]. The transmission curve is structureless from 1070 to 2080 eV [1, 2]. YB_{66} is white with a pink cast in reflected white light. In the near IR it is opaque and the reflection data are featureless. In the far IR, absorption measurements show that YB_{66} transmits light only below 80 cm^{-1} for 0.01 cm thin samples [3]. The absorption coefficient vs. photon energy between 6 and 110 cm^{-1} at 4.2 K shows no peaks of the Y–B vibrations, see the figure given in the paper [10].

Chemical Behavior

YB_{66} is chemically stable except that concentrated HNO_3 attacks at room temperature [1, 2].

References:

[1] Tanaka, T.; Otani, S.; Ishizawa, Y. (J. Less-Common Metals **117** [1986] 293/5).
[2] Tanaka, T.; Otani, S.; Ishizawa, Y. (J. Cryst. Growth **73** [1985] 31/6).
[3] Oliver, D. W.; Brower, G. D. (J. Cryst. Growth **11** [1971] 185/90).
[4] Oliver, D. W.; Brower, G. D.; Horn, F. H. (J. Cryst. Growth **12** [1972] 125/31).
[5] Richards, S. M.; Kasper, J. S. (Acta Cryst. B **25** [1969] 237/51).
[6] Naslain, R.; Etourneau, J.; Kasper, J. S. (J. Solid State Chem. **3** [1971] 101/11).
[7] Kasper, J. S. (J. Less-Common Metals **47** [1976] 17/21).
[8] Naslain, R. (in: Matkovich, V. I., Boron and Refractory Borides, Springer, Berlin – New York 1977, pp. 139/202, 143/9).
[9] Slack, G. A.; Oliver, D. W.; Horn, F. H. (Phys. Rev. [3] B **4** [1971] 1714/20).
[10] Slack, G. A.; Oliver, D. W.; Brower, G. D.; Young, J. D. (J. Phys. Chem. Solids **38** [1977] 45/9).

[11] Bilir, N. (Diss. Stanford Univ. 1974, Palo Alto, Calif., from [10]; Diss. Abstr. Intern. B **35** [1974/75] 2963/4).
[12] Golikova, O. A.; Tadzhiev, A. (J. Non-Cryst. Solids **87** [1986] 64/9).
[13] Richards, S. M. (Diss. Rensselaer Polytech. Inst. 1966, pp. 1/227, 23; Diss. Abstr. B **27** [1967] 3894).
[14] Türkes, P. R. H.; Swartz, E. T.; Pohl, R. O. (AIP [Am. Inst. Phys.] Conf. Proc. No. 140 [1986] 346/61, 350).

32.1.8.3 Lanthanum Borides

Additional data for LaB_4 and LaB_6 are found in Sections 32.1.4 and 32.1.5, on pp. 6 and 21, respectively.

32.1.8.3.1 The La–B System

Phase Diagram

The phase diagram of the system La–B is given in **Fig. 60** from Spear [1] and is a revision of the one given by Johnson, Daane [2], see also Elliot [6], near LaB_6 and the B-rich region. A further revision is shown in **Fig. 61**, according to which the liquidus curve near LaB_6 is very flat, McKelvy et al. [3]. The La transition points are not influenced by B. The solubility of B in La and of La in B seems to be very low. Unlike other M–B systems, di- and dodecaborides do not exist. LaB_4 melts incongruently at 1800±15°C [2], ~1850°C (taken from the figure) [3]. The peritectic melt composition is 67 at% B, and LaB_4 has a narrow homogeneity region [2]. LaB_6

melts congruently at 2715°C [1]; the composition range is narrow on the B side at low temperature and widens on increasing temperature, Storms, Mueller [4], see also [3]. It exists between 85.5 and 88 at% B. Near the B side, a eutectoid reaction at ~2360°C was assumed involving a solid solution of La in B (which is stable at high temperature only) leading to LaB_6 and B [2]. However, Fig. 60 shows a peritectic at 2350°C. The existence of LaB_{66} may also be possible between 2200 and 2350°C [1]. In a further survey, a eutectic was assumed at 2025°C, Spear [5]. Recently, thermal arrests at 2052±7°C and 2007±10°C (2325 and 2280 K) were interpreted as eutectic $S \rightarrow LaB_6 + B$ and peritectoid LaB_6 (purple) $+ 3B \rightarrow LaB_9$ (blue), respectively. The eutectic composition is B/La=12.5. Ordered La vacancies are assumed in the blue LaB_9 [4]. However, according to detailed transmission electron microscopic, X-ray, and NMR studies, this blue boride is a randomly La-deficient hexaboride, instead, with an overall composition $LaB_{6.13\pm0.03}$ within the homogeneity range of LaB_6. The compound LaB_9 may exist between 2007 and 1627°C or less [3].

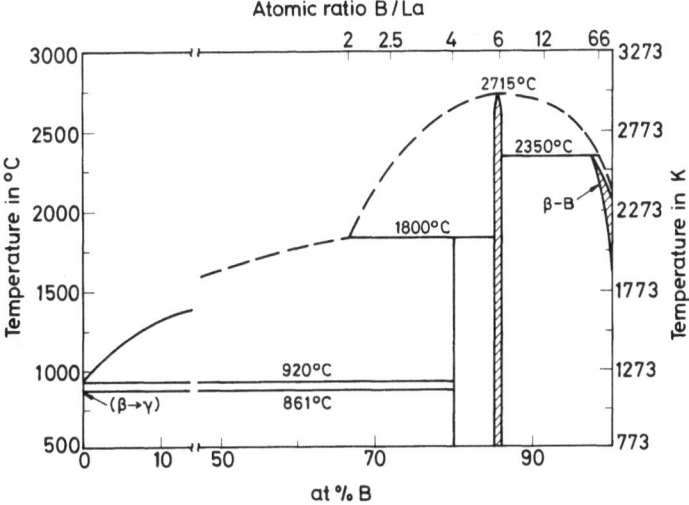

Fig. 60.   The La–B phase diagram.

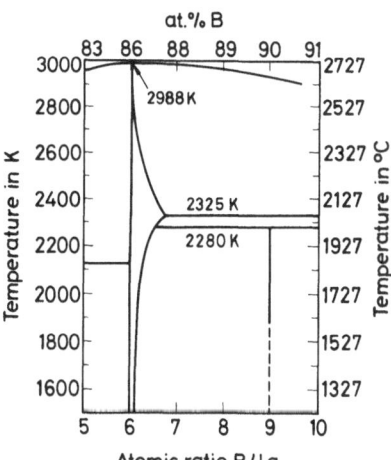

Fig. 61.   Partial La–B phase diagram.

Preparation and Properties

Samples in the composition range La to LaB_4 were prepared by heating La and LaB_4 in the appropriate amounts in a Ta crucible under vacuum at 1773 K for 15 min in an induction furnace. LaB_4 was obtained by arc melting of pressed B powder and a two- to threefold excess of La. The LaB_6 was prepared from LaB_4 and B in vacuum at 1873 K in 15 min. The compositions LaB_6 to B were arc melted from LaB_6 and previously arc-melted B [2]. After arc melting, the borides with compositions from LaB_6 to B were broken to < 200 mesh. Fe contaminations were leached out with HCl and the powder was then annealed at ~2000 K for 15 h [4]. Arc melting was also used by [3]. Nonequilibrium glassy films of nominal $LaB_{1.9}$ were evaporated at 5×10^{-10} Torr ($\cong 0.07$ μPa) from LaB_6 onto polycrystalline Ta substrate at room temperature. On annealing up to 920 K, crystalline La, LaB_4, and LaB_6 are formed, Ociepa [9].

Material with 32 to 59 wt% B ($\cong$ compositions between $LaB_4 + LaB_6$ to $LaB_6 + B$) has a low temperature coefficient of resistance and an extended range of resistance ratings, Shulishova, Shcherbak [7]. The preparation of directionally solidified composite eutectics, the most promising being LaB_6 needles in B matrix, and its potential use as composite field emission cathode material, is described by Goodrum [8]. The thermionic properties of compositions $LaB_{5.9}$, $LaB_{6.01}$, and $LaB_{8.5}$ were studied. For details see p. 242, Jaskie [10].

References:
[1] Spear, K. E. (in: Alper, A. M., Refractory Materials, Phase Diagrams, Vol. 4, Academic, New York 1976, pp. 91/159, 142/3).
[2] Johnson, R. W.; Daane, A. H. (J. Phys. Chem. **65** [1961] 909/15).
[3] McKelvy, M. J.; Eyring, L.; Storms, E. K. (J. Phys. Chem. **88** [1984] 1785/90).
[4] Storms, E.; Mueller, B. (J. Phys. Chem. **82** [1978] 51/9).
[5] Spear, K. E. (in: Matkovich, V. I., Boron and Refractory Borides, Springer, Berlin – New York 1977, pp. 439/56, 446/7).
[6] Elliott, R. P. (Constitution of Binary Alloys, 1st Suppl., McGraw-Hill, New York 1965, pp. 123/4).
[7] Shulishova, O. I.; Shcherbak, I. A. (USSR 688014 [1978/80] from C. A. **92** [1980] No. 225264).
[8] Goodrum, J. W. (AD-A 101703 [1981] 1/125, 4/48; C. A. **96** [1982] No. 44786).
[9] Ociepa, J. G. (Thin Solid Films **120** [1984] 123/31).
[10] Jaskie, J. E. (DOE-ET-15419-T 1 [1981] 1/121, 32; C. A. **99** [1983] No. 80946; Diss. Arizona State Univ. 1981, pp. 1/123, 32; C. A. **97** [1982] No. 32162).

32.1.8.3.2 Nonconfirmed Borides La_3B, La_2B, and LaB

For a composition La_3B with $a = 4.368$ Å, the superconducting critical temperature $T_c = 27$ to 33.4 K was predicted by various methods, Wang [1].

The existence of a lower boride La_2B, in addition to the known LaB_4 and LaB_6, has been reported by Markovskii, Vekshina [2], who prepared it by sintering pressed La and B powder mixtures at 1073 K. The compound is less stable chemically than LaB_4 and LaB_6 and is readily decomposed by HCl (1:10) with the evolution of large amounts of boranes. The existence has not been confirmed, see for instance Spear [3].

The heat of formation of hypothetic $LaB = -63$ kJ/mol was predicted with the help of electronegativity and density parameters of the components, Miedema et al. [4].

References:

[1] Wang, R. (Phys. Status Solidi A **94** [1986] 445/52).

[2] Markovskii, L. Ya.; Vekshina, N. V. (Zh. Prikl. Khim. **38** [1965] 1945/9; J. Appl. Chem. [USSR] **38** [1965] 1908/11).

[3] Spear, K. E. (in: Alper, A. M., Refractory Materials, Phase Diagrams, Vol. 4, Academic, New York 1976, pp. 91/159, 99).

[4] Miedema, A. R.; de Boer, F. R.; Boom, R. (Proc. 12th Rare Earth Res. Conf., Vail, Colo., 1976, Vol. 2, pp. 838/46).

32.1.8.3.3 LaB$_4$

LaB$_4$ is an incongruently melting (at 2073 ±15 K) phase in the La–B system, Johnson, Daane [1], see p. 162. The homogeneity range of LaB$_4$ is rather narrow [1]. The wide range of homogeneity supposed by Felten et al. [2] is questioned by Samsonov [3, p. 68].

Formation. Preparation

Powders

LaB$_4$ forms as the major component in the reaction of boron and lanthanum in an atomic ratio B/La between two and four at about 1573 K, Felten et al. [2]. Pure LaB$_4$ (at least comparatively) was obtained at an atomic ratio La/B = 1/2 at 1273 K and treatment of the products with dilute hydrochloric acid, Markovskii, Vekshina [4]. No excess of a reaction component is mentioned by Etourneau et al. [5] in their direct syntheses at 1773 K in sealed Mo or Ta tubes or by Buschow, Creyghton [6], who slowly heated prereacted (by arc melting) samples to 1873 K within sealed Mo containers followed by annealing for approximately 1 h.

Reduction of La$_2$O$_3$ by boron under 10^{-6} Torr ($\hat{=}$133 µPa) at temperatures between 1773 and 2473 K yielded LaB$_4$ with admixtures of hexaboride or La metal, their amounts varying with temperature and reaction time, Etourneau et al. [7]. The reduction according to La$_2$O$_3$ + 2 B$_4$C + C = 2 LaB$_4$ + 3 CO at temperatures between 1473 and 1773 K gave carbon-containing products, up to 8.5 wt% C, and varying ratios of La/B. The product with La/B nearest to LaB$_4$ and 3.7 wt% C was obtained at 1673 K, Neshpor [8], see also Samsonov [3, p. 68], Etourneau et al. [9, p. 118], [10, p. 272].

Crystal Growth

Large LaB$_4$ crystals can be grown by controlled cooling of boron-lanthanum melts containing 40 at% B from 2053 K to about 1573 K in tantalum crucibles (according to the phase diagram, LaB$_4$ is in equilibrium with La at all temperatures below its peritectic decomposition, see p. 163). For example, after equilibration at 1973 K, the sample was cooled at 16 K/h to about 1623 K at which point the furnace was switched off. The La matrix was etched away in 1:1 dilute hydrochloric acid for 16 h, which, after washing with distilled water, left the LaB$_4$ crystals. Most of the good crystals were 5 to 8 mm in diameter; sizes were only limited by the container dimensions, Deacon, Hiscocks [11]. This procedure was developed from the method of Johnson, Daane [1]. These authors prepared pure, black LaB$_4$ crystals with sharp edges and smooth shiny faces by arc melting pressed mixtures of boron powder with a two- to threefold excess of La, and subsequent treatment in dilute hydrochloric acid [1]. Black single-crystal platelets were obtained from mixtures of La and B with an atomic ratio of 1 : 0.65 in sealed Ta tubes, which were heated to 1973 K for 10 min and then cooled to 1273 K over a period of 10 min; the treatment of the excess La, which acted as flux, is not described, Fisk et al. [12].

Crystallographic Properties

LaB_4 is isomorphous with the other rare earth tetraborides: tetragonal UB_4 type; the structure is described on p. 8. The following values, in Å, are reported for the room temperature lattice parameters a and c, respectively, from powder patterns: 7.3240 ± 0.0005, 4.1811 ± 0.0006, Johnson, Daane [1], 7.328 ± 0.004, 4.175 ± 0.005, Zhuravlev et al. [13], 7.327, 4.182, Buschow, Creyghton [6], 7.30, 4.17, Felten et al. [2]; from single crystals: 7.324 ± 0.001, 4.181 ± 0.001, Fisk et al. [12], confirmed by Kato et al. [14], 7.32462 ± 0.00003, 4.18091 ± 0.00002 at 293 K, Deacon, Hiscocks [11] by X-ray diffractometer (no further data were given to justify this high precision). No variation of the lattice parameters, due to possible deviations from ideal stoichiometry, was found [1, 2].

An X-ray single-crystal structure determination with $R = 0.035$ showed the short La–B distances to range between 2.818(4) and 2.969(6) Å with two additional La–B distances at 3.155(9) Å. The short B–B distances range from 1.775(11) to 1.850(12) Å, Kato et al. [14]. For thermal expansion coefficients parallel and perpendicular to c from X-ray diffraction, see p. 167.

The growth facets of the crystals are, in the order of decreasing size, {001}, {110}, and {100} [11]. A prismatic shape of a crystal bounded predominantly by {110} is mentioned by Kato et al. [14].

^{11}B Nuclear Magnetic Resonance

The nuclear magnetic resonance (NMR) signal of ^{11}B in polycrystalline LaB_4, recorded at 4.2 K and 17.910 MHz, should be separated into the contributions from the three different boron sites. However, the three central magnetic transitions (leading to the Knight shifts) strongly overlap and cannot be analyzed, but it is possible to derive the Knight shifts from the satellites, which are separated from each other by their different (electric) quadrupole interaction. The so-derived Knight shifts for LaB_4 are zero with respect to $Na_2B_2O_4$: isotropic magnetic shift $K_{iso} = \frac{1}{3}(K_x + K_y + K_z) = 0.00(2)\%$ for all the sites, and the axial magnetic shift is $K_{ax} = K_z - K_{iso} = 0.00(3)\%$ for site 4e and 0.00(2)% for the other sites; the asymmetry parameter ε of the shift remained undetermined. The results for the quadrupole interaction H_Q, in G, or the equivalent ν_Q, in kHz, and for the asymmetry parameter η of the quadrupole interaction are:

boron site	point symmetry	H_Q	ν_Q	η
4e	4	251(4)	343(4)	0
8j	m	302(3)	412(4)	0.53(1)
4h	mm2	398(3)	544(4)	0.045(20)

Within the experimental accuracy, these values are independent of temperature between T = 4 and 300 K. Small differences between the observed and the calculated spectra were ascribed to a second phase (3 to 4 wt% LaB_6), the presence of which was also indicated in X-ray diagrams. The impossibility of observing any difference in the Knight shift for the three boron sites rules out the possibility of resolving the question whether a nonuniform polarization model (demanded for the RKKY model) gives a better description of the experimental data than a more simple model of uniform polarization, Creyghton et al. [15].

Mechanical and Thermal Properties

The X-ray density is given as $D_{calc} = 5.396$ g/cm^3, Kato et al. [14] or 5.44 g/cm^3 [3] from the lattice parameters of Felten et al. [2]. The microhardness on {001} at 50 g load is 2190 kg/mm^2 ($\cong 21.6$ GPa) [11].

The mean thermal expansion coefficients are $(8.36 \pm 1.03) \times 10^{-6}$ and $(7.17 \pm 1.16) \times 10^{-6}$ K^{-1} parallel and perpendicular to c, respectively, derived from X-ray diffraction between 298 and 773 to 973 K, Zhuravlev et al. [13]. An approximate value of 4×10^{-6} K^{-1} parallel to c, obviously for room temperature, was given by Deacon, Hiscocks [11].

The Gibbs free energy of formation at 1700 K at the high boron phase boundary was related as $\Delta G = -41$ kcal/mol ($\triangleq -172$ kJ/mol), Storms [16].

Magnetic and Electrical Properties

LaB$_4$ is diamagnetic at room temperature, $\chi_g = -8.3 \times 10^{-8}$ cm^3/g, Buschow, Creyghton [6].

LaB$_4$ is a metallic conductor [5]. The electrical resistivity at 292 K is $\varrho = (22 \pm 4) \times 10^{-6}$ Ω·cm, Deacon, Hiscocks [11]. The resistivity decreases with decreasing temperature from $\varrho \approx (24 \pm 12) \times 10^{-6}$ Ω·cm at room temperature to roughly half that value at 83 K, Johnson, Daane [1]. The resistivity ϱ in the (001) plane decreases from $\sim 17.5 \times 10^{-6}$ Ω·cm at 300 K to below 1×10^{-6} Ω·cm at 4.2 K, as read from a figure in Kasaya et al. [17]. The metallic conductivity agrees with the proposed bonding situation (La^{3+}, strong covalent B–B bonds, no covalent La–B bonds, one conduction electron per La), see pp. 17/8 [1].

The Hall constant, $R_H = -2.8 \times 10^{-3}$ cm^3/C at room temperature (I in (001) plane), shows only weak temperature dependence. This might be due to unequal temperature dependences of the mobilities of electrons and holes, Kasaya et al. [17].

The electron-emitting properties of LaB$_4$ are qualitatively similar to those of LaB$_6$. Color changes after electron-emission tests may indicate a change to LaB$_6$ at the sample surface by loss of La [11].

X-Ray Spectra

The B K$_\alpha$ emission points to strong covalent B–B bonds, see p. 18. The La L$_{III}$ absorption spectra of LaB$_4$, stoichiometric or with an imperfect lattice, show no change. The observed long-wave shift of the principal absorption maximum, 0.3 eV relative to the oxide, is only slightly more than the experimental error of ± 0.2 eV. This shift is 0.8 eV for LaB$_6$; the difference was interpreted as a small valency change of La attributed to the change in structure and nearest neighborhood, Vainshtein et al. [18].

Chemical Behavior and Use

LaB$_4$ is stable against cold, dilute HCl [1, 4, 11], it can be etched by concentrated sulfuric acid [11], and is readily soluble in nitric acid, and decomposes in concentrated hydrochloric acid on boiling [4].

The transition region at the pn junction in a LaB$_4$ single crystal, partly doped with Ca, has been shown to serve as a laser material at $\lambda = 9300$ Å if a current higher than 5000 A/cm^2 passes the pn junction, Vickery [19].

References:

[1] Johnson, R. W.; Daane, A. H. (J. Phys. Chem. **65** [1961] 909/15).

[2] Felten, E. J.; Binder, I.; Post, B. (J. Am. Chem. Soc. **80** [1958] 3479).

[3] Samsonov, G. V. (Tugoplavkie Soedineniya Redkozemel'nykh Metallov s Nemetallami, Metallurgiya, Moscow 1964; High-Temperature Compounds of Rare Earth Metals with Nonmetals, Consultants Bureau, New York 1965, pp. 1/280, 1/100).

[4] Markovskii, L. Ya.; Vekshina, N. V. (Zh. Prikl. Khim. **38** [1965] 1945/9; J. Appl. Chem. [USSR] **38** [1965] 1908/11).

[5] Etourneau, J.; Mercurio, J.-P.; Berrada, A.; Hagenmuller, P.; Georges, R.; Bourezg, R.; Gianduzzo, J. C. (J. Less-Common Metals **67** [1979] 531/9).

[6] Buschow, K. H. J.; Creyghton, J. H. N. (J. Chem. Phys. **57** [1972] 3910/4).

[7] Etourneau, J.; Mercurio, J.-P.; Naslain, R.; Hagenmuller, P. (Compt. Rend. C **274** [1972] 1688/91).

[8] Neshpor, V. S. (Vysokotemp. Metallokeram. Mater. **1962** 96/101; N.S.A. **19** [1965] No. 5912).

[9] Etourneau, J.; Mercurio, J.-P.; Hagenmuller, P. (in: Matkovich, V. I., Boron and Refractory Borides, Springer, Berlin – New York 1977, pp. 115/38).

[10] Etourneau, J. (J. Less-Common Metals **110** [1985] 267/81).

[11] Deacon, J. A.; Hiscocks, S.E.R. (J. Mater. Sci. **6** [1971] 309/12).

[12] Fisk, Z.; Cooper, A. S.; Schmidt, P. H.; Castellano, R. N. (Mater. Res. Bull. **7** [1972] 285/8).

[13] Zhuravlev, N. N.; Belousova, I. A.; Manelis, R. M.; Belousova, N. A. (Kristallografiya **15** [1970] 836/8; Soviet Phys.-Cryst. **15** [1970/71] 723/4).

[14] Kato, K.; Kawada, I.; Oshima, C.; Kawai, S. (Acta Cryst. B **30** [1974] 2933/4).

[15] Creyghton, J. H. N.; Locher, P. R.; Buschow, K. H. J. (Phys. Rev. [3] B **7** [1973] 4829/43).

[16] Storms, E. K. (J. Phys. Chem. **85** [1981] 1536/40).

[17] Kasaya, M.; Takegahara, K.; Yanase, A.; Kasuya, T. (Cryst. Electr. Field Eff. f-Electron Magn. Proc. 4th Intern. Conf., Wroclaw 1981 [1982], pp. 95/9; C.A. **98** [1983] No. 64284).

[18] Vainshtein, E. E.; Blokhin, S. M.; Paderno, Yu. B. (Fiz. Metal. Metalloved. **18** [1964] 450/1; Phys. Metals Metallog. [USSR] **18** No. 3 [1964] 132/3).

[19] Vickery, R. C. (U.S. 3340108 [1963/67] 1/3; C.A. **68** [1968] No. 74014).

32.1.8.3.4 LaB$_6$

32.1.8.3.4.1 Homogeneity Range

LaB$_6$ exhibits a homogeneity range, see the partial phase diagram from [1] on p. 163. The existence and extent of the homogeneity range were controversial for a long time, McKelvy et al. [1], see also the review articles [2 to 4]. The homogeneity range at room temperature was determined in a most recent careful study [1] to reach from B/La = 6 to 6.13 ± 0.03 with a change of color from purple to blue at about B/La = 6.07. The samples were studied by high resolution transmission electron microscopy (HRTEM), back scattered electron imaging, quantitative characteristic X-ray analysis, ^{11}B NMR, density based separation (sink float method), and other techniques [1]. Earlier either broader homogeneity ranges were established: B/La = 6 to 7.8 by Johnson, Daane [5], and ~6 to 6.5 by Zhuravlev et al. [6] from sample density and microscopical studies or the existence of a homogeneity range was refuted on the basis of lattice spacing measurements of different compositions (B/La = 5.8 to 6.6) by Mamedov et al. [7]. In contrast an extremely wide range of homogeneity was claimed from experiments on sintering LaB$_6$ with B: La$_x$B$_6$ with $0.33 \leqq x \leqq 1.0$ ($\hat{=}$ LaB$_{18}$ to LaB$_6$) at 1873 K. At higher temperatures the value of x is said to be confined to $x \geqq 0.77$, Badzian [23]. The blue boride, which was earlier postulated to be a new phase, "LaB$_9$", with ordered La vacancies, Storms, Mueller [8], Zhuravlev et al. [9], was determined to be a randomly La-deficient La hexaboride with composition LaB$_{6.13 \pm 0.03}$, the boron-rich end of the homogeneity region [1], see also p. 162. Diffraction patterns of preparations with various boron contents ranging from LaB$_6$ to LaB$_{12}$ showed only the normal LaB$_6$ unit cell with no significant variation of lattice parameter [6, 10, 11, 26]. Crystals of β-rhombohedral boron were recognized by transmission electron microscopy in a LaB$_6$ sample analyzed as La$_{0.81}$B$_6$ (63.1 wt% La, 36.4 wt% B) which in X-ray diffraction showed only the LaB$_6$ powder pattern. The β-boron crystals were found mostly joined to LaB$_6$ crystals but without actual intergrowth in a ratio less than 1:20, Olsson [24]. There are some doubts whether the composition range of La hexaboride is caused by vacancies at the La or B positions or both. Chemical analyses of zone-refined single crystals gave B/La ratios which were stoichiometric within experimental error [17 to 19, 21] and also values significantly lower than 6. Values as low as B/La = 5.71 ± 0.05 [13], 5.74 [15], 5.75 [12, 14, 22] were reported. Metallographic analyses of single crystals revealed occasional inclusions of LaB$_4$, Aida, Fukazawa [27], Paderno et al. [19], Takagi, Ishii [28], Hohn et al. [29]. However, as Noack, Verhoeven [12] point out, a LaB$_6$/LaB$_4$ mixture with overall ratio B/La = 5.75 would have enough LaB$_4$ to be easily detected in optical, scanning electron microscopic, and X-ray diffraction studies. The authors conclude that the samples studied were single phase with B/La less than stoichiometric [12]. Vacancies at boron positions were indicated, although not conclusively established, from an X-ray diffraction study of three different LaB$_6$ crystals prepared from an Al melt containing excess boron or lanthanum. The structure was refined to R values of 1.10 to 1.36% and in the three crystals boron occupancies of 96.4(7), 97.2(5), and 98.2(5)% were found, corresponding to B/La = 5.78 ± 0.13, 5.83 ± 0.09, and 5.89 ± 0.09, respectively, Korsukova et al. [16]. Vacancies in both the La and the B sublattice, were proposed from the comparison of pycnometric and X-ray densities of a sample analyzed as LaB$_{5.9}$. This led to the formula La$_{0.97}$B$_{5.72}$, Aivazov et al. [20].

The homogeneity region of LaB$_6$ broadens appreciably above 2000 K. The maximum of boron content (B/La ≈ 6.8, see Fig. 61 on p. 163) is reported to be at 2325 K, the temperature of the LaB$_6$–B eutectic [1, 8]. The previously proposed phase diagram [8] was adjusted to a congruent melting composition very close to stoichiometric [1] (B/La ≈ 6.04 as read from Fig. 61 on p. 163, instead of 6.2 in [8]) which is also close to the congruently vaporizing composition (CVC) of B/La = 6.034 and 6.042 at 1500 and 1700 K given by Storms, Mueller [25], see also p. 205 for additional also disagreeing data. The homogeneity range at temperatures above the LaB$_6$–B eutectic was found to be narrower [1] than was given earlier [8].

References:

[1] McKelvy, M. J.; Eyring, L.; Storms, E. K. (J. Phys. Chem. **88** [1984] 1785/90).

[2] Lundström, T. (Z. Anorg. Allgem. Chem. **540/541** [1986] 163/8).

[3] Korsukova, M. M.; Gurin, V. N. (Usp. Khim. **56** [1987] 3/28; Russ. Chem. Rev. **56** [1987] 1/15).

[4] Korsukova, M. M.; Gurin, V. N. (Current Topics Mater. Sci. **11** [1984] 390/439).

[5] Johnson, R. W.; Daane, A. H. (J. Phys. Chem. **65** [1961] 909/15).

[6] Zhuravlev, N. N.; Manelis, R. M.; Gramm, N. V.; Stepanova, A. A. (Poroshkovaya Metal. **7** No. 2 [1967] 95/101; Soviet Powder Met. Metal Ceram. **1967** 158/62).

[7] Mamedov, F. G.; Meerson, G. A.; Zhuravlev, N. N.; Umanskii, Ya. S. (Izv. Akad. Nauk SSSR Neorgan. Materialy **3** [1967] 950/6; Inorg. Materials [USSR] **3** [1967] 851/6).

[8] Storms, E.; Mueller, B. (J. Phys. Chem. **82** [1978] 51/9).

[9] Zhuravlev, N. N.; Manelis, R. M.; Belousova, I. A.; Nurmukhamedov, V. Kh. (Izv. Akad. Nauk SSSR Neorgan. Materialy **9** [1973] 1162/5; Inorg. Materials [USSR] **9** [1973] 1033/5).

[10] Post, B.; Moskowitz, D.; Glaser, F. W. (J. Am. Chem. Soc. **78** [1956] 1800/2).

[11] Neshpor, V. S. (Vysokotemp. Metallokeram. Mater. **1962** 96/101; N.S.A. **19** [1965] No. 5912).

[12] Noack, M. A.; Verhoeven, J. D. (J. Cryst. Growth **49** [1980] 595/99).

[13] Davis, P. R.; Swanson, L. W.; Hutta, J. J.; Jones, D. L. (J. Mater. Sci. **21** [1986] 825/36).

[14] Paderno, Yu. B.; Lazorenko, V. I.; Buryak, N. I.; Kovalev, A. V.; Matvienko, A. A.; Galasun, A. P. (Poroshkovaya Metal. **1983** No. 1, pp. 56/60; Soviet Powder Met. Metal Ceram. **22** [1983] 50/3).

[15] Swanson, L. W.; Gesley, M. A.; Davis, P. R. (Surf. Sci. **107** [1981] 263/89).

[16] Korsukova, M. M.; Gurin, V. N.; Lundström, T.; Tergenius, L.-E. (J. Less-Common Metals **117** [1986] 73/81).

[17] Tanaka, T.; Bannai, E.; Kawai, S.; Yamane, T. (J. Cryst. Growth **30** [1975] 193/7).

[18] Verhoeven, J. D.; Gibson, E. D.; Noack, M. A.; Conzemius, R. J. (J. Cryst. Growth **36** [1976] 115/20).

[19] Paderno, Yu. B.; Lazorenko, V. I.; Kovalev, A. V. (Poroshkovaya Metal. **1981** No. 10, pp. 60/5; Soviet Powder Met. Metal Ceram. **20** [1981] 717/21).

[20] Aivazov, M. I.; Aleksandrovich, S. V.; Zinchenko, K. A.; Mkrtchyan, V. S. (Fiz. Metal. Metalloved. **46** [1978] 1171/5; Phys. Metals Metallog. [USSR] **46** No. 6 [1978] 39/42).

[21] Olsen, G. H.; Cafiero, A. V. (J. Cryst. Growth **44** [1978] 287/90).

[22] Noack, M. A. (IS-T-869 [1979] 1/90; C.A. **92** [1980] No. 138 783).

[23] Badzian, A. R. (Proc. Conf. Phys. [Wroclaw] **3** [1981] 244/51; C.A. **96** [1982] No. 26 971).

[24] Olsson, P.-O. (J. Solid State Chem. **76** [1988] 301/12).

[25] Storms, E. K.; Mueller, B. A. (J. Appl. Phys. **50** [1979] 3691/8).

[26] Paderno, Yu. B.; Lundström, T. (Acta Chem. Scand. A **37** [1983] 609/12).

[27] Aida, T.; Fukazawa, T. (J. Cryst. Growth **80** [1987] 9/16).

[28] Takagi, K.; Ishii, M. (J. Cryst. Growth **40** [1977] 1/5).

[29] Hohn, F. J.; Chang, T. H. P.; Broers, A. N.; Frankel, G. S.; Peters, E. T.; Lee, D. W. (J. Appl. Phys. **53** No. 3, Pt. 1 [1982] 1283/96).

32.1.8.3.4.2 Preparation

Introduction

Lanthanum hexaboride can be prepared by the general methods used for the preparation of borides of the rare earth elements such as reduction of La_2O_3 at elevated temperature mostly by B or B_4C in a solid state reaction finally yielding powders, by B in an Al flux or electrolytically in a molten mixture of salts, yielding crystals. LaB_6 may be deposited from the gas phase after reducing vapors of La and B halides by hydrogen. Lanthanum and boron also combine directly at higher temperatures. Growth of single crystals, together with a purification, is accomplished by recrystallization, mainly using the floating-zone technique. Larger parts of LaB_6 for technical application must be produced by powder metallurgical methods: pressing at normal or elevated temperature and/or sintering.

The calculation of Gibb's free energy of several possible formation reactions of LaB_6 from La_2O_3 indicated LaB_6 to be formed in any case if the reaction temperature is above 1617 K, Peshev [1].

Preparation of Polycrystalline Samples

Preparation from the Elements

La- (or La-hydride) powder or filings were mixed with amorphous boron powder, pressed into bars, and heated (covered by boron powder) in graphite crucibles under pure hydrogen for 1 h at 1650 K, followed by 20 min at 2073 K after removal of the boron powder. The crushed boride was washed in hydrochloric acid and again after dry milling in a steel ball mill to remove any free metal, metal oxide, and boron oxide. Small amounts of free boron were removed by stirring the powder in water and decanting the liquid, making use of differences in sedimentation rate, Lafferty [2].

La and B powders, at a B excess of 2% relative to stoichiometric composition, were heated under pressure up to 800 kg/cm^2 ($\cong$80 MPa) during 4 h at 1273 K with subsequent annealing for 2 h at 2123 K for crystallization. The purple products had a density of 4.00 g/cm^3 and showed only the LaB$_6$ X-ray diffraction lines; slightly different reaction parameters gave a density of 4.19 g/cm^3, Schmitt, Setton [3]. The same procedure, except for only 0.2 h annealing at 2073 K, gave samples with a density of 4.59 g/cm^3 ($\cong$0.98 D$_{theor}$) [4]. Arc melting of La and B with subsequent homogenization of the pulverized melt buttons, either by cold pressing and sintering at 1700 K, or hot pressing at 2100 K in Ta-lined graphite dies, was used by Storms [5], cf. Storms, Mueller [6]. Direct synthesis with intermediate formation of LaB$_4$ was reported by Johnson, Daane [7].

If pure boron is available, direct synthesis is regarded as preferable to the reaction of B$_4$C and La$_2$O$_3$, for pure LaB$_6$, Rabenau et al. [8]. For preparation of LaB$_6$ from the elements see also "Single-Crystal Growth" on p. 179.

Borothermal Reduction of La$_2$O$_3$

The reaction of La$_2$O$_3$ with boron under vacuum at temperatures between 1773 and 2073 K for 30 to 60 min was investigated with emphasis on obtaining high-purity LaB$_6$. The starting materials were used according to the equation La$_2$O$_3$ + 15 B → 2 LaB$_6$ + $\frac{3}{2}$ B$_2$O$_2$ and also with 5 or 10% excess boron. The equation is only nominal, as B$_2$O$_3$ also forms. The materials were carefully mixed for 8 h under alcohol in a rubber-lined steel ball mill with rubber coated balls. Other organic wetting agents are said to increase the content of carbon in the product. The dried mixture was sieved and briquetted. Products from reaction at 1773 and 1923 K were nonuniform in color, the core of the briquets having a bluish tinge and the periphery purple-violet. Only reaction at 2073 K gave uniform purple-violet samples. A stoichiometric product needed 5% excess of boron. Reaction times long enough to ensure complete removal of B$_2$O$_2$ and excess boron were 0.5 to 1.5 h, depending on the sample size. When reduced at 1923 and 1773 K, stoichiometric charges gave B/La ratios <6, whereas in charges containing excess boron, this excess was not volatilized during the reaction. The reduction under vacuum decreased the content of some impurities such as Si, Mg, Al, Ca. Various mill and mortar materials used in grinding the LaB$_6$ had an effect on purity, Meerson, Mamedov [9]. A later paper by the same group, in addition, specified an optimum briquetting pressure, 0.4 to 0.5 t/cm^2 ($\cong$40 to 50 MPa), giving a porosity of 53 to 55%. This, together with heating from 1573 to 2173 K within 30 min, favored liberation of reaction products and complete reduction within 1 h at 2173 K. This temperature was regarded as optimum, giving an average grain size of 4 to 6 μm. Higher temperatures gave larger values. Lower heating rates gave a dense crust on the sample surface. A moisture content of larger than 10% in the La$_2$O$_3$ is said to break the tablets. Ta was used as crucible material, Meerson et al. [10], see also Zhuravlev et al. [11]. The formation of La-borates as impurities in LaB$_6$ on borothermal reduction of La$_2$O$_3$ was indicated. Acid treatment did not completely remove them, Kovalev et al. [114].

From literature data, a calculation of Gibb's free energy for the reaction formulated above indicated the reaction to start at >1900 K for equilibrium conditions at 1 atm. The onset temperature of the reaction is lowered at reduced pressure to 1670, 1500, and 1100 K for pressures of 10^{-1}, 10^{-3}, and 10^{-5} Torr ($\cong$13, 0.13, and 1.3×10^{-3} Pa), respectively, Meerson et al. [12]. These values are said [12] to agree to within 200 K with the experimental data of [10], where the reaction was described to start at 1673 to 1773 K at 10^{-3} Torr.

Single-phase LaB_6 formed within 1 h under vacuum at temperatures of 1773 K up to 2273 K, from briquetted mixtures of $La_2O_3 + 15$ B in graphite crucibles. At lower temperature the reaction was incomplete. Stoichiometric LaB_6 was obtained at 1773 to 1873 K; higher temperatures gave higher B/La ratios, Samsonov et al. [13]. The preparation by borothermal reduction at 1773 K in induction-heated graphite crucibles under vacuum on a pilot-plant scale gave a product with $D = 4.65 \pm 0.04$ g/cm³ in 90% yield, Breusov et al. [14]. For the preparation of LaB_6 by borothermal reduction of La_2O_3 see also p. 23.

Boron Carbide Method

LaB_6 begins to form from La_2O_3 and B_4C, mixed in a mole ratio of 1:3 and pressed to tablets, above 1473 K (vacuum, graphite crucibles). But only at temperatures of 1873 K and higher was complete reaction obtained within 1 h. The content of carbon in the product decreased with increasing reaction temperature, from 1.3 wt% at 1873 K to 0.3 wt% at 2273 K, Samsonov et al. [13]. To obtain carbon-free products, any free carbon in the B_4C should be taken into account in the initial charge. This may be done preferably by increasing the La_2O_3 content appropriately, or by adding free boron or B_2O_3, Neshpor [15].

Methods and apparatus for LaB_6 production on a pilot plant scale were developed by Breusov et al. [14]: B_4C and La_2O_3, 30% in excess, were ball milled and heated for 3 h at 1873 K and 0.1 Torr ($\triangleq$ 13 Pa). The charge was in a graphite crucible heated by induction within a Cu lined steel vacuum reactor, thermally insulated by a ZrO_2 layer. After washing with HCl solution, a yield of 85% was obtained. The product density was $D = 4.44 \pm 0.03$ g/cm³ [14].

Reactions with Boron Nitride

LaB_6 was synthesized by the reaction of powdered hexagonal BN in a BN crucible with calcined La citrate. Optimum conditions were an initial B/La ratio of 5.0 to 6.0 and heating under vacuum at 1753 to 1873 K with repeated powdering and reheating. Lanthanum citrate hydrate was calcined at 1273 K in a N_2 atmosphere, Shiota et al. [16]. The reaction of La_2O_3 and BN in a mole ratio of 1 to 10 at 2023 K forms a lanthanum boride, O'Connor [17].

Electrolytic Deposition of LaB_6

For the growth of LaB_6 single crystals by molten salt electrolysis see also p. 26.

Following the early research of Andrieux [18], LaB_6 was prepared by electrolytic reduction of mixtures of La_2O_3 and B_2O_3 in molten salts. Experimental data are collected in the following table:

melt composition in mol%	T in K	potential in V	materials of cathode	anode	crucible	Ref.
a) 3.7 La_2O_3, 21.8 $Li_2B_4O_7$, 74.5 Li_3AlF_6	1123	-0.3 c)	Mo	cru. d)	graphite	[19]
5.49 La_2O_3, 21.94 CaB_4O_7, 72.57 $CaCl_2$	1273	1.7 to 3.3	W	cru. d)	graphite	[20]
2.2 La_2O_3, 33.5 B_2O_3, 31.2 Li_2O, 33.1 LiF	1073	1.85 to 2.1	Au	Au	Pt	[21]
b) 12.4 La_2O_3, 40.7 B_2O_3, 46.8 LiF	1273	1.3 to 2	graphite	cru. d)	pyrolytic graphite	[22]
$La_2(B_4O_7)_3$, $CaCl_2$	1400 ± 25	—	—	—	—	[23]
$LaCl_3$, $Li_2B_4O_7$, LiCl	923	—	cru.	graphite	graphite	[24]

a) Calculated from 7.1, 21.8, and 71.1 wt%, respectively. – b) Calculated from 50, 35, and 15%, respectively, presumably by weight. – c) Relative to the LaB$_6$ reference electrode. – d) Crucible served as anode.

The cell reaction was formulated as $2 La_2O_3 + 12 B_2O_3 \rightarrow 4 LaB_6 + 21 O_2$, Zubeck et al. [21]. The formation of LaB$_6$ from melts comprised of La$_2$O$_3$, Li$_2$B$_4$O$_7$, and Li$_3$AlF$_6$ was concluded to proceed by a primary deposition of elemental boron followed by discharge of La ions at a slightly higher potential, Uchida [19]. Some CaB$_6$, co-deposited with the LaB$_6$ from La$_2$O$_3$–CaB$_4$O$_7$–CaCl$_2$ melts, was removed by heavy-liquid separation, Uchida, Shiota [20]. The electrolysis cells were flushed continuously with inert gas to remove the oxygen generated at the anode [19, 21], Scholz et al. [22].

Densification

The preparation of larger dense bodies of LaB$_6$ was studied because of the importance in special technical applications. Owing to the high values of the hardness, cold pressing and sintering at ~2073 K for 15 min in hydrogen or under vacuum yields only ~50% of the theoretical density D$_{theor}$, Lafferty [2]; higher densities may be achieved either by sintering under pressure or by additives that act as binders and lubricants.

Sintered parts made from LaB$_6$ powder of technical purity were found to decompose again into LaB$_6$ powder after storage at ambient atmosphere for 2 to 180 d. In dry air the parts were stable. This behavior is attributed to hydrolysis of impurity phases such as carbides, boron carbides, oxides, or borides. The impurities could be removed by washing with (warm [25]) H$_2$O and with dilute hydrochloric acid, Paderno et al. [25, 26]. Only after purification, hot-pressed parts resisted corrosion during 100 h at 593 to 613 K under 1.5×10^5 Pa of H$_2$O vapor [25]. Disintegration was observed for parts from technical grade LaB$_6$ powder that had been sintered below 2273 K, Kuznetsova et al. [27]. Bulk LaB$_6$ samples with 98% of D$_{theor}$ were obtained from powder at temperatures between 2273 and 2323 K under a pressure of 500 kg/cm^2 ($\triangleq$50 MPa) in a graphite matrix for 10 min. Below 2273 K the reaction is not sufficiently fast, above 2323 K the graphite reacts, Rabenau et al. [8]. High densities, up to 99% D$_{theor}$, for hot-pressed parts were reached for powder with a mean grain diameter of 10 μm or less, prepared from technical grade B$_4$C containing several hundreds ppm by weight of impurities (Mo, Fe, Ti). Purer LaB$_6$ powder with impurities less than 100 ppm almost defied hot pressing, Colton [28]. Hot pressing in graphite dies, 60 min, 300 kg/cm^2 ($\triangleq$30 MPa), of LaB$_6$ powder with a grain size of 5 to 8 μm gave densities of 0.73, 0.93, and 0.98 D$_{theor}$ at 1900, 2200, and 2400 K, respectively, (read from a figure), Scholz [29].

Nearly zero porosity was obtained by hot pressing at ~35 kbar ($\triangleq$3.5 GPa) and ~2073 K for 2 min. The influence of the sintering conditions on composition and microstructure was studied, Bochko et al. [30]. Weight losses, residual porosities, and chemical compositions are tabulated and the microstructure was studied on samples, obtained by cold pressing (0.8 GPa) of LaB$_6$ powders with various particle size distributions (a solution of glycerin in alcohol was added) and with heating between 2073 and 2473 K in vacuum or H$_2$ and between 2073 and 2723 K in Ar atmosphere, Meerson, Mamedov [36]. For the microstructure and the fracture pattern of LaB$_6$ parts sintered at 2.5 to 3 MPa by an electric discharge of between 4 and 12.5 MW for 0.001 s, giving estimated 4273 K at 10.6 MW discharge, see Volkogon et al. [31].

The blue phase "LaB$_9$" was described as a decomposition product during hot pressing LaB$_6$ at 2573 K and 8.5 GPa, Zhuravlev et al. [11, 32], cf. p. 168.

By addition of Ni to the LaB$_6$ powder, the same densification during hot pressing was achieved at 100 to 150 K lower than without Ni. In both cases, densification is a result of creep of material, controlled by dislocation slip. The activation energies without and with Ni were

References for 32.1.8.3.4.2 on p. 186

10.6 and 9.0 eV, respectively. In addition to LaB_6 a second phase in the sintered parts contained La, Ni, and B; it occurred mainly at grain boundaries, Kondrashov et al. [33]. Previously, additives such as W, Re, Ta, Mo, Ti, Ir, Pt, Os, or Ni in amounts of 1 to 5% gave no significant lowering of the hot-pressing temperature, Kuznetsova et al. [27]. Various mixtures were used as additives to promote compacting, see Table 1 from Pastor [115], which summarizes the sintering characteristics of LaB_6. The table was supplemented by some data.

Addition of up to 4% Y_2O_3 activated shrinkage during sintering of LaB_6 parts; an amount of 2% gave the largest effect. Al_2O_3 addition deactivated sintering, Shlyuko et al. [41].

Several papers deal with the effects of a pretreatment of LaB_6 powder on the sintering behavior. Shock waves with amplitudes up to 350 kbar ($\cong 35$ GPa) affect many properties of LaB_6 powder. Shock-wave treated powder gave pellets with, for example, 0.95 D_{theor} after 1 h vacuum sintering at 2323 K, compared to 0.68 D_{theor} untreated. The microhardness increased to 4000 kg/cm^2 from 2800 kg/cm^2 ($\cong 4$ to 2.8 GPa) untreated. The effect of ultrasound is smaller than that of shock waves, Meerson et al. [42], see also Anan'in et al. [43]. After shock-wave treatment, a lowering by 200 to 250 K was observed for the hot-pressing temperature necessary to achieve the same density, Volkogon et al. [44]. The fraction of particles below 1 to 2 μm was increased to 30 to 60% from the initial 10% by one and two shock-wave treatments, respectively, Strashinskaya et al. [45].

Single-Crystal Growth and Zone Refining

General Remarks

The most important method for crystal growth of LaB_6 is the recrystallization from the melt, mainly by containerless floating-zone melting. Other possibilities are crystallization from molten aluminium, chemical transport, or chemical vapor deposition (CVD), and electrolytic deposition. For comprehensive reviews see Korsukova, Gurin [46, 47].

Zone Melting

Data on preparation conditions and the characteristics of LaB_6 single crystals obtained by zone melting are collected in Table 2, p. 176 from Korsukova, Gurin [47], which was supplemented by further data. Only the zone-melting method was used for growing LaB_6 single crystals crucible-free which avoids contact between the hexaboride melt and any container material. The quality of the single crystals is largely determined by composition and purity of the starting material. Most experimenters used commercially available LaB_6 powder, which was pressed in rods or bars and sintered, see for example [48 to 52]. Various heating methods were used to generate the molten zone, see Table 2, p. 176. In addition to the data, mentioned in the table, r.f. induction heating was also used by Windsor [53], Curtis, Graffenberger [54], electron-beam heating, arc- and laser-beam heating, by Feigelson et al. [55], Gibson, Verhoeven [56].

Regarding simplicity, low cost, and the possibility of oriented growth, the arc floating-zone method was considered preferable to other floating-zone techniques [51]. Dissociation and evaporation of LaB_6 was avoided by carrying out the zone-melting operation in an atmosphere of a protective gas, predominantly Ar, see Table 2, p. 176. The effect of different protective gas types, N_2, Ar, H_2, He on the monocrystallinity and phase content of the zone-melted product was studied by Hafner [57]. No preferred orientation of single crystals at spontaneous growth was observed. Most often the growth axis was close to $\langle 001 \rangle$, $\langle 111 \rangle$, $\langle 103 \rangle$, $\langle 113 \rangle$, and $\langle 123 \rangle$. Directions near $\langle 011 \rangle$ were less frequent, and $\langle 012 \rangle$ and $\langle 112 \rangle$ were not observed, Lazorenko et al. [50]. Oriented seed crystals proved to be successful in pulling LaB_6 single crystals in the desired direction [50, 52]. Dislocation density in LaB_6 crystals was closely related to crystal diameter [63], and the subgrain density to rotation rate during crystal growth [64].

Table 1
Sintering Characteristics of LaB$_6$.

binder content in wt%	compacting p in GPa	heating conditions T in K	time in min	atmosphere or vacuum V p in Torr	relative density in %	remarks	Ref.
—	—	1648	15	—	—	—	[2]
aq. soln. 10% PVA [a]	0.4 to 0.5	2680	30	V: (1 to 3)×10^{-5}	82	wt. loss: 15%	[34]
aq. soln. 2% PVA [a]	0.6 to 0.7	2573 to 2623	60	V: (5 to 10)×10^{-3}	99.2	LaB$_6$ powder packing	[35]
aq. soln. 3% PVA [a]	0.15 to 0.25	2623 to 2673	60 to 90	V: —	100	wt. loss, grain growth	[27]
glycerin + alcohol	0.8	2273	120	V: 10^{-5}	81	wt. loss: 9%	[36] [c]
glycerin + alcohol	0.8	2273	120	Ar: 798 to 836	89.5	5%	[36] [c]
glycerin + alcohol	0.8	2273	120	H$_2$: 798 to 836	100	6%	[36] [c]
85 paraffin + 15 beeswax	(0.1×10^{-3}) [b]	2573 to 2623	60	V: (5 to 10)×10^{-3}	95 to 97	LaB$_6$ powder packing	[38,39]
PhOH+HCOH+B	0.1 to 0.5	2423 to 2473	5 to 60	Ar: 750	99.3	—	[40]

[a] PVA: polyvinyl alcohol. – [b] Hot casting at 5 to 8 atm. – [c] See also [37].

Table 2
Preparation of LaB_6 Single Crystals by Zone Melting (Floating-Zone Techniques).

preparation conditions				characteristics of single crystals			Ref.
zone generation	atmosphere p in Torr	pulling rate V in mm/h and directions	starting composition, impurity content in wt%	formula [a]	impurities in wt%	orientation [b], size in mm	
r.f. heating	Ar (1.1×10^4)	20 down	LaB_6 (99.9)	LaB_6 (within limits of chemical analysis)	undetected by qualitative emission spectral analysis (triple zone passage)	undetected l = 60 d = 7	[48]
electron-beam heating	N_2 (20)	$300 < V < 420$ down, sample rotations rpm: $40 < R < 60$	$LaB_6 + (1.6 \text{ to } 2)B$ (61.6 to 62.7 La), $LaB_6 > 99\%$ B: 95 to 97% La_2O_3 99.99	$La_xB_{6.0}$ $0.85 < x < 1.0$	Ti, Mo <0.1 (microprobe analysis), Na, Mg, Cr, Fe, Cu, Zr, Zn, Sn by qualit. SIMS [c] analysis (single zone passage)	undetected l = 25 d = 3.6	[57]
arc heating Ta electrode	Ar (20)	180 zone displacement downwards	$LaB_{6 \pm 0.03}$, total impurity content: 0.49 to 1.9, C: 0.063, $LaB_{5.99}$, $LaB_{5.81}$	$LaB_{6.00 \pm 0.15}$	total amount: 45×10^{-4}, main impurities: Fe, Pr, Nd, W: $(30 \text{ to } 2) \times 10^{-4}$, by mass spectral analysis (triple zone passage), C: 6×10^{-3}	l = 26 d = 0.25 to 1.5 for seed $\langle 110 \rangle$, l = 23 d = 1 for seed $\langle 100 \rangle$ orientation preserved within 1°	[52, 58] [d]

method	atmosphere (pressure)	rate / notes	starting material / impurity	product	impurity data	dimensions	Ref.
laser-beam heating CO$_2$ laser 10.6 μm 400 W	Ar (760)	58 feed rod 95 seed crystal down	LaB$_6$, total impurity content: 0.485	data unavailable	total amount: 0.043, main impurities: Mg, Si: 0.01, Ca, Cu, Fe, Ti <0.005, by emission spectral analysis (double zone passage)	l = 60 d = 1.1	[59]
r.f. heating	Ar, H$_2$ (750)	—	—	LaB$_6$	total amount: 0.015, principal impurity: C: 0.01	l = 120 d = 4 to 16	[60, 61]
laser-beam heating CO$_2$ laser 10.6 μm 500 W	Ar (~760)	51 down (pulling zone: 2:1 diameter attenuation)	LaB$_{6.3}$ total impurity content: 0.23	LaB$_{6.1}$	total amount: 0.05 Al, Ca, Fe, Mo, Si by d.c. plasma emission spectroscopy; significant reduction in Fe and Ca	l > 150 d = 1	[62]
arc heating LaB$_6$ electrode	Ar	50 down	LaB$_{6.2}$ total impurity content: >3.2	LaB$_{6.09}$ starting materials with B/La <6.2 produce a B/La gradient in two-pass zone refined specimens	total amount <0.01 (double zone passage) main impurities: Ni, Cl, S, Si, Al by spark mass-spectrometry	⟨100⟩ d = 1 to 3	[51]
IR heating in xeron arc image furnace	Ar (760) (300 L/h)	10 rotations 0 to 40 rpm	total impurity content: <0.16	—	total amount <0.02	⟨100⟩ d = 2 to 5	[63 to 65]

References for 32.1.8.3.4.2 on p. 186

Table 2 (continued)

preparation conditions			starting composition, impurity content in wt%	characteristics of single crystals			Ref.
zone generation	atmosphere p in Torr	pulling rate V in mm/h and directions		formula[a]	impurities in wt%	orientation[b], size in mm	
r.f. heating	—	20 to 30 down specimen rotation 20 rpm	total impurity content: 1.5 to 2	—	total amount <0.01	⟨100⟩, ⟨110⟩, ⟨111⟩ l=55 to 60 grown on single-crystal oriented seeds orientation preserved within 1°	[50]
r.f. heating	Ar, H₂	—	—	—	—	l=10 d=5 to 6	[66, 67]
r.f. heating	Ar (760)	—	total impurity content: 1.6	—	total amount of impurities: (triple zone passage) 0.01 the amounts of main impurities fell by two or three orders	⟨111⟩ orientation preserved within 5°	[49, 68]

a) According to chemical analysis. – b) Or pulling direction. – c) Secondary ion mass spectroscopy. – d) Additional analytical data in [58].

Experiments on r. f. induction-heated melting of LaB$_6$ in compact, near-eutectic mixtures with boron revealed, after directional solidification, small areas of aligned LaB$_6$ fibers in the boron matrix. The result suggested a possible design of ordered arrays of LaB$_6$ fibers for electron emission facilities. None of the investigated crucible materials were compatible with the LaB$_6$–B mixture and suitable for induction-heated melting, Logan [69].

Floating-zone crystal growth is very effective in lowering impurity concentration according to analyses by spark source mass spectrometry (SSMS) [51, 52, 63], emission spectroscopy [48, 59, 62, 68], and special analyses for C, N, and O. This is due to both the normal refining process and to a vaporization effect, Verhoeven et al. [52]. Analyses for the light elements after two zone passages, for example, in two samples from different sources indicate oxygen contents of 240 and 10 ppmw and carbon contents of 40 and 60 ppmw, respectively [51], after three passages C content <40 ppmw [48], and after four passages C content = 55 ppmw, O content = 20 ppmw; nitrogen and hydrogen were not detected [58]. For the other impurities after three passes the overall concentration was lowered to below the detection limit, Tanaka et al. [48], Paderno et al. [68], or by a factor of 315 for concentrations in atomic ppm, whereas one passage gave a factor of 70 [52], or one to two orders of magnitude [68]. The total impurity content, from SSMS, was below 30 ppm after three passages [58], below 100 ppm after two passages [51]. A final purity of 99.99% is estimated by [63] and of 99.9% by [68]. Data on the actual concentrations of the impurities are reported in [51, 52, 59, 62, 63, 68].

The importance of the vaporization for the refining is evident from the high content of impurities in condensed vapor, especially for Al, Fe, Si, and C [63]. According to the effect of a single impurity on the melting point of LaB$_6$, or to the respective solubility in the solid or the melt, the impurity distribution along the crystal is different: the concentrations of Hf, Al, Si, Sn, Ca, N, O, and C were found to decrease, whereas Mg, Ta, Mn, Fe, Zr, Nb, Mo, and Pb were driven to the end of the crystal [68].

Al-Flux Method

LaB$_6$ single crystals were prepared by this method as follows: known weights of boron and lanthanum or La$_2$O$_3$ were added to the Al flux in alumina crucibles, heated in an Ar atmosphere to the required temperature for several hours to homogenize the melt, followed by controlled cooling. Then the Al ingots were dissolved in dilute HCl, NaOH, or KOH to isolate the LaB$_6$ crystals, see for example, the review of Korsukova, Gurin [46]. In Table 3, p. 180, from [46] are summarized preparation data of LaB$_6$ single crystals by the Al-flux method. The table was supplemented by some data.

Needles, plates, and cubes occur in the ratio 76:23:1 at a cooling rate of 75 K/h. At any of the applied cooling rates down to 6.2 K/h, the needles were most numerous and their thickness decreased roughly exponentially with increasing cooling rate [71]. Longer equilibration times were said to favor needle- and plate-like crystals, whereas shorter or no equilibration gave intergrown cubes [76]. Increasing concentration of La + 6 B in the Al solution favored formation of plate-like crystals, Korsukova, Gurin [46].

For dependences of crystal shape and size on the B/La ratio in raw material and on reaction time, see the figures in the original paper of Okada et al. [78].

Preliminary data on the dependence of the growth rate of LaB$_6$ needles on the given cation/anion ratio in the molten Al solution were said to fit a theoretical expression for the kinetics of dislocation growth of two-component crystals, Moshkin, Nardov [79].

Table 3
Preparation of LaB$_6$ Single Crystals by the Al-Flux Method.

initial components Al purity (in wt%); weight in g	purity of B and La (in wt%); B/La ratio	wt% LaB$_6$ [a]	preparative conditions			characteristics of single crystals		impurity conc. in wt%	Ref.
			T in K	time [b] in h	rate [c] in K/h	shape and size of crystals in mm	method of analysis; formula		
Al (99.9999)	B (99.99), La (99.9); 6:1	~2 [f]	1773	0.16	25	needles 5.0 to 7.0×0.1×0.1; plates 2.00×3.00×0.05; cubes [g]	—	0.01	[70]
Al (99.9); 90 to 140	B (99.99), La (99.9); 4.98:1	2.4 to 3.6	1473	5 to 10	6.2 to 75	needles 5.0×0.5×0.5; plates; cubes	—	0.045	[71]
Al (99.8); 58	B (99.8), La (99.5); 5.8:1	3.06	1773	8	30	needles; plates; cubes	—	0.005 to 0.01	[72]
Al (99.99)	B (>99), La (>99); 6:1 or 12:1	5 to 10	1573 to 1673	2 to 10	—	needles 10.0×0.5×0.4; plates 6.00×7.00×0.15; cubes	chemical analysis [d]: LaB$_{5.86}$ (31.0 wt% B); LaB$_{6.04}$ (32.0 wt% B)	0.056 (Al)	[46] [e]
Al (99.99)	B (>99), La (99.7); 5:1 to 12:1	—	1673 to 2073	4	—	—	LaB$_{5.67}$ to LaB$_{6.82}$	—	[75]
Al (99.5)	B (99.5), La (99.5)	—	1500	16	5	cubes; needles; plates	—	—	[76]
Al; 41	B (99.9), La$_2$O$_3$; (4 to 8):1	~10	1773	120	5 to 20	needles 5×1×1; plates 2.0×2.0×0.1; cubes 1×1×1	microprobe; (30.00 ± 0.08 wt% B)	—	[77]

[a] Concentration of LaB$_6$ in solution. — [b] Holding time. — [c] Cooling rate. — [d] Theoretical boron content of LaB$_6$ is 31.8 wt%. — [e] See also [73, 74]. — [f] 13.35 wt% reported by [46]. — [g] Reported by [46].

Single crystals, La^{10}B$_6$ and La^{11}B$_6$, grown by the Al-flux method showed the same morphology (needles, plates) as LaB$_6$, Zhukova et al. [80]. Al impurities were found up to 0.1 wt%, sometimes in the form of Al-rich lamellae, Futamoto et al. [71]; Al inclusions were below 0.02 wt%, Gurin et al. [73], 50 to 10 ppm [72]. Other impurities were less than 0.01 wt% per species [71], or ranged from 50 to 100 ppm for Fe, Mn, Si, which is less than in the starting materials, Aita et al. [72].

The LaB$_6$ crystallization from Al flux was discussed in the quasi-binary section LaB$_6$–Al of the ternary system La–B–Al by Korsukova, Gurin [46, p. 5], [47, p. 408].

Vapor Phase Methods

LaB$_6$ single crystals could be grown either by chemical transport through the vapor phase by suitable transport agents [81 to 83] or by chemical reaction (synthesis of LaB$_6$) in the vapor phase [66, 84, 85]. Data on the preparation of LaB$_6$ single crystals by the vapor phase method are listed in Table 4, p. 182, from [47], which was supplemented by further data.

Electrocrystallization from Molten Salts

For electrodeposition of LaB$_6$, see also p. 26. LaB$_6$ single crystals were electrodeposited from baths, containing in mol%: 2.2 La$_2$O$_3$, 33.5 B$_2$O$_3$, 31.2 Li$_2$O, and 33.1 LiF, at 1073 K under He. The electrolysis cell consisted of a platinum crucible, gold anode and cathode. A constant voltage between 1.85 and 2.1 V was found most effective for crystal growth. The proper magnitude of cathodic current density for good crystal growth (20 to 40 mA/cm^2) was adjusted by variation of the immersed areas of cathode and anode. A growth period of 300 h produced clusters of crystallites 4 mm on a side. Deposition on LaB$_6$ seeds produced crystals 6 × 5 × 5 mm in 200 h. The stoichiometry varied from La$_{0.67}$B$_6$ to La$_{0.87}$B$_6$ on varying the La$_2$O$_3$ content of the bath from 0.1 to 4.0 mol% [21], Elwell et al. [88]. Galvanostatic electrolysis, using (in wt%): 7.1 La$_2$O$_3$, 21.8 Li$_2$B$_4$O$_7$, and 71.1 Li$_3$AlF$_6$, a graphite crucible serving as anode and a Mo rod as cathode, at an initial current density of 250 mA/dm^2 for 96 h gave an intergrowth of single crystals below 1 mm in size [19]. Crystals of 2 mm size could be deposited from a bath with (in wt%): 50 La$_2$O$_3$, 35 B$_2$O$_3$, and 15 LiF using the gradient-reversal technique. The general property of this technique is the decrease and control of nucleation rate by periodic variation of the potential height. Growth periods and dissolution (switching off) periods, both of the order of several minutes, were alternated for 5 to 20 h. The size of the crystals was increased from 10 μm (constant potential) to 2 mm [22]. The rate of crystal growth of LaB$_6$ depends on interface kinetic and charge-transfer processes as was found from the resistance, measured during electrocrystallization, Elwell et al. [89].

Preparation of Films and Coatings

In order to make use of the favorable electron emitting properties of LaB$_6$ in place of standard W filaments in a variety of electron sources, the coating of filaments by LaB$_6$ and the filament properties were studied as well as flat films on a variety of other substrates. Films of lanthanum hexaboride were prepared by condensation of electron-beam evaporated LaB$_6$, by cathode sputtering of LaB$_6$, by co-deposition of La and B vapors, and by chemical vapor deposition (CVD). Cataphoretic coating by LaB$_6$ powders was also used, or painting with a LaB$_6$/amyl acetate paint.

References for 32.1.8.3.4.2 on p. 186

Table 4

Preparation of LaB$_6$ Single Crystals by Vapor Phase Methods.

preparation technique	preparation conditions			characteristics of the crystals		Ref.
	T in K	time in h	other conditions	shape and size in mm	color	
ΔT transport: agent Br$_2$	1423 1173	—	starting material: LaB$_6$ powder, Ar	quadratic prisms $0.15 \times 0.15 \times 2$	violet	[81]
agents H$_2$ + BBr$_3$ or H$_2$ + Br$_2$	1273 1575	—	starting material: LaB$_6$; substrate: sintered LaB$_6$; single-crystalline LaB$_6$: max. size of single-crystal grains: 1 mm^2; coating: Au; d up to ~400 nm	whiskers: $\langle 100 \rangle$; conical and pyramidal crystals	—	[82, 83]
chemical reaction: La$_2$O$_3$ + 21 H$_2$ + 12 BCl$_3$ → 2 LaB$_6$ + 36 HCl + 3 H$_2$O	A zone: 1273 B zone: 1623 to 1723	3	starting material: pressed La$_2$O$_3$; gas flow rate: H$_2$ 900 mL/min; BCl$_3$ 0.28 g/min; substrate: graphite	cubes {100}; cubo-octahedra {111}, $1 \times 1 \times 1$	purple	[66, 84]
chemical reaction (CVD): LaCl$_3$(g) + 6 BCl$_3$(g) + (21/2)H$_2$(g) = LaB$_6$(s) + 21 HCl(g)	1373 to 1623	0.5 to 2	BCl$_3$/LaCl$_3$ = 10:1; flow rate (BCl$_3$ + LaCl$_3$): 0.02 to 0.2 mL/s; p = 760 Torr; substrate: graphite	whiskers, growth direction $\langle 100 \rangle$, $l \approx 2$ to 5, 1 to 20 µm thick; pyramids, growth direction $\langle 100 \rangle$, rounded tips, 0.01 to 0.1 µm	purple-blue	[85]

| chemical reaction:
$LaCl_3(g) + 6\,BCl_3(g) + (21/2)H_2(g) = LaB_6(s) + 21\,HCl(g)$ | — | 1373 to 1523 | substrate: Cu, Ag, Au, Fe, Co, Ni, Pd, Pt, Si, SiO$_2$, mullite porcelain | — | growth direction: ⟨100⟩ with square cross section; ⟨111⟩ on Fe with hexagonal cross section; growth rate is highest on Pt: ~2.4 mm/h | [87]*) |

*) See also [86].

Electron-Beam Evaporation

LaB$_6$ films were prepared by evaporation under vacuum of ~10^{-7} Torr (~13 µPa) from a bulk polycrystalline LaB$_6$ plate heated to between 1873 and 2473 K by electron bombardment on the rear side. The vapor was condensed onto an air-cleaved NaCl(100) surface at a rate of ~200 nm/min for a source temperature of 2473 K. Films deposited at substrate temperatures below 623 K were amorphous or microcrystalline, at substrate temperatures between 623 and 873 K the films were polycrystalline. No other phases than LaB$_6$ were found, suggesting a nearly stoichiometric evaporation of LaB$_6$, Oshima et al. [90], see also Bessaraba et al. [91]. LaB$_6$ films, deposited by electron-beam evaporation onto vacuum cleaved MgO crystals at substrate temperatures between 473 and 1123 K, were found to grow epitaxially above 1003 K at 0.02 nm/s and above 1063 K at 0.1 nm/s with orientation LaB$_6$(001)||MgO(001); [100]||[100], which has only a very small misfit (1.2%). After heating to 1173 K for 12 h other orientations were also observed: LaB$_6$(001)||MgO(001); [110]||[100], LaB$_6$(011)||MgO(001); [01$\bar{1}$]||[100], and LaB$_6$(011)||MgO(001); [100]||[100]. Air-cleaved MgO had to be heated for 12 h at 1473 K before deposition in order to obtain epitaxial growth, Muranaka, Kawai [92]. The following parameters were given to produce 10 to 200 nm thick LaB$_6$ films on Pyrex and quartz by electron-beam evaporation: 10^{-6} Torr ($\triangleq$133 µPa), target: molten LaB$_6$ pill, substrate temperature: 373 to 1273 K, deposition rate 1.5 to 100 nm/min, Peschmann et al. [93].

The composition of films prepared by electron-beam evaporation onto Mo at room temperature, Ociepa, Mroz [94], or Ta, Ociepa, Mroz [95] inferred from intensity in Auger spectra corresponds essentially to LaB$_6$ only for the highest source temperatures T$_s$, up to 2273 K [94]; they were richer in La at lower T$_s$ [94, 95]. On annealing, the La-rich films crystallized at above 573 to 673 K with segregation into La and LaB$_4$. The LaB$_4$ was in the outer regions of the films according to the Auger spectra [95]. Previous results of Ociepa, Mroz [96] were withdrawn because of impurities from the vacuum system [95]. The annealing behavior of a film with composition LaB$_{1.9}$ is described by Ociepa [97]; here La is assumed to segregate at the outer region, the La boride in the inner region, in contrast to the earlier statements [95].

Cathode Sputtering

The essential of this method is the deposition of LaB$_6$, sputtered from an LaB$_6$ cathode by Ar ions, onto an appropriate substrate. Thus, LaB$_6$ films were deposited with a rate of 1.5 to 4 nm/min onto amorphous optical and quartz glass and monocrystalline sapphire and quartz employing a d.c.-triode sputtering system. The Ar pressure was 5×10^{-3} Torr ($\triangleq$670 mPa), the target was single-crystalline LaB$_6$, target voltage −2000 V, current 2 mA, substrate temperature 373 to 1523 K. Prior to deposition, the substrates were cleaned and etched. The 0.01 to 1.00 µm thick layers were amorphous at substrate temperatures below 473 K, and polycrystalline above this temperature, Winsztal et al. [98]. Similar process parameters for deposition onto Pyrex and quartz were given by Peschmann et al. [93]. For preparation of LaB$_6$ films on Ta in a d.c.-diode sputtering system see Li, Wan [99].

Cataphoretic Deposition

Cataphoretic coating was performed with a mixture of 1 g finely powdered LaB$_6$ in 10 mL of methanol plus one drop of concentrated aqueous HCl as activator. This gave uniform coatings on clean Re wire used as cathode; the anode was of Ta. At ~90 V the current density was 0.03 A/cm^2. The coating was sintered by flashing under vacuum, 10^{-6} Torr ($\triangleq$133 µPa), at 1873 to 1973 K for 10 min or alternatively in pure H$_2$, He, or Ar at atmospheric pressure. Interstitial diffusion of B atoms into Re was not observed, Favreau [100]. Occasional flaking off of the coating from the Re wire was found to be due to impurities, boron or boron oxide, in the LaB$_6$. These could be eliminated by vacuum firing the LaB$_6$ powder to 2073 K for 2 h and repeated (up to 5 times) elutriation in ethanol, Favreau, Koenig [101]. The obtained coatings were described as loosely stacked with moderate adhesion to the Re substrate using ethanol as liquid, and HNO$_3$ as activator. This made the electrodes good halide ion sources. Replacing HNO$_3$ by NH$_4$NO$_3$ gave dense, smooth coatings with excellent adhesion to the Re substrate, which made excellent electron emitters, Delmore [102].

Cataphoresis of a freshly stirred suspension of 0.5 g of technical grade LaB$_6$ powder, grain size 5 to 8 µm, in acetone together with few mg of La(NO$_3$)$_3$ as activator gave, after 15 s at 95 V, a uniform coating of 0.1 mm thickness. Re is preferable as filament material, yet also W is satisfactory at operation below 1400 K, having life times of 500 to 1000 h, Hayward, Taylor [103].

Prior to cataphoretic deposition of LaB$_6$ (10 g LaB$_6$, 10 mg LaCl$_3 \cdot 7$H$_2$O, 100 mL absolute ethanol, 10 mL H$_2$O) onto Ta, the substrate was carburized by heating to 1573 K in a CH$_4$ atmosphere, Nasini, Redaelli [104].

Miscellaneous Methods

La boride films, 10 to 1000 nm thick, could be prepared by successively depositing vapors of La, from a Ta boat, and of B, from a graphite boat, onto glass or NaCl single crystals and annealing at 623 to 823 K. On the NaCl substrate the films were crystallized to a higher degree. Electron diffraction indicated the presence of the hexaboride with impurities of B$_4$C and La$_2$O$_3$. These are attributed to the crucible and to poor vacuum, 10^{-5} Torr ($\triangleq$1.3 mPa), Paderno et al. [105].

Pyrolysis of lanthanum boranes at about 673 K is a method by which complex surfaces can be covered by La boride, Frank et al. [106]. Layers of vapor-deposited La reacted with B$_2$H$_2$ at temperatures above 473 K to form LaB$_6$ after formation of intermediates, presumably lanthanum hydride, Knauff [107].

For deposition by painting, the LaB$_6$ powder was mixed with ethanol or amyl acetate to a paste, painted on carburized Ta and heat-treated in vacuum to 1873 to 2073 K, Lafferty [2], Muranaka, Kawai [108]. For operating temperatures of 1023 to 1073 K tungsten wire was coated by LaB$_6$ powder with an organic binder. After exposure to the atmosphere, the filament can be reactivated by short flashing to 1673 K. The organic binder introduced hydrocarbon into the UHV system, Margoninski et al. [109].

LaB$_6$ can be bonded to electric devices or, for example, to BeO substrates by an adherent coating, consisting of a tin-vanadium alloy (15 to 2% V). The composite is heated in an atmosphere of CO at approximately 1273 K, Intrater, Bertoldo [110].

LaB$_6$ was described to remain unchanged (55.33 wt% La and 27.68 wt% B) after being plasma-sprayed in air, Lécrivain, Provost [111]. LaB$_6$ coatings on Mo were deposited electrolytically by Bogacz et al. [113] from an La$_2$O$_3$–Li$_2$B$_4$O$_7$–Li$_3$AlF$_6$ electrolyte, Uchida [19].

Barrier Layers

Barrier layers were recommended to prevent diffusion of boron from the LaB$_6$ into the metal substrate. Barrier layers of Ta carbide on Ta were generated by carburizing the substrate in charcoal at 2573 K by [2], or by painting with a mixture of TaC in ethanol and heating [108].

0.4 to 0.6 µm thick barrier layers of TiC, ZrC, TiB$_2$, or ZrB$_2$ reduce the boron diffusion into the Mo substrate but could not prevent it totally. The ratios of boron diffusion coefficients without and with barrier layers were 4 to 6 for the carbides and TiB$_2$ and about 10 for ZrB$_2$, Shaginyan et al. [112].

References for 32.1.8.3.4.2 on p. 186

References:

[1] Peshev, P. (Rev. Intern. Hautes Temp. Refract. **4** [1967] 289/96; C. A. **69** [1968] No. 5740).
[2] Lafferty, J. M. (J. Appl. Phys. **22** [1951] 299/309).
[3] Schmitt, J.; Setton, R. (Verres Refractaires **18** [1964] 319/25; C. A. **62** [1965] 7480).
[4] Compagnie Francaise Thomson-Houston (Neth. Appl. 6507182 [1965] 1/6; C. A. **64** [1966] No. 13803).
[5] Storms, E. K. (J. Appl. Phys. **52** [1981] 2961/5).
[6] Storms, E. K.; Mueller, B. A. (J. Appl. Phys. **50** [1979] 3691/8).
[7] Johnson, R. W.; Daane, A. H. (J. Phys. Chem. **65** [1961] 909/15).
[8] Rabenau, A.; Kauer, E.; Klotz, H. (Colloq. Int. Centre Natl. Rech. Sci. [Paris] No. 157 [1967] 495/8).
[9] Meerson, G. A.; Mamedov, F. G. (Izv. Akad. Nauk SSSR Neorgan. Materialy **3** [1967] 802/7; Inorg. Materials [USSR] **3** [1967] 717/21).
[10] Meerson, G. A.; Manelis, R. M.; Nurmukhamedov, V. Kh. (Izv. Akad. Nauk SSSR Neorgan. Materialy **6** [1970] 1219/23; Inorg. Materials [USSR] **6** [1970] 1070/3).

[11] Zhuravlev, N. N.; Manelis, R. M.; Nikanorova, I. A. (Bor: Poluch. Strukt. Svoistva Mater. 4th Mezhdunar. Simp. Boru, Tbilisi 1972 [1974], Vol. 2, pp. 162/71; C. A. **83** [1975] No. 125357).
[12] Meerson, G. A.; Nurmukhamedov, V. Kh.; Manelis, R. M. (Izv. Akad. Nauk SSSR Neorgan. Materialy **9** [1973] 2133/5; Inorg. Materials [USSR] **9** [1973] 1884/5).
[13] Samsonov, G. V.; Paderno, Yu. B.; Kreingol'd, S. U. (Zhur. Prikl. Khim. **34** [1961] 10/5; J. Appl. Chem. [USSR] **34** [1961] 8/13).
[14] Breusov, O. N.; Korotkevich, M. N.; Nazarov, V. I. (Prom. Khim. Reakt. Osobo Chist. Veshchestv No. 10 [1967] 9/11 from C. A. **70** [1969] No. 79588).
[15] Neshpor, V. S. (Vysokotemp. Metallokeram. Mater. **1962** 96/101; N.S.A. **19** [1965] No. 5912).
[16] Shiota, M.; Tsutsumi, M.; Uchida, K. (J. Mater. Sci. **15** [1980] 1987/92).
[17] O'Connor, T. E.; E. I. Du Pont de Nemours & Co. (Brit. 951280 [1961/64] 1/17).
[18] Andrieux, L. (Ann. Chim. [Paris] [10] **12** [1929] 345/507).
[19] Uchida, K. (Surf. Technol. **7** [1978] 137/43).
[20] Uchida, K.; Shiota, M. (Surf. Technol. **7** [1978] 299/304).

[21] Zubeck, I. V.; Feigelson, R. S.; Huggins, R. A.; Pettit, P. A. (J. Cryst. Growth **34** [1976] 85/91).
[22] Scholz, H.; Bauhofer, W.; Ploog, K. (Solid State Commun. **18** [1976] 1539/42).
[23] Kuroda, T. (Denki Shikensho Kenkyu Hokoku No. 561 [1957] 1/103 from C. A. **1958** 8795).
[24] Wold, A. (AD-660615 [1967] 1/32, 5; C. A. **68** [1968] No. 90511).
[25] Paderno, Yu. B.; Tkach, A. V.; Dudnik, E. M.; Masyuk, T. V.; Zaitseva, Z. A. (Poroshkovaya Metal. **1985** No. 5, pp. 61/5; Soviet Powder Met. Metal Ceram. **24** [1985] 394/8).
[26] Paderno, Yu. B.; Dudnik, E. M.; Zaitseva, Z. A.; Nazarchuk, T. N.; Strashinskaya, L. V.; et al. (Poroshkovaya Metal. **1981** No. 4, pp. 56/60; Soviet Powder Met. Metal Ceram. **20** [1981] 282/6).
[27] Kuznetsova, G. M.; Kudintseva, G. A.; Likhacheva, T. K.; Suchkov, L. K. (Poroshkovaya Metal. **1972** No. 12, pp. 46/51; Soviet Powder Met. Metal Ceram. **11** [1972] 810/4).
[28] Colton, E. (Proc. 12th Rare Earth Res. Conf., Vail, Colo., 1976, Vol. 2, pp. 1026/31).
[29] Scholz, S. (Spec. Ceram. No. 2 [1962/63] 293/305).
[30] Bochko, A. V.; Volkogon, V. M.; Karyuk, G. G.; Paderno, V. N.; Paderno, Yu. B.; Pilyankevich, A. N. (Goryachee Pressovanie Dokl. 3rd Vses Nauchno Tekhn. Konf. Goryachee Pressovanie Poroshk. Metall., Novocherkassk, USSR, 1976 [1977], pp. 3/9; C. A. **88** [1978] No. 175957).

[31] Volkogon, V. M.; Paderno, V. N.; Martynenko, A. N.; Paderno, Yu. B.; Dubovka, G. V. (Poroshkovaya Metal. **1984** No. 2, pp. 36/41; Soviet Powder Met. Metal Ceram. **23** [1984] 121/5).

[32] Zhuravlev, N. N.; Manelis, R. M.; Belousova, I. A.; Nurmukhamedov, V. Kh. (Izv. Akad. Nauk SSSR Neorgan. Materialy **9** [1973] 1162/5; Inorg. Materials [USSR] **9** [1973] 1033/5).

[33] Kondrashov, A. I.; Paderno, Yu. B.; Paderno, V. N. (Poroshkovaya Metal. **1982** No. 6, pp. 16/21; Soviet Powder Met. Metal Ceram. **21** [1982] 437/41).

[34] Meerson, G. A.; Manelis, R. M.; Telyukova, T. M. (Izv. Akad. Nauk SSSR Neorgan. Materialy **2** [1966] 291/8; Inorg. Materials [USSR] **2** [1966] 250/5).

[35] Shlyuko, V. Ya.; Chernyak, L. V. (Poroshkovaya Metal. **1967** No. 7, pp. 41/5; Soviet Powder Met. Metal Ceram. **1967** 969/72).

[36] Meerson, G. A.; Mamedov, F. G. (Izv. Akad. Nauk SSSR Neorgan. Materialy **3** [1967] 942/9; Inorg. Materials [USSR] **3** [1967] 844/50).

[37] Kudintseva, G. A.; Kuznetsova, G. M.; Mamedov, F. G.; Meerson, G. A.; Tsarev, B. M. (Izv. Akad. Nauk SSSR Neorgan. Materialy **4** [1968] 49/53; Inorg. Materials [USSR] **4** [1968] 38/42).

[38] Medvedev, O. G.; Trunov, G. V.; Chernyak, L. V.; Shlyuko, V. Ya. (Poroshkovaya Metal. **1971** No. 3, pp. 101/2; Soviet Powder Met. Metal Ceram. **1971** 250/1).

[39] Bilyk, I. I.; Kresanov, V. S.; Medvedev, O. G. (Poroshkovaya Metal. **1980** No. 12, pp. 90/2; C.A. **94** [1981] No. 88533).

[40] Knoch, H.; Bechler, E.; Lipp, A. (Ger. Offen. 3516955 [1986] 1/12; C.A. **106** [1987] No. 22291).

[41] Shlyuko, V. Ya.; Morozov, V. V.; Besov, A. V.; Chernyak, L. V.; Guzenko, L. V. (Poroshkovaya Metal. **1974** No. 7, pp. 29/33; Soviet Powder Met. Metal Ceram. **13** [1974] 544/6).

[42] Meerson, G. A.; Nurmukhamedov, V. Kh.; Manelis, R. M.; Dremin, A. N.; Adadurov, G. A.; Breusov, O. N.; Pershin, S. V.; Tatsii, V. F. (Fiz. Khim. Obrab. Mater. **1974** No. 5, pp. 140/4; C.A. **82** [1975] No. 63396).

[43] Anan'in, A. V.; Breusov, O. N.; Dremin, A. N.; Ivanova, V. B.; Pershin, S. V.; Tatsii, V. F.; Fekhretdinov, F. A. (Poroshkovaya Metal. **1974** No. 8, pp. 74/9; Soviet Powder Met. Metal Ceram. **13** [1974] 662/6).

[44] Volkogon, V. M.; Paderno, Yu. B.; Chernyavskii, Yu. A.; Martynenko, A. N. (Vliyanie Vys. Davlenii Veshchestvo Mater. 2nd Resp. Nauchn. Semin., Kiev 1976 [1977] 67/70 from C.A. **92** [1980] No. 10179).

[45] Strashinskaya, L. V.; Dudnik, E. M.; Kopylova, L. I.; Savvakin, G. I. (Poroshkovaya Metal. **1986** No. 5, pp. 79/83; Soviet Powder Met. Metal Ceram. **25** [1986] 424/7).

[46] Korsukova, M. M.; Gurin, V. N. (Usp. Khim. **56** [1987] 3/28; Russ. Chem. Rev. **56** [1987] 1/15).

[47] Korsukova, M. M.; Gurin, V. N. (Current Topics Mater. Sci. **11** [1984] 390/439).

[48] Tanaka, T.; Bannei, E.; Kawai, S.; Yamane, T. (J. Cryst. Growth **30** [1975] 193/7).

[49] Paderno, Yu. B.; Lazorenko, V. I.; Kovalev, A. V. (Poroshkovaya Metal. **1981** No. 10, pp. 60/5; Soviet Powder Met. Metal Ceram. **20** [1981] 717/21).

[50] Lazorenko, V. I.; Lotsko, D. V.; Platonov, V. F.; Kovalev, A. V.; Galasun, A. P.; Matvienko, A. A.; Klinkov, A. E. (Poroshkovaya Metal. **1987** No. 3, pp. 51/7; Soviet Powder Met. Metal Ceram. **26** [1987] 229/33).

[51] Davis, P. R.; Swanson, L. W.; Hutta, J. J.; Jones, D. L. (J. Mater. Sci. **21** [1986] 825/36).

[52] Verhoeven, J. D.; Gibson, E. D.; Noack, M. A.; Conzemius, R. J. (J. Cryst. Growth **36** [1976] 115/20).

[53] Windsor, E. E. (Proc. Inst. Elec. Eng. [London] **116** [1969] 348/50).

[54] Curtis, B. J.; Graffenberger, H. (Mater. Res. Bull. **1** [1966] 27/31).

[55] Feigelson, R. S.; Kway, W. L.; Route, R. K. (Opt. Eng. **24** [1985] 1102/7).
[56] Gibson, E. D.; Verhoeven, J. D. (J. Phys. E **8** [1975] 1003/4).
[57] Hafner, P. V. (Diss. ETH 5710 Zürich 1976, pp. 1/117).
[58] Noack, M. A.; Verhoeven, J. D. (J. Cryst. Growth **49** [1980] 595/9).
[59] Takagi, K.; Ishii, M. (J. Cryst. Growth **40** [1977] 1/5).
[60] Morozov, V. V. (Vestn. Kievsk. Politekhn. Inst. Mashinostr. No. 21 [1984] 48/52; C.A. **101** [1984] No. 142229).

[61] Morozov, V. V.; Mal'nev, V. I.; Dub, S. N.; Loboda, P. I.; Kresanov, V. S. (Izv. Akad. Nauk SSSR Neorgan. Materialy **20** [1984] 1421/3; Inorg. Materials [USSR] **20** [1984] 1225/6).
[62] Hohn, F. J.; Chang, T. H. P.; Broers, A. N.; Frankel, G. S.; Peters, E. T.; Lee, D. W. (J. Appl. Phys. **53** Pt. 1 [1982] 1283/96).
[63] Aida, T.; Fukazawa, T. (J. Cryst. Growth **78** [1986] 263/73).
[64] Aida, T.; Fukazawa, T. (J. Cryst. Growth **80** [1987] 9/16).
[65] Aida, T.; Fukazawa, T. (Japan. Kokai Tokkyo Koho 86-83698 [1986] 1/3 from C.A. **105** [1986] No. 144063).
[66] Niemyski, T.; Kierzek-Pecold, E. (J. Cryst. Growth **3/4** [1968] 162/5).
[67] Niemyski, T.; Pračka, I.; Jun, J.; Winsztal, S.; Majewska-Minor, H.; Wisniewska, M. (Bor: Poluch. Strukt. Svoistva Mater. 4th Mezhdunar. Simp. Boru, Tbilisi 1972 [1974], Vol. 2, pp. 146/54; C.A. **83** [1975] No. 124155).
[68] Paderno, Yu. B.; Lazorenko, V. I.; Buryak, N. I.; Kovalev, A. V.; Matvienko, A. A.; Galasun, A. P. (Poroshkovaya Metal. **1983** No. 1, pp. 56/60; Soviet Powder Met. Metal Ceram. **22** [1983] 50/3).
[69] Logan, K. V. (Mater. Res. Soc. Symp. Proc. **12** [1982] [In Situ Compos. 4] 195/204; C.A. **99** [1983] No. 26803).
[70] Arko, A. J.; Crabtree, G.; Karim, D.; Mueller, F. M.; Windmiller, L. R.; Ketterson, J. B.; Fisk, Z. (Phys. Rev. [3] B **13** [1976] 5240/7).

[71] Futamoto, M.; Aita, T.; Kawabe, U. (Japan. J. Appl. Phys. **14** [1975] 1263/6).
[72] Aita, T.; Kawabe, U.; Honda, Y. (Japan. J. Appl. Phys. **13** [1974] 391).
[73] Gurin, V. N.; Korsukova, M. M.; Nikanorov, S. P.; Smirnov, I. A.; Stepanova, N. N.; Shul'man, S. G. (J. Less-Common Metals **67** [1979] 115/23).
[74] Stepanov, N. N.; Zyuzin, A. Yu.; Shul'man, S. G.; Gurin, V. N.; Korsukova, M. M.; Nikanorov, S. P.; Smirnov, I. A. (Fiz. Tverd. Tela [Leningrad] **20** [1978] 935/8; Soviet Phys.-Solid State **20** [1978] 542/3).
[75] Korsukova, M. M.; Gurin, V. N.; Lundström, T.; Tergenius, L. E. (J. Less-Common Metals **117** [1986] 73/81).
[76] Inoue, T.; Nakada, M.; Kozumi, T.; Sugata, E. (J. Vac. Sci. Technol. **21** [1982] 952/6).
[77] Olsen, G. H.; Cafiero, A. V. (J. Cryst. Growth **44** [1978] 287/90).
[78] Okada, S.; Imai, Y.; Atoda, T. (Yogyo Kyokaishi **90** [1982] 73/82; C.A. **96** [1982] No. 113669).
[79] Moshkin, S. V.; Nardov, A. V. (J. Cryst. Growth **52** [1981] 816/9).
[80] Zhukova, T. B.; Korsukova, M. M.; Nardov, A. V.; Gurin, V. N. (Izv. Akad. Nauk SSSR Neorgan. Materialy **17** [1981] 353/4; C.A. **94** [1981] No. 130669).

[81] Klotz, H. (Naturwissenschaften **52** [1965] 451).
[82] Givargizov, E. I.; Obolenskaya, L. N. (J. Cryst. Growth **51** [1981] 190/4).
[83] Givargizov, E. I.; Obolenskaya, L. N. (J. Less-Common Metals **117** [1986] 97/103).
[84] Niemyski, T.; Appenheimer, S.; Cinak, J. (Pol. 62515203 [1971] 1/2; C.A. **75** [1971] No. 142390).
[85] Motojima, S.; Takahashi, Y.; Sugiyama, K. (J. Cryst. Growth **44** [1978] 106/9).

[86] Hagimura, A.; Kato, A. (Koen Yoshishu 24th Jinko Kobutsu Toronkai, Sendai, 1979, pp. 71/2 from C.A. **92** [1980] No. 138776).

[87] Hagimura, A.; Kato, A. (Nippon Kagaku Kaishi **1980** No. 7, pp. 1108/13 from C.A. **93** [1980] No. 104935).

[88] Elwell, D.; Zubeck, I. V.; Feigelson, R. S.; Huggins, R. A. (J. Cryst. Growth **29** [1975] 65/8).

[89] Elwell, D.; Demattei, R. C.; Zubeck, I. V.; Feigelson, R. S.; Huggins, R. A. (J. Cryst. Growth **33** [1976] 232/8).

[90] Oshima, C.; Horiuchi, S.; Kawai, S. (Japan. J. Appl. Phys. **13** Suppl. 2, Pt. 1 [1974] 281/4).

[91] Bessaraba, V. I.; Vereshchak, V. M.; Ivanchenko, L. A.; Kugai, L. N.; Paderno, Yu. B. (Poroshkovaya Metal. **1976** No. 7, pp. 69/73; Soviet Powder Met. Metal Ceram. **15** [1976] 552/5).

[92] Muranaka, S.; Kawai, S. (Japan. J. Appl. Phys. **15** [1976] 587/94).

[93] Peschmann, K. R.; Calow, J. T.; Knauff, K. G. (J. Appl. Phys. **44** [1973] 2252/6).

[94] Ociepa, J. G.; Mroz, S. (Acta Univ. Wratislav. Mat. Fiz. Astron. No. 37 [1980] 25/30; C.A. **94** [1981] No. 93805).

[95] Ociepa, J. G.; Mroz, S. (Thin Solid Films **85** [1981] 43/51).

[96] Ociepa, J. G.; Mroz, S. (Vacuum **29** [1979] 241/4).

[97] Ociepa, J. G. (Thin Solid Films **120** [1984] 123/31).

[98] Winsztal, S.; Majewska-Minor, H.; Wisniewska, M.; Niemyski, T. (Mater. Res. Bull. **8** [1973] 1329/35).

[99] Li, X.; Wan, Y. (Zhejiang Daxue Xuebao **20** [1986] 53/61; C.A. **105** [1986] No. 33859).

[100] Favreau, L. J. (Rev. Sci. Instrum. **36** [1965] 856/7).

[101] Favreau, L. J.; Koenig, D. F. (Rev. Sci. Instrum. **38** [1967] 841).

[102] Delmore, J. E. (Rev. Sci. Instrum. **54** [1983] 158/60).

[103] Hayward, D. O.; Taylor, N. (J. Sci. Instrum. **43** [1966] 762/3).

[104] Nasini, M.; Redaelli, G. (Rev. Sci. Instrum. **42** [1971] 1765/7).

[105] Paderno, Yu. B.; Ivanchenko, L. A.; Bessaraba, V. I.; Vereshchak, V. M. (Poroshkovaya Metal. **1975** No. 6, pp. 106/8; Soviet Powder Met. Metal Ceram. **14** [1975] 515/6).

[106] Frank, B.; Lydtin, H.; Gärtner, G. (Ger. Offen. 3148441 [1981] 1/47; C.A. **106** [1987] No. 77289).

[107] Knauff, K. G. (Basic Probl. Thin Film Phys. Proc. Intern. Symp., Clausthal-Zellerfeld and Göttingen 1965 [1966], pp. 207/11; Metallurg. Abstr. [3] **2** [1967] 751/2).

[108] Muranaka, S.; Kawai, S. (Japan. J. Appl. Phys. **15** [1976] 1809/10).

[109] Margoninski, Y.; Wolky, S. P.; Zdanuk, E. J. (Vacuum **11** [1961] 287/8).

[110] Intrater, J.; Bertoldo, G. (U.S. 4426423 [1984] 1/4; C.A. **100** [1984] No. 125821).

[111] Lécrivain, L.; Provost, G. (Ber. Deut. Keram. Ges. **45** [1968] 347/51).

[112] Shaginyan, L. R.; Vasil'ev, A. M.; Bessaraba, V. I.; Chernyaev, V. N.; Blokhin, V. G. (Poroshkovaya Metal. **1984** No. 2, pp. 70/4; Soviet Powder Met. Metal Ceram. **23** [1984] 150/3).

[113] Bogacz, A.; Los, P.; Szklarski, W.; Josiak, J. (Rudy Metale Niezelazne **28** No. 4 [1982] 134/9).

[114] Kovalev, A. V.; Minakov, V. N.; Dudnik, E. M.; Zaitseva, Z. A. (Poroshkovaya Metal. **1988** No. 9, pp. 46/50; Soviet Powder Met. Metal Ceram. **27** [1988] 717/20).

[115] Pastor, H. (in: Matkovich, V. I., Boron and Refractory Borides, Springer, Berlin – New York 1977, pp. 457/93).

32.1.8.3.4.3 Crystallographic Properties

Habit. Epitaxy

For the habit of single crystals inclusive whiskers derived from vapor phase methods, see Table 4, p. 182.

The plate- and cube-like crystals grown from an Al flux have large {100} faces, Aita et al. [1], see also Zhukova et al. [2], Swanson, Dickinson [3]. Generally, needles are elongated in the [100] direction, Inoue et al. [4], and exhibit {100} faces [2], see also Korsukova, Gurin [5]. The growth direction [110] and a well-developed (1$\bar{1}$0) face was observed by Futamoto et al. [6]. The LaB$_6$ rods grown spontaneously by the floating-zone technique with induction heating most often had a growth axis close to the direction $\langle 001 \rangle$, $\langle 111 \rangle$, $\langle 103 \rangle$, $\langle 113 \rangle$, or $\langle 123 \rangle$ and less frequently to $\langle 011 \rangle$. The directions $\langle 012 \rangle$ and $\langle 112 \rangle$ were not observed. A preferred orientation was not observed, Lazorenko et al. [7]. Selected orientations on seed crystals can be grown by this technique with the different methods of heating, see pp. 28, 176 to 178. For the thermal stability of various planes, see p. 254.

Various ways of epitaxial growth are observed for LaB$_6$ films on MgO, Muranaka, Kawai [8], see p. 184.

Crystal Structure

Analogous to the other rare earth hexaborides, crystals of LaB$_6$ are cubic, with the CaB$_6$ structure type, space group Pm$\bar{3}$m-O$_h^1$ (No. 221), Z = 1, see p. 33. An X-ray structure refinement with a single crystal grown from Al flux gave x = 0.19957(15) for the positional parameter of the boron atom, position 6(f): x, ½, ½; La is in position 1(a): 0,0,0. Refinement of only the high-angle reflections gave no significant difference. The values of x for two other crystals from different preparation conditions were identical within the range of the standard deviation. R values for the three crystals ranged between 1.10 and 1.36%. With the lattice parameter a = 4.1566(1) Å, the interatomic distances are: 1.766 Å between B atoms within the B$_6$ octahedra and 1.659 Å between the octahedra, 3.054 Å between La and B. Remarkably, the refinement of site occupation factors revealed full occupancy of the lanthanum position for the three crystals and an occupancy between only 96.4(7) and 98.2(5)% for the boron position, which is equivalent to B/La = 5.78 and 5.89, respectively; chemical analyses on these charges indicated B/La = 5.67 and 6.82, respectively, Korsukova et al. [9, 10].

The results are essentially identical to those of Eliseev et al. [11], see p. 33. For other investigations on the crystal structure, see also Barantseva, Paderno [12], and for bond lengths B–B also La Placa [13].

Lattice Constant

At Room Temperature

The lattice constant of LaB$_6$ remains essentially unchanged across the homogeneity range and the blue phase, regarded as B-rich (see p. 168), has the same lattice constant as the purple phase, within the limit of accuracy, McKelvy et al. [15], Johnson, Daane [16]: a (in Å) = 4.1564 ± 0.0003 for the carbon-free LaB$_6$ in the whole single-phase region, Lundström [17, 18], and a = 4.1564(1) for individual single crystals prepared by different methods [17], Noack, Verhoeven [19], Futamoto et al. [6]. The values a = 4.1561 ± 0.0003 [16], 4.1565 ± 0.0003, Storms, Mueller [20], and 4.1566 ± 0.0005, Mamedov et al. [21] were all observed for both violet and blue polycrystalline samples. Seven Al-flux grown single crystals of different composition had values between 4.1566(1) and 4.1571(1), without trend (mean value 4.1569), Korsukova et al. [10]. For further individual determinations yielding values in the range 4.1561 to 4.1569, see [2, 19, 22 to 26] and for a = 4.1573 ± 0.0010, Zhuravlev et al. [27], similarly [28]. Lattice constants as small as 4.1558(2) [19], Noack [29], and 4.1540 [17] have been related to carbon contaminations. For values in this range see also [30 to 34].

A La^{11}B₆ single crystal had the same lattice constant (a = 4.1566 ± 0.0001 Å) as one with natural isotope distribution. Crystals of La^{10}B₆ had a = 4.1567 to 4.1571 Å, both ± 0.0001 Å, Zhukova et al. [2].

The lattice parameter of LaB₆ changes only very little on addition of Ta, W, Re, and Hf, see a figure in Bondarenko et al. [35]. Additions of Ni increase the lattice constant, Kondrashev et al. [31].

LaB₆ films evaporated onto graphite had a = 4.153 Å, identical to that of the bulk material, Gert et al. [36]; a ≈ 4.12 Å for a film on a NaCl substrate, Oshima et al. [37]. A film prepared from the elements on a glass or NaCl substrate had a = 4.15 Å, Paderno et al. [38]. Purple and blue whiskers from chemical vapor growth had a = 4.151 and 4.155 Å, respectively, Motojima et al. [39].

Temperature Dependence

On cooling, the lattice constant decreases from 4.1567 Å at 296 K to 4.1529 Å at 77 K, and 4.1528 Å at 4.6 K, Etourneau et al. [40]. A similar, relatively strong decrease at the higher temperatures and a smooth one at the lower temperatures is also observed in studies between 4.3 and 300 K, see a figure in Alekseev et al. [41].

The increase of a from 4.1554 ± 0.0003 Å at 294 K to 4.1993 ± 0.0005 Å at 1573 K may be fitted by a = 4.1551(1 + 7.20 × 10^{-6} t + 8.0 × 10^{-10} t^2) with t in °C, Aivazov et al. [33].

Lattice Defects

For inclusions of Al in single crystals grown from an Al flux and of LaB₄ and B in floating-zone refined rods, see p. 181 and below.

The most perfect single crystals are grown from an Al flux, with the smallest subgrain misalignment, the lowest dislocation density, and the possibility of stoichiometric composition, see the review by Korsukova, Gurin [5].

Subgrains occur in single crystals grown from an Al flux, with a maximum misorientation of 2° in platelets and only ca. 10′ in needles, Zhukova et al. [2]. They are relatively large in rods grown by the floating-zone technique without crystal rotation during growth, Aida, Fukazawa [42]. For example, on use of induction heating and a pulling rate of 1 to 2 cm/h, single crystals (7 mm in diameter, 60 mm in length) had subgrains of 1 mm³ in size that were misoriented by several seconds in the center to a few degrees in the outer part of the crystal boules, Tanaka et al. [43]. Similar rods had subgrains of cross section 0.6 to 1 mm², misoriented by < 0.5° in the center and by < 2° in the periphery, Morozov et al. [44]. For the observation of subgrains with laser heating, see Takagi, Ishii [45]. Studies with IR heating (i.e., Xe arc image furnace) at a pulling rate of 1 cm/h showed that, as the rotation rate is increased from zero to 40 rpm, the subgrain density increases from ca. <10 cm^{-2} to 1000 cm^{-2} and the subgrain size decreases. In addition, the distribution of subgrains on cleaved (001) surfaces becomes more uniform compared to that observed at 10 rpm crystal rotation, with subgrains mainly concentrated in the periphery. At this high density (1000 cm^{-2}), subgrains are cubes of 0.2 to 0.5 mm in length, generated parallel and perpendicular to the [100] growth axis, with misorientations of up to 1°. Some subboundaries with a large width of ca. 5 μm contain a yellow second phase, probably LaB₄, and a higher impurity level (La-rich composition in the molten zone). Subgrains were considered to originate from two mechanisms. One is a polygonization boundary composed of dislocations gathered in a straight line, the other mechanism is a cellular structure, in which the subboundaries are produced by constitutional supercooling [42].

Cleaved surfaces of zone-refined rods prepared without crystal rotation [46] or at low rotation rate of 10 rpm [42] reveal a higher dislocation density D_{disl} in the center than at the periphery. Studies with IR heating show that this difference increases with increasing crystal diameter ($d = 2$ to 5 mm), see a figure in the paper. A dependence of D_{disl} on the pulling rate (r) was not obvious [47], whereas crystal rotation of 40 rpm during growth led to a more uniform distribution [42]. Values for D_{disl} observed on crystals grown from an Al flux are compared with those on cleaved surfaces of rods grown by the floating-zone method (fl.z.) using various heatings (h = heating, c = center, p = periphery):

method	D_{disl} in cm^{-2}	d in mm	r in cm/h	Ref.
Al flux	10^3; square pits on plates, rectangular pits on needles	—	—	[6]
fl.z., laser h	5×10^5 to 10^7 (c)	1	5.1	[46]
	1.6×10^6 (c) on (100); square pits	1	9.5	[45]
fl.z., IR h	$\sim 5 \times 10^6$ (c), $\sim 5 \times 10^5$ (p) } on (100); mainly	5	1	[47]
	$\sim 2 \times 10^4$ (c, p) } pyramid-like pits	2	1	[47]
fl.z., induction h	10^6 (10^8, may be with crystal rotation)	?	2 to 3 ($\geqq$3)	[7]
	$\geqq 10^6$	?	0.8 to 2	[48]
fl.z., electric arc	$\sim 2.5 \times 10^7$ (1-pass); 10^6 (2-pass) on (100); square pits	?	5	[49]

In the zone-melted crystals prepared on use of IR heating, the surfaces of the pyramid-like etch pits consist of {110} planes. At $D_{disl} = 10^5$ cm^{-2}, the etch pits were inclined to line up in small groups along $\langle 100 \rangle$. At densities of $\sim 10^4$ cm^{-2} a characteristic pattern appeared with the dislocations lined up in parallel, plus isolated dislocations. They were identified as some kind of spiral and pure screw dislocations, respectively. The slip plane and direction of the dislocation are {100}/$\langle 110 \rangle$, which is a slip system of a simple cubic lattice. The experimental results on the dislocation distribution could be explained by Tsivinsky's temperature gradient model giving the axial temperature gradient $\Delta T_n = 3.9 \times 10^4$ K/cm in the core and $\Delta T_n = 7.0 \times 10^3$ K/cm in the periphery [47]. Plate-like (100) crystals grown from an Al flux, presumably at a cooling rate of 4.8 K/h from 1500 to 700 K, exhibit cubic-like and pyramid-like etch pits. The latter are associated with a stratified growth along $\langle 100 \rangle$. The pyramid faces are inclined by ca. 30° from the surface, Inoue et al. [4].

For the existence of vacancies in the La and B sublattices, see pp. 168 and 190. Vacancy densities of 4.5×10^{20} cm^{-3} (La) and 4.1×10^{21} cm^{-3} (B), corresponding to the composition $La_{0.97}B_{5.71}$, were postulated for a sample of analytical composition $LaB_{5.9}$ from a comparison of pycnometric and X-ray density, Aivazov et al. [50].

Surface Structure

For the surface structure at ambient and elevated temperatures in various atmospheres, see p. 262 and p. 265.

Surface Energy. Surface Tension

An estimate for the effective surface energy $\gamma_{eff} = 10$ J/m^2 ($\hat{=} 1 \times 10^4$ erg/cm^2) was evaluated from the frequency and the length of cracks at the various loads, Paderno et al. [51].

For molten LaB$_6$ a surface tension of 200 dyn/cm ($\triangleq 0.2$ J/m^2) was estimated from the falling-drop method applied in floating-zone melting experiments, Curtis, Graffenberger [48].

References:

[1] Aita, T.; Kawabe, U.; Honda, Y. (Japan. J. Appl. Phys. 13 [1974] 391).
[2] Zhukova, T. B.; Korsukova, M. M.; Nardov, A. V.; Gurin, V. N. (Izv. Akad. Nauk SSSR Neorgan. Materialy 17 [1981] 353/4).
[3] Swanson, L. W.; Dickinson, T. (Appl. Phys. Letters 28 [1976] 578/80).
[4] Inoue, T.; Nakada, M.; Uozumi, T.; Sugata, E. (J. Vac. Sci. Technol. 21 [1982] 952/6).
[5] Korsukova, M. M.; Gurin, V. N. (Usp. Khim. 56 [1987] 3/28; Russ. Chem. Rev. 56 [1987] 1/15, 6/7).
[6] Futamoto, M.; Aita, T.; Kawabe, U. (Japan. J. Appl. Phys. 14 [1975] 1263/6).
[7] Lazorenko, V. I.; Lotsko, D. V.; Platonov, V. F.; Kovalev, A. V.; Galasun, A. P.; Matvienko, A. A.; Klinkov, A. E. (Poroshkovaya Metal. 1987 No. 3, pp. 51/7; Soviet Powder Met. Metal Ceram. 26 [1987] 229/34).
[8] Muranaka, S.; Kawai, S. (Japan. J. Appl. Phys. 15 [1976] 587/94).
[9] Korsukova, M. M.; Lundström, T.; Gurin, V. N.; Tergenius, L.-E. (Z. Krist. 168 [1984] 299/306).
[10] Korsukova, M. M.; Gurin, V. N.; Lundström, T.; Tergenius, L.-E. (J. Less-Common Metals 117 [1986] 73/81).

[11] Eliseev, A. A.; Efremov, V. A.; Kuz'micheva, G. M.; Konovalova, E. S.; Lazorenko, V. I.; Paderno, Yu. B.; Khlyustova, S. Yu. (Kristallografiya 31 [1986] 803/5; Soviet Phys.-Cryst. 31 [1986] 476/7 [incomplete]).
[12] Barantseva, I. G.; Paderno, Yu. B. (Vysokotemp. Boridy Silitsidy 1982 8/12; C.A. 101 [1984] No. 51504).
[13] La Placa, S. (from [14]).
[14] Naslain, R.; Etourneau, J.; Hagenmuller, P. (in: Matkovich, V. I., Boron and Refractory Borides, Springer, Berlin – New York 1977, pp. 262/92, 285).
[15] McKelvy, M. J.; Eyring, L.; Storms, E. K. (J. Phys. Chem. 88 [1984] 1785/90).
[16] Johnson, R. W.; Daane, A. H. (J. Phys. Chem. 65 [1961] 909/15).
[17] Lundström, T. (Z. Anorg. Allgem. Chem. 540/541 [1986] 163/8).
[18] Lundström, T. (private communication to the Gmelin Institute, 1989).
[19] Noack, M. A.; Verhoeven, J. D. (J. Cryst. Growth 49 [1980] 595/9).
[20] Storms, E.; Mueller, B. (J. Phys. Chem. 82 [1978] 51/9).

[21] Mamedov, F. G.; Meerson, G. A.; Zhuravlev, N. N.; Umanskii, Ya. S. (Izv. Akad. Nauk SSSR Neorgan. Materialy 3 [1967] 950/6; Inorg. Materials [USSR] 3 [1967] 851/6).
[22] Paderno, Yu. B.; Lundström, T. (Acta Chem. Scand. A 37 [1983] 609/12).
[23] Uchida, K.; Shiota, M. (Surf. Technol. 7 [1978] 299/304).
[24] Blum, P.; Bertaut, F. (Acta Cryst. 7 [1954] 81/6).
[25] Meerson, G. A.; Manelis, R. M.; Nurmukhamedov, V. Kh. (Izv. Akad. Nauk SSSR Neorgan. Materialy 6 [1970] 1219/23; Inorg. Materials [USSR] 6 [1970] 1070/3).
[26] Olsen, G. H.; Cafiero, A. V. (J. Cryst. Growth 44 [1978] 287/90).
[27] Zhuravlev, N. N.; Belousova, I. A.; Manelis, R. M.; Belousova, N. A. (Kristallografiya 15 [1970] 836/8; Soviet Phys.-Cryst. 15 [1970] 723/4).
[28] Zhuravlev, N. N.; Manelis, R. M.; Gramm, N. V.; Stepanova, A. A. (Poroshkovaya Metal. 1967 No. 2, pp. 95/101; Soviet Powder Met. Metal Ceram. 1967 158/62).
[29] Noack, M. A. (IS-T-869 [1979] 1/90, 65; C.A. 92 [1980] No. 138783).

[30] Stepanov, N. N.; Zyuzin, A. Yu.; Shul'man, S. G.; Gurin, V. N.; Korsukova, M. M.; Nikanorov, S. P.; Smirnov, I. A. (Fiz. Tverd. Tela [Leningrad] 20 [1978] 935/8; Soviet Phys.-Solid State 20 [1978] 542/3).

[31] Kondrashev, A. I.; Paderno, Yu. B.; Paderno, V. N. (Poroshkovaya Metal. 1982 No. 6, pp. 16/21; Soviet Powder Met. Metal Ceram. 21 [1982] 437/41).

[32] Okada, S.; Imai, Y.; Atoda, T. (Yogyo Kyokaishi 90 No. 2 [1982] 74/82; C.A. 96 [1982] No. 113669).

[33] Aivazov, M. I.; Evseev, B. A.; Tsarev, O. M. (Izv. Akad. Nauk SSSR Neorgan. Materialy 15 [1979] 1296/7; Inorg. Materials [USSR] 15 [1979] 1015/6).

[34] Shelykh, A. I.; Sidorin, K. K.; Karin, M. G.; Bobrikov, V. N.; Korsukova, M. M.; Gurin, V. N.; Smirnov, I. A. (J. Less-Common Metals 82 [1981] 291/6).

[35] Bondarenko, V. P.; Morozov, V. V.; Chernyak, L. V. (Poroshkovaya Metal. 1971 No. 1, pp. 73/8; Soviet Powder Met. Metal Ceram. 10 [1971] 57/61).

[36] Gert, L. M.; Minashkin, V. I.; Babad-Zakhryapin, A. A. (Izv. Akad. Nauk SSSR Neorgan. Materialy 5 [1969] 2200/1; Inorg. Materials [USSR] 5 [1969] 1880/1).

[37] Oshima, C.; Horiuchi, S.; Kawai, S. (Japan. J. Appl. Phys. Suppl. 2, Pt. 1 [1974] 281/4).

[38] Paderno, Yu. B.; Ivanchenko, L. A.; Bessaraba, V. I.; Vereshchak, V. M. (Poroshkovaya Metal. 1975 No. 6, pp. 106/8; Soviet Powder Met. Metal Ceram. 14 [1975] 515/6).

[39] Motojima, S.; Takahashi, Y.; Sugiyama, K. (J. Cryst. Growth 44 [1978] 106/9).

[40] Etourneau, J.; Mercurio, J.-P.; Naslain, R.; Hagenmuller, P. (J. Solid State Chem. 2 [1970] 332/42).

[41] Alekseev, P. A.; Konovalova, E. S.; Lazukov, V. N.; Lyukshina, S. I.; Paderno, Yu. B.; Sadikov, I. P.; Udovenko, E. V. (Fiz. Tverd. Tela [Leningrad] 30 [1988] 2024/31; Soviet Phys.-Solid State 30 [1988] 1167/71).

[42] Aida, T.; Fukazawa, T. (J. Cryst. Growth 80 [1987] 9/16).

[43] Tanaka, T.; Bannai, E.; Kawai, S.; Yamane, T. (J. Cryst. Growth 30 [1975] 193/7).

[44] Morozov, V. V.; Mal'nev, V. I.; Dub, S. N.; Loboda, P. I.; Kresanov, V. S. (Izv. Akad. Nauk SSSR Neorgan. Materialy 20 [1984] 1421/3; Inorg. Materials [USSR] 20 [1984] 1225/6).

[45] Takagi, K.; Ishii, M. (J. Cryst. Growth 40 [1977] 1/5).

[46] Hohn, F. J.; Chang, T. H. P.; Broers, A. N.; Frankel, G. S.; Peters, E. T.; Lee, D. W. (J. Appl. Phys. 53 [1982] 1283/96).

[47] Aida, T.; Fukazawa, T. (J. Cryst. Growth 78 [1986] 263/73).

[48] Curtis, B. J.; Graffenberger, H. (Mater. Res. Bull. 1 [1966] 27/31).

[49] Davis, P. R.; Swanson, L. W.; Hutta, J. J.; Jones, D. L. (J. Mater. Sci. 21 [1986] 825/36).

[50] Aivazov, M. I.; Aleksandrovich, S. V.; Zinchenko, K. A.; Mkrtchyan, V. S. (Fiz. Metal. Metalloved. 46 [1978] 1171/5; Phys. Metals Metallog. [USSR] 46 No. 6 [1978] 39/42).

[51] Paderno, V. N.; Paderno, Yu. B.; Pilyankevich, A. N.; Lazorenko, V. I.; Bulchev, S. I. (J. Less-Common Metals 67 [1979] 431/6).

32.1.8.3.4.4 Lattice Vibrations

Introduction

The three external acoustical phonon modes and its dispersion were determined by inelastic neutron scattering. Factor group analysis for the hexaboride structure predicts three Raman (R) active and two IR active modes. The modes classify at Γ: $1 A_{1g}$ (R) $+ 1 E_g$ (R) $+ 1 F_{1g}$ (inactive) $+ 1 F_{2g}$ (R) $+ 2 F_{1u}$ (IR) $+ 1 F_{2u}$ (inactive). The three Raman active modes can be attributed to the internal vibrations of the boron octahedron. The rare earth ions cannot contribute to the Raman spectra because they occupy centers of symmetry, see e.g., Schell et al. [1]; see also p. 39.

In the following, the Raman spectra that give the frequencies for optical modes at Γ are described first. Then the phonon dispersion for the acoustical and one low-lying optical mode (F$_{1u}$), as derived from the neutron scattering data, are treated; from the dispersion curves, the phonon density of states (lattice spectrum) can be evaluated. Thereafter, theoretical studies of lattice dynamics leading to force constants and theoretical results of phonon dispersion and phonon density of states are considered. Next, Debye and Einstein temperatures are given; above 200 K, the Einstein model is more appropriate for the description of the vibrations of the lanthanum ion than the Debye model, whereas the vibrations of the boron lattice are correctly described by the Debye model at all temperatures. Finally, results for the electron-phonon interaction are given, which come, to some extent, from point contact spectroscopy (see p. 233).

Raman Spectra. Inelastic Neutron Scattering Spectra

Raman spectra on single-crystal LaB$_6$ were recorded with excitation by polarized light, on the (100) face, λ = 488.0 and 457.9 nm, Scholz et al. [2], on (100) and (111) faces, λ = 488.0 and 514.5 nm, Ishii et al. [3], on the (100) face, λ = 488.0 and 514.5 nm, Betsch, White [4], on the (100) face, λ = 514.0 nm, Mörke et al. [5]. Consistently, three intense peaks were observed and attributed to three Raman active modes F$_{2g}$, E$_g$, and A$_{1g}$ derived from the factor group analysis (in cm^{-1}):

F$_{2g}$	E$_g$	A$_{1g}$	Ref.
675	1125	1245	[2]
682	1120	1258	[3]
685	1138	1256	[4]
688	1138	1262	[5]

The A$_{1g}$ mode corresponds to a deformation of the boron lattice, whereas the other two modes are B–B stretching modes [3]. There was reasonable agreement between the experimental Raman frequencies and those from a simple lattice dynamical model using B–B bond stretching as single parameter; from the frequency of the F$_{2g}$ mode, a force constant of the B–B bond of 14.6×10^4 dyn/cm ($\triangleq$146 N/m) was evaluated [2]. For more detailed models on the lattice vibrations, see p. 197.

Weak additional lines that obey the same selection rules as the A$_{1g}$ mode were observed at 1390 cm^{-1} [4] and at 1385, 1160, and 205 cm^{-1} [3]. A weak line at 215 cm^{-1} increased and broadened with decreasing temperature [5]. The lines were tentatively attributed to second-order Raman scattering [3], induced by defects [5]. The interpretation of the 1390 cm^{-1} line as the second harmonic of the 685 cm^{-1} F$_{2g}$ mode is in contrast to its appearance in the polarized spectra (not observed for crossed polarizer and analyzer); it was suggested that the extra line could be the F$_{1u}$ mode, an odd-parity infrared active mode [4].

Phonon dispersion curves were measured by inelastic neutron scattering at room temperature using a single crystal of ^{11}B enriched LaB$_6$; the considerable absorption due to the actual ^{10}B content, 1.5%, caused difficulties in measuring some of the phonon modes. The measured phonon frequencies are shown in **Fig. 62**, p. 196. The triply degenerate optical mode at about 5.8 THz ($\triangleq$24 meV$\triangleq$193 cm^{-1}) at Γ represents the La atoms vibrating out-of-phase with the rigid B$_6$ group, F$_{1u}$, at Γ$_{15}$. The internal modes of the B$_6$ group are expected not to mix with the external modes because of the very high frequencies of the former indicated by the Raman scattering data. The rapid flattening of both the longitudinal and transversal acoustic modes to a value of about 3 THz for wave vector $\zeta = 0.25$ to 0.5 is rather unusual; it is attributed to the decoupled vibration of the La atom and may be described in the Einstein model, Smith et al. [6]. A flat phonon dispersion curve at ~140 K ($\triangleq 2.9$ THz $\triangleq 12$ meV) over a large part of the Brillouin zone was also found by inelastic neutron scattering by Rossat-Mignod et al. [7].

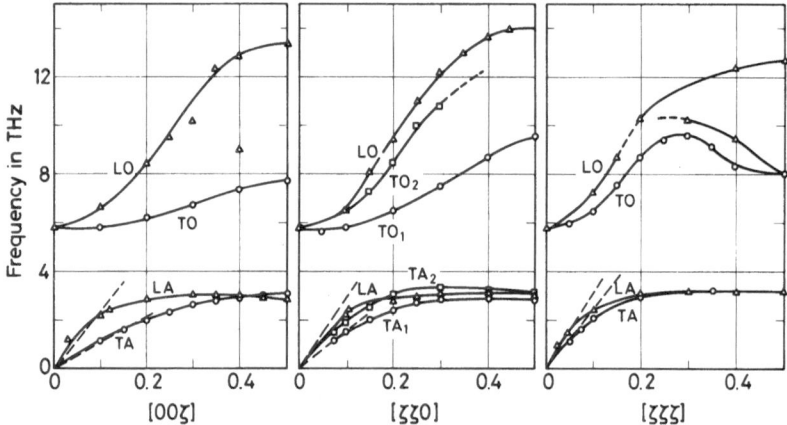

Fig. 62. Phonon dispersion curves of La^{11}B$_6$ from inelastic neutron scattering; the dashed lines are expected slopes of the dispersion curves extrapolated from the measured [15] sound velocity.

Earlier, inelastic neutron scattering experiments on polycrystalline ^{11}B enriched LaB$_6$ gave the phonon density of states shown in **Fig. 63**. The peak at 13.0 meV ($\triangleq 3.1$ THz $\triangleq 105$ cm^{-1}) was attributed to the translational acoustic mode F_{1u}; the peak near 38 meV (9.2 THz $\triangleq 306$ cm^{-1}) to the rotational mode B$_6$–B$_6$, and the peaks at 24 and 55 meV to optical modes, Schell et al. [1], which, except for the B$_6$–B$_6$ rotational mode, is in reasonable agreement with Fig. 64, p. 198 [6]. The phonon density of states between 60 and 120 meV was attributed to unresolved deformation modes of the B$_6$ octahedron [1]. For preliminary results see also Gompf [8], Schell et al. [9]. Phonon states in the range of 40 meV are attributed to a rotational mode at R$_{15'}$, Takegahara, Kasuya [10], see also p. 197; the intensity of a rotational mode at the Γ point is expected to be much weaker than the optical modes, F_{1u} [6].

The lattice dynamics of La in LaB$_6$ has also been studied between 34 and 500 K by Rutherford backscattering for 0.9 MeV protons along $\langle 100 \rangle$ of a single-crystalline disk, cut perpendicular to $\langle 111 \rangle$. The behavior of boron was studied by analysis of α particles from the nuclear reaction ^{11}B (p, α)^{8}Be. The temperature dependence of the angular half-width of the channeling curves was compared with predictions from the Debye and the Einstein models at various characteristic temperatures. The thermal vibration of the lanthanum ions follows the Einstein model above about 200 K; anharmonic terms in the potential, if any, are small. The Einstein temperature is $\Theta_E = 150 \pm 10$ K. At sufficiently high temperature, the La vibration is decoupled from the boron lattice. Below ~200 K the Einstein model is less suitable and below 150 K the Einstein-like phonon modes are exponentially depopulated; only the usual long-wave acoustic part with $\omega = v \cdot q$, where q is the wave vector, is thermally excited in the center of the Brillouin zone. The Debye model with $\Theta_D = 404$ K becomes more and more appropriate as the temperature is lowered. The vibration of the boron lattice is correctly described by the Debye model with both $\Theta_D = 404$ and 600 K. It appears that the channeling is mainly governed by the B$_6$ cluster acoustic modes rather than by its internal vibrations, Peysson et al. [11].

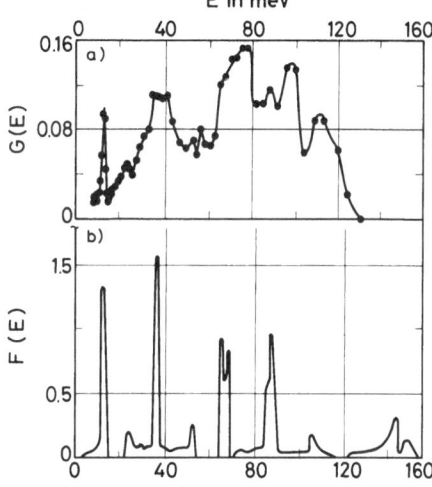

Fig. 63. Measured a) and calculated b) phonon density of states of LaB$_6$.

Theoretical Studies of Lattice Dynamics

Theoretical studies of the lattice dynamics were initiated for LaB$_6$ in order to understand several properties of the rare earth hexaborides in general; namely, the Raman spectra, the small value of the elastic constant c_{12}, and the specific heat anomaly below 10 K of some non-magnetic hexaborides. Central force interactions were considered up to the third neighbors for B–B interactions, and the first and second neighbors for La–B interactions. To account for the small c_{12}, a three-body force (Keating force) and a volume-dependent force acting between the lanthanum atoms were discussed, but only the latter gave favorable results. The force defined is of the long-range type and thus has an effect only for small wave vectors k. In addition, a central force between nearest-neighbor La atoms only was used. Within this model the phonon dispersion curves were calculated and adjusted to the observed Raman frequencies and to the optical F_{1u} mode at 24 meV. The experimental elastic constants, see p. 202, the phonon dispersion curve, and the peak at ~110 meV ($\triangleq$ 890 cm^{-1}), interpreted as upper F_{1u} mode, could be reasonably reproduced, see **Fig. 64**, p. 198. Without the long-range volume-dependent force, c_{12} is too large and is nearly equal to c_{44}. Negative values of c_{12} are obtained by appropriate choice of this force; the constant c_{11} is adjusted to the experimental value by the force between first neighbor La atoms, Takegahara, Kasuya [10], see also [12]; for the obtained force constants, see the paper. The volume-dependent force is explained by the properties of the electronic band structure: The occupied part of the conduction band is derived from a bonding orbit made by the first antibonding orbit of the B$_6$ molecule and the La 5d(e_g) orbit; cf. p. 214. At compression, the most significant effect is the strong energy shift of the antibonding B$_6$ orbital, which causes the conduction band to shift to higher energies [10, 12].

Similar phonon dispersion curves were previously calculated by Schell et al. [1, 9] using central forces between nearest neighbor and next-nearest neighbor boron atoms, and a central force between La and the nearest neighbor boron atoms. Instead of addition of further central forces, it proved necessary to introduce a three-body force (Keating force) acting between boron atoms in order to get reasonable agreement with the observed density of states curve [1], see Fig. 63. These calculations, however, have been criticized as failing to describe the observed frequency of the E$_g$ mode in the Raman spectra, the elastic constants, and the actually observed behavior of the rotational mode [10]. Also, calculations of Yoshimatsu [13], in an attempt to reproduce the Raman data and the optical F_{1u} mode (~ 24 meV) by a short-range central force model, gave only unsatisfactory results [10].

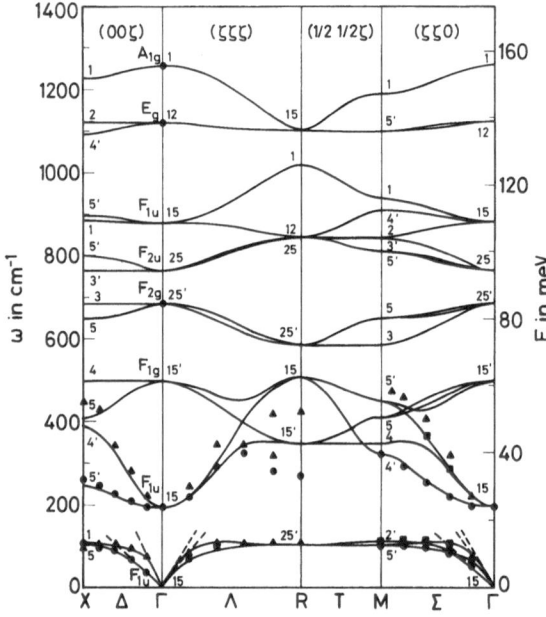

Fig. 64. Calculated phonon dispersion curves for LaB$_6$. The dashed lines were derived from the observed sound velocities; frequencies measured by the Raman and neutron scattering experiments are indicated by solid marks.

Debye and Einstein Temperatures

The values of the Debye temperature attributed to LaB$_6$ cover a considerable range, depending on the method of evaluation. In addition, studies of the lattice dynamics by ion channeling and inelastic neutron scattering, see p. 196, showed that above 200 K the Einstein model is more appropriate for the description of the vibration of lanthanum than the Debye model, which is correct for the boron sublattice at all temperatures studied.

The overall value from the single-crystal X-ray data $\Theta_D = 491$ K corresponds to individual values $\Theta_D(B) = 732(30)$ K and $\Theta_D(La) = 417(4)$ K, Korsukova et al. [14]. The boron value is in good agreement with $\Theta_D = 773$ K estimated from the elastic constants, Tanaka et al. [15], with the values from the thermal expansion [16], and the temperature dependence of the X-ray diffraction intensity [17]. A different value $\Theta_D = 404$ K was given by Smith et al. [6] as corresponding to the elastic constants of [15]. A re-evaluation, by considering all phonon modes, gave $\Theta_D = 468$ K from the temperature dependence of the electrical resistivity below 300 K [15] instead of the previously given $\Theta_D = 245$ K, obtained for the acoustic modes only [18]. The ion channeling results were consistent with $\Theta_D = 404$ K and 600 K for boron, and the results for La below ~ 100 K were between the predictions of the Debye and Einstein models with $\Theta_D = 404$ K and $\Theta_E = 150$ K, Peysson et al. [11]. From calorimetric data $\Theta_D = 438$ K, Viswanathan [19]. Lower values were obtained from other low-temperature specific heat data: $\Theta_D = 212$ K from data below 6 K and $\Theta_D = 250$ K above 6 K, Mercurio et al. [20], Etourneau et al. [21].

The decoupled thermal vibration of La above 200 K is appropriately described by the Einstein model with $\Theta_E = 150 \pm 10$ K [6] and from the X-ray data at room temperature $\Theta_E = 139 \pm 2$ K, Korsukova et al. [22]. The values of the Einstein frequency near 3 THz ($\hat{=} 100$ cm^{-1}) [1, 6] correspond to $\Theta_E = 150$ K [22].

A contribution to scattering of electrons in the electrical resistivity indicated an Einstein temperature $\Theta_E = 920$ K for the polar optical phonons [18]. The heat capacity of LaB$_6$ between 200 and 2200 K was described by a theoretical formula implying $\Theta_D = 1110$ K, from the coefficient $\gamma = 2.66$ mJ·mol^{-1}·K^{-2} of the specific heat and $\Theta_E = 6310$ K ($\hat{=}$ Einstein frequency 4385 ± 20 cm^{-1}) for internal oscillations of the B$_6$ groups, Gordienko [23].

Electron-Phonon Interaction

The electron-phonon interaction parameter λ (McMillan parameter) can be obtained by different methods. From the superconduction transition temperature T$_c$ (see p. 232) the value $\lambda = 0.33$ was obtained, Schell et al. [1]. Other experiments on the electron-phonon coupling (point contact spectroscopy (PCS), see p. 233) suggested still lower values, $\lambda = 0.21$ for PCS along [100] and $\lambda = 0.1$ for [111], Samuely et al. [24]; whereas estimates from the cyclotron masses or the specific heat suggested $\lambda = 1.14$ and 2.53, respectively, Arko et al. [25].

From the point contact spectra, the Eliashberg coupling function $\alpha^2(\omega)F(\omega)$ was derived [24] and compared with that derived theoretically by [1] within the rigid-muffin-tin approximation. From this function λ can be obtained.

References:

[1] Schell, G.; Winter, H.; Rietschel, H.; Gompf, F. (Phys. Rev. [3] B **25** [1982] 1589/99).

[2] Scholz, H.; Bauhofer, W.; Ploog, K. (Solid State Commun. **18** [1976] 1539/42).

[3] Ishii, M.; Tanaka, T.; Bannai, E.; Kawai, S. (J. Phys. Soc. Japan **41** [1976] 1075/6).

[4] Betsch, R. J.; White, W. B. (Proc. 12th Rare Earth Res. Conf., Vail, Colo., 1976, Vol. 2, pp. 534/41).

[5] Mörke, I.; Dvorak, V.; Wachter, P. (Solid State Commun. **40** [1981] 331/4).

[6] Smith, H. G.; Dolling, G.; Kunii, S.; Kasaya, M.; Liu, B.; Takegahara, K.; Kasuya, T.; Goto, T. (Solid State Commun. **53** [1985] 15/9).

[7] Rossat-Mignod, J.; Effantin, J. M.; Vettier, C.; Kunii, S.; Kasuya, T. (from Peysson, Y.; et al., J. Phys. [Paris] **47** [1986] 113/9).

[8] Gompf, F. (KfK-2670 [1978] 17/9; INIS **10** [1979] No. 438945).

[9] Schell, G.; Winter, H.; Rietschel, H. (Supercond. d-f-Band Metals Proc. Conf., San Diego 1979 [1980], Vol. 3, pp. 465/71; C.A. **93** [1980] No. 124363).

[10] Takegahara, K.; Kasuya, T. (Solid State Commun. **53** [1985] 21/5).

[11] Peysson, Y.; Daudin, B.; Dubus, M.; Benenson, R. E. (Phys. Rev. [3] B **34** [1986] 8367/71).

[12] Takegahara, K.; Kasaya, M.; Goto, T.; Kasuya, T. (Physica B + C **130** [1985] 49/51).

[13] Yoshimatsu, H. (Diss. Tohoku Univ., Sendai 1982 from Takegahara, K., Kasuya, T. [10]).

[14] Korsukova, M. M.; Lundström, T.; Tergenius, L. E.; Gurin, V. N. (Solid State Commun. **63** [1987] 187/9).

[15] Tanaka, T.; Yoshimoto, J.; Ishii, M.; Bannai, E.; Kawai, S. (Solid State Commun. **22** [1977] 203/5).

[16] Zhuravlev, N. N.; Stepanova, A. A.; Paderno, Yu. B.; Samsonov, G. V. (Kristallografiya **6** [1961] 791/4; Soviet Phys.-Cryst. **6** [1961] 636/8).

[17] Dutchak, Ya. I.; Fedyshin, Ya. I.; Paderno, Yu. B.; Vadets, D. I. (Izv. Vysshikh Uchebn. Zavedenii Fiz. **16** No. 1 [1973] 154/6; Soviet Phys. J. **16** [1973] 133/6).

[18] Tanaka, T.; Akahane, T.; Bannai, E.; Kawai, S.; Tsuda, N.; Ishizawa, Y. (J. Phys. C **9** [1976] 1235/41).

[19] Viswanathan (private communication to Tanaka, T.; et al. [15]).

[20] Mercurio, J.-P.; Etourneau, J.; Naslain, R.; Bonnerot, J. (Compt. Rend. B **268** [1969] 1766/9).

[21] Etourneau, J.; Mercurio, J.-P.; Naslain, R.; Hagenmuller, P. (J. Solid State Chem. **2** [1970] 332/42).

[22] Korsukova, M. M.; Gurin, V. N.; Nikanorov, S. P.; et al. (Neutron and X-Ray Diffraction Studies of the Crystal Structure of Rare Earth Hexaborides, Preprint No. 1188, Acad. Sci. USSR, Phys. Tech. Inst., Leningrad 1987, pp. 1/18).

[23] Gordienko, S. P. (Poroshkovaya Metal. **1981** No. 1, pp. 83/5; Soviet Powder Met. Metal Ceram. **20** [1981] 66/8).

[24] Samuely, P.; Reiffers, M.; Flachbart, K.; Akimenko, A. I.; Yanson, I. K.; et al. (Japan. J. Appl. Phys. **26** Suppl. 3 [1987] 647/8).

[25] Arko, A. J.; Crabtree, G.; Ketterson, J. B.; Mueller, F. M.; Walch, P. F.; et al. (Intern. J. Quantum Chem. Symp. No. 9 [1975] 569/78).

32.1.8.3.4.5 Nuclear Magnetic Resonance

For additional data see p. 42.

^{11}B NMR measurements on a powdered sample of LaB_6 were performed for comparison with SmB_6 at temperatures between 100 and 480 K against a H_3BO_3 solution as reference, Bose et al. [1]. In frequency variation studies (range 2 to 15 MHz), second-order quadrupolar splitting was observed below about 5 MHz; above this range only a single line was obtained, indicating that quadrupolar and magnetic interactions are interwoven. The actual transition frequency was difficult to determine, owing to an unfavorable signal-to-noise ratio. The data were analyzed using the method of Jones et al. [2], giving an anisotropic Knight shift at 300 K with $K_{iso} = +0.03 \pm 0.02\%$ and anisotropic component $a = K_{ax}/(1 + K_{iso}) = +0.153 \pm 0.02\%$, evaluated from the separate shifts of the two halves of the split central transition, and $a = 0.11\%$ from the width of the central transition. The quadrupole interaction parameter at 300 K is $\nu_Q = 0.492$, 0.495, and 0.46 MHz, all ± 0.01 MHz ($\nu_Q = 3e^2qQ/[2I(2I-1)\cdot h]$), evaluated from the separate shifts of the split central transition, and of the width, respectively. $\nu_Q = 0.509$ MHz was obtained from the satellite separation, Bose et al. [1].

An anisotropic Knight shift, $K_{\parallel} = 0.067\%$ and $K_{\perp} = 0.16\%$, together with the quadrupole coupling constant $b = 50 \pm 10$ kHz, were derived from ^{11}B NMR, Malyuchkov, Povitskii [3]. Knight shifts of $+0.031$ and $+0.018\%$, relative to ^{11}B in a sodium metaborate solution were measured for two samples of LaB_6 (commercial sample as supplied and after reaction with B, respectively) at room temperature, at $H = 5.8$ kG, McNiff, Shapiro [4].

The relaxation rates in LaB_6 are more than one order of magnitude slower than those in SmB_6 at all temperatures according to ^{11}B NMR. The upper boundary for the relaxation rate $1/T_1 \approx 0.07$ s^{-1} at 77 K, and $\lesssim 0.004$ s^{-1} at 4.2 K, Peña et al. [5].

The Knight shift of ^{139}La in LaB_6 is $-0.038 \pm 0.01\%$ (no further details given), Bose et al. [1].

References:

[1] Bose, M.; Roy, K.; Basu, A. (J. Phys. C **13** [1980] 3951/9).

[2] Jones, W. H., Jr.; Graham, T. P.; Barnes, R. G. (Phys. Rev. [2] **132** [1963] 1898/909).

[3] Malyuchkov, O. T.; Povitskii, V. A. (Fiz. Metal. Metalloved. **13** [1962] 933/4; Phys. Metals Metallog. **13** No. 6 [1962] 124/5).

[4] McNiff, E. J., Jr.; Shapiro, S. (J. Phys. Chem. Solids **24** [1963] 939/45).

[5] Peña, O.; Lysak, M.; MacLaughlin, D. E.; Fisk, Z. (Solid State Commun. **40** [1981] 539/41).

32.1.8.3.4.6 Mechanical Properties

Density in g/cm^3

The calculated density is $D_{calc} = 4.715$, Okada et al. [1], 4.711, Winter et al. [2], see also p. 45. The measured density is $D_{exp} = 4.71 \pm 0.01$ for a zone-refined sample with B:La = 5.75 containing 60 ppm C as the main impurity, Noack, Verhoeven [3], $D_{exp} = 4.55$ to 4.71 for La$_x$B$_6$ with x = 0.85 to 1, Hafner [4], $D_{exp} = 4.72$ (corresponding to 99.4% of the theoretical density) for a sample with 0.44 wt% impurities, Sanders, Probst [5]. $D_{exp} = 4.68 \pm 0.02$ for crystals grown in an Al flux [1].

Hardness. Strength

Single Crystals

The following Knoop microhardness values H$_K$ in GPa (1 kg/mm$^2 \triangleq 9.8 \times 10^6$Pa) were measured under a load of 5 N at room temperature on melt-grown single crystals: 18.1 ± 0.9 on (001), 15.2 ± 0.6 on (110), and 14.6 ± 0.65 on (111). The Vickers microhardness H$_V$ is higher (e.g. H$_V \approx 21$) on (001). With decreasing load, H$_K$ and H$_V$ increase, and at 0.5 N both H$_K$ and H$_V$ are ~27, see the given figure for the (001) plane, Morozov et al. [6]. The marked load dependence is related to the formation of cracks at higher loads, Korsukova, Gurin [7]. Cracks were observed at loads as low as ≥ 0.176 N [8]. H$_V = 25.7 \pm 0.8$ at 0.98 N on (110) planes of needles and (100) planes of plates grown from Al flux, Gurin et al. [9] and H$_V = 19.8$ at 2.94 N, Hafner [4]. H$_V = 33$ to 36 at 0.49 N on (100) of a crystal grown from the vapor phase, Motojima et al. [10]. H$_K$ on (100) is anisotropic and decreases from 19.42 ± 1.27 to 16.2 ± 1.5 with the long direction of the indentor along [010] and [011], respectively, at 1.96 N for flux-grown crystals, Futamoto et al. [11]. Similarly H$_K = 21.4$ to 25 and 15.0 to 17.9 along [010] and [011] on the (100) face at 0.49 N, Okada et al. [1].

For data on the fracture toughness and the Lawn-Marshall brittleness factor of single crystals, see Morozov et al. [6].

Polycrystalline Samples

The microhardness of polycrystalline samples is H$_V \approx 36.3$ GPa ($\triangleq 3700$ kg/mm^2) at 0.196 to 0.392 N ($\triangleq 20$ to 40 g) load and H$_V \approx 29.4$ GPa ($\triangleq 3000$ kg/mm^2) at 0.49 to 0.98 N ($\triangleq 50$ to 100 g) load. Error limits are large. However, the general tendency of H$_V$ to fall with increasing load is shown. This phenomenon is associated with the great brittleness and the crack formation in the impressions, Pilyankevich, Paderno [12]. Single data of other authors do not agree with this tendency:

H$_V$ in GPa (kg/mm^2)	load in N(g)	Ref.
22.5 (2300)	0.294 and 0.49 (30 and 50)	[13]
24.7 (2520)	0.49 (50)	[5]
25.5 (2600)	0.588 (60)	[14]
31.9 (3250)	0.98 (100)	[15]

Additional values (in GPa) without specification of load: 26 ± 0.6, Ordan'yan et al. [16], 27.2 (2770 ± 160 kg/mm^2), Samsonov, Grodshtein [17], 27.7, Ordan'yan et al. [18], 28.9 ± 2, Kondrashov et al. [19], Samsonov et al. [20]. The hardness at ~1900 K is 3.6 (371 kg/mm^2) at a load of 24.5 N (2500 g) [5].

Under hydrostatic pressure of 230 MPa (2300 kg/cm^2) some ductility appeared. The plastic deformation reached 3.5% before fracture [13].

The bending strength of cold-pressed and sintered LaB_6 increases with the densification of the sample on increasing sintering temperature. It reaches a maximum of 83 to 88 MPa (850 to 900 kg/cm²) at 2680 K and then drops [15].

Elastic Properties

The elastic constants of LaB_6 were evaluated from measurements of the sound velocities at 300 K in LaB_6 crystals from floating-zone crystal growth. The velocities of propagation for longitudinal waves along [001] and [111] are 9793 and 7779 m/s, respectively; for shear waves in the [001] and [110] direction with polarization [1$\bar{1}$0] they are 4362 and 6830 m/s, respectively, and in the [110] direction with polarization [001] 4335 m/s. The accuracy was ±1%. From these values, the following elastic constants (in GPa) were calculated by a least-square method: $c_{11} = 453.3 \pm 1.1$, $c_{12} = 18.2 \pm 1.7$, and $c_{44} = 90.1 \pm 0.5$. The anisotropy factor is $2c_{44}/(c_{11} - c_{12}) = 0.41$, Tanaka et al. [21].

The temperature dependence of the elastic constants is shown in **Fig. 65**, derived on crystals from floating-zone crystal growth by measurement of sound velocities using the pulse-echo overlap method. The error in the c_{ij} is estimated to be about 1%. Below 20 K all c_{ij} are temperature-independent; c_{44} shows an anomaly between 50 and 150 K, Winter et al. [2]. The transverse mode c_{44} increases nearly linearly on cooling from $\sim 7.85 \times 10^{11}$ erg/cm³ ($\hat{=} 78.5$ GPa) at 80 K to $\sim 7.87 \times 10^{11}$ erg/cm³ ($\hat{=} 78.7$ GPa) at 4.2 K (read from a figure) [22].

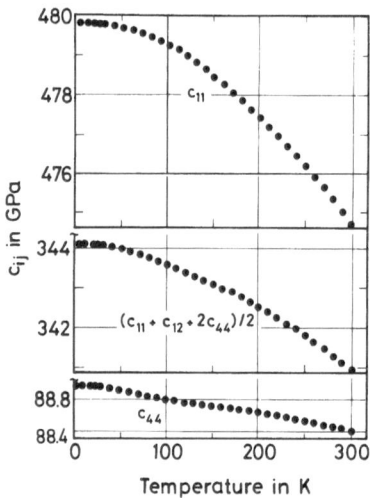

Fig. 65.   Elastic constants c_{11}, c_{44}, and ½ ($c_{11} + c_{12} + 2c_{44}$) of LaB_6 as a function of temperature.

The experimental elastic constants of [21] could be obtained within the model of the lattice dynamics of LaB_6 except for $c_{44} = 95.2$ GPa, which is slightly larger than the experimental value, Takegahara, Kasuya [23], see p. 197. The elastic properties of LaB_6 are understood mainly in terms of the character of the boron sublattice alone [21].

The Young's modulus $E = 451.8$ GPa along [100] was calculated from the c_{ij} and $E = 400$ GPa was estimated from the Raman frequencies with assumed force constants, $K_{inter} = 2.18$ mdyn/Å and $K_{intra} = 1.28$ mdyn/Å ($\hat{=} 218$ and 128 N/m, respectively) for inter- and intra-octahedra boron–boron bonds, respectively, Tanaka et al. [21]. From the dependence of the indentation depth on the load during unloading in microhardness studies, the Young's modulus $E \approx 40\,000$ kg/mm² ($\hat{=} 392$ GPa) was estimated, Paderno et al. [8]. A theoretical calculation gave E≈48 000 kg/mm^2 (≙ 470 GPa), Samsonov, Vinitsky (from [8]); E = 398 GPa, Samsonov et al. [24].

The bulk modulus K$_0$ = 191 GPa was derived from X-ray diffraction data (i.e. change of the lattice parameter) under hydrostatic pressure up to 6 GPa and K$_0$ = 163.2 GPa was calculated from the elastic constants of [21] by King et al. [25].

The compressibility $\varkappa$ = 1/K = (0.58 ± 0.03) × 10^{-11} Pa^{-1} (≙K = 172 GPa) was evaluated from X-ray diffraction investigations under a uniaxial load up to 8 GPa by Lundström et al. [26].

Magneto-Mechanical Properties

The de Haas-van Alphen (dHvA)-like oscillations (acoustic dHvA effect) of the c$_{44}$ and the (c$_{11}$-c$_{12}$)/2 mode were studied at 4.2 K in fields up to 8 T, Suzuki et al. [27, 28] and at 1.3 K in fields up to 8.5 T, Goto et al. [22], Ewert et al. [29], see p. 221.

The relative change of crystal dimensions, Δ l/l, along [100] was found to oscillate in a dHvA-like way as a function of an external magnetic field parallel to [100]. This oscillatory magneto-striction was studied up to 8.5 T at temperatures between 1.39 and ~8 K; Δ l/l had an order of magnitude of up to 2 × 10^{-7}, Sera et al. [30] cf. p. 220.

References:

[1] Okada, S.; Imai, Y.; Atoda, T. (Yogyo Kyokaishi **90** [1982] 73/82; C.A. **96** [1982] No. 113669).

[2] Winter, K. M.; Lenz, D.; Schmidt, H.; Ewert, S.; Blumenröder, S.; Zirngiebel, E.; Winzer, K. (Solid State Commun. **59** [1986] 117/2).

[3] Noack, M. A.; Verhoeven, J. D. (J. Cryst. Growth **49** [1980] 595/9).

[4] Hafner, P. V. (Diss. E. T. H. Zürich 1976, pp. 1/117, 68).

[5] Sanders, W. A.; Probst, H. B. (J. Am. Ceram. Soc. **49** [1966] 231/2).

[6] Morozov, V. V.; Mal'nev, V. I.; Dub, S. N.; Loboda, P. I.; Kresanov, V. S. (Izv. Akad. Nauk SSSR Neorgan. Materialy **20** [1984] 1421/3; Inorg. Materials [USSR] **20** [1984] 1225/6).

[7] Korsukova, M. M.; Gurin, V. N. (Current Top. Mater. Sci. **11** [1984] 389/439, 423/5).

[8] Paderno, V. N.; Paderno, Yu. B.; Pilyankevich, A. N.; Lazorenko, V. I.; Bulychev, S. I. (J. Less-Common Metals **67** [1979] 431/6).

[9] Gurin, V. N.; Korsukova, M. M.; Nikanorov, S. P.; Smirnov, I. A.; Stepanova, N. N.; Shul'man, S. G. (J. Less-Common Metals **67** [1979] 115/23).

[10] Motojima, S.; Takahashi, Y.; Sugiyama, K. (J. Cryst. Growth **44** [1978] 106/9).

[11] Futamoto, M.; Aita, T.; Kawabe, K. (Mater. Res. Bull. **14** [1979] 1329/34).

[12] Pilyankevich, A. N.; Paderno, V. N. (Fiz. Khim. Mekh. Mater. **11** No. 2 [1975] 60/5; Soviet Mater. Sci. **11** [1975] 184/8).

[13] Martynov, E. D.; Beresnev, B. I.; Baranov, I. A.; Mezis, V. Ya.; Fokin, A. E.; et al. (Fiz. Metal. Metalloved. **24** [1967] 522/7; Phys. Metals Metallog. [USSR] **24** No. 3 [1967] 138/43).

[14] Bondarenko, V. P.; Morozov, V. V.; Chernyak, L. V. (Poroshkovaya Metal. **1971** No. 1, pp. 73/8; Soviet Powder Met. Metal Ceram. **10** [1971] 57/61).

[15] Meerson, G. A.; Manelis, R. M.; Telyukova, T. M. (Izv. Akad. Nauk SSSR Neorgan. Materialy **2** [1966] 291/8; Inorg. Materials [USSR] **2** [1966] 250/5).

[16] Ordan'yan, S. S.; Paderno, Yu. B.; Khoroshilova, I. K.; Nikolaeva, E. E. (Poroshkovaya Metal. **1984** No. 2, pp. 79/81; Soviet Powder Met. Metal Ceram. **23** [1984] 157/9).

[17] Samsonov, G. V.; Grodshtein, A. E. (Zh. Fiz. Khim. **30** [1956] 379/81).

[18] Ordan'yan, S. S.; Paderno, Yu. B.; Khoroshilova, I. K.; Nikolaeva, E. E.; Maksimova, E. V. (Poroshkovaya Metal. **1983** No. 11, pp. 87/90; Soviet Powder Met. Metal Ceram. **22** [1983] 946/8).

[19] Kondrashov, A. I.; Paderno, Yu. B.; Paderno, V. N. (Poroshkovaya Metal. **1982** No. 6, pp. 16/21; Soviet Powder Met. Metal Ceram. **21** [1982] 437/41).

[20] Samsonov, G. V.; Kondrashov, A. I.; Okhremchuk, L. N.; Podchernyaeva, I. A.; Siman, N. I.; Fomenko, V. S. (J. Less-Common Metals **67** [1979] 415/8).

[21] Tanaka, T.; Yoshimoto, J.; Ishii, M.; Bannai, E.; Kawai, S. (Solid State Commun. **22** [1977] 203/5).

[22] Goto, T.; Suzuki, T.; Ohe, Y.; Fujimura, T.; Tamaki, A. (J. Magn. Magn. Mater. **76/77** [1988] 305/11).

[23] Takegahara, K.; Kasuya, T. (Solid State Commun. **53** [1985] 21/5).

[24] Samsonov, G. V.; Koval'chenko, M. S.; Ogorodnikov, V. V.; Krainii, A. G. (At. Energiya SSSR **24** [1968] 191/2; Soviet At. Energy **24** [1968] 232/3).

[25] King, H. E., Jr.; LaPlaca, S. J.; Penney, T.; Fisk, Z. (Valence Fluctuations Solids St. Barbara Inst. Theor. Phys. Conf., Santa Barbara, Calif., 1981, pp. 333/6).

[26] Lundström, T.; Lönnberg, B.; Törmä, B.; Etourneau, J.; Tarascon, J. M. (Phys. Scr. **26** [1982] 414/6).

[27] Suzuki, T.; Goto, T.; Fujimura, T.; Kunii, S.; Suzuki, T.; Kasuya, T. (J. Magn. Magn. Mater. **52** [1985] 261/3).

[28] Suzuki, T.; Goto, T.; Sakatsume, S.; Tamaki, A.; Kunii, S.; Kasuya, T.; Fujimura, T. (Japan. J. Appl. Phys. **26** Suppl. 3 [1987] 511/2).

[29] Ewert, S.; Guo, S.; Lemmens, P.; Lenz, D.; Sander, W.; Thalmeier, P.; Winzer, K. (Japan. J. Appl. Phys. **26** Suppl. 3 [1987] 537/8).

[30] Sera, M.; Kunii, S.; Kasuya, T. (J. Phys. Soc. Japan **57** [1988] 13/5).

32.1.8.3.4.7 Thermal Properties

32.1.8.3.4.7.1 Thermal Expansion and Melting Point

The linear thermal expansion coefficient of stoichiometric LaB_6 at room temperature is $\alpha = 7.22 \times 10^{-6}$ K^{-1}; it increases to 8.00×10^{-6} K^{-1} at 1273 K and to 8.24×10^{-6} K^{-1} at 1573 K according to X-ray diffraction studies in this temperature range, Aivazov et al. [1]. Similarly, a mean value of $\alpha = (7.4 \pm 0.5) \times 10^{-6}$ K^{-1} was derived from studies between 298 and 973 K, Zhuravlev et al. [2]. Additional values are $\alpha = (6.0 \pm 0.1) \times 10^{-6}$ K^{-1} at 300 K, Kondrashov et al. [3] and $\alpha = (4.9 \pm 0.3) \times 10^{-6}$ K^{-1}, Samsonov, Grodshtein [4].

The melting point is $T_m = 2803 \pm 40$ K, Ordan'yan et al. [5], essentially equal to 2820 ± 60 K, Ordan'yan et al. [6, 7]; $T_m = 2877$ K, Noack, Verhoeven [8]; $T_m = 2988$ K, Mordovin et al. [9]. A temperature of 2773 K is given as the lower limit for the melting point, Johnson, Daane [10].

References:

[1] Aivazov, M. I.; Evseev, B. A.; Tsarev, O. M. (Izv. Akad. Nauk SSSR Neorgan. Materialy **15** [1979] 1296/7; Inorg. Materials [USSR] **15** [1979] 1015/6).

[2] Zhuravlev, N. N.; Belousova, I. A.; Manelis, R. M.; Belousova, N. A. (Kristallografiya **15** [1970] 836/8; Soviet Phys.-Cryst. **15** [1970] 723/4).

[3] Kondrashov, A. I.; Paderno, Yu. B.; Paderno, V. N. (Poroshkovaya Metal. **1982** No. 6, pp. 16/21; Soviet Powder Met. Metal Ceram. **21** [1982] 437/41).

[4] Samsonov, G. V.; Grodshtein, A. E. (Zh. Fiz. Khim. **30** [1956] 379/82).

[5] Ordan'yan, S. S.; Paderno, Yu. B.; Khoroshilova, I. K.; Nikolaeva, E. E.; Maksimova, E. V. (Poroshkovaya Metal. **1983** No. 11, pp. 87/90; Soviet Powder Met. Metal Ceram. **22** [1983] 946/8).

[6] Ordan'yan, S. S.; Nikolaeva, E. E.; Kozlovskii, L. V. (Izv. Akad. Nauk SSSR Neorgan. Materialy **20** [1984] 1821/4; Inorg. Materials [USSR] **20** [1984] 1580/3).

[7] Ordan'yan, S. S.; Paderno, Yu. B.; Khoroshilova, I. K.; Nikolaeva, E. E. (Poroshkovaya Metal. **1984** No. 2, pp. 79/81; Soviet Powder Met. Metal Ceram. **23** [1984] 157/9).

[8] Noack, M. A.; Verhoeven, J. D. (J. Cryst. Growth **49** [1980] 595/9).

[9] Mordovin, O. A.; Timofeeva, E. N. (Zh. Neorgan. Khim. **13** [1968] 3155/8; Russ. J. Inorg. Chem. **13** [1968] 1627/9).

[10] Johnson, R. W.; Daane, A. H. (J. Phys. Chem. **65** [1961] 909/15).

32.1.8.3.4.7.2 Vaporization. Sublimation

For additional data see p. 89, for data on field evaporation see p. 253.

The results on vapor composition and vapor pressures of La and/or B above LaB$_6$ do not agree, which is attributed, for example, to effects of the wall materials in Knudsen effusion experiments, Nordine, Schiffman [1], Storms, Mueller [2].

Type of Evaporation. Vapor Composition

Solid LaB$_6$ vaporizes congruently under 4000 Pa (30 Torr) of Ar, Nordine, Schiffman [1] and under vacuum of 10^{-6} Torr ($\triangleq$133 µPa), Etourneau et al. [3]. Congruent vaporization was also inferred from the composition of films from the LaB$_6$ vapor deposition, Oshima et al. [4] and is believed to occur at the highest temperatures used by Ociepa, Mroz [5]; cf. p. 184. Free evaporation was studied from polycrystalline samples with B/La = 5.3 to 8.7 in the mass spectrometer. From the dependence of the vapor composition on the surface composition at 1500 and 1700 K (see a figure in the original paper) congruently vaporizing compositions LaB$_{6.034}$ and LaB$_{6.042}$, respectively, are indicated, Storms, Mueller [6]. Congruent vaporization was observed only at ~2000 K, whereas below 1700 K the B/La atomic ratio was about 3, Swanson, Dickinson [7]. Incongruent vaporization (the La evaporation rate is five times that of B) was observed during Langmuir evaporation by Goldstein, Szostak [8] and also during electron-beam heated evaporation under 20 Torr ($\triangleq$2.7 kPa) of inert gas. The composition of the deposited vapors depends on the size of the inert atoms or molecules, Hafner [9, p. 59]. Dissociation according to LaB$_6$(s) = La(g) + 6 B(s) was inferred from studies under equilibrium conditions between 2045 and 2300 K in a mass spectrometer. B/La never exceeded 4:1 in the vapor, even at 2300 K, Gordienko et al. [10].

The vapor phase over clean LaB$_6$ faces consists only of atomic B and La, Swanson et al. [11], see also Gordienko et al. [10], Torshina et al. [12]. LaO$^+$, in addition, was found even at 10^{-10} Torr ($\triangleq$13 nPa) owing to adsorption of O$_2$ whether at room temperature or at 1273 K. The existence of BO$^+$ was not observed even though evaporation was from a surface on which both B and La have been oxidized; possibly it dissociated just prior to evaporation, Goldstein, Szostak [8]. However, both LaO$^+$ and BO$^+$ were observed in the presence of O$_2$ impurities, Storms, Mueller [2].

Evaporation Rate

The experimental evaporation rates of the (100) face of a LaB$_{6.09}$ crystal are shown in **Fig. 66**, p. 206, at 1450 to 1950 K. The mass loss rate R$_m$ in g·cm^{-2}·s^{-1} at <1.33 µPa was given by $R_m = 1.13 \times 10^6 \exp(-60524/T)$; the evaporation is isotropic from the various exposed crystal surfaces. As shown in the figure, R_m is strongly dependent on the residual gases, mainly at low temperature, by formation of volatile oxides. At O_2 pressures of >66.5 µPa, the oxygen-enhanced volatility is anisotropic and increases in the order $(110)<(111)<(100)$ as was inferred from facet formation on conical LaB_6 cathodes at 1700 and 1750 K, Davis et al. [13]. From weight losses of freely evaporating polycrystalline samples, R_m between 2015 and 2248 K for $LaB_{6.01}$ was given by $\log R_m = -29997/T + 7.604$ ($\hat{=} R_m = 4.0 \times 10^7 \exp(-69071/T)$) and for $LaB_{5.9}$ between 2019 and 2132 K by $\log R_m = -30646/T + 8.151$ ($\hat{=} 1.4 \times 10^8 \exp(-70565/T)$). Mass loss rates were also calculated from the vapor composition by mass spectrometry and agree reasonably well with the data from weight loss measurements, see the paper for data on more boron-rich compositions, Storms, Mueller [2].

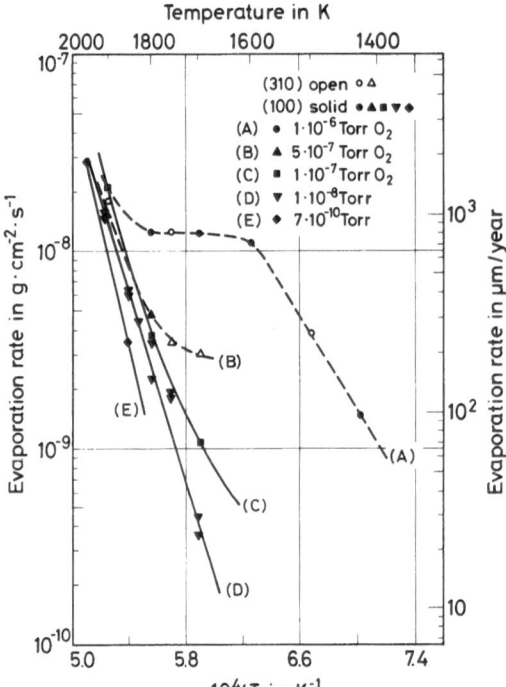

Fig. 66. Evaporation rate of a $LaB_{6.09}$ single crystal versus temperature under vacuum and in various controlled pressures of oxygen.

From the degradation of a LaB_6 single-crystal cathode between 773 and 2173 K, an evaporation rate of $4 \times 10^{14} \exp(-570 \times 10^3/RT)$ in µm/h ($\hat{=} 5 \times 10^7 \exp(-68700/T)$ in $g \cdot cm^{-2} \cdot s^{-1}$) was derived. The evaporation rate was nearly constant below $\sim 10^{-4}$ Pa ($\hat{=} 10^{-6}$ Torr) but increased for poorer vacuum, for example by a factor of ~ 30 at $\sim 10^{-1}$ Pa at 1923 K, Futamoto et al. [14]. The La evaporation rate from experiments with a Knudsen-like geometry was 2.89×10^{-8}, 2.79×10^{-7}, and 2.66×10^{-6} $g \cdot cm^{-2} \cdot s^{-1}$ at 1953, 2063, and 2183 K, Lafferty [15]. A high value of $\sim 5.5 \times 10^{-5}$ $g \cdot cm^{-2} \cdot s^{-1}$ at 2123 K was observed by Gordienko et al. [10], whereas their value $\sim 20.6 \times 10^{-5}$ $g \cdot cm^{-2} \cdot s^{-1}$ for 2208 K is consistent with data from [2, 13], see the criticism by [2]. Similarly, material loss of 0.25 µm/h ($\hat{=} 3.3 \times 10^{-8}$ $g \cdot cm^{-2} \cdot s^{-1}$) was observed at 1823 K from single-crystal LaB_6; it was increased for higher impurity levels, Hagiwara et al. [16], similarly found on hot-pressed LaB_6 by Vogel [17]. For polycrystalline LaB_6 cathodes, mass loss rates of 1.51×10^{-8} and 2.03×10^{-8} $g \cdot cm^{-2} \cdot s^{-1}$ at 1810 and 1880 K, respectively, were observed at $(0.5$ to $2) \times 10^{-6}$ Torr $((\sim 67$ to $266) \times 10^{-6}$ Pa$)$ N_2 equivalent pressure, Hafner [9]. Lower values in g·cm^{-2}·s^{-1}, are <1.85×10^{-8} at 1873, 3.9×10^{-8} at 2273, and 7.3×10^{-7} at 2373 K, given by Wagner et al. [18]. For sintered LaB$_6$ cathodes with impurities of La-richer and B-richer phases loss rates of 6×10^{-10} and 7×10^{-11} g·cm^{-2}·s^{-1}, respectively, were observed at 1700 K, Jacobson, Storms [19]. For a discussion of the previous results see [1], cf. p. 209.

Vapor Pressure

The equation log p_{La}[atm] = (8.539 ±0.003) − (29340 ±500)/T was derived for the La partial pressure over LaB$_6$ from equilibrium vaporization (Knudsen cell with Mo and Ta) between 2045 and 2270 K. The data show good agreement with values evaluated from weight loss determinations, Gordienko et al. [10]. Boron vapor pressures about 20 times that of pure boron are implied in the data of Ames, McGrath [20] according to [2].

The dependence of the La activity over a sample LaB$_{6.02}$ is given by log a_{La} = 0.22 −4810/T, the boron activity by log a_B = −1.97 +1300/T. The data were derived between 1400 and 2100 K from samples with bulk ratios B/La = 4.24 to 29.2. There is a decrease of the La activity by four orders of magnitude on going from the La-rich side of LaB$_6$ to the boron-rich side, Storms, Mueller [2].

Thermodynamic Data of Vaporization

The vaporization enthalpy, for the reaction 1/7 LaB$_6$(s) = 1/7 La(g) + 6/7 B(g) at 298 K was evaluated as $\Delta H^\circ_{v,298}$ = 602.9 ± 5.9 kJ/mol from the intensity of laser-induced fluorescence (LIF) in the wake of an aerodynamically levitated LaB$_6$ sphere heated to 2530 K by a continuous-wave CO$_2$ laser (for details see the paper). This value is consistent with that calculated from the formation enthalpy determined by Topor, Kleppa [21], however, it is inconsistent with that calculated from the formation enthalpy given by [2], see p. 209, Nordine, Schiffman [1].

From the temperature dependence of vapor pressure between 1273 and 2073 K, the sublimation enthalpy is 105 ±12 kcal/mol (≙439 ± 50 kJ/mol), Torshina et al. [12]. ΔH_{subl} = 169 kcal/mol (≙707 kJ/mol) from La evaporation rates from LaB$_6$ between 1953 and 2183 K, Lafferty [15, 22]; ΔH_{subl} = 707 kJ/mol, Futamoto et al. [14]. At 2173 K the vaporization enthalpy ΔH° =134±2 kcal/mol (≙561±8 kJ/mol) and the entropy ΔS° =39.4±2 cal·mol^{-1}·K^{-1} (≙165±8 J·mol^{-1}·K^{-1}) assuming dissociation according to LaB$_6$(s) = La (g) +6B(s) were derived from vapor pressure measurements between 2045 and 2270 K, Gordienko [10], for previous reports see also [23 to 25]. However, see p. 209.

Vaporization studies by mass spectrometry of vaporizing species from various faces of zone-refined crystals LaB$_{5.74}$ and LaB$_{5.86}$ between 1750 and 2080 K, Swanson et al. [11], see also [7], and for free evaporation of Al-flux grown LaB$_6$, Goldstein, Szostak [8] give the following partial vaporization enthalpy ΔH of boron and lanthanum in eV (or kJ/mol in parentheses):

composition ...	LaB$_{5.74}$ [11]		LaB$_{5.86}$ [11]		LaB$_6$ [8]
crystal face	(110)	(100)	(110)	(346)	(100)
ΔH_B	6.40±0.05(617)	7.3±0.2(704)	5.8±0.1(559)	6.3±0.1(608)	5.6(540)
ΔH_{La}	4.40±0.1(424)	4.4±0.1(424)	5.3±0.05(511)	5.4±0.1(520)	5.6(540)

Vaporization of polycrystalline samples with ratios B/La from 4.24 to 29.2 between 1400 and 2100 K from Knudsen cells of presaturated W analyzed by mass spectrometry, gave ΔH_B =133.2 ±0.9 and ΔH_{La} =122.7 ±0.3 kcal/mol (≙544.8 and 513.4 kJ/mol) for LaB$_{6.02}$. Whereas ΔH_B is nearly independent of sample composition, there was a strong increase of ΔH_{La} for the boron-rich samples. By comparison with data from free vaporization, there are some inconsistencies that were attributed to a concentration gradient and a boron evaporation coefficient below unity [2].

From a lanthanum boride film deposited on a tungsten substrate, desorption energies of 6.14 ± 0.2 eV for La and 5.8 ± 0.2 eV for B were determined, Okuno et al. [26]. A very low value of 3.25 eV for the mean activation energy of LaB_6 evaporation was observed for a LaB_6-covered Re thermionic cathode, Buckingham [27].

References:

[1] Nordine, P. C.; Schiffman, R. A. (High Temp. Sci. **20** [1985] 1/20).
[2] Storms, E. K.; Mueller, B. (J. Phys. Chem. **82** [1978] 51/9).
[3] Etourneau, J.; Mercurio, J.-P.; Naslain, R. (Compt. Rend. C **275** [1972] 273/6).
[4] Oshima, C.; Horiuchi, S.; Kawai, S. (Japan J. Appl. Phys. **13** Suppl. 2, Pt. 1 [1974] 281/4).
[5] Ociepa, J. G.; Mroz, S. (Acta Univ. Wratislaw. Mat. Fiz. Astron. No. 37 [1980] 25/30; C. A. **94** [1981] No. 93805).
[6] Storms, E. K.; Mueller, B. A. (J. Appl. Phys. **50** [1979] 3691/8).
[7] Swanson, L. W.; Dickinson, T. (Appl. Phys. Letters **28** [1976] 578/80).
[8] Goldstein, B.; Szostak, D. J. (Surf. Sci. **74** [1978] 461/78).
[9] Hafner, P. V. (Diss. ETH Zürich 1976, pp. 1/117).
[10] Gordienko, S. P.; Guseva, E. A.; Fesenko, V. V. (Teplofiz. Vys. Temp. **6** [1968] 821/5; High Temp. [USSR] **6** [1968] 785/9).

[11] Swanson, L. W.; Gesley, M. A.; Davis, P. R. (Surf. Sci. **107** [1981] 263/89).
[12] Torshina, V. V.; Smolina, G. N.; Dobychin, S. Ya. (Zh. Neorgan. Khim. **10** [1965] 1275/6; Russ. J. Inorg. Chem. **10** [1965] 691/5).
[13] Davis, P. R.; Schwind, G. A.; Swanson, L. W. (J. Vac. Sci. Technol. B **4** [1986] 112/5).
[14] Futamoto, M.; Nakazawa, M.; Usami, K.; Hosoki, S.; Kawabe, U. (J. Appl. Phys. **51** [1980] 3869/76).
[15] Lafferty, J. M. (J. Appl. Phys. **22** [1951] 299/309).
[16] Hagiwara, H.; Hiraoka, H.; Terasaki, R.; Ishii, M.; Shimizu, R. (Scanning Electron Microsc. **1982** 473/83).
[17] Vogel, S. F. (Rev. Sci. Instrum. **41** [1970] 585/7).
[18] Wagner, S.; Albrecht, H. E.; Kotsch, H. (Neue Hütte **9** [1964] 278/81).
[19] Jacobson, D. L.; Storms, E. K. (IEEE Trans. Pharma. Sci. **6** No. 2 [1978] 191/9).
[20] Ames, L. L.; McGrath, L. (High Temp. Sci. **7** [1975] 44/54).

[21] Topor, L.; Kleppa, O. J. (J. Chem. Thermodyn. **16** [1984] 993/1002).
[22] Lafferty, J. M. (Phys. Rev. [2] **79** [1950] 1012).
[23] Gordienko, S. P.; Samsonov, G. V.; Fesenko, V. V. (Poroshkovaya Metal. **1965** No. 8, pp. 70/3; Soviet Powder Met. Metal Ceram. **1965** 661/3).
[24] Fesenko, V. V.; Bolgar, A. S.; Gordienko, S. P. (Rev. Intern. Hautes Temp. Refract. **3** [1966] 261/71).
[25] Fesenko, V. V.; Bolgar, A. S. (Poroshkovaya Metal. **3** No. 1 [1963] 17/25; Soviet Powder Met. Metal Ceram. **1963** 11/7).
[26] Okuno, K.; Sasaki, T.; Kim, H.; Inoue, T.; Sugata, E. (Japan. J. Appl. Phys. **17** [1978] 719/20).
[27] Buckingham, J. D. (Brit. J. Appl. Phys. **16** [1965] 1821/32).

32.1.8.3.4.7.3 Enthalpy of Formation

The standard molar enthalpy of formation of LaB$_6$ was determined calorimetrically from the reaction of LaB$_6$ and La + 6 B with solid Pt at temperatures near 1373 K, yielding a liquid alloy. A value H$^\circ_{f,298}$ = −(400.4 ±11.9) kJ/mol was obtained, Topor, Kleppa [1]. A consistent value ΔH$^\circ_{f,298}$ = −430 ±86 kJ/mol was calculated from the measured enthalpy of evaporation, Nordine, Schiffman [2]. From the experimental La activities in the vapor phase above LaB$_6$ and the B activities calculated by the Gibbs-Duhem technique, the standard enthalpy of formation is H$^\circ_{f,298}$ = −254 ±8 kJ/mol (the value −287 kJ/mol given in [3] is incorrect), Storms [4]. However, this value seems to suffer from B reaction effects with the Knudsen cell [2]. This is also suggested by [3] for the values ΔH$^\circ_{f,298}$ = −30.7 ±4 kcal/mol ($\triangleq$ −128.4 kJ/mol) from 2nd law thermodynamic calculations and ΔH$^\circ_{f,298}$ = −24.9 ±1.5 kcal/mol ($\triangleq$ −104.2 kJ/mol) from 3rd law thermodynamic calculations, based on studies on both vaporization and heat capacity, Gordienko et al. [5]. From the temperature dependence of the reaction of La$_2$O$_3$ and 3 B$_4$C, the enthalpy of formation of LaB$_6$ was estimated as −112.3 ±6.5 kcal/mol ($\triangleq$ −470 kJ/mol), Samsonov et al. [6].

References:

[1] Topor, L.; Kleppa, O. J. (J. Chem. Thermodyn. **16** [1984] 993/1002).
[2] Nordine, P. C.; Schiffman, R. A. (High Temp. Sci. **20** [1985] 1/20).
[3] Storms, E. K.; Mueller, B. (J. Phys. Chem. **82** [1978] 51/9).
[4] Storms, E. K. (private communication to Nordine, Schiffman [2]).
[5] Gordienko, S. P.; Guseva, E. A.; Fesenko, V. V. (Teplofiz. Vys. Temp. **6** [1968] 821/5; High Temp. [USSR] **6** [1968] 785/9).
[6] Samsonov, G. V.; Paderno, Yu. B.; Kreingol'd, S. U. (Zh. Prikl. Khim. **34** [1961] 10/5; J. Appl. Chem. [USSR] **34** [1961] 8/13).

32.1.8.3.4.7.4 Heat Capacity. Thermodynamic Functions

At Low Temperature

The low-temperature heat capacity is strongly sample dependent; the following table gives some experimental values for the parameters γ and β in C$_p$ = γT + βT^3 (zm = zone melted):

temperature range in K	2 to 6	6 to 12	2 to 8	<6
γ in mJ·mol^{-1}·K^{-2}	2.6	6.4	2.3	~3.4
β in mJ·mol^{-1}·K^{-4}	0.21	0.13	0.0305	—
sample	zm	zm	single cryst.	zm
Ref.	[1, 2]	[1, 2]	[3]	[4]

Remarkably, two regions with different slopes in the otherwise linear C$_p$/T vs. T^2 plot were observed by Mercurio et al. [1], Etourneau et al. [2], cf. Fig. 19, p. 51. This was not observed, however, by von Molnar et al. [3], whose results are considerably lower than those of [1] and lower than C$_p$ = 0.1 to 1.3 J·mol^{-1}·K^{-1} for T = 10 to 20 K of Lee et al. [5]. A careful remeasurement by Takegahara, Kasuya [6] showed that no such anomaly exists in very pure LaB$_6$; it was tentatively attributed to a defect-induced softening of a phonon mode [6] rather than to the rotational motion of B$_6$ as previously proposed by Kasuya et al. [4].

The heat capacity of LaB$_6$ from ~10 to 80 K, Fujita et al. [7], cf. [4] is shown in **Fig. 67**, p. 210, from Smith et al. [8], together with curves calculated within a simple model from the total phonon density of states and from the acoustic modes only (solid and dashed lines, respectively), as observed by inelastic neutron scattering and Raman spectroscopy (see p. 195). In doing so for small wave vectors, the acoustic modes are treated by the triply degenerate Debye model with $\Theta_D = 404$ K, whereas in the flat part, the Einstein model is assumed with the Einstein frequency chosen to be 100 cm^{-1} ($\hat{=}$ 3 THz). The dispersion of the optical modes is assumed as $\omega = \omega_0 + Mq^2$ with q the wave vector, Smith et al. [8]. Heat capacity data from 10 to 200 K with quantitative agreement between 10 and 60 K with those of [7] were measured by Peysson et al. [10], cf. [11].

Earlier C_p measurements from Westrum et al. [12, 13] between 5 and 350 K gave the following values (converted from cal/mol and cal·mol^{-1}·K^{-1}):

T in K	25	50	100	200	298.15
C_p in J·mol^{-1}·K^{-1}	2.845	13.56	28.85	64.77	96.90
$H_T^\circ - H_0^\circ$ in J/mol	14.10	223.0	1292	5933	13937
S_T° in J·mol^{-1}·K^{-1}	0.703	6.058	20.27	51.00	83.14
$-(G_T^\circ - H_0^\circ)/T$ in J·mol^{-1}·K^{-1} ..	0.1389	1.5979	7.355	21.34	36.39

At High Temperature

Fig. 68 shows the temperature dependence of the heat capacity from 150 to 1000 K [8]. The experimental data are from [7, 9]; the calculated data (dashed line) are from the phonon density of states, see above.

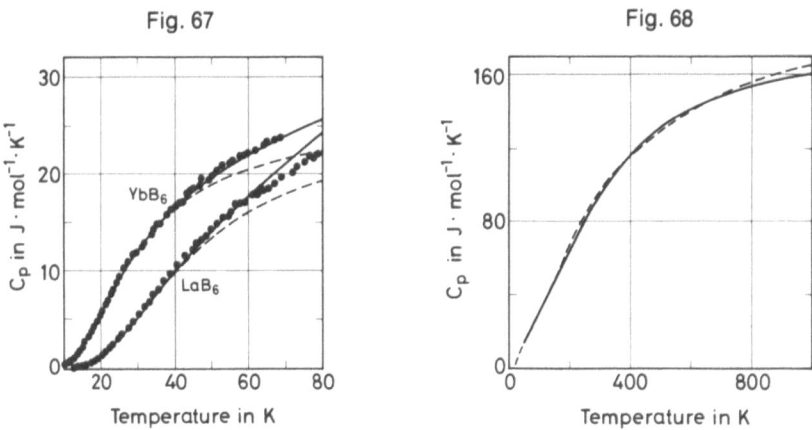

Heat capacity of LaB$_6$ and YbB$_6$ at low temperatures (Fig. 67) and of LaB$_6$ up to 1000 K (Fig. 68) compared with theoretical data, see text.

At high temperature, from 1100 to 2223 K, the enthalpy $H_T - H_{298.15}^\circ$ was described within ±0.7% by $H_T - H_{298}^\circ$ (in kJ/kg) = $0.4902 \cdot T + 16.150 \times 10^{-5} \cdot T^2 - 1997.2 \cdot T^{-1} - 152.11$, or (in J/mol) = $99.89 \cdot T + 0.032909 \cdot T^2 - 4.0697 \times 10^5 \cdot T^{-1} - 3.09957 \times 10^4$, samples were from borothermal reduction with an impurity content of 0.2%, Kapyrina et al. [14]. Similarly, the measured enthalpy change between 1341 and 2018 K was fitted by $H_T - H_{298}^\circ$ (in cal/mol) = $26.3 \cdot T + 67.367 \times 10^{-4} \cdot T^2 - 9018$, or $H_T - H_{298}^\circ$ (in J/mol) = $110 \cdot T + 0.028186 \cdot T^2 - 3.773 \times 10^4$, Gordienko et al. [15].

From these data, the entropy change and the heat capacity were extrapolated down to 298 K and up to 2500 K, taking $S_{298}^\circ = 19.88$ cal·mol^{-1}·K^{-1} ($\hat{=}$ 83.17 J·mol^{-1}·K^{-1}), Gordienko et al. [15], see the table in the paper. Combining the data from [12, 14] the following equations were derived by Storms, Mueller [16]:

$H_T - H^\circ_{298} =$

$6.6221 \times 10^5 \cdot T^{-1} - 1.052 \times 10^4 + 24.8 \cdot T + 1.109 \times 10^{-2} \cdot T^2 - 3.35 \times 10^{-6} \cdot T^3 + 8.009 \times 10^{-10} \cdot T^4$ cal/mol

$2.7707 \times 10^6 \cdot T^{-1} - 4.402 \times 10^4 + 104 \cdot T + 4.640 \times 10^{-2} \cdot T^2 - 1.40 \times 10^{-5} \cdot T^3 + 3.351 \times 10^{-9} \cdot T^4$ J/mol.

Selected numerical values (converted from cal/mol and cal$\cdot$mol$^{-1}\cdot$K^{-1}) are:

T in K	298.15	600	900	1200	1500	1800	2100
C_p in J$\cdot$mol$^{-1}\cdot$K^{-1}	96.90	139.3	159.4	175.7	192.5	211.7	236.8
$H^\circ_T - H^\circ_{298}$ in J/mol	0	36970	82015	132353	187535	248069	315198
S°_T in J$\cdot$mol$^{-1}\cdot$K^{-1}	83.14	167.2	227.7	275.9	316.9	353.6	388.0
$-(G^\circ_T - H^\circ_{298})/T$ in J$\cdot$mol$^{-1}\cdot$K^{-1}	83.14	105.5	136.6	165.6	191.8	215.8	237.9

Storms, Mueller [16]. The heat capacity equation $C_p = 21.73 + 0.0204 \cdot T$ in cal$\cdot$mol$^{-1}\cdot$K^{-1} ($\triangleq 90.92 + 0.0854 \cdot T$ in J$\cdot$mol$^{-1}\cdot$K^{-1}) was reported by Samsonov et al. [17]. The difference $C_p - C_v$ is estimated as 1 J$\cdot$mol$^{-1}\cdot$K^{-1} at 1000 K, Smith et al. [8].

The experimental heat capacities above about 1000 K are substantially larger than values extrapolated from low-temperature specific heat data, implying an estimated $\Theta_D = 1110$ K and $\gamma = 2.66$ mJ$\cdot$mol$^{-1}\cdot$K^{-2}. Thus, in addition to the usual contributions of lattice and electrons to C_p, contributions from internal vibrations of the B$_6$ octahedra are postulated that are characterized by the Einstein temperature $\Theta_E = 6310$ K, Gordienko [18].

References:

[1] Mercurio, J.-P.; Etourneau, J.; Naslain, R.; Bonnerot, J. (Compt. Rend. B **268** [1969] 1766/9).

[2] Etourneau, J.; Mercurio, J.-P.; Naslain, R.; Hagenmuller, P. (J. Solid State Chem. **2** [1970] 332/42).

[3] von Molnar, S.; Theis, T.; Benoit, A.; Briggs, A.; Flouquet, J.; Ravex, J. (Valence Instab. Proc. Intern. Conf., Zürich, Switz.,1982, pp. 389/95).

[4] Kasuya, T.; Takegahara, K.; Fujita, T.; Tanaka, T.; Bannai, E. (J. Phys. Colloq. [Paris] **40** [1979] C5-308/C5-313).

[5] Lee, K. N.; Bachmann, R.; Geballe, T. H.; Maita, J. P. (Phys. Rev. [3] B **2** [1970] 4580/5).

[6] Takegahara, K.; Kasuya, T. (Solid State Commun. **53** [1985] 21/5).

[7] Fujita, T.; Suzuki, M.; Komatsubara, T.; Kunii, S.; Kasuya, T.; Ohtsuka, T. (Solid State Commun. **35** [1980] 569/72).

[8] Smith, H. G.; Dolling, G.; Kunii, S.; Kasaya, M.; Liu, B.; Takegahara, K.; Kasuya, T.; Goto, T. (Solid State Commun. **53** [1985] 15/9).

[9] (Rept. Natl. Inst. Res. Inorg. Materials [Japan] No. 17 [1978] from Smith et al. [8]).

[10] Peysson, Y.; Ayache, C.; Rossat-Mignod, J.; Kunii, S.; Kasuya, T. (J. Phys. [Paris] **47** [1986] 113/9).

[11] Peysson, Y.; Ayache, C.; Salce, B.; Rossat-Mignod, J.; Kunii, S.; Kasuya, T. (J. Magn. Magn. Mater. **47/48** [1985] 63/5).

[12] Westrum, E. F., Jr.; Clever, H. L.; Andrews, H. L.; Andrews, J. T. S.; Flick, G. (Proc. 4th Conf. Rare Earth Res., Phoenix, Ariz., 1964 [1965], pp. 597/605).

[13] Westrum, E. F., Jr. (Colloq. Intern. Centre Natl. Rech. Sci. [Paris] No. 180 [1970] 443/50).

[14] Kapyrina, V. Y.; Prilepskii, V. N.; Timofeev, V. A.; Timofeeva, E. N.; Trubitsyn, A. Ya. (Teplofiz. Vys. Temp. **6** [1968] 193/4; High Temp. [USSR] **6** [1968] 188/9).

[15] Gordienko, S. P.; Guseva, E. A.; Fesenko, V. V. (Teplofiz. Vys. Temp. **6** [1968] 821/5; High Temp. [USSR] **6** [1968] 785/9).

[16] Storms, E.; Mueller, B. (J. Phys. Chem. **82** [1978] 51/9).

[17] Samsonov, G. V.; Paderno, Yu. B.; Kreingol'd, S. U. (Zh. Prikl. Khim. **34** [1961] 10/5; J. Appl. Chem. [USSR] **34** [1961] 8/13).

[18] Gordienko, S. P. (Poroshkovaya Metal. **1981** No. 1, pp. 83/5; Soviet Powder Met. Metal Ceram. **20** [1981] 66/8).

32.1.8.3.4.7.5 Thermal Conductivity

At Low Temperature. Influence of a Magnetic Field

The thermal conductivity λ of LaB_6 was measured between 1.2 and 200 K on a crystal (obtained by the floating-zone method) parallel to [111]. The conductivity λ increases from ~0.24 $W \cdot cm^{-1} \cdot K^{-1}$ at 1.5 K to a maximum of 5.0 $W \cdot cm^{-1} \cdot K^{-1}$ near 17 K and decreases to a nearly constant value of ~1.3 above 100 K. The increase up to 17 K was proportional to $T^{1.1}$, thus the Wiedemann-Franz law holds quite well in this range, despite some deviations due to inelastic scattering. The Lorenz number, calculated from the residual electrical resistivity of this sample, $\varrho_0 = (7.5 \pm 0.2) \times 10^{-8} \, \Omega \cdot cm$, was close to the value for the free electron gas model $L_0 = 2.45 \times 10^{-8} \, W \cdot \Omega \cdot K^{-2}$ for temperatures below 2 K and above 70 K, but negative deviations occurred between 2 and 70 K, up to 60% at 25 K. This indicates that the heat transport is mainly by conduction electrons. When the electronic conductivity λ_{el} or the electronic thermal resistivity W_{el}, are expressed by $\lambda_{el} = 1/W_{el} = 1/(aT^{-1} + bT^2)$, the terms in the denominator represent the scattering by static lattice defects (first term) and by phonons (second term). With $a = 2.9 \, K^2 \cdot cm \cdot W^{-1}$ and $b = 3 \times 10^{-4} \, cm \cdot W^{-1} \cdot K^{-1}$, derived from the low- and high-temperature data, the maximum at $T_M = (a/2b)^{1/3}$ was correctly obtained at 17 K. A theoretical value for b, within the Bloch-Grüneisen approximation of the electron-phonon interaction, was roughly half the experimental value. The deviation is said to be due to neglect of the d character of the conduction electrons in the approximations used and to the large deviation from the Debye model, Peysson et al. [1], cf. [2].

Thermal conductivity data between 0.7 and 25 K for a zone-refined single crystal (residual resistance ratio = 175) by Flachbart et al. [3], see **Fig. 69**, showed only qualitative agreement; maximum $\lambda \approx 10 \, W \cdot cm^{-1} \cdot K^{-1}$ at 15 K. Variation of λ was in proportion to T^2 between 4 and 12 K and to T^3 below 3 K. This would suggest the phonon thermal conductivity to be the dominant component of the heat transport. However, the reduced thermal conductivity in a longitudinal magnetic field of 3 T indicated a major role in heat transport by electronic thermal conductivity [3]; but no effect of a longitudinal magnetic field up to 7 T on the thermal conductivity λ, both along [111], was found by [1].

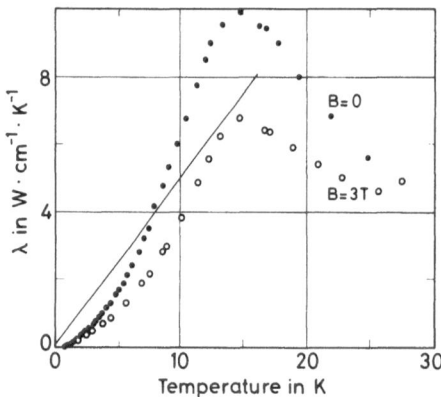

Fig. 69. Thermal conductivity of LaB_6. The solid line is the electronic thermal conductivity according to the Wiedemann-Franz law $\lambda_{el} = L_0 T / \varrho_0$.

In order to study again the phonon contribution λ_{ph} at these low temperatures, the electronic component was reduced by magnetic fields perpendicular to the heat flow (Corbino method), in this case from the center to the edge of a disk parallel to (111), Peysson et al. [4]. In zero field, the measurements at 1.5 to 15 K gave satisfactory agreement with the results obtained in standard geometry [1]. In magnetic fields, λ was strongly reduced, by a factor of 12 for B = 7 T. In the high-field limit, the electronic contribution perpendicular to the field, λ_{el} (B), is expected to decrease as B^{-2}; this regime, however, was not achieved. Thus only an upper limit for the phonon contribution λ_{ph} was estimated by extrapolation to infinite field. The ratio $\lambda_{ph}/\lambda_{el} = 1:25$ seems to be rather large compared to usual metals [4]. At 200 K, half of the heat current was thought to be due to the phonons [1].

At High Temperature

The thermal diffusivity $a = \lambda/(C_p \cdot D)$ was measured between 1300 and 2000 K for a hot-pressed rod of LaB$_6$, relative density 92.2%, and the thermal conductivity λ was calculated from a. The decrease of λ from 0.54 W·cm^{-1}·K^{-1} at 1300 K to 0.40 W·cm^{-1}·K^{-1} at 2000 K was proportional to 1/T. The Wiedemann-Franz law was not obeyed; the actual L value deviates from L$_0$. Assuming the electronic contribution λ_{el} to follow the Wiedemann-Franz law gives the phonon thermal conductivity $\lambda_{ph} = 3.5 \times 10^{-2}$ cal·cm^{-1}·s^{-1}·K$^{-1} \cong 0.146$ W·cm^{-1}·K^{-1} at 1300 K, in good agreement with a theoretical estimate assuming a Debye temperature of $\Theta_D = 1000$ K, Tanaka [5].

A consistent thermal diffusivity on a hot-pressed sample with zero porosity was measured between 1300 and 2100 K; values for a polycrystalline, fused sample are higher. The change of λ is very nearly proportional to 1/T with $\lambda \approx 0.84$ and 0.69 W·cm^{-1}·K^{-1} at 1300 and 2100 K, respectively, for the fused sample. Assuming the electronic thermal conduction to be temperature-independent gives $\lambda_{el} = 0.45$ W·cm^{-1}·K^{-1} for both samples, Neshpor et al. [6].

References:

[1] Peysson, Y.; Ayache, C.; Salce, B.; Kunii, S.; Kasuya, T. (J. Magn. Magn. Mater. 59 [1986] 33/40).
[2] Peysson, Y.; Ayache, C.; Rossat-Mignod, J.; Kunii, S.; Kasuya, T. (J. Phys. [Paris] 47 [1986] 113/9).
[3] Flachbart, K.; Reiffers, M.; Janos, S.; Paderno, Yu. B.; Lazorenko, V. I.; Konovalova, E. S. (J. Less-Common Metals 88 [1982] L3/L6).
[4] Peysson, Y.; Salce, B.; de Göer, A. M. (Springer Ser. Solid-State Sci. 68 [1986] 281/3).
[5] Tanaka, T. (J. Phys. C 7 [1974] L177/L180).
[6] Neshpor, V. S.; Fridlander, B. A.; Paderno, V. N.; Paderno, Yu. B.; Nerus, M. A. (Teplofiz. Vys. Temp. 14 [1976] 903/6; High Temp. [USSR] 14 [1976] 802/4).

32.1.8.3.4.8 Magnetic Properties

LaB$_6$ is diamagnetic, the susceptibility $\chi = -0.3 \times 10^{-6}$ cm^3/g is temperature invariant between 77 and 500 K, Etourneau et al.

The de Haas-van Alphen effect and other dHvA-like magnetic oscillations were measured in connection with studies of the Fermi surface, see p. 218; for data on the magnetoresistance including the Shubnikov-de Haas effect see p. 233.

Reference:

Etourneau, J.; Mercurio, J.-P.; Naslain, R.; Hagenmuller, P. (J. Solid State Chem. 2 [1970] 332/42).

32.1.8.3.4.9 Electrical Properties

32.1.8.3.4.9.1 Electronic Structure and Related Spectra

Core Levels

The B 1s level was observed at 188 eV below E_F from XPS with AlK_α radiation, Aono et al. [1] referring to [2], at 187.9 ± 0.1 eV, Duc et al. [3].

The level La $3d_{5/2}$ was observed at a binding energy ~833 eV (from a figure) by XPS with AlK_α radiation, Berrada et al. [4]; the levels La $4p_{3/2}$ at 198 eV [2], La $4d_{3/2}$ at 107.0 eV, La $4d_{5/2}$ at 104.0 eV, La 5s at 35.3 eV, La $5p_{1/2}$ at 20.2 eV, and La $5p_{3/2}$ at 18.0 eV below E_F from photoelectron spectra on the (001) face, Aono et al. [1]; cf. "Seltenerdelemente" B 4, 1976, p. 335.

Electron Energy Bands

The band structure of LaB$_6$ has been calculated by various methods: quite recently the full-potential linearized augmented plane wave (FLAPW) method was employed by Kubo, Asano [5], and with inclusion of the spin-orbit interaction, Harima et al. [6]. To evaluate the effect of the spin-orbit interaction, a self-consistent calculation by means of a relativistic Korringa-Kohn-Rostoker (RKKR) method with the muffin-tin approximation was made [5].

The early self-consistent APW calculation with the one-electron potential constructed by the X_α method by Hasegawa, Yanase [7] is still used for reference. The calculations were later improved by using the local-spin-density approximation (LDA) for the exchange term in the one-electron potential and including relativistic effects, Yanase, Hasegawa [8], cf. [9]. A self-consistent calculation using the fast-symmetrized-cluster approach with muffin-tin potentials similar to those of [7] was made by Schell et al. [10]. Using the ASW method (augmented-spherical wave, a descendant of the linear muffin-tin orbital technique) the band structure of LaB$_6$ was calculated by van der Heide et al. [11]. A discrete variational method using non-muffin-tin potentials was used by Walch et al. [12], see also [13, 14]. The LCAO approach was used by Perkins et al. [15]. For previous or more general work, see Section 32.1.5.11.1, p. 57.

The results of the calculation of Hasegawa, Yanase [7] along the ΓM, ΓX, and XM lines are shown in **Fig. 70** (solid lines) together with experimental data from angle-resolved photoemission, Aono et al. [16], see p. 215. The size of the APW spheres in the calculation was set to give La–La contacts along $\langle 100 \rangle$ and to allow B to touch its five nearest neighbors. The 4f states were omitted in the calculation. The results for full Slater exchange, $\alpha = 1.0$, and for $\alpha = 2/3$ do not differ in their essential features, but the Fermi surface for $\alpha = 1.0$ agrees better with experiment. The lowest three bands shown in Fig. 70 arise from La 5p states. The next band, the lowest of the valence bands, has large B 2s components but also has fairly large B 2p admixtures into the Bloch states near the Brillouin zone boundaries. The separation of this band from the higher valence bands is attributed to the crystal field. The higher bands, 2nd to 10th of the valence band, consist mainly of the B 2sp states and the La 6s state and are completely occupied. The total width of these bands is about 0.6 Ry ($\triangleq 8$ eV). The Γ_{12} state and the $\Gamma_{25'}$ state in the conduction band, at 1.101 and 1.300 Ry, respectively, are interpreted as belonging to the La 5d bands, with B 2sp admixtures due to strong hybridization with B 2s and B 2p states of the B$_6$ octahedron, which together form d-like orbitals about the center of the octahedron. This interaction also leads to a large dispersion of the Δ_2, S$_2$, and Z$_3$ branches, along the ΓX, XR, and XM directions, respectively, in the vicinity of the Fermi level. This large wave vector dependence of the d bands introduces light effective masses of conduction electrons, in qualitative agreement with experiments, Hasegawa, Yanase [7], see pp. 221 and 258. In general, the calculation agrees with the experimental result from angle-resolved photo-emission (ARUPS), see Fig. 70, to within ~0.5 eV [16].

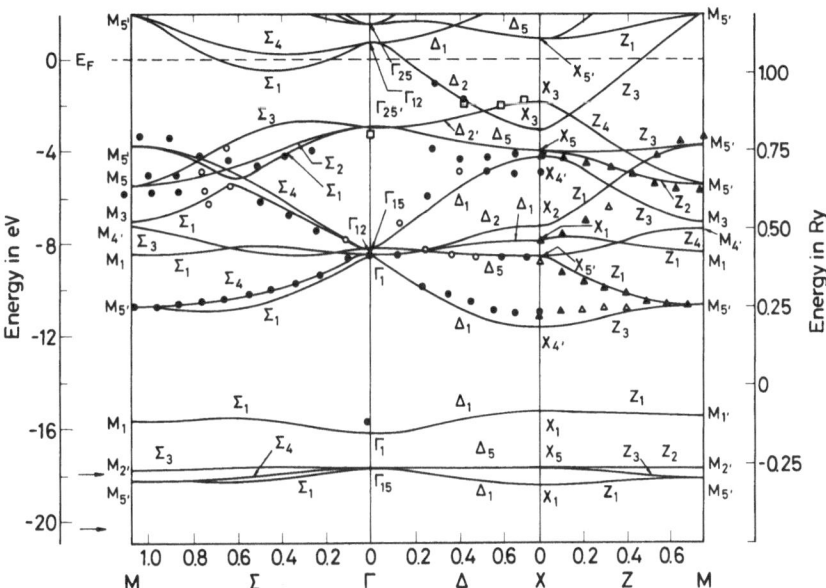

Fig. 70. Energy band structure of LaB$_6$. Solid lines
represent the results of the APW calculation [7]; circles
and triangles correspond to experimental data from
angle-resolved UPS on the (001) face (open circles or
triangles indicate weak peaks or shoulders, squares are
based on a contaminated surface) [16].

The inclusion of the La 4f states into the calculation yields an appreciable band width for the
4f band and a distortion of the conduction band owing to a La 4f–B 2p interaction. Thus at the
Γ point of the Brillouin zone, the conduction band state Γ_{25} (at 1.157 Ry, see Fig. 70), formed
mainly from B 2p states, goes down towards E_F owing to hybridization with a Γ_{25} state of the 4f
bands, Hasegawa, Yanase [8, 9]. The Σ_4 state was below the Σ_1 [9] according to [6]. The band
structure obtained in this calculation agrees well with the angle-resolved photoemission
measurements [16], interpreted using a direct transition model with a free-electron approximation for the final state. The following table gives calculated energies E_{calc} in Ry [8] and
transformed to eV (taking 0 eV at 1.331 Ry) and experimental binding energies E_B from Aono
et al. [16] for selected symmetry points of the occupied valence band (^{u} = upper, ^{l} = lower
state):

state	Γ_1^l	Γ_1^u	Γ_{15}	Γ_{12}	$\Gamma_{25'}$	$M_{5'}^l$	M_5	Ref.
E_{calc} in Ry	0.1543	0.6805	0.6972	0.7081	1.0965	0.5455	0.9205	[8]
E_{calc} in eV	−16.0	−8.9	−8.6	−8.5	−3.2	−10.7	−5.6	[8]
E_B in eV	−15.8		−8.6		−3.3	−10.8	−5.8	[16]

state	$M_{5'}^u$	$X_{4'}^l$	$X_{5'}$	X_1^u	X_5	$X_{4'}^u$	Ref.
E_{calc} in Ry	1.0330	0.4576	0.6985	0.7415	0.9747	1.0141	[8]
E_{calc} in eV	−4.1	−11.9	−8.6	−8.0	−4.8	−4.3	[8]
E_B in eV	−3.4	−11.2	−8.8	−8.0	−5.0	−4.2	[16]

The FLAPW calculation by Harima et al. [6] shows the band, cutting the Fermi level, to have mainly La 5d character near the X point and mainly B 2p character near the Γ point. The corresponding Fermi surface shows nearly-spherical bellies centered at X, see p. 218. The band that cuts the Fermi level on the Σ axis (Σ_4) corresponds to necks connecting the bellies. Details of this result, however are inconsistent with experimental results on the Fermi surface (FS), which may be due to approximations used in the calculation. To account for this, the empty 4f levels (at 1.3 to 1.4 Ry) were tentatively shifted upward by 0.10 Ry, see **Fig. 71**. The Σ_1 and Σ_4 branches nearly degenerate with Σ_1 slightly lower; both of them cut the Fermi level. This corresponds to short necks (Σ_1) connecting the main parts of the FS and electron pockets (Σ_4) within these necks, Harima et al. [6].

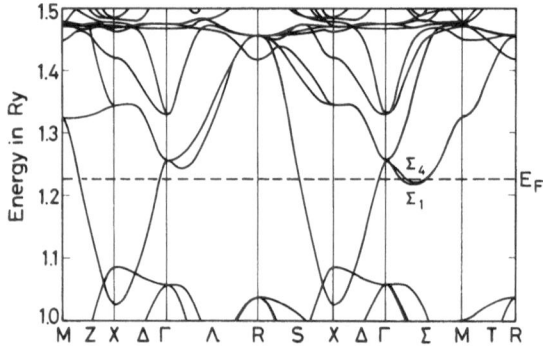

Fig. 71.   FLAPW band structure of LaB$_6$
including 4f states, see text.

The FLAPW calculation by Kubo, Asano [5] gives results roughly comparable to that of [7] (i.e. the topologies of the corresponding FS are similar), for a figure of the band structure see the paper. The effect of inclusion of the spin-orbit interaction (calculation by the RKKR method) is mainly a removing of band degeneracies and band crossings. The effect on the FS topology is not large, however the character of the electron states at E_F is somewhat modified, Kubo, Asano [5].

The results by van der Heide et al. [11] are in very close agreement with those of [7]. The Fermi level (at ∼0.83 Ry, see the figure in the paper) is positioned in a band of complex mixed La d and B p character. A doublet of La d(e_g) symmetry occurs at Γ. The La 4f bands are at about 1 Ry [11]. The results of Schell et al. [10] show general agreement with those of [7]. The results of the calculation by Walch et al. [12], cf. [13, 14] are similar to those of [7] near E_F in that the Fermi level cuts a single band and in the wave vector dependence of this band. Discrepancies occur in the assignment of the symmetry properties of the Bloch states near the cutting of this band with the valence bands [12]. The band structure calculation within an LCAO approach neglects the La d states, Perkins et al. [15] and leads to a Fermi surface quite different from the experimental one. For previous or more general work, see p. 57.

Density of States

The calculated density of states (DOS) is shown in **Fig. 72** from Hasegawa, Yanase [7]; it is in good agreement with photoemission spectra, UPS by Aono et al. [1], angle-resolved UPS by [17], XPS by Chazalviel et al. [18 to 20], and X-ray B K-emission spectra by Okusawa et al. [21], Lyakhovskaya et al. [22]. The density of states is broad with two dense parts at 0.70 and 0.35 Ry

(10 and 5 eV) below E$_F$ [7]. The DOS curve from the calculation by Schell et al. [10] shows general agreement with that of [7]. Also the DOS calculated by Perkins et al. [15], showing peaks at 5, 9, 11, 15, and 20 eV below E$_F$ is in fair agreement.

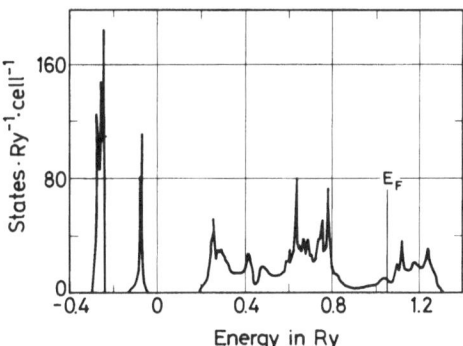

Fig. 72. Theoretical density of electronic states of LaB$_6$.

According to the calculation by Schell et al. [10] the band near −0.3 Ry (~1.3 Ry below E$_F$) is due to La p states and the band at −0.05 Ry corresponds to states of A$_{1g}$ symmetry with intraoctahedron binding character. The minimum in the DOS at 0.85 Ry is due to the separation between states of T$_{2g}$ (below) and T$_{2u}$ symmetry. The lanthanum states lie above ~0.85 Ry and are mainly of d character, but near E$_F$ (at 1.0 Ry) there is considerable admixture of f states to the DOS; this f contribution leads to a peak at 1.14 Ry [10]; preliminary results in [23], see also p. 57.

Photoemission spectra (UPS) show filled valence bands formed from La 5d and 6s and B 2s and 2p valence orbitals to extend from E$_F$ to ~12 eV below E$_F$ with emission maxima at ~5 and ~10 eV, Aono et al. [1], see Fig. 78, p. 225. At the same energies, two weak broad maxima were found in the otherwise almost featureless valence band spectra from XPS by AlK$_\alpha$, Chazalviel et al. [18 to 20]. The valence band (~12 eV wide) is characterized by three main bands located at ~2, ~5, and ~10 eV below E$_F$ with additional subsidiary structures; from XPS by AlK$_\alpha$ and UPS by HeI, II, and NeII radiation. The comparison of XPS and UPS shows that mainly the La 5d-6s and B 2s density of states are mapped by XPS, whereas for hν ≲ 25 eV the excitation probability of B 2p states is comparatively higher. Thus the boron 2p states contribute to the whole valence band with a prominent maximum at ~5.5 eV, whereas the lower edge of the valence band was attributed to B 2s levels, and the states in the vicinity of E$_F$ are mainly from La orbitals, Duc et al. [3]. By comparison with the results of Walch et al. [12] the conduction band was found to be about 3 eV wide and to have predominant La 5d-6s hybrid character with admixture of B 2p states. In addition, strong covalent overlap was demonstrated to take place in the bonding bands, whereas B 2s states are more localized, Duc et al. [3].

For the density of states at the Fermi level N(E$_F$), the following results were obtained in the calculations, in states·Ry^{-1}·spin^{-1}·cell^{-1}: 9.57 for α = 1.0 and 8.78 for α = 2/3 [7], 13.3 [8, 9], 15.32 [5], 10.86 [10], 5.01 [12]. These values have to be compared with experimental data: 13.8 [8] (from the specific heat data of Kasuya et al. [24]), 15.0 and 19.6 [7] (from analogous data of Etourneau et al. [25] see p. 209), 10.28 after correcting for the electron-phonon mass-enhancement factor λ = 0.33 [10] (from the data of [25]). For conclusions on the density of states near E$_F$ from the composition dependence of the work function Φ by Storms [26], see p. 242.

References for 32.1.8.3.4.9.1 on p. 228

Fermi Surface and Magnetic Oscillations

Topology of the Fermi Surface

Experimentally, the first models of the Fermi surface (FS) of LaB_6 have been derived from the de Haas-van Alphen (dHvA) measurements by Arko et al. [13, 14] and Ishizawa et al. [27, 28]; for a preliminary report see [29]. Models for the Fermi surface have also been constructed from energy band structure calculations, Arko et al. [13], Hasegawa, Yanase [7], Perkins et al. [15], and quite recently by Harima et al. [6] and Kubo, Asano [5]. The FS obtained from the APW calculation [7] shows three equivalent electron surfaces centered at the X points, which are closed along the Δ, S, and Z axes, see **Fig. 73** a. Since the Fermi energy cuts the Σ_1 branch, these electron surfaces are connected by necks which intersect the Σ axes. Compared to the experiment, these necks are too large [7]; for new results concerning the necks see below. Later, modified calculations gave roughly the same FS [9] or a similar topology, but with apparently no intersection of an energy band and the Fermi level on the Σ axis [5].

An interpretation comprehending recent theoretical calculations [6], see p. 216, and results from dHvA experiments [30, 31], describes the Fermi surface to consist of nearly spherical bellies at X points with a diameter of $\sim 1.1\,\text{Å}^{-1}$ (comparable to the previous model), which, however, are connected by "short" necks. In addition, small electron pockets are postulated, corresponding to the Σ_4 branch in the recalculated electronic band structure, see p. 216. The pockets have the shape of a flat ellipsoid ("surfboard" [31]) with the longest axis along [110], the shortest along [$\bar{1}$10], and the third along [001], see Fig. 73b. This model was adjusted to the frequencies observed for the ϱ branch in the dHvA measurements and the magnetoacoustic quantum oscillations (oscillatory elastic properties); using frequency data from [28] gave Fermi wave vectors of 0.033, 0.0063, and 0.018 Å^{-1}, respectively, along the three principal axes. The model results in a fairly good agreement between the observed and calculated dHvA frequencies and intensities, Harima et al. [6], see also Kasuya et al. [32].

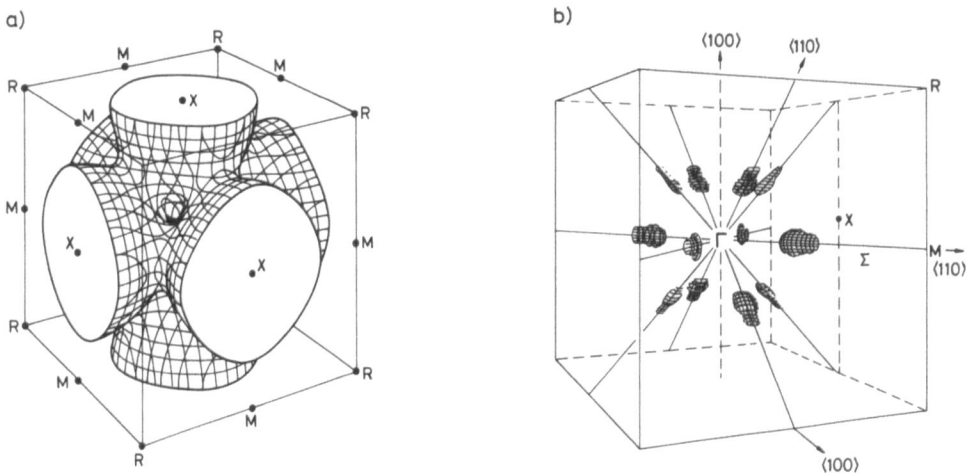

Fig. 73.   Fermi surface of LaB_6.
a) Gross feature, showing the position of the multiply-connected spherical electron bellies [7].
b) Position of the small electron pockets [6].

The assumption that the main Fermi surface is centered at the X points rather than at the M points was confirmed by a study of the momentum distribution of electrons by the two-dimensional angular correlation of positron annihilation radiation, see p. 222, Tanigawa et al. [33].

The volume of the FS corresponds to an electron density of 1.39×10^{22} cm^{-3} which is equivalent to 1 electron per unit cell [27].

De Haas-van Alphen Effect

The de Haas-van Alphen effect has been studied by the field modulation method at 1.5 K. In addition to the dominant dHvA frequencies with the order of 1000 T, frequencies in the order of 100 T and 10^4 T (the latter are weak) were also observed in limited angular ranges, Ishizawa et al. [27].

The most prominent of the observed dHvA frequencies is the α frequency which has three branches α_1, α_2, α_3 in the (010) plane that degenerate into two in the $(1\bar{1}0)$ plane. This α orbit was assigned to an electron orbit around X, see **Fig. 74**. The existence of three branches for the α frequency indicates that the α branches correspond to at least three or six equivalent Fermi surface pieces. The γ orbit is a hole-like orbit around M and thus the observed angular region is limited to a narrow one. The β orbit was assigned to an electron orbit centered at Γ and about the six α surfaces. The ϵ orbit is a hole-like orbit around Γ. Further orbits (at frequencies of ~100 T) are a ζ orbit, a hole-like orbit with a triangular shape surrounded by three α surfaces, and a δ branch, possibly a hole-like orbit almost around the ΓM axis. Additional weak high dHvA frequencies were tentatively attributed to electron orbits through the neck: ν and μ orbit, see Fig. 74, ξ and λ orbit, see the paper [27].

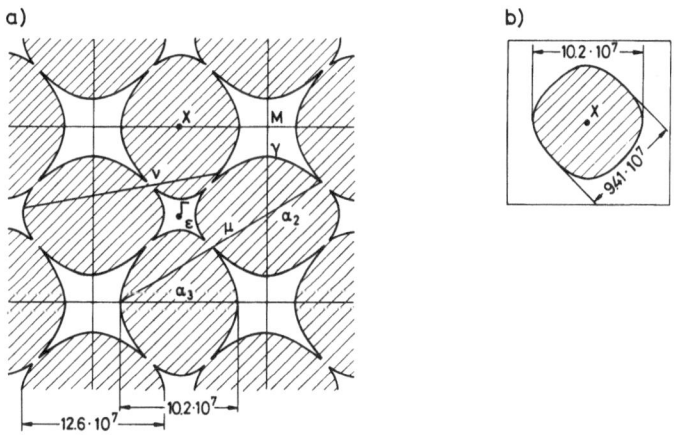

Fig. 74.   Cross sections of the Fermi surface in a) the ΓXM plane
and b) the XRM plane. The dimension is in cm^{-1}.

The somewhat older measurements of Arko et al. [13] by the field modulation technique were made in magnetic fields up to $B = 13.2$ T and at temperatures down to 0.30 K. There is a rough agreement of this study with the data from [27], especially concerning the χ, μ, γ, and ν orbits (α, γ, ϵ, and δ, respectively, in the notation of [27]), however the χ branch was observed over a broader angular range and the high-field data at $B \geq 7.2$ T in this study had not been found by [27]. The frequencies observed by [13] for high fields B along nearly [001] were attributed to magnetic breakdown through intraband transitions by Ishizawa et al. [28] from an estimate of the magnetic breakdown field $B_b \approx 15.0$ T. A low-frequency branch ν (area $\cong 0.015$ to 0.1 a.u.) observed for a range of angles near B∥ [110], was attributed to the extremal orbit

References for 32.1.8.3.4.9.1 on p. 228

around the neck of the Fermi surface [13]; this branch (now labeled δ) was assigned differently to a hole-like orbit almost around the ΓM axis [27]. Already by [13] the appearance of some branches was interpreted by magnetic breakdown. Later measurements in fields between 30 and 35 T by van Deursen et al. [34] gave results fully consistent with [13].

The observed frequencies F for the magnetic field B along high symmetry directions from [27] are given below, together with data from Sera et al. [35], and values taken from figures in Arko et al. [13], and Suzuki et al. [30]; when necessary for comparison, data were transformed from the extremal cross-sectional area S.

B direction	[001]			[110]			[111]			Ref.
orbit	α	γ	ε	$\alpha_{1,2}$	α_3	δ	α	β	ζ	[27]
orbit	χ	μ	γ	$\chi_{2,3}$	χ_1	ν	—	—	—	[13]
F in 10^3 T	7.89	3.22	0.845	9.21	8.30	0.525	8.51	8.02	0.213	[27]
F in 10^3 T	7.89	3.2	0.86	9.35	8.23	0.51	—	—	—	[13]
S in a.u.	0.211	0.086	0.023	0.25[a]	0.22[a]	0.015	—	—	—	[13]
F in 10^3 T	7.9	—	0.84	9.3	—	0.52	8.4	8.2	0.23	[30]
S[a] in Å^{-2}	0.75	—	0.80	0.89	—	0.050	0.81	0.78	0.022	[30]
F in 10^3 T	—	3.1	0.78	—	—	0.51	—	—	—	[35]
F in 10^3 T	7.89[b]	3.24	0.83	—	—	—	—	—	—	[34]

[a] Read from figures in the paper. – [b] The two other branches of the χ orbit were observed at frequencies of 10.03×10^3 T.

In a later study by the torque method from Ishizawa et al. [28], dHvA frequencies in the order of 10 T were found in fields between 0.3 and 1.2 T. The observed frequencies for this branch labeled ϱ at special directions of the magnetic field are listed below together with equivalent data (all read from figures) from oscillatory magnetostriction [35] and oscillatory elastic properties (acoustic dHvA effect) [30].

B direction	[001]	[101]	[111]		Ref.
F in T	5.2	3.8	6.0	4.2	[28]
F in T	5.2	4.2	—	—	[35]
F in T	5.4	—	6.1	4.4	[30]
S in 10^{-4} Å^{-2}	5.2	—	5.8	4.2	[30]

This low-field dHvA oscillation, ϱ branch, was attributed to the neck orbit; a diameter of 0.021 Å^{-1} and a length of 0.044 Å^{-1}, i.e., a slender neck was derived from the angular variation of the frequency [28], an interpretation superseded by more recent results, see p. 218.

Oscillatory Magnetostriction

The effect was measured in magnetic fields up to B = 8.5 T at temperatures between 1.39 and ~8 K. The relative variation of the crystal length could be registered by the three-terminal capacitance method down to a limit of $\Delta l/l = 3 \times 10^{-9}$. With the magnetic field along [001], three dHvA-like frequencies were observed at fields above ~0.3, above ~3, and above ~6 T and were attributed to the neck orbit ϱ, the ε orbit, and the γ orbit, respectively. The results are reasonably consistent with those of the usual dHvA experiments, see the tables above, but inconsistent with the acoustic measurements, Sera et al. [35].

Magnetoacoustic Quantum Oscillations

The acoustic de Haas-van Alphen effect was studied by measuring the sound velocity change of the $(c_{11}-c_{12})/2$ and c_{44} modes at 70 and 30 MHz, respectively, at 4.2 K in magnetic fields up to 8 T, Suzuki et al. [36]. A later improvement of the resolution to 0.1 ppm permitted the observation of the whole of the orbits. The longitudinal c_{11} mode was measured; for results see the lower table on p. 220, Suzuki et al. [30]. This method revealed additional data of the ϱ branch from measurements of the transverse c_{44} mode. These new data, in both amplitude and frequency, confirm the interpretation of the ϱ branches as the small electron pockets and the short thick necks predicted in the band-structure calculation [6], Goto et al. [31], see also Suzuki et al. [37].

The cross-sectional areas of the ϱ branch exhibit individual angular dependences on the different ultrasonic modes, c_{11}, $(c_{11}-c_{12})/2$, or c_{44}; this is attributed to the effect of selection rules for the interaction of the band electrons and the elastic strain. A further point is the anisotropic behavior of the c_{44} mode, which may be related to the anisotropy of the Landau state [30].

A frequency F = 6.3 T was observed on investigation of the c_{44} mode ($k \| \langle 110 \rangle$). The frequency is temperature-independent between 0.5 and 5 K and essentially the same holds for the effective Dingle temperature $\Theta_D \approx 1$ K, Ewert et al. [38]. For earlier, preliminary studies of the magnetoacoustic quantum oscillations see also Suzuki et al. [36] and Lüthi et al. [39].

Oscillations of the magnetoresistance (Shubnikov-de Haas effect) are described on p. 234.

Effective Electron Masses

The cyclotron masses of electrons m* in the various orbits were derived from the temperature dependence of the amplitudes of the magnetic oscillations, and are compared to mass values from band structure calculations, in units of the free electron mass m_0 (ac. = acoustic, mag.str. = magnetostriction, band str. = band structure):

method	T in K	α_3	$\alpha_{1,2}$, α_3	ε	ϱ	Ref.
dHvA	1.5	0.64[1]	0.67[2]	—	—	[27]
dHvA	2.5 to 4.4	—	—	—	0.046[1], 0.050[2], 0.040[4]	[28]
dHvA	≧1.2	0.610[1]	0.650[3]	—	—	[13]
ac. dHvA	0.5 to 5	—	—	—	0.066	[38]
ac. dHvA	4.2	0.59[1]	—	0.44[1]	0.058[1]	[30]
mag.str.	1.39 to 8	—	—	0.49[1]	0.045[1]	[35]
band str.	—	0.44	—	0.23	—	[7]
band str.	—	0.40	—	0.37	—	[13]

[1] B $\|$ [001]. — [2] B $\|$ [111]. — [3] B $\|$ [110]. — [4] B at 45° away from [001] in the (1$\bar{1}$0) plane.

For the χ ($\hat{=}\alpha$) branch m*(χ_1) = 0.6 ± 0.1, m*($\chi_{2,3}$) = 0.7 ± 0.1 was derived from dHvA studies by [34].

Adjusting the parameters of the (ellipsoidal) electron pocket model, see p. 218, to the observed frequency and effective mass values of the ϱ branch, using the relation $E_F = \hbar^2/2m_0 \cdot (a_1 k_1^2 + a_2 k_2^2 + a_3 k_3^2)$, gives $1/a_1 = 0.32$, $1/a_2 = 0.012$, $1/a_3 = 0.091$, $E_F = 1.016$ mRy, $k_1 = 3.3 \times 10^{-2}$ Å^{-1}, $k_2 = 0.63 \times 10^{-2}$ Å^{-1}, and $k_3 = 1.8 \times 10^{-2}$ Å^{-1}; a_i are the inverse of the effective mass ratios and k_i are the wave vectors on the three principal axes along [110], [$\bar{1}$10], and [001], respectively, Harima et al. [6]. Assuming a cylindrical shape for the ϱ-FS, a cyclotron mass about the cylindrical axis of $\sim$0.03 m_0 was deduced [28]. For optical masses see p. 258.

References for 32.1.8.3.4.9.1 on p. 228

Electron Momentum Distribution

The electron momentum distribution can be evaluated from the two-dimensional angular correlation of positron annihilation radiation. First measurements were made by Tanigawa et al. [33]. Later measurements of Tanigawa et al. [40] are so far only preliminarily published in Kubo, Asano [5]. There is good agreement between the three-dimensional Lock-Crisp-West (LCW) folded momentum densities (3D LCW FMD's) reconstructed from the two-dimensional angular correlation of positron annihilation radiation and those from the FLAPW calculation. The basic structures of the 3D LCW FMD's are predominantly formed by contributions from the conduction bands; the Fermi surface topology plays an important role. However, contributions from the valence band cannot be neglected. For explanations of discrepancies between theory and experiment, the effects of spin-orbit interaction, correlation between f electrons, electron-positron enhancement, and electron-electron correlation were discussed, Kubo, Asano [5]; for plots of the 3D LCW FMD's and further details see the paper.

Surface Electron States

Surface electron states on LaB_6 (001), Aono et al. [17, 41], LaB_6 (110) and (111), Nishitani et al. [42] were studied on clean surfaces using angle-resolved UV photoelectron spectroscopy (ARUPS) with unpolarized light. The surfaces were checked for impurities: absence of O1s and C1s peaks in XPS and of oxygen derived O2p peaks in UPS; a LEED pattern was clearly observed [17, 42].

The surface states were interpreted as due mainly to the boron 2p orbitals that run along $\langle 100 \rangle$ directions, that is, by the dangling bonds left by disrupting the bonds between the boron octahedra, Nishitani et al. [42].

(001) Surface

Spectra from the (001) surface were recorded with incident beam (at $\alpha = 20°$ off the surface normal) and registered direction of emitted electrons both in the (010) plane or the ($1\bar{1}0$) plane; the angle Θ is between the surface normal and the registering direction. Whereas most of the peaks in the spectra were due to bulk energy bands, a sharp peak at ~ 2 eV below E_F was attributed to a surface state because: (1) this peak disappeared at oxygen exposure; (2) it was fairly intense, even at large Θ; (3) no bulk peak was expected in this range. This surface state was observed independently of the energy of the exciting light, NeI and NeII, HeI and HeII, $h\nu = 16.8$ to 40.8 eV. Electron angular distribution curves, obtained after correcting for the direction of the incident light, give the maximum of emitted electrons at $\Theta = 10°$ in the ($1\bar{1}0$) plane; a relative minimum at $\Theta = 0°$ in the ($1\bar{1}0$) plane is the intensity maximum in the (010) plane. The two-dimensional energy band formed by the surface state was constructed from the Θ dependence of the energy position of the surface state peaks, assuming that the electron momentum parallel to the surface is conserved before and after photoemission, see **Fig. 75**. The figure shows in addition a second energy band, because the observed surface state peak really consists of two peaks; in most of the spectra one of the two peaks is predominant, Aono et al. [17]. The upper surface state with the minimum at $\bar{\Gamma}$ and a dispersion of the order of 0.45 eV was also found in a more recent measurement using ArI ($h\nu = 11.83$ eV). Electron intensity variation at variation of the angle of incidence α, for example zero intensity in the normal direction at incidence angle $\alpha = 0°$ and large intensity at $\alpha = 60°$, indicates Δ_1 symmetry at $\bar{\Gamma}$ ($k_\parallel = 0$) and even parity of the surface state with respect to both the (010) and the ($1\bar{1}0$) planes, Nishitani et al. [42]. A surface state at 2.2 to 2.8 eV below E_F and a surface state at just below E_F that overlaps an occupied conduction band (thus surface resonance) were observed by UPS excited at $hv = 28$ eV on (001), Aono et al. [16].

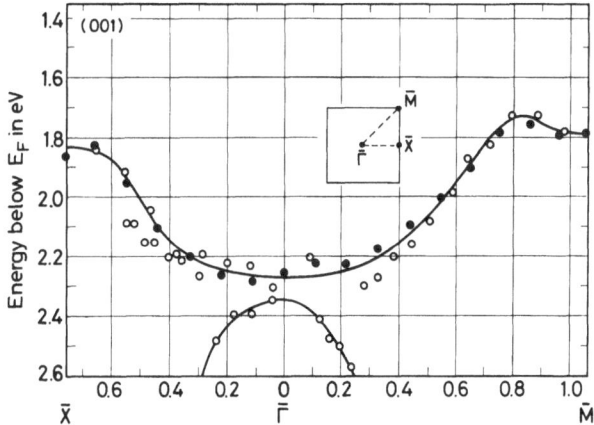

Fig. 75.   Brillouin zone and surface states of the LaB₆ (001) surface. Experimental points shown by open circles have been reduced from higher Brillouin zones to the first zone.

Based on the model with La atoms as the outer layer of the (001) surface of LaB₆, the upper state in Fig. 75 was supposed to arise mainly from dangling-bond electrons of the outermost boron atoms (p_z orbitals) [17, 42]. However, the assignment of the observed surface state to dangling bonds (Shockley surface states) was disputed by Tomášek, Pick [43, 44]. They expect Shockley surface states at lower energies (>4 eV below E_F) based on projections of the bulk energy bands onto the (001) surface Brillouin zone. Instead, they attribute the states just below the Fermi level to so-called "hetero atomic surface states", derived from La states, which are split from the bottom of the La bands by the surface formation. The lower surface states, ~2 eV below E_F, are attributed to Tamm states stemming from B atoms [43]. Tamm states arise in metallic or ionic systems if the energy of a localized state is shifted by potential difference between bulk and surface, Tomášek, Koutecky [45].

(110) Surface

The analogous studies for the (110) surfaces also using Ar I light ($hv = 11.83$ eV) and varying the angle of the incident light in the range $-60° < \alpha < 85°$ resulted in a surface state at ~1.8 eV below E_F having even parity with respect to the ($1\bar{1}0$) plane and odd parity with respect to the (001) plane, and a surface state at ~3.0 eV below E_F having even parity with respect to both the ($1\bar{1}0$) and (001) planes, see **Fig. 76**, p. 224. The level at ~3.0 eV was attributed to the surface state into which all surface p orbitals from the dangling bonds contribute with the same sign. The state at ~1.8 eV is constituted from p orbitals, the signs of which change from one octahedron to the next along [001]. The surface states on (110) are associated with the occurrence of the c(2 × 2) surface structure, cf. p. 262. The effect on the energy levels of the surface states from the coupling of the boron dangling bonds with the surface La atoms is not clear at present (no calculation of the surface electronic structure is available), but is regarded as rather small because of the small band dispersion of about 0.2 eV [42].

References for 32.1.8.3.4.9.1 on p. 228

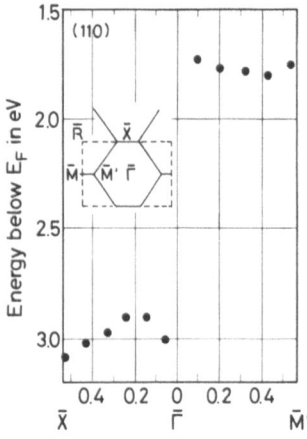

Fig. 76. Surface states on the (110) surface. The inset shows the surface Brillouin zone of the c(2 × 2) structure (solid line) and the unreconstructed surface (dashed line).

(111) Surface

The energy band dispersion of the surface states on (111) is shown in **Fig. 77**. In addition to that around 2 eV, a surface state at ~1.5 eV below E_F is observed in the (1$\bar{1}$0) plane, giving high electron emission at high angle of incidence. The state at ~2 eV has Λ_1 symmetry at $k_{\parallel} = 0$ and was supposed to correspond to a state into which the p orbitals of the three dangling bonds per octahedron contribute with the same sign. The dispersion is comparatively large, 0.9 eV [42].

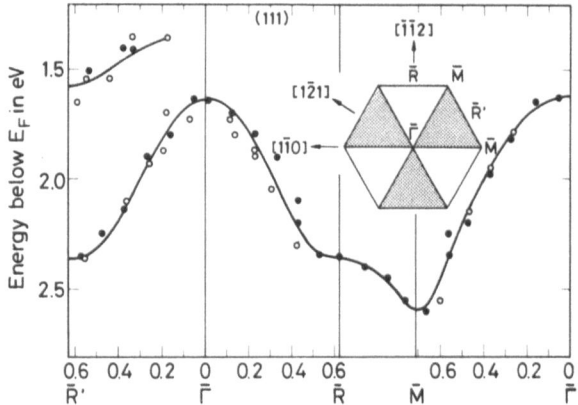

Fig. 77. Energy dispersion of the surface states on the (111) surface of LaB_6. The inset shows the surface Brillouin zone.

(210) Surface

A surface state at ~2.5 eV below E_F was observed on (210) by UPS, $h\nu = 21.2$ eV, with the incident light in the ($\bar{1}$20) plane at $\alpha = 25°$ and normal emission, Oshima et al. [46].

Surface States at Oxygen Exposure

1 L (Langmuir) $= 10^{-6}$ Torr·s.

The LaB_6 surface states are quenched at exposure to oxygen, Nishitani et al. [47, 48]. The surface state on (100) at ~2 eV below E_F disappeared at an oxygen exposure of ~1.4 L [47]; earlier 7 L was given by [17]. New peaks appear in UPS, a first peak at ~6.6 eV below E_F, which is overcome at higher exposures by a peak at ~6.0 eV below E$_F$ [47]. The surface states on (110) disappeared at ~0.4 L of oxygen exposure, that on (111) disappeared at ~2 L and similar to the (100) face, oxygen-derived states evolved [48].

Photoelectron Spectra

Angle-Integrated Spectra

Photoemission spectra for the (001) face of LaB$_6$ excited by photon energies hν ranging from 97.5 to 118.0 eV are shown in **Fig. 78**. The spectra for hν = 97.5 and 100 eV are very similar. Spectra at higher exciting energy show characteristic differences, especially in the extra peaks (A, B, C in Fig. 78), which shift to lower initial energies for higher exciting hν, Aono et al. [1].

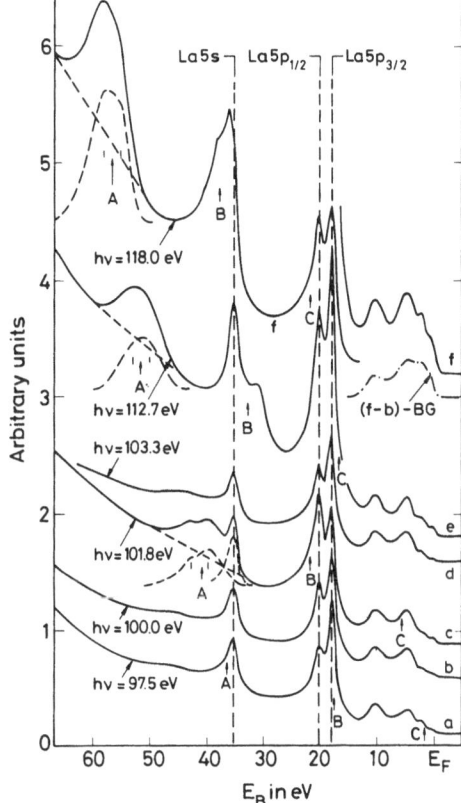

Fig. 78. Photoemission spectra of LaB$_6$(001) excited by various photon energies, see text.

The yield of secondary electrons with a kinetic energy of 11 eV as a function of the exciting photon energies hν from 90 to 140 eV shows two weak, narrow maxima at 97.5 and 101.8 eV and a broad, strong feature from about 110 to 140 eV with a maximum at 118.5 eV. These are due to La 4d-4f resonant excitations from the 4d^{10} ground state $^1S_0$ to 4d^9 4f^1 excited states $^3P_1$, $^3D_1$, and $^1P_1$, respectively. The observed intensity ratios agree reasonably well with the dipole transition selection rules. The photoemission spectra of Fig. 78 for hν = 97.5, 101.8, and 118 eV, thus, correspond to the maxima of the photoemission partial yield. The strong enhancement of the La 5s and 5p core level peaks in the spectra at hν = 118 and 112.7 eV can be interpreted by direct recombination decay processes of the excited configuration $4d^9 4f^1$, accompanied by emission from the La 5s and 5p core levels. A weaker enhancement is observed for the spectra at $h\nu = 101.8$ and 97.5 eV. The difference spectrum between spectra at $h\nu = 118$ and 100 eV with a linear background subtracted ($\triangleq(f-b)-BG$) reflects the stronger enhancement of the upper part of the conduction and valence band compared to lower part because of the contribution of the La 5d and 6s orbitals. Three extra peaks with fine structure appear in the spectra except for spectra at 100 and 103.3 eV. The peaks A, B, C in Fig. 78, p. 225, (A consists of two peaks, separated by 3 eV, due to spin-orbit splitting) are shifted from one spectrum to another, in such a way that their kinetic energies are nearly independent of $h\nu$ (the arrows indicate positions of constant kinetic energy: 61, 80, and 96 eV, respectively). They are assigned to Auger deexcitations of the excited states $4d^9 4f^1$: $N_{4,5}O_1V$, $N_{4,5}O_{2,3}V$, and $N_{4,5}VV$ (for A, B, and C, respectively), V representing a La-derived valence orbital. These transitions are regarded as not ordinary Auger transitions because of a strong exchange interaction between the $4f^1$ electron and the 4d hole in the $4d^9 4f^1$ excited state. The exchange interaction results in the creation of a 4d hole in the form of $4d^{10} 4f^1$ at the $^3D_1$ and $^3P_1$ resonant photon energies $h\nu = 101.8$ and 97.5 eV already below the one-electron 4d threshold of 104 eV (see the La 4d core level energies, p. 214). At other, off-resonant photon energies below 104.0 eV, there is no mechanism to excite 4d holes. A shift by ~ 0.5 eV of the kinetic energy of the lowest extra peak in the spectrum at 101.8 eV compared to those at 112.7 and 118 eV is interpreted as indicating the Auger deexcitation channel to be the main decay channel for the $^3D_1$ excited state at $h\nu$ below the one-electron core-level threshold. For the $^1P_1$ excited state at $h\nu$ above the threshold, the Auger deexcitation and direct recombination decay channels are comparable in strength, Aono et al. [1].

For the giant 4d photoabsorption of La, both the total photoabsorption spectrum and the $N_{4,5}$-derived Auger emission intensity spectrum increase significantly above the photon energy $h\nu \cong 112$ eV, with spectral peaks at $h\nu = 118$ and 119 eV, respectively. However, the predominant 4d photoemission partial cross section shows a delayed onset of ~ 4 eV, with a peak at $h\nu = 121$ eV, whereas the 5s, 5p, and 5d partial cross sections all show a strong resonant enhancement at lower energies, with spectral peaks at $h\nu = 116.6$ eV. In analogy to results from a many-body calculation of the photoabsorption for Ce, the explanation for this observation is that electrons excited from the 4d subshell into the continuum on the low-energy side of the resonance are not directly emitted as photoelectrons but are localized owing to the $4d^9 \varepsilon f\ ^1P$ resonantly localized continua; consequently, direct recombination decay of the localized states enhances the 5s, 5p, and 5d photoemission. In contrast, at higher energies autoionization/direct emission into the continuum occurs and a predominant 4d partial cross section is observed, Aono et al. [49], for further details see the paper.

Photoelectric yield spectra in the vicinity of the B K absorption edge, excited by the bremsstrahlung spectrum of a tungsten anode, showed two peaks at 194.2 and 200.5 eV with shoulders at 191.5 and 198.1 eV, respectively [22].

The comparison of XPS and UPS leads to some conclusions on the contributions of the B 2s, B 2p, and La 5d states to the density of states, see p. 217, Duc et al. [3].

Angle-Resolved Spectra

Angle-resolved photoelectron spectra have been recorded of various surfaces of single crystalline LaB_6 in order to derive the surface structure, both topographic and electronic, and to permit an understanding of the low work function of LaB_6 and the influence of chemisorbed oxygen. Angle-resolved UPS was interpreted for bulk electron energy band dispersion, Aono et al. [16], see p. 215, for results on surface states, see p. 222. Studies on (001) by XPS (MgKα)

are reported by Aono et al. [2,50]; all three surfaces (001), (110), and (111) are discussed by Nishitani et al. [42, 51]; the (210) surface by Oshima et al. [46].

An observed structure in the emission intensity from La 4d and B 1s for variation of the polar angle ϑ (from the surface normal), ϑ up to 85°, were tentatively attributed to diffraction effects. The ratio of the La 4d peak height to the B 1s peak height for the (001) and (110) surfaces tends to increase strongly for $\vartheta > 60°$ but on (110) not as rapidly as for the (001) face. This was found for analyzer positions in both the (1$\bar{1}$0) and the {001} planes [51]. The ratio of the La 4d and B 1s emission was nearly independent of ϑ for the (111) face for all ϑ [51] and for the (210) face for $\vartheta < 60°$; it increased strongly for $\vartheta > 70°$, independent of analyzer positions in the (001) or (1$\bar{2}$0) planes [46]. Within a simple model the results on all faces were interpreted to indicate La as outermost layer on the surfaces with surface relaxation of the La atoms [51], see p. 262.

X-Ray Emission and Absorption Spectra

Boron K Spectra

The B K emission and absorption spectra of Okusawa et al. [21] are reported on p. 85. The previous B K emission spectra of Lyakhovskaya et al. [22] are consistent with the data of [21]: The spectra are characterized by a strong peak at 184.5 eV starting at 188.5 eV with shoulders at 187.4, 183.4, and 179.0 eV and a weak peak at 173.1 eV; the resolution was 0.2 eV [22]. The calculated density of states, see pp. 216/7, is consistent with the B K emission of [21], Schell et al. [10].

Lanthanum L Spectra

In La L$_{III}$ absorption spectra of LaB$_6$, no significant shift of the "white line" (i.e., principal absorption maximum dominated by $2p \rightarrow 5d$ transitions) at $h\nu = 5503.2 \pm 0.1$ eV relative to La$_2$O$_3$ was observed. A shoulder (at 5507 eV) was shifted to lower energies by 2 eV (see "Seltenerdelemente" B 4, 1976, p. 407, Fig. 47), Troneva [52]. A shift of this maximum by 0.8 eV to lower energy relative to that of La$_2$O$_3$ was measured by Vainshtein et al. [53]. The L$_{II}$ absorption spectra were fully analogous to the L$_{III}$ absorption spectra. The trivalent state of La was confirmed, Vainshtein et al. [54]. Thus, the X-ray absorption structure of LaB$_6$ above the L$_{III}$ edge, which is due to multiple scattering of the photoelectron on the neigboring atomic shells, is like that of an integral valent compound characterized by a single peak (at ~ 25 eV above the white line), Beaurepaire et al. [55]. Additional data see on p. 86.

Lanthanum N$_{4,5}$ Spectra

La N$_{4,5}$ emission spectra were excited in LaB$_6$ by electrons with energies from 1 to 3 keV. The samples were prepared in situ by electron beam evaporation onto a Cu target. The spectra show a broad and intense emission peak, around 117.5 eV, which is above the N$_{4,5}$ threshold energy. The intensity of this peak decreases with increasing energy of the exciting electrons. Below the N$_{4,5}$ threshold there are sharp peaks around 80 eV (intense), 90 eV (medium), and 102 eV (weak). The peak at 80 eV is attributed to an electron transition $5p_{1/2, 3/2}$ to $4d_{3/2, 5/2}$. The peak at 90 eV is explained by a second-order diffraction of the boron K-emission line. The other two peaks coincide with peaks in the absorption spectra; that at 102 eV is attributed to a radiative decay from the ^3D state of the excited configuration $4d^9 4f^1$, and that at 117.5 eV is analogously attributed to the ^1P state of the $4d^9 4f^1$ configuration. A very weak peak observed at ~ 97 eV was attributed to valence band emission. From the expected emissions corresponding to transitions 4d to $4p_{3/2}$ (92.5 eV) and 4d to $4p_{1/2}$ (106.1 eV), the first overlaps with the peak at 90 eV and the other is recognized as a shoulder around 106 eV in the 1 keV excited spectrum, Ichikawa et al. [56].

References for 32.1.8.3.4.9.1 on p. 228

15*

La $N_{4,5}$ absorption spectra exhibit sharp absorption at ~97.4 and 101.8 eV and a broad (~40 eV) absorption starting at about 110 eV with a maximum at 118 eV. The absorptions were assigned to the same transitions (multiplet states of the $4d^9 4f^1$ configuration in the final state) as the secondary electron yield spectra, Aita et al. [57], see p. 225.

Auger Spectra

Auger electron spectra (AES) of LaB_6 were used to investigate any variation of the surface composition with crystallographic orientation, temperature, method of cleaning or the presence of impurities, see for example, Goldstein, Szostak [58], Swanson et al. [59], Berrada et al. [4], Oshima, Kawai [60]. The observed AES spectra showed peaks at 59, 75, 95 eV attributed to La [4]; the electron energy of the $La(N_{4,5}O_{2,3}O_{2,3})$ transition was 78 eV, the $La(M_5N_{4,5}N_{4,5})$ transition was at 625 eV, the $B(KL_1L_1)$ transition at 179 eV, Chambers, Swanson [61], Swanson et al. [59]. For the heavily oxidized LaB_6 surface the La(MNN) emission changed to a doublet characteristic of the surface oxide, the B(KLL) emission shifted to lower energy, Goldstein, Szostak [58], from 179 to 171 eV, Oshima, Kawai [60], see also [4]. For La Auger transitions involving valence band states, $N_{4,5}O_1V$ at 61 eV, $N_{4,5}O_{2,3}V$ at 80 eV, $N_{4,5}VV$ at 96 eV, see pp. 225/6.

AES data at room temperature and at 1600 K on various single-crystal faces of crystals with different chemical composition $LaB_{5.86}$ and $LaB_{5.74}$ (the densities, however, indicating $LaB_{6.0}$), taken after initial thermal cleaning and long-time annealing (~1800 K), indicated that the La-NOO and B-KLL Auger peak heights varied with temperature whereas the La-MNN peak was mostly unchanged, suggesting that only an outer layer participates in any reversible temperature dependent compositional change. Above about 1200 K the temperature dependence of the AES peak ratios was negligible. For annealed crystals, the B/La ratios increased in the sequence (100) < (110) < (111), Swanson et al. [59].

Angle-resolved AES on (001) showed an increase of the ratio of the La (MNN) to B (KLL) Auger intensities for shallow take-off angles, $\vartheta \gtrsim 60°$ that agree with angle-resolved photoemission data (see p. 226), indicating La to be the outermost atom layer, Chambers, Swanson [61]. On the (110) plane there was a similar but very slight increase of this ratio; a maximum was observed for electrons emitted along the $\langle 111 \rangle$ direction [61]; cf. p. 262.

References:

[1] Aono, M.; Chiang, T. C.; Knapp, J. A.; Tanaka, T.; Eastman, D. E. (Phys. Rev. [3] B **21** [1980] 2661/5).
[2] Aono, M.; Oshima, C.; Tanaka, T.; Bannai, E.; Kawai, S. (J. Appl. Phys. **49** [1978] 2761/4).
[3] Duc, T. M.; Hollinger, G.; Jugnet, Y.; Mercurio, J.-P.; Berrada, A.; et al. (Conf. Ser. Inst. Phys. [London] No. 37 [1977/78] 134/40; C.A. **89** [1978] No. 82656).
[4] Berrada, A.; Mercurio, J.-P.; Etourneau, J.; Alexandre, F.; Theeten, J. B.; Duc, T. M. (Surf. Sci. **72** [1978] 177/88).
[5] Kubo, Y.; Asano, S. (Phys. Rev. [3] B **39** [1989] 8822/31).
[6] Harima, H.; Sakai, S.; Kasuya, T.; Yanase, A. (Solid State Commun. **66** [1988] 603/7).
[7] Hasegawa, A.; Yanase, A. (J. Phys. F **7** [1977] 1245/60).
[8] Yanase, A.; Hasegawa, A. (in: Moriya, T., Electron Correlation and Magnetism in Narrow Band Systems, Springer, Berlin – New York 1981, pp. 230/6).
[9] Hasegawa, A.; Yanase, A. (J. Magn. Magn. Mater. **15/18** [1980] 887/8).
[10] Schell, G.; Winter, H.; Rietschel, H.; Gompf, F. (Phys. Rev. [3] B **25** [1982] 1589/99).

[11] van der Heide, P. A. M.; ten Cate, H. W.; ten Dam, L. M.; de Groot, R. A.; de Vroomen, A. R. (J. Phys. F **16** [1986] 1617/23).
[12] Walch, P. F.; Ellis, D. E.; Mueller, F. M. (Phys. Rev. [3] B **15** [1977] 1859/66).

[13] Arko, A. J.; Crabtree, G.; Karim, D.; Mueller, F. M.; Windmiller, L. R.; et al. (Phys. Rev. [3] B **13** [1976] 5240/7).

[14] Arko, A. J.; Crabtee, G.; Ketterson, J. B.; Mueller, F. M.; Walch, P. F.; Windmiller, L. R.; Fisk, Z.; et al. (Intern. Quantum Chem. Symp. No. 9 [1975] 569/78).

[15] Perkins, P. G.; Armstrong, D. R.; Breeze, H. (J. Phys. C **8** [1975] 3558/70).

[16] Aono, M.; Chiang, T. C.; Knapp, J. A.; Tanaka, T.; Eastman, D. E. (Solid State Commun. **32** [1979] 271/4).

[17] Aono, M.; Tanaka, T.; Bannai, E.; Oshima, C.; Kawai, S. (Phys. Rev. [3] B **16** [1977] 3489/92).

[18] Chazalviel, J. N.; Campagna, M.; Wertheim, G. K.; Schmidt, P. H. (Phys. Rev. [3] B **14** [1976] 4586/92).

[19] Chazalviel, J. N.; Campagna, M.; Wertheim, G. K.; Schmidt, P. H.; Longinotti, L. D. (Physica B + C **86/88** [1977] 237/8).

[20] Chazalviel, J. N.; Campagna, M.; Wertheim, G. K.; Schmidt, P. H. (Solid State Commun. **19** [1976] 725/8).

[21] Okusawa, M; Ichikawa, K.; Matsumoto, T.; Tsutsumi, K. (J. Phys. Soc. Japan **51** [1982] 1921/6).

[22] Lyakhovskaya, I. I.; Zimkina, T. M.; Fomichev, V. A. (Fiz. Tverd. Tela [Leningrad] **12** [1970] 174/80; Soviet Phys.-Solid State **12** [1970] 138/43).

[23] Schell, G.; Winter, H.; Rietschel, H. (Supercond. d-f-Band Metals Proc. Conf., San Diego, Calif., 1979 [1980], pp. 465/71).

[24] Kasuya, T.; Takegahara, K.; Fujita, T.; Tanaka, T.; Bannai, E. (J. Phys. Colloq. [Paris] **40** [1979] C5-308/C5-313).

[25] Etourneau, J.; Mercurio, J.-P.; Naslain, R.; Hagenmuller, P. (J. Solid State Chem. **2** [1970] 332/42).

[26] Storms, E. K. (J. Appl. Phys. **52** [1981] 2961/5).

[27] Ishizawa, Y.; Tanaka, T.; Bannai, E.; Kawai, S. (J. Phys. Soc. Japan **42** [1977] 112/8).

[28] Ishizawa, Y.; Nozaki, H.; Tanaka, T.; Nakajima, T. (J. Phys. Soc. Japan **48** [1980] 1439/42).

[29] Ishizawa, Y.; Tanaka, T.; Kawai, S.; Bannai, E. (Proc. 14th Intern. Conf. Low Temp. Phys., Otaniemi, Finl., 1975, Vol. 3, pp. 137/40).

[30] Suzuki, T.; Goto, T.; Sakatsume, S.; Tamaki, A.; Kunii, S.; Kasuya, T.; Fujimura, T. (Japan. J. Appl. Phys. **26** Suppl. 3 [1987] 511/2).

[31] Goto, T.; Suzuki, T.; Ohe, Y.; Fujimura, T.; Tamaki, A. (J. Magn. Magn. Mater. **76/77** [1988] 305/11).

[32] Kasuya, T.; Sakai, O.; Harima, H.; Ikeda, M. (J. Magn. Magn. Mater. **76/77** [1988] 46/52).

[33] Tanigawa, S.; Terakado, S.; Iwase, Y.; Suzuki, R.; Komatsubara, T.; Onuki, Y. (J. Magn. Magn. Mater. **52** [1985] 313/6).

[34] van Deursen, A. P. J.; Fisk, Z.; de Vroomen, A. R. (Solid State Commun. **44** [1982] 609/12).

[35] Sera, M.; Kunii, S.; Kasuya, T. (J. Phys. Soc. Japan **57** [1988] 13/5).

[36] Suzuki, T.; Goto, T.; Fujimura, T.; Kunii, S:; Suzuki, T.; Kasuya, T. (J. Magn. Magn. Mater. **52** [1985] 261/3).

[37] Suzuki, T.; Goto, T.; Ohe, Y.; Fujimura, T.; Kunii, S. (J. Phys. Colloq. [Paris] **49** [1988] C8-799/C8-800).

[38] Ewert, S.; Guo, S.; Lemmens, P.; Lenz, D.; Sander, W.; Thalmeier, P.; Winzer, K. (Japan. J. Appl. Phys. **26** Suppl. 3 [1987] 537/8).

[39] Lüthi, B.; Blumenröder, S.; Hillebrands, B.; Zirngiebl, E.; Güntherodt, G.; Winzer, K. (Z. Physik B **58** [1984] 31/8).

[40] Tanigawa, S.; Kurihara, T.; Osawa, M.; Komatsubara, T.; Onuki, Y. (Proc. 8th Intern. Conf. Positron Annihilation, Belgium 1988, unpublished, from [5]).

[41] Aono, M; Tanaka, T.; Bannai, E.; Kawai, S. (Appl. Phys. Letters 31 [1977] 323/5).

[42] Nishitani, R.; Aono, M.; Tanaka, T.; Kawai, S.; Iwasaki, H.; Oshima, C.; Nakamura, S. (Surf. Sci. 95 [1980] 341/58).

[43] Tomášek, M.; Pick, S. (Surf. Sci. 100 [1980] L454/L456).

[44] Tomášek, M.; Pick, S. (Czech. J. Phys. B 29 [1979] 557/61).

[45] Tomášek, M.; Koutecky, J. (Intern. J. Quantum Chem. 3 [1969] 249/67).

[46] Oshima, C.; Aono, M.; Tanaka, T.; Nishitani, R.; Kawai, S. (J. Appl. Phys. 51 [1980] 997/1000).

[47] Nishitani, R.; Kawai, S.; Iwasaki, H.; Nakamura, S.; Aono, M.; Tanaka, T. (Surf. Sci. 92 [1980] 191/200).

[48] Nishitani, R.; Oshima, C.; Aono, M.; Tanaka, T.; Kawai, S.; Iwasaki, H.; Nakamura, S. (Surf. Sci. 115 [1982] 48/60).

[49] Aono, M.; Chiang, T. C.; Himpsel, F. J.; Eastman, D. E. (Solid State Commun. 37 [1981] 471/4).

[50] Aono, M.; Nishitani, R.; Oshima, C.; Tanaka, T.; Bannai, E.; Kawai, S. (Surf. Sci. 86 [1979] 631/7).

[51] Nishitani, R.; Aono, M.; Tanaka, T.; Oshima, C.; Kawai, S.; Iwasaki, H.; Nakamura, S. (Surf. Sci. 93 [1980] 535/49).

[52] Troneva, N. V. (Izv. Akad. Nauk SSSR Ser. Fiz. 27 [1963] 403/8; Bull. Acad. Sci. USSR Phys. Ser. 27 [1963] 410/4).

[53] Vainshtein, E. E.; Blokhin, S. M.; Paderno, Yu. B. (Fiz. Metal. Metalloved. 18 [1964] 450/1; Phys. Metals Metallog. [USSR] 18 No. 3 [1964] 132/3).

[54] Vainshtein, E. E.; Staryi, I. B.; Blokhin, S. M.; Paderno, Yu. B. (Zh. Strukt. Khim. 3 [1962] 200/7; J. Struct. Chem. [USSR] 3 [1962] 185/91).

[55] Beaurepaire, E.; Kappler, J. P.; Krill, G. (Solid State Commun. 57 [1986] 145/9).

[56] Ichikawa, K.; Nisawa, A.; Tsutsumi, K. (Phys. Rev. [3] B 34 [1986] 6690/4).

[57] Aita, O.; Ichikawa, K.; Kamada, M.; Okusawa, M.; Nakamura, H.; Tsutsumi, K. (J. Phys. Soc. Japan 56 [1987] 649/54).

[58] Goldstein, B.; Szostak, D. J. (Surf. Sci. 74 [1978] 461/78).

[59] Swanson, L. W.; Gesley, M. A.; Davis, P. R. (Surf. Sci. 107 [1981] 263/89).

[60] Oshima, C.; Kawai, S. (Appl. Phys. Letters 23 [1973] 215/6).

[61] Chambers, S. A.; Swanson, L. W. (Surf. Sci. 131 [1983] 385/402).

32.1.8.3.4.9.2 Electrical Transport Properties

LaB$_6$ behaves like a monovalent metallic conductor, Mori et al. [1], with a carrier concentration in the order of 1 electron per La atom according to the Hall constant, see p. 234, and the volume of the Fermi surface, see p. 218.

Electrical Conductivity

At Room Temperature

At 293 K polycrystalline sintered samples of LaB$_6$ had an electrical conductivity $\sigma = 67 \times 10^3$ $\Omega^{-1} \cdot cm^{-1}$, Aivazov et al. [3], equivalent to a resistivity $\varrho = 15\ \mu\Omega \cdot cm$, Bondarenko et al. [4]. Other measured values for sintered samples: $\varrho = 12.0 \pm 0.4\ \mu\Omega \cdot cm$, Kondrashev et al. [5], $\varrho = 13\ \mu\Omega \cdot cm$, Rabenau et al. [6], $\varrho = 17.4\ \mu\Omega \cdot cm$, Samsonov, Grodshtein [7], and $\varrho \approx 10\ \mu\Omega \cdot cm$,

Berrada et al. [8]. Addition of small amounts of Ni as sintering activator during hot pressing increases ϱ, Kondrashev et al. [5]. The resistivity of a zone-refined single crystal was 8.90 μΩ·cm at 300 K, Tanaka et al. [2].

Temperature Dependence

Below Room Temperature. Near room temperature the resistivity decreases with temperature in proportion to $T^{1.5}$ and below about 70 K at increased rate, Tanaka et al. [2], see **Fig. 79**, similarly stated by Peysson et al. [9], Flachbart et al. [10], see also Etourneau et al. [11], Shulishova, Shcherbak [12]. Below about 20 K, the residual resistivity is reached that may be as low as 2×10^{-8} Ω·cm, equivalent to a residual resistance ratio (RRR) of 450 [2]. Other values for the residual resistivity of single crystals are $\varrho_{res} = (7.5 \pm 0.2) \times 10^{-8}$ Ω·cm [9] and 5×10^{-8} Ω·cm ≙ RRR = 175 [10], RRR = 750, Ishizawa et al. [13], or RRR = 60 and 20 for the center and edge of a zone-melted crystal [11]; in all these cases, the resistivity remained constant down to 1.5 K or lower. In contrast, a slight increase between 0.6 and 1.1% on cooling to 1.7 K was observed by Shulishova, Shcherbak [12].

The temperature dependence cannot be fitted by only the Bloch-Grüneisen term (for scattering by acoustic phonons) and the impurity scattering term. Inclusion of scattering by polar optical phonons characterized by the Einstein temperature $\Theta_E = 920$ K and taking $\Theta_D = 245$ K for the acoustic phonons resulted in the solid line in Fig. 79. The polar optical phonon scattering becomes appreciable above about 100 K. At room temperature the optical phonon part is about 30% of the total resistivity, Tanaka et al. [2]. This interpretation was confirmed by [10].

A mean temperature coefficient $d\varrho/dT = 4 \times 10^{-8}$ Ω·cm·K^{-1} between 77 and 300 K was given for polycrystalline LaB$_6$ [8].

Above Room Temperature. The electrical resistivity ϱ of a LaB$_6$ single crystal increased linearly with temperature from ~8 μΩ·cm at 293 K to ~28 μΩ·cm at 773 K, Sato et al. [14]. The resistivity of remelted LaB$_6$ from room temperature to 2300 K is shown in **Fig. 80**; above about 1200 K there is a steeper than linear increase of the resistivity, Neshpor et al. [15]. The same behavior was observed for sintered LaB$_6$ between 80 and 1700 K (the values for ϱ, however, were higher), Lafferty [16]. Higher resistivity values with a linear increase of ϱ from 100 K (ϱ ≈ 6 μΩ·cm) to 1000 K (ϱ ≈ 41 μΩ·cm) were found by [6, 17] and from 1100 K (ϱ ≈ 65 μΩ·cm)

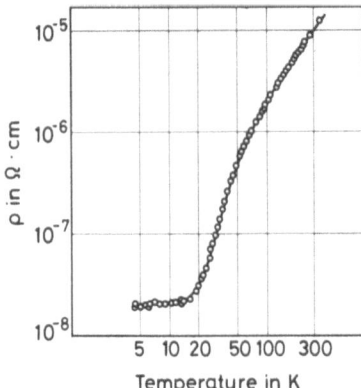

Fig. 79. Electrical resistivity ϱ of a LaB$_6$ single crystal below room temperature.

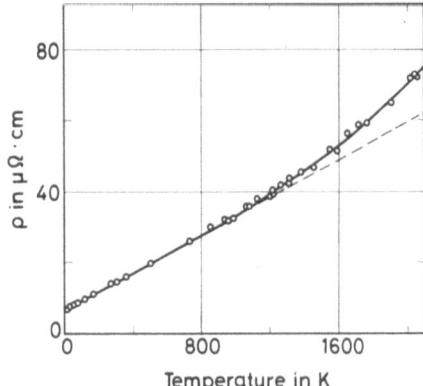

Fig. 80. Electrical resistivity ϱ of a melted LaB$_6$ sample at high temperatures.

to 1850 K ($\varrho \approx 115$ $\mu\Omega \cdot$cm) by Tanaka [18], similarly between 300 and 2273 K by Paderno et al. [19]; read from figures in [17, 18].

A nearly linear increase of ϱ with the temperature from 1100 to 2200 K was observed, independent of the sample density D, between 0.60 and 0.90 of the theoretical density D_{th}. The relation $\varrho = A \cdot d^n \cdot T$ was found to be a good approximation for the dependence of ϱ (in $\Omega \cdot$cm) on temperature (in K) and density ratio $d = D/D_{th}$ with $A = 4.1 \times 10^{-8}$ $\Omega \cdot$cm$\cdot$K^{-1} and $n = -8/3$, Williams et al. [20]. Earlier at constant temperature, $\varrho = \varrho_0 \cdot (1-P)^{-3}$ was proposed with P = sample porosity, Paderno et al. [19].

Pressure Dependence

Under hydrostatic pressure, studied up to 3.0 GPa, the room temperature resistivity of a LaB$_6$ single crystal was found to decrease linearly with pressure up to $\sim$1.8 GPa, from $\varrho = 10.9$ $\mu\Omega \cdot$cm at normal pressure to $\varrho = 10.3$ $\mu\Omega \cdot$cm, values read from a figure. At higher pressures ϱ deviates from the linear relation to lower resistivities with about -5% deviation at 3.0 GPa. The dependence of ϱ at 3.0 GPa on the temperature, measured down to 80 K is similar to that at normal pressure, Mori et al. [1].

A decrease of ϱ (at room temperature) with pressure was measured on powdered LaB$_6$ up to 10 GPa. However, this decrease was regarded as insignificant for experimental reasons, Leger et al. [21]. Previously, the resistivity of Al-flux grown crystals at room temperature $\varrho = 17$ $\mu\Omega \cdot$cm was observed to be practically independent on hydrostatic pressures up to 1.6 GPa, Stepanov et al. [22].

Films

The electrical resistivity of LaB$_6$ films deposited from the vapor phase depended on both the substrate temperature and the final film thickness. Quasi-amorphous films, deposited below 600 K, had resistivities $\varrho > 100$ $\mu\Omega \cdot$cm. Fine-grained polycrystalline material had ϱ values about 1 order of magnitude higher than that of bulk material. Films deposited above about 1000 K are comparable to bulk material. At a film thickness of 0.2 μm the resistivity was about 1000 $\mu\Omega \cdot$cm and ϱ approached the values of bulk LaB$_6$ at film thickness above $\sim$1 μm, Bessaraba et al. [23]. A lower value $\varrho = 600$ $\mu\Omega \cdot$cm for a film of 0.06 μm was measured for a film deposited from electron beam heated LaB$_6$ onto glass at 573 K, Hara et al. [24]. A decrease of the film resistivity at high-temperature annealing (after cold deposition) was reported by Ryan, Roberts [25].

Superconductivity

LaB$_6$ becomes superconducting at a critical temperature $T_c = 0.122$ K according to inductive measurements. Less-pure samples had lower and broader transitions, Arko et al. [26]. The remarks on superconductivity, $T_c \approx 0.1$ K, by Sobzcak, Sienko [27] and Hiebl, Sienko [28] apparently refer to this measurement. A higher value $T_c = 0.45$ K was related by Schell et al. [29] and Gompf [30], however without reference to any measurement. The earlier report of $T_c = 5.7$ K, Matthias et al. [31] is apparently superseded.

Making use of their calculations on the electronic structure, see p. 216, and the phonon model of LaB$_6$, see p. 197, Schell et al. [29] solved the Eliashberg functions $\alpha^2 F(\omega)$ numerically, treating the electron-phonon coupling in the rigid-muffin-tin approximation. They obtained $T_c = 0.78$ K. The superconductivity is mainly due to the boron sublattice according to the contributions of the individual phonon modes to the McMillan parameter λ. Both LaB$_6$ and YB$_6$ are weak-coupling superconductors. The difference between the values of T_c for LaB$_6$ and YB$_6$

($T_c \approx 7$ K) is large, even though they have almost identical lattice parameters. The difference was attributed to differences in λ, which was as low as $\lambda = 0.33$ for LaB$_6$ [29].

Point Contact Spectroscopy

Point contact spectra (PCS) were recorded at 1.8 K for point contacts between the cleaved (100) surface of LaB$_6$ and a LaB$_6$ needle tip. The differential resistance dU/dI versus the bias voltage U showed a flat bottom within ± 10 mV, followed by a weak rise due to electron-phonon interaction. The second derivative d^2U/dI^2 showed prominent peaks at ~12 and 24 meV, a broad peak near 38 meV, and a broad structure around 50 meV. There is a good correlation with the results of inelastic neutron scattering experiments, see p. 195, except for the peak at 24 meV. Several explanations were discussed, Kunii [40], cf. [41].

Study of 80 point contacts (in the [100] and [111] directions) revealed a relatively strong anisotropy, Samuely et al. [39]. The point contacts in the [100] direction agree with the data by [41], but in addition, show structures up to ~100 meV. All the phonon modes, also accounting for the anisotropy, could be identified in agreement with the previous inelastic neutron scattering und Raman scattering experiments and the model of Schell et al. [29] (see pp. 195 and 197) [39]. The results from the point contact spectra could be transformed in an approximate way to the Eliashberg function; a comparison with the Eliashberg function derived by Schell et al. [29], showed qualitative agreement; also the observed temperature dependence of the electrical resistivity and of the heat capacity could be reproduced reasonably well [39].

Earlier studies by Frankowski, Wachter [42, 43] gave different results, showing structures in the second derivative curve at ~35 and ~68 meV [42], which might be due to insufficient cleaning of the surface [40]. Intentionally oxidized surfaces gave different PCS. The bias-voltage dependence of the differential resistance is reversed compared to that of the clean surface. The analysis of the metal-insulator-metal tunnel current gave a barrier width $d = 20.3$ Å and a barrier height $\Phi = 0.45$ V, Kunii [40].

Magnetoresistance

No effect of magnetic fields parallel to the electric current on ϱ of single crystals was observed for fields up to 7T, Peysson et al. [9].

The high-field magnetoresistance $\Delta\varrho/\varrho_0$, measured down to 1.2 K, saturated for general directions of the current I and field B. This saturation is expected for an uncompensated metal either with closed orbits in the Fermi surface or for a current perpendicular to the open orbit direction (not considering magnetic breakdown). For I$\|$[100] and rotation of the magnetic field, 6.2 T, in the (100) plane, the magnetoresistance showed minima for B$\|\langle001\rangle$ and $\langle011\rangle$. At these minima, the magnetoresistance tended to saturation, yet was not saturated at 6.5 T. In the other directions, the high magnetoresistance showed nonsaturating behavior, increasing approximately as B$^{1.5}$ characteristic of open orbits along [100], the direction of I. However, neither a true B^2 dependence nor saturation were observed for any direction of B in the (100) plane. On variation of the angle ϑ between I and B, the magnetoresistance showed a sharp maximum for $\vartheta = 90°$. This behavior was attributed to magnetic breakdown, resulting in topological changes of the orbits, Arko et al. [32].

For I$\|$[1$\bar{1}$0] and B (up to 6.0T) in (1$\bar{1}$0) the transverse magnetoresistance $\Delta\varrho/\varrho_0$ saturated or tended towards saturation for general field directions except for about 3° and 27° from [001]. At B$\|$[001] $\Delta\varrho/\varrho_0$ saturated, indicating [001] to be a singular field direction. The nonsaturating magnetoresistance at the mentioned exceptions was attributed to open orbits on the Fermi surface along [1$\bar{1}$0]. The experimental results are consistent with the proposed Fermi surface described on p. 218. The field dependence of $\Delta\varrho/\varrho_0$ is lower than quadratic, $\sim B^{1.1}$ for B at 27° away from [001] in the (1$\bar{1}$0) plane, and $\sim B^{1.3}$ for B at 20° away from [001] in the (010) plane, Ishizawa et al. [33].

Superimposed on the monotonic magnetoresistance there is a low-frequency oscillatory magnetoresistance at magnetic fields B>1.0 T. This was observed for all regions in the (010) plane and at angles of 10°, 27° to 49°, and near 90° (i.e. [110]) away from [001] in the (1$\bar{1}$0) plane. These oscillations are considered to be due to the Shubnikov-de Haas effect and their frequencies are estimated to be in the range from 3.5 to 6.5 T, which corresponds to the low-frequency branch observed in the de Haas-van Alphen effect, Ishizawa et al. [33]. Field dependent low-frequency oscillations of $\Delta\varrho/\varrho_0$ were superimposed on the background magnetoresistance for fields below 5.0 T, and weaker high-frequency oscillations were observed at fields greater than 5.0 T, Arko et al. [32].

Hall Effect. Charge Carrier Concentration and Mobility

For effective cyclotron masses of electrons see p. 221, effective electron masses from optical data see p. 258.

The Hall constant of a single crystal at room temperature was $R_H = -4.5 \times 10^{-10}$ m³/C, Tanaka et al. [34]. Other consistent room temperature values: $R_H = -4.68 \times 10^{-10}$ m³/C, Paderno et al. [35], $R_H = -4.9 \times 10^{-10}$ m³/C, Kondrashev et al. [5], Bondarenko et al. [4]; a deviating value of -7.7×10^{-10} m³/C was measured for a sintered sample with a high porosity, Lafferty [16].

From Hall data of LaB$_6$ single crystals at room temperature, a charge carrier concentration n = 1 per La atom (the La density is 1.393×10^{22} cm^{-3}) was obtained, Tanaka et al. [34], in reasonable agreement with n = 0.93 per La atom, Paderno et al. [35], n = 1.45×10^{22} cm^{-3} in sintered polycrystalline LaB$_6$, Kauer [17], n = 1.4×10^{22} cm^{-3}, Aivazov et al. [3], n = 0.92 per La, Bondarenko et al. [4]. Essentially one negative charge carrier per La was inferred from Hall data, Lafferty [16]. Hall data in the temperature range between that of liquid air and 373 K indicate no change in the carrier concentration, Kauer [17]. The carrier mobility at room temperature was 32 cm²·V^{-1}·s^{-1}, Kauer [17], Bondarenko et al. [4].

Thermoelectric Properties

At low temperature, 1.5 to 5 K, the Seebeck coefficient S of a single crystal is close to zero (±1 µV/K) and gradually increases to ~5 µV/K at 30 K, Peysson et al. [9], partly consistent with previous data on a polycrystalline sample between 2 and 20 K, Ali, Woods [36] who report a negative maximum S = −0.28 µV/K near 5.5 K, and S>0 above 12 K, see Fig. 28, p. 69 [36]. Between about 403 and 873 K the increase of S with temperature is fairly linear from ~6.7 µV/K at 423 K to ~10.0 µV/K at 773 K. The thermal EMF was fitted by the following relation: E = $5.18 \times 10^{-3} \cdot t + 5.20 \times 10^{-6} \cdot t^2 - 0.96 \times 10^{-9} \cdot t^3$ with E in mV and t in °C, Kondo et al. [37].

The following negative values of S were given for sintered samples at room temperature: S = -2.9 ± 0.06 µV/K, Kondrashev et al. [5], -2.1 µV/K, Aivazov et al. [3], -5 µV/K, Etourneau et al. [11], but S = $+23.6$ µV/K for a Ni-containing hot-pressed sample was reported by Kondrashev et al. [5].

The thermoelectric power against Cu was 4.0 and 6.5 µV/K at 373 and 673 K, respectively, for a LaB$_6$ whisker from chemical vapor growth, Motojima et al. [38].

References:

[1] Mori, N.; Sato, N.; Kasuya, T. (Solid State Phys. Pressure Recent Advan. Anvil Devices **1985** 259/62).

[2] Tanaka, T.; Akahane, T.; Bannai, E.; Kawai, S.; Tsuda, N.; Ishizawa, Y. (J. Phys. C **9** [1976] 1235/41).

[3] Aivazov, M. I.; Aleksandrovich, S. V.; Evseev, B. A.; Zinchenko, K. A.; Mkrtchyan, V. S. (Izv. Akad. Nauk SSSR Neorgan. Materialy **15** [1979] 61/3; Inorg. Materials [USSR] **15** [1979] 46/8).

[4] Bondarenko, V. P.; Kovenskaya, B. A.; Morozov, V. V. (Izv. Vysshikh Uchebn. Zavedenii Fiz. **13** No. 2 [1970] 12/5; Soviet Phys. J. **13** [1970] 147/9).

[5] Kondrashev, A. I.; Paderno, Yu. B.; Paderno, V. N. (Poroshkovaya Metal. **1982** No. 6, pp. 16/21; Soviet Powder Met. Metal Ceram. **21** [1982] 437/41).

[6] Rabenau, A.; Kauer, E.; Klotz, H. (Colloq. Intern. Centre Natl. Rech. Sci. [Paris] No. 157 [1967] 495/8).

[7] Samsonov, G. V.; Grodshtein, A. E. (Zh. Fiz. Khim. **30** [1956] 379/81).

[8] Berrada, A.; Mercurio, J.-P.; Etourneau, J.; Hagenmuller, P.; Shroff, A. M. (J. Less-Common Metals **59** [1978] 7/25, 18).

[9] Peysson, Y.; Ayache, C.; Salce, B.; Kunii, S.; Kasuya, T. (J. Magn. Magn. Mater. **59** [1986] 33/40).

[10] Flachbart, K.; Reiffers, M.; Janos, S.; Paderno, Yu. B.; Lazorenko, V. I.; Konovalova, E. S. (J. Less-Common Metals **88** No. 2 [1982] L3/L6).

[11] Etourneau, J.; Mercurio, J.-P.; Naslain, R.; Hagenmuller, P. (J. Solid State Chem. **2** [1970] 332/42).

[12] Shulishova, O. I.; Shcherbak, I. A. (Izv. Akad. Nauk SSSR Neorgan. Materialy **3** [1967] 1495/7; Inorg. Materials [USSR] **3** [1967] 1304/6).

[13] Ishizawa, Y.; Tanaka, T.; Bannai, E.; Kawai, S. (J. Phys. Soc. Japan **42** [1977] 112/8).

[14] Sato, N.; Kunii, S.; Oguro, I.; Komatsubara, T.; Kasuya, T. (J. Magn. Magn. Mater. **47/48** [1985] 86/8).

[15] Neshpor, V. S.; Fridlender, B. A.; Paderno, V. N.; Paderno, Yu. B.; Nerus, M. A. (Teplofiz. Vys. Temp. **14** [1976] 903/6; High Temp. [USSR] **14** [1976] 802/4).

[16] Lafferty, J. M. (J. Appl. Phys. **22** [1951] 299/309).

[17] Kauer, E. (Phys. Letters **7** [1963] 171/3).

[18] Tanaka, T. (J. Phys. C **7** [1974] L177/L180).

[19] Paderno, Yu. B.; Samsonov, G. V.; Fomenko, V. S. (Fiz. Metal. Metalloved. **10** [1960] 633/40; Phys. Metals Metallog. **10** No. 4 [1960] 143/5).

[20] Williams, M. D.; Jackson, L. T.; Kippenhan, D. O.; Leung, K. N.; West, M. K.; Crawford, C. K. (Appl. Phys. Letters **50** [1987] 1844/5).

[21] Leger, J. M.; Percheron-Guegan, A.; Loriers, C. (Phys. Status Solidi A **60** [1980] K23/K26).

[22] Stepanov, N. N.; Zyuzin, A. Yu.; Shul'man, S. G.; Gurin, V. N.; Korsukova, M. M.; Nikanorov, S. P.; Smirnov, I. A. (Fiz. Tverd. Tela [Leningrad] **20** [1978] 935/8; Soviet Phys.-Solid State **20** [1978] 542/3).

[23] Bessaraba, V. I.; Vereshchak, V. M.; Ivanchenko, L. A.; Kugai, L. N.; Paderno, Yu. B. (Poroshkovaya Metal. **1976** No. 7, pp. 69/73; Soviet Powder Met. Metal Ceram. **15** [1976] 552/5).

[24] Hara, T.; Shinmi, A.; Fukui, M.; Hajimoto, Y.; Shirato, Y. (Japan. Kokai Tokkyo Koho 79-65394 [1977/79] 1/3 from C.A. **91** [1979] No. 100955).

[25] Ryan, J. G.; Roberts, S. (Thin Solid Films **135** [1986] 9/19).

[26] Arko, A. J.; Crabtree, G.; Ketterson, J. B.; Mueller, F. M.; Walch, P. F.; et al. (Intern. J. Quantum Chem. Symp. No. 9 [1375] 569/78).

[27] Sobzak, R. J.; Sienko, M. J. (J. Less-Common Metals **67** [1979] 167/71).

[28] Hiebl, K.; Sienko, M. J. (Inorg. Chem. **19** [1980] 2179/80).

[29] Schell, G.; Winter, H.; Rietschel, H.; Gompf, F. (Phys. Rev. [3] B **25** [1982] 1589/99).

[30] Gompf, F. (KfK-2670 [1978] 17/9; INIS **10** [1979] No. 438945).

[31] Matthias, B. T.; Geballe, T. H.; Andres, K.; Corenzwit, E.; Hall, G. W.; Maita, J. P. (Science **159** [1968] 530).

[32] Arko, A. J.; Crabtree, G.; Karim, D.; Mueller, F. M.; Windmiller, L. R.; Ketterson, J. B.; Fisk, Z. (Phys. Rev. [3] B **13** [1976] 5240/7).

[33] Ishizawa, Y.; Tanaka, T.; Bannai, E. (J. Phys. Soc. Japan **49** [1980] 557/61).

[34] Tanaka, T.; Bannai, E.; Kawai, S.; Yamane, T. (J. Cryst. Growth **30** [1975] 193/7).

[35] Paderno, Yu. B.; Garf, E. S.; Niemyskii, T.; Pračka, I. (Poroshkovaya Metal. **1969** No. 10, pp. 55/8; Soviet Powder Met. Metal Ceram. **1969** 821/3).

[36] Ali, N.; Woods, S. B. (Solid State Commun. **46** [1983] 33/5).

[37] Kondo, I.; Iwasa, M.; Kinoshita, M. (Osaka Kogyo Gijutsu Shikensho Kiho **37** No. 1 [1986] 31/5; C.A. **105** [1986] No. 31790).

[38] Motojima, S.; Takahashi, Y.; Sugiyama, K. (J. Cryst. Growth **44** [1978] 106/9).

[39] Samuely, P.; Reiffers, M.; Flachbart, K.; Akimenko, A. I.; Yanson, I. K.; et al. (Japan. J. Appl. Phys. **26** Suppl. **3** [1987] 647/8).

[40] Kunii, S. (J. Phys. Soc. Japan **57** [1988] 361/6).

[41] Kunii, S. (J. Magn. Magn. Mater. **63/64** [1987] 673/6).

[42] Frankowski, I.; Wachter, P. (Solid State Commun. **41** [1982] 577/80).

[43] Frankowski, I.; Wachter, P. (Valence Instab. Proc. Intern. Conf., Zürich, Switz., 1982, pp. 309/12).

32.1.8.3.4.9.3 Electron Emission

32.1.8.3.4.9.3.1 Introduction

The electron emission of LaB_6 is a thoroughly studied property. The interest started from the research of Lafferty [1, 2] on sintered, polycrystalline LaB_6 samples with 50% of the theoretical density. A very low thermionic Richardson work function $\Phi_R = 2.66$ eV, emission constant $A_R = 29$ $A \cdot cm^{-2} \cdot K^{-2}$ in the temperature range 1100 to 1300 K was obtained for this refractory compound. This made LaB_6 a promising cathode material for many technical applications, see p. 255. Research on single crystals showed the work function to be anisotropic, see **Fig. 81** from Gesley, Swanson [3], cf. [4].

In the following, experimental values of Φ from different methods will be given. Then other data on the thermionic properties (i.e., dependence of brightness and emission patterns on shape and orientation of the cathodes in electron emitting systems, "electron guns", etc.) are given on p. 247. Finally, data on the electron field emission (including thermal field emission) and the secondary electron emission are given. The photoelectron spectra are treated on p. 225 in connection with the electronic structure; Auger spectra have mainly been employed in numerous studies for surface analysis, see pp. 228, 261.

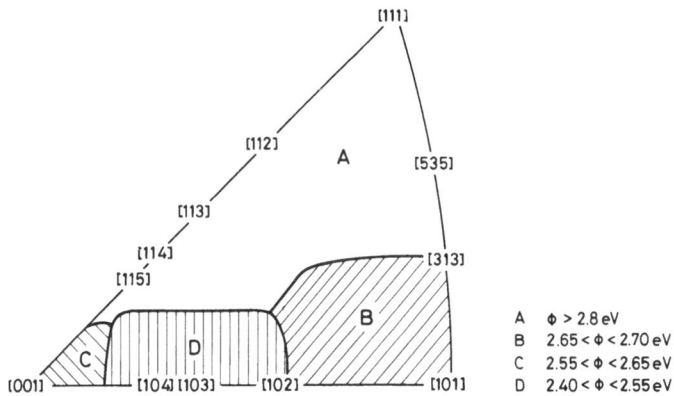

Fig. 81. Crystallographic map of the work
function variation (for 300<T<1600 K). The
directions indicate the surface normal.

References:

[1] Lafferty, J. M. (J. Appl. Phys. **22** [1951] 299/309).

[2] Lafferty, J. M. (Phys. Rev. [2] **79** [1950] 1012).

[3] Gesley, M.; Swanson, L. W. (Surf. Sci. **146** [1984] 583/99).

[4] Gesley, M. A.; Davis, P. R.; Swanson, L. W. (Proc. 29th Intern. Field Emiss. Symp., Goeteborg 1982, pp. 121/7).

32.1.8.3.4.9.3.2 Work Function

General Remarks

The work function was derived by various methods: at room temperature from the field emission retarding potential (FERP), from the contact potential difference relative to a reference material (Kelvin method, K), from the cut-off energy of secondary electron emission and the kinetic energy at the Fermi level in photoelectron spectra (PES), from the photoelectric yield, and from field emission (FE); at high temperature the work function was mainly derived from thermionic emission current densities (TE), but also from the energy spectrum of the thermal electrons and from X-ray bremsstrahlung isochromates.

The temperature dependence of the current density I of thermally emitted electrons is usually described by the Richardson equation $I = A_R \cdot T^2 \cdot \exp(-\Phi_R/kT)$. Plots of $\ln(I/T^2)$ vs. $1/T$ give the Richardson work function Φ_R (from the slope) and the Richardson emission constant A_R at $1/T = 0$. Using the theoretical value $A = 120 \ A \cdot cm^{-2} \cdot K^{-2}$ from the free-electron model a so-called effective work function Φ_{eff} may be calculated from the emission current at a given temperature. A temperature dependence of Φ_{eff} which implies a deviation of the emission constant from the theoretical value is usually described by $\Phi_{eff} = \Phi_0 + (d\Phi/dT) \cdot T$. To allow for the electric field dependence of the emission current (Schottky effect) the zero-field currents were obtained from plots of $\ln I$ vs. $(voltage)^{1/2}$. The data from thermionic emission given in the following were derived in diode systems with cathode (LaB$_6$) and collecting anode parallel within a vacuum system.

                    References for 32.1.8.3.4.9.3.2 on p. 244

Ultrahigh vacuum (UHV) is appropriate for the measurements, for example, 5×10^{-9} Torr [1 to 3]. For reproducible results of work function measurements, a highly defined surface state is important. For procedures for preparation of a clean surface, see p. 262.

Significant differences of experimental results from thermionic measurements for a sample between initial and later runs occurred; at the same time, the ratio of the AES peaks of B and La decreased. This was attributed to a deviation of the initial surface composition from the congruently vaporizing composition that was approached during high-temperature treatment [4], cf. [5]. Similarly nonlinear Richardson plots were obtained in the initial state of heating; straight lines were obtained after at least 20 h at 1474 K, giving reproducible results for Φ [6].

Work Function of Single Crystals

At Room Temperature

The work function Φ at room temperature was determined by the FERP method to ± 0.05 eV, Gesley, Swanson [7], Gesley et al. [8], Swanson et al. [4], cf. [5], by the Kelvin method, Oshima et al. [6], and from PES to ± 0.1 eV, Nishitani et al. [1, 9], see also [10, 11]. The resulting values Φ for the investigated clean crystal faces of zone-melted crystals are (B/La = bulk composition):

(100)	(110)	Φ in eV for the crystal face (111)	(321)	(346)	(210)	(310)	(211)	B/La	method	Ref.
2.60	2.65	2.90	—	2.90	—	—	—	5.86	FERP	[4, 7]
2.68	2.84	—	3.02	—	—	—	—	5.74	FERP	[4]
2.69	—	—	—	—	2.55	2.50	3.05	6.09	FERP	[7]
2.3	2.5	3.3	—	—	2.2	—	—	—	PES	[1, 12]
—	—	3.6	—	—	—	—	—	stoich.	Kelvin	[6]

A value $\Phi = 2.7$ eV on (100) by FERP was measured by Inoue et al. [13]; $\Phi \approx 2.7$ eV from field emission data at room temperature of a single-crystal tip (no orientation given) by Fowler-Nordheim plots, Windsor [14].

At High Temperature

The following work function values Φ in eV for various crystal faces were derived from the emission current density I at a given temperature (Φ_{eff}) or in a temperature range (Richardson work function Φ_R, emission constant A_R in $A \cdot cm^{-2} \cdot K^{-2}$; see p. 239).

T in K	B/La	Φ_{eff} for the crystal face (100)	(110)	(111)	(210)	(310)	(321)	(346)	Ref.
1600	5.86[a]	2.52	2.64	2.90	—	—	—	2.41	[4, 7]
1600	5.74[a]	2.71	2.75	—	—	—	2.80	—	[4]
1600	6.09	—	—	2.57	2.46	2.42	—	—	[7]
1700	6 to 7	2.70	2.98	3.35	—	—	—	—	[15]
1600	—	2.75	2.89	2.98	—	—	—	—	[16]
1000	—	2.5	2.7	~2.9	—	—	—	~2.9	[17][b]

[a] Composition data from chemical analysis; the densities indicated B/La near 6.0 [4]. – [b] Read from a figure.

| temp. range | Φ_R (A_R) for the crystal face | | | | Ref. |
(in K)	(100)	(110)	(111)	(321)	
1200 to 1650	2.86(82)	2.68(57)	3.4 ± 0.2(71)	—	[6]
1500 to 1650	2.4(—)	—	—	2.3(—)	[18]
1300 to 1950	2.91(510)	3.09(251)	2.85(3.84)	—	[15]
1200 to 1700	—	2.74(14)	—	—	[19]

Also, electron emission data from LaB$_6$ cathodes in electron emitting devices were analyzed using the Richardson function, however, the assignment of values of Φ to crystal faces is problematic. A value of $\Phi_R \gtrsim 3$ eV and A_R (always in $A \cdot cm^{-2} \cdot K^{-2}$) = 100 was obtained for a pointed LaB$_6$ rod with [100] the rod axis and $\Phi_R = 2.5$ eV and $A_R = 60$ with the rod axis [110] at 1373 to 1623 K, Schmidt et al. [20], cf. [21]; a lower value $\Phi_R = 2.6 \pm 0.04$ eV and $A_R = 49$ to 150 was evaluated for a [100] oriented LaB$_6$ tip, Zaima et al. [22], analogously $\Phi_R = 2.3 \pm 0.1$ eV, Futamoto et al. [23], and $\Phi_{eff} = 2.86 \pm 0.04$ eV at 1873 K for a bar-like LaB$_6$ tip showing {100} faces, except for the slightly rounded edges, Futamoto et al. [24].

Except for the data by Oshima et al. [6], there is agreement that for thermally equilibrated single-crystal faces the work function increases in the order $\Phi(100) < \Phi(110) < \Phi(111)$, cf. Fig. 81, p. 237. Differences in the experimental values may be due to differing bulk stoichiometries and thermal history, Swanson et al. [4], thus a 2% increase in the bulk B/La stoichiometry causes a ~6% reduction of Φ for the (110) and (100) planes; the minimum of Φ is supposed for interior compositions near the congruently vaporizing composition [4].

The (210) face [7, 12] and the (310) face [7] have even lower work functions than the (100) or (110) faces. The work function of the (310) face, both from FERP and thermionic measurements, remained unchanged by heating at 1800 K for 100 h, indicating the stability of this face at these conditions [7].

Whereas the low-index planes are thermally stable under UHV (see also p. 206), the work function measurements on high-index faces may suffer from macroscopic surface reconstruction [7]. Under poor vacuum (~10^{-6} Torr, $\triangleq 133$ µPa), facet formation ({111} on a (110) plane), Oshima et al. [6], or an emission pattern typical for polycrystalline cathodes, Oshima et al. [25], were observed during operation at 1673 K or higher temperature.

Temperature Dependence

A comparison of the room temperature FERP values, Φ, and the high temperature thermionic emission data (TE) for the work function, Φ_{eff}, shows the TE values generally to be smaller by ~0.1 eV for a given plane, Gesley, Swanson [7]. Similarly, an approximate equivalence of the work function on (100) at room temperature, $\Phi = 2.4$ eV, and $\Phi = 2.6$ eV at 1073 K (read from a figure; values derived from the energy spectrum of secondary electrons and thermal electrons) was found by Aono et al. [10]; a measured increase of the apparent work function especially at higher temperature, up to ~4 eV at ~1800 K, was explained as an effect of the space charge that could not be avoided in the given experimental setup. No strong variation of the work function with temperature was expected because of the observed temperature independence of the composition of clean LaB$_6$ surfaces [10].

The variation of the thermionic effective work function Φ_{eff} with increasing temperature T, see **Fig. 82**, p. 240, for (100), (210), and (310) faces [7] and analogous data for the (100), (110), (111), and (346) faces, shown in a previous paper [4], indicate temperature-independent or slightly increasing Φ_{eff} values up to ~1600 K and a stronger increase at higher temperatures [4, 7]. See also Fig. 30, p. 74.

References for 32.1.8.3.4.9.3.2 on p. 244

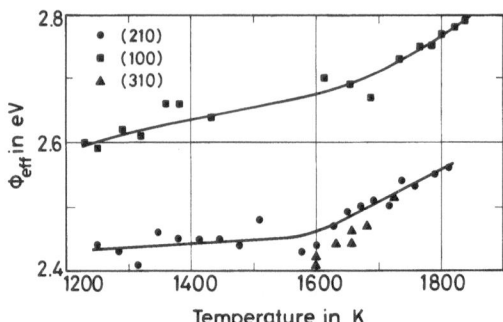

Fig. 82.   Temperature dependence of the
work function Φ_{eff} for three faces of a
LaB$_{6.09}$ single crystal.

Theoretical Considerations

The low work function Φ of LaB$_6$ was at first interpreted as due to the presence of lanthanum at the surface, as inferred from the activation behavior, Lafferty [26]. The study of the temperature dependence of the surface composition, however, revealed that the low value of Φ is a property of the material itself, Goldstein, Szostak [3], Berrada et al. [2]. Φ may be separated into two contributions, the bulk chemical potential of the electrons relative to the mean electrostatic potential in the material and the electrostatic barrier at the surface. Estimates within a jellium model give low values for the bulk chemical potential, Oshima (as reported by Nishitani et al. [1]). A value $\Phi = 2.31 \pm 0.05$ eV was calculated within the jellium model by Yamauchi et al. [18].

The anisotropy of the work function is attributed to differences of the surface dipole moment for the various faces [1, 7]. The value $\Phi = 3.3$ eV for the (111) plane at room temperature was supposed to reflect the bulk property. The values for the (100) and (110) faces are lowered by dipoles produced from the positive charge of the La ions after surface relaxation [1].

A calculation of the dipolar contribution to the work function, assuming unreconstructed (100), (110), and (111) surfaces or the surface reconstructions (proposed by [1]), resulted in differences in the work functions that were larger by at least one order of magnitude than the measured difference. From this, moderately reduced ionic charges compared to those in the bulk were proposed in order to harmonize the model with the experimental data, Watson, Perlman [27]. In spite of the inability to specify properly the surface ionic charge, the dipole model predicts the right sequence of the low-index faces including the (210) face, even though the assumed displacement of 2.08 Å for some surface La atoms might seem rather large, Gesley, Swanson [7]; cf. p. 262.

Work Function of Polycrystalline Samples

The mean work function of polycrystalline LaB$_6$, both as bulk material and as coatings on a substrate was determined by a variety of methods, see p. 237. The experimental values depend on the bulk composition, Kudintseva et al. [28] or, more precisely, on the surface composition, Storms, Mueller [29], cf. [30]. Without information on the bulk composition comparison of work functions was regarded as impossible. In addition, the emission current density depends on other factors, such as thermal history, purity of sample and surrounding gases, and average crystallographic orientation of the surface [29].

For bulk LaB$_6$ work function mean values, in eV, Φ_{eff}, Φ_R, from thermionic emission in diodes, or Φ from other methods are compiled below together with the Richardson emission constant A_R (in $A \cdot cm^{-2} \cdot K^{-2}$), the B/La ratio, and the percentage of the theoretical density D.

Φ, Φ_{eff}	Φ_R	A_R	T in K	B/La	D (%)	Ref.
2.91	2.86[a]	79	1500	6.021[b]	73	[29]
2.91	3.15[a]	881	1500	6.032[b]	66	[29]
2.93	3.12[a]	568	1500	6.047[b]	66	[29]
2.94	2.88[a]	76	1500	6.068[b]	72	[29]
—	2.36	120	1050 to 1350	—	62	[31]
2.92	—	—	1373	—	—	[32]
2.62	—	—	1773	6.0	85	[28]
2.56	—	—	1773	5.8	85	[28]
2.92	—	—	1773	6.4	100	[28]
—	2.7	25 to 29	~1800 to 2073	—	94 to 99	[33]
2.67[c]	—	—	300	—	50	[26]
2.77[c]	—	—	300	—	—	[34]
2.6[d]	—	—	873, 1473	—	100	[35]

[a] From emission current densities between 1300 and 1500 K. – [b] Surface compositions, cf. Fig. 83, p. 242. – [c] From photoelectric yield. – [d] From field emission.

From the Richardson plot (1740 to 1960 K) of the LaB$_6$ rod cathode in an electron source $\Phi_R = 2.4$ eV and $A_R = 40$ $A \cdot cm^{-2} \cdot K^{-2}$, Ahmed, Broers [36], and from the influence of the work function on the starting voltage of X-ray bremsstrahlung isochromates, the value $\Phi = 2.68 \pm 0.07$ eV was derived, Merz [37].

Using a thermoemission microscope the thermionic current from small portions (10 µm in diameter) of the surface could be measured or the surface could be visualized on a fluorescent screen. On initial heating of the sample to ~1300 K, thermoemission occurred only on a few "bright" spots which, however, grew in number and size with time, resulting in a statistical mean $\Phi = 3.16$ eV. During a 5 h-activation at 1773 K, the work function decreased, leading to a mean value of 2.92 eV at 1373 K. Measurement of local Φ values gave minimum $\Phi = 2.75$ and maximum $\Phi = 3.05$ eV; the degree of "blackness" then is 0.8. The activation was interpreted as migration of La to the surface, Karetnikov et al. [32], which is in contrast to the conclusions from the single crystal data.

For polycrystalline coatings on various substrates, the work function values and the emission constants A_R in $A \cdot cm^{-2} \cdot K^{-2}$ from thermionic emission are:

Φ_{eff} in eV	Φ_R in eV	A_R	T in K	substrate	coating method	Ref.
2.64	—	—	1500	W	spark alloying	[38]
2.74	—	—	1800	Mo or Ta	—	[39]
—	2.70	30	1473 to 1573	Re	cataphoresis	[40]
—	2.7 ± 0.2	—	1450 to 1700	Re	cataphoresis	[41]
—	2.61 ± 0.05	41	1250 to 1430	Ta	vapor deposition	[42]
—	2.10 to 2.55	—	1173 to 1773	W or Ta	sputtering	[43]

A photoelectric work function $\Phi = 2.67$ eV was evaluated from the energy distribution of photoelectrons of a LaB_6 coating on Ta activated by heating, $\Phi \cong 3$ eV before activation, Kul'varskaya [44]. Φ of sputtered LaB_6 films (2 to 3 μm) on W substrates was found to be lower by 0.20 eV, on the average, than on Ta substrates. Under otherwise identical conditions, Φ decreased with film thickness and deposition temperature; the films contained some LaB_4, Shaginyan et al. [43].

The effect of composition on Φ is shown in **Fig. 83** where Richardson plots for sintered samples with bulk compositions ranging from atom ratio B/La = 5.3 to 8.7 are given together with surface compositions inferred from the vapor composition above the sample. The data indicate a strong increase of the work function Φ_{eff} with increasing B content, Storms, Mueller [29]; for previous data see also Jacobsen, Storms [45], Jaskie [46]. Similarly the lowest work function was observed for a bulk composition $LaB_{5.8}$, Kudintseva et al. [28]. Previously, the work function of a pressed LaB_6 cathode with average characteristics was reported to decrease during vapor deposition of metallic La at a rate of $\sim 1 \times 10^{-9}$ g·cm^{-2}·s^{-1}, giving a minimum $\Phi = 2.3$ eV in the range 1140 to 1230 K. The base pressure was (1 to 5) $\times 10^{-9}$ Torr ($\cong 133$ to 665 nPa). Increase of temperature did not increase the emission current appropriately, giving $\Phi = 2.52$ eV for 1600 K, Kudintseva et al. [47].

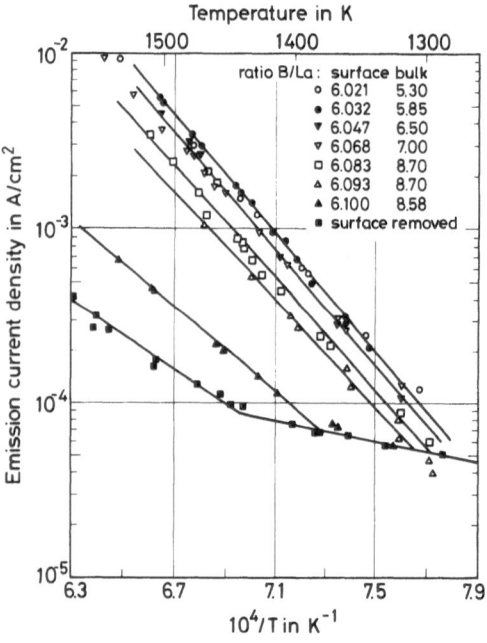

Fig. 83. Emission current densities vs. 1/T for sintered polycrystalline lanthanum boride cathodes.

The experimental data on the composition dependence of the work function and the emission current in the composition range B/La = 6.00 to 6.11 [29], were explained by assuming the Fermi level at a position of a large energy dependence of the density of states. By further assuming the presence of only La vacancies, $\Phi = 3.041 + 9.92 \times 10^{-5}\, T^2 \exp(-0.11/v)$ was obtained by a fit to the observed emission current densities; Φ in eV, T in K, and the vacancy concentration v in vacancies per La site, Storms [30]. These experimental results may need a re-discussion in terms of heterogeneous LaB_4–LaB_6 and LaB_6–B or –LaB_x mixtures when the LaB_6–B phase diagram in the high-temperature range is unambiguously established, compare p. 168.

An influence of the diode voltage on the emission current was observed for samples with a high boron content but not for samples low in boron. The explanation may be the low ionization potential and large ionization cross section of La atoms compared to boron. This could lead to a flux of evaporated La$^+$ ions back to the cathode thereby affecting the surface composition [29].

An increase of the cathode porosity from 0 to 20% decreased the work function by 0.2 eV. From samples of different porosity, a value $\Phi_{eff} = 2.76$ eV at 1773 K was extrapolated for extra-pure LaB$_6$ of zero porosity, Kudintseva et al. [28]. The influence of the porosity on the observed emission current density and therefore the work function was attributed to an increased surface area of porous cathodes and additionally to a changed surface composition because of altered diffusion and vaporization rates, Storms, Mueller [29].

Influence of Oxygen Exposure

1 L (Langmuir) = 10^{-6} Torr·s. For additional data on the effect of impurities on thermo-cathode properties, see p. 248.

In general, the work function Φ of LaB$_6$ is increased by adsorption of oxygen. The increase of Φ at room temperature, derived from the cut-off energy of the photoexcited secondary electrons and the kinetic energy of the Fermi level in UPS spectra, is shown in **Fig. 84**, p. 244, for the three low-index faces; the oxygen pressure was $\sim 1 \times 10^{-9}$ Torr ($\triangleq 133$ nPa), Nishitani et al. [9], see also [48]. The change in the work function, $\Delta\Phi$, on (100) increases linearly with increasing amount of surface oxygen up to $\Delta\Phi \approx 1.0$ eV at an exposure of 1.4 L and with a decreased slope at higher oxygen contents. This was explained by a dependence of the surface dipole moment on the coverage Θ; the dipole moment was estimated as 0.55 or 0.25 D (D $\cong 3.34 \times 10^{-30}$ C·m) for coverages lower or higher than $\Theta = 0.5$, respectively; $\Theta = 0.5$ was defined as coverage at 1.4 L of oxygen exposure, in accordance with the different sites of adsorption in the two regimes [48], see also [49]. A different interpretation was given by Chambers et al. [50]. A linear relation between $\Delta\Phi$ and the intensity of the oxygen Auger signal on LaB$_6$ (100) was also found up to $\Delta\Phi = 1.1$ eV by Goldstein, Szostak [3]. A constant work function of ~ 3.8 eV, from field emission retarding potential measurements, was obtained for relative coverage, $\Theta > 1$ on LaB$_6$ (100), Inoue et al. [13].

On the (110) face, the work function increases linearly with the amount of surface oxygen. The work function on the (111) face, on the other hand, showed an intermediate maximum of $\Phi \cong 4.4$ eV at an exposure of 2 L and decreased again to saturate at ~ 3.8 eV at oxygen exposures of above 100 L [9]. This again was related to the different effect on the surface dipole moment for the different sites of adsorption [9], see pp. 265/7. The work function on the (210) face increased monotonously from 2.2 eV with increasing exposure; saturation occurred at 3.3 eV at an oxygen exposure of 15 L, Oshima et al. [12].

For oxygen pressures above 10^{-8} Torr ($\triangleq 1.3$ μPa) a marked decrease of the thermionic emission appears on all crystal planes studied (mainly the low-index faces). At lower oxygen pressure, however, a remarkable enhancement of the thermionic emission was found for the (210) face giving $\Phi_{R,ox} = 2.27$ eV compared to $\Phi_R = 2.45$ eV for the clean face in 10^{-10} Torr ($\triangleq 13$ nPa). A similar, though smaller, effect was measured for the (110) face. The emission currents follow the Richardson plot, Shimizu et al. [51].

In a study on the effect of an oxygen partial pressure, up to 1×10^{-6} Torr, on the work function of LaB$_6$ (100) at increased temperature, 1073 to 1800 K, the observed change of the thermionic current from the clean surface was attributed to a space-charge effect, V_{sc}, rather than to an increase of the work function Φ. At oxygen exposure the combined effects of work function change and V_{sc} give a large difference compared to the clean surface at low

References for 32.1.8.3.4.9.3.2 on p. 244   16* temperature. At higher temperature, the difference compared to the behavior of clean LaB$_6$ becomes zero at a temperature T$_c$. For temperatures above T$_c$, the sum of $\Phi + V_{sc}$ was independent of the oxygen pressure, whereas below T$_c$ the work function was strongly increased by the oxygen. The value of T$_c$, which depends on the oxygen pressure, corresponds to the minimum temperature at which the surface oxide evaporates fast enough to leave a clean surface; T$_c \approx$ 1623 K at p$_{O_2}$ = 1×10^{-6} Torr ($\triangleq$ 133 µPa) [10, 52]. For consistent results from a LaB$_6$ single-crystal rod cathode under oxygen exposure, see also [23], Nakazawa et al. [53].

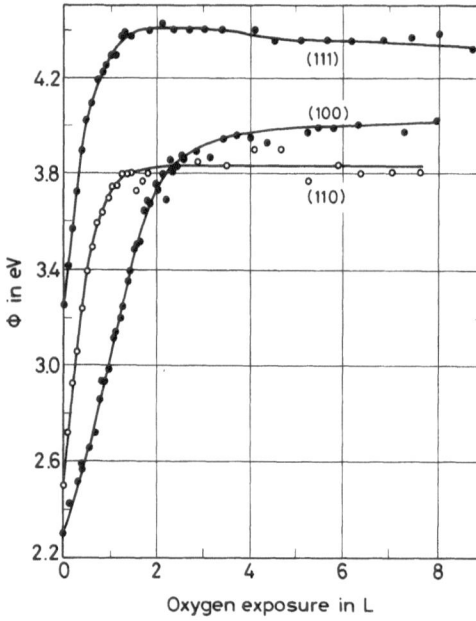

Fig. 84. Influence of oxygen exposure on the work function for the LaB$_6$ (100), (110), and (111) surfaces.

Influence of Caesium

Adsorption of Cs on the clean LaB$_6$ (100) face reduced the work function (measured by FERP) from Φ = 2.77 eV to a minimum of 1.97 eV and a saturation value of 2.07 eV at a monolayer of Cs, Chambers et al. [50], Davis et al. [54]. A decrease of Φ by 0.6 eV with no intermediate minimum was measured by the CDP method on (100) by Gorodetskii et al. [55, 56]. The minimum was Φ = 1.88 eV on the (110) face with saturation at Φ = 2.00 eV, Chambers et al. [50].

Adsorption of Cs onto the surface with preadsorbed oxygen (saturated) reduced the work function to Φ = 1.35 eV on (100) and to 1.47 eV on (110), Chambers et al. [50]. Similarly a decrease of Φ by 1.1 eV for Cs adsorption on the oxygen-saturated (100) face was found by Gorodetskii et al. [55]; a reduced coverage (partially desorbed oxygen) gave a smaller decrease [55]. For further details see the papers [50, 54, 56 to 59].

For older data on the thermionic work functions of LaB$_6$ cathodes in the presence of Cs, see [39, 60, 61]; for the work function in the presence of BaO, see [34].

References:

[1] Nishitani, R.; Aono, M.; Tanaka, T.; Oshima, C.; Kawai, S.; Iwasaki, H.; Nakamura, S. (Surf. Sci. **93** [1980] 535/49).

[2] Berrada, A.; Mercurio, J.-P.; Etourneau, J.; Alexandre, F.; Theeten, J. B.; Duc, T. M. (Surf. Sci. **72** [1978] 177/88).

[3] Goldstein, B.; Szostak, D. J. (Surf. Sci. **74** [1978] 461/78).

[4] Swanson, L. W.; Gesley, M. A.; Davis, P. R. (Surf. Sci. **107** [1981] 263/89).

[5] Swanson, L. W.; Dickinson, T. (Appl. Phys. Letters **28** [1976] 578/80).

[6] Oshima, C.; Bannai, E.; Tanaka, T.; Kawai, S. (J. Appl. Phys. **48** [1977] 3925/7).

[7] Gesley, M. A.; Swanson, L. W. (Surf. Sci. **146** [1984] 583/99).

[8] Gesley, M. A.; Davis, P. R.; Swanson, L. W. (Proc. 29th Intern. Field Emiss. Symp., Goeteborg 1982, pp. 121/7).

[9] Nishitani, R.; Oshima, C.; Aono, M.; Tanaka, T.; Kawai, S.; Iwasaki, H.; Nakamura, S. (Surf. Sci. **115** [1982] 48/60).

[10] Aono, M.; Nishitani, R.; Oshima, C.; Tanaka, T.; Bannai, E.; Kawai, S. (J. Appl. Phys. **50** [1979] 4802/7).

[11] Aono, M.; Oshima, C.; Tanaka, T.; Bannai, E.; Kawai, S. (J. Appl. Phys. **49** [1978] 2761/4).

[12] Oshima, C.; Aono, M.; Tanaka, T.; Nishitani, R.; Kawai, S. (J. Appl. Phys. **51** [1980] 997/1000).

[13] Inoue, T.; Nakada, M.; Uozumi, T.; Sugata, E. (J. Vac. Sci. Technol. **21** [1982] 952/6).

[14] Windsor, E. E. (Proc. Inst. Elec. Eng. [London] **116** [1969] 348/50).

[15] Hafner, P. V. (Diss. ETH Zürich 1976, pp. 1/117).

[16] Morozov, V. V. (Vestn. Kievsk. Polytekhn. Inst. Mashinostr. No. 21 [1984] 48/52).

[17] Lazorenko, V. I.; Lotsko, D. V.; Platonov, V. F.; Kovalev, A. V.; Galasun, A. P.; Matvienko, A. A.; Klinkov, A. E. (Poroshkovaya Metal. **1987** No. 3, pp. 51/7; Soviet Powder Met. Metal Ceram. **26** [1987] 229/33).

[18] Yamauchi, H.; Takagi, K.; Yuito, I.; Kawabe, U. (Appl. Phys. Letters **29** [1976] 638/40).

[19] Berrada, A.; Mercurio, J.-P.; Etourneau, J.; Hagenmuller, P.; Shroff, A. M. (J. Less-Common Metals **59** [1978] 7/25).

[20] Schmidt, P. H.; Longinotti, L. D.; Joy, D. C.; Ferris, S. D.; Leamy, H. J.; Fisk, Z. (J. Vac. Sci. Technol. **15** [1978] 1554/60).

[21] Schmidt, P. H.; Joy, D. C.; Longinotti, L. D.; Leamy, H. J.; Ferris, S. D.; Fisk, Z. (Appl. Phys. Letters **29** [1976] 400/1).

[22] Zaima, S.; Sase, M.; Adachi, H.; Shibata, Y.; Oshima, C.; Tanaka, T.; Kawai, S. (J. Phys. D **13** [1980] L47/L49).

[23] Futamoto, M.; Nakazawa, M.; Usami, K.; Hosoki, S.; Kawabe, U. (J. Appl. Phys. **51** [1980] 3869/76).

[24] Futamoto, M.; Nakazawa, M.; Kawabe, U. (Surf. Sci. **100** [1980] 470/80).

[25] Oshima, C.; Aono, M.; Tanaka, T.; Kawai, S.; Shimizu, R.; Hagiwara, H. (J. Appl. Phys. **51** [1980] 1201/6).

[26] Lafferty, J. M. (J. Appl. Phys. **22** [1951] 299/309).

[27] Watson, R. E.; Perlman, M. L. (Surf. Sci. **122** [1982] 371/82).

[28] Kudintseva, G. A.; Kuznetsova, G. M.; Mamedov, F. G.; Meerson, G. A.; Tsarev, B. M. (Izv. Akad. Nauk SSSR Neorgan. Materialy **4** [1968] 49/53; Inorg. Materials [USSR] **4** [1968] 38/42).

[29] Storms, E. K.; Mueller, B. A. (J. Appl. Phys. **50** [1979] 3691/8).

[30] Storms, E. K. (J. Appl. Phys. **52** [1981] 2961/5).

[31] Pelletier, J.; Pomot, C. (Appl. Phys. Letters **34** [1979] 249/51).

[32] Karetnikov, D. V.; Koryukin, V. A.; Obrezumov, V. P. (Poverkhnost **1982** No. 9, pp. 50/3; Phys. Chem. Mech. Surf. **1982** 2607/14).

[33] Wagner, S.; Albrecht, H. E.; Kotsch, H. (Neue Hütte **9** [1964] 278/81).

[34] Miroshnichenko, L. S. (Radiotekhn. Elektron. **6** [1961] 673; Radio Eng. Electron. Phys. [USSR] **6** [1961] 314/6).

[35] Elinson, M. I.; Kudintseva, G. A. (Radiotekhn. Elektron. **7** [1962] 1511/8; Radio Eng. Electron. Phys. [USSR] **7** [1962] 1417/23).

[36] Ahmed, H.; Broers, A. N. (J. Appl. Phys. **43** [1972] 2185/92).

[37] Merz, H. (Phys. Status Solidi A **1** [1970] 707/13).

[38] Podchernyaeva, I. A.; Siman, N. I.; Verkhoturov, A. D.; Chiplik, V. N.; Morozov, V. Ya.; Isaeva, L. P.; Paramonov, A. M. (Poroshkovaya Metal. **1984** No. 2, pp. 50/3; Soviet Powder Met. Metal Ceram. **23** [1984] 132/5).

[39] Kul'varskaya, B. S.; Rekov, A. I.; Serebrennikova, V. E.; Nikolaeva, V. A.; Kan, K. S. (Radiotekhn. Elektron. **13** [1968] 1304/7; Radio Eng. Electron. Phys. [USSR] **13** [1968] 1131/4).

[40] Buckingham, J. D. (Brit. J. Appl. Phys. **16** [1965] 1821/32).

[41] Blais, J. C.; Bolbach, G. (Intern. J. Mass Spectrom. Ion Phys. **24** [1977] 413/27).

[42] Oshima, C.; Horiuchi, S.; Kawai, S. (Japan. J. Appl. Phys. **13** Suppl. 2, Pt. 1 [1974] 281/4).

[43] Shaginyan, L. R.; Chernyaev, V. N.; Kondrashin, A. A.; Bessaraba, V. I. (Poroshkovaya Metal. **1981** No. 9, pp. 88/91; Soviet Powder Met. Metal Ceram. **20** [1981] 659/61).

[44] Kul'varskaya, B. S. (Zh. Tekhn. Fiz. **41** [1971] 1481/4; Soviet Phys.-Tech. Phys. **16** [1971] 1165/7).

[45] Jacobsen, D. L.; Storms, E. K. (IEEE Trans. Plasma Sci. PS-6 No. 2 [1978] 191/9; C.A. **89** [1978] No. 98437).

[46] Jaskie, J. E. (Diss. Arizona State Univ. 1981, pp. 1/123; Diss. Abstr. Intern. B **42** [1982] 4098; C.A. **97** [1982] No. 32162).

[47] Kudintseva, G. A.; Kuznetsova, G. M.; Nikulov, V. V. (Radiotekhn. Elektron. **12** [1967] 857/61; Radio Eng. Electron. Phys. [USSR] **12** [1967] 798/802).

[48] Nishitani, R.; Kawai, S.; Iwasaki, H.; Nakamura, S.; Aono, M.; Tanaka, T. (Surf. Sci. **92** [1980] 191/200).

[49] Aono, M.; Tanaka, T.; Bannai, E.; Kawai, S. (Appl. Phys. Letters **31** [1977] 323/5).

[50] Chambers, S. A.; Davis, P. R.; Swanson, L. W.; Gesley, M. A. (Surf. Sci. **118** [1982] 75/92).

[51] Shimizu, R.; Onoda, H.; Hashimoto, H.; Hagiwara, H. (J. Appl. Phys. **55** [1984] 1379/87).

[52] Kawai, S. (Mem. Inst. Sci. Ind. Res. Osaka Univ. **40** [1983] 47/61; C.A. **98** [1983] No. 226394).

[53] Nakazawa, M.; Futamoto, M.; Hosoki, S. (Japan. J. Appl. Phys. **19** [1980] 1267/75).

[54] Davis, P. R.; Chambers, S. A.; Swanson, L. W. (Proc. Intersoc. Energy Convers. Eng. Conf. **15** [1980] 2327/30; C.A. **94** [1981] No. 124554).

[55] Gorodetskii, D. A.; Koshelyuk, A. S.; Nonik, V. P.; Tskhakaya, V. K.; Shchudlo, Yu. G.; et al. (Poverkhnost **1983** No. 10, pp. 79/82; Phys. Chem. Mech. Surf. **2** [1984/85] 2991/9).

[56] Gorodetskii, D. A.; Shchudlo, Yu. G.; Yas'ko, A. A.; Koshelyuk, A. S.; Mel'nik, Yu. P. (Poverkhnost **1985** No. 3, pp. 123/6; Phys. Chem. Mech. Surf. **4** [1986] 877/84).

[57] Chambers, S. A.; Davis, P. R.; Swanson, L. W. (Surf. Sci. **118** [1982] 93/102).

[58] Gorodetskii, D. A.; Tskhakaya, V. K.; Shchudlo, Yu. G.; Yarygin, V. I.; Yas'ko, A. A. (Izv. Akad. Nauk SSSR Ser. Fiz. **46** [1982] 1224/9; Bull. Acad. Sci. USSR Phys. Ser. **46** No. 7 [1982] 6/10).

[59] Gorodetskii, D. A.; Tskhakaya, V. K.; Shchudlo, Yu. G.; Yarygin, V. I.; Yas'ko, A. A. (Poverkhnost **1983** No. 1, pp. 29/35; Phys. Chem. Mech. Surf. **2** [1984/85] 54/67).

[60] Mikhailovskii, B. I. (Izv. Akad. Nauk SSSR Ser. Fiz. **28** [1964] 1504/7; Bull. Acad. Sci. USSR Phys. Ser. **28** [1964] 1404/7).

[61] Mikhailovskii, B. I. (Ukr. Fiz. Zh. **7** [1962] 75/7; C.A. **57** [1962] 10629/30).

32.1.8.3.4.9.3.3   Thermionic Emission. Thermionic Cathodes

Introduction

Values for thermionic work functions Φ_{eff} and Φ_R and the Richardson emission constant A_R are given in the preceding chapter.

LaB$_6$ is a preferred cathode material with three advantages over other thermal electron emitters: (1) The availability of high peak emission density (>100 A/cm^2); (2) the long lifetime as compared to other metallic emitters (more than thousand hours of operation in high brightness mode); (3) LaB$_6$ cathodes can be operated in standard high-vacuum systems (10^{-6} to 10^{-7} Torr $\cong 133$ to13 µPa), according to a recent review on LaB$_6$ cathode properties by Hohn [1].

For a review on mounting methods with indirect heating (Broers type) or direct heating with carbon block mounting (Vogel type), and other designs, see Crawford [2], or Nakagawa [3]. For early designs of thermionic electron beam sources using polycrystalline LaB$_6$, see Broers [4, 5] and Vogel [6]. A brightness of 6.4×10^4 A·cm^{-2}·sr^{-1} at 1773 K and 41.1×10^4 A·cm^{-2}·sr^{-1} at 1973 K at an acceleration voltage of 12 kV was obtained for the indirectly heated cathode [5]; a brightness up to 28×10^4 A·cm^{-2}·sr^{-1} was measured for the directly heated cathode [6].

Emission Patterns. Crossover Distribution. Brightness

As already mentioned for the work function, the thermionic emission of LaB$_6$ single crystals is anisotropic. This has been observed directly using a hemispherical LaB$_6$ cathode and fluorescent screen as anode. Immediately after activation for 90 s at 1773 K, the emission pattern from the $\langle 110 \rangle$ oriented LaB$_6$ rod at 1323 K in $\sim 2 \times 10^{-9}$ Torr ($\cong 266$ nPa) indicated emission mainly from the $\{100\}$ and $\{210\}$ faces. The $\{100\}$ spots became weaker with time and nearly vanished compared to the $\{210\}$ spots, which became brighter and determined the stable final pattern. Without initial activation, emission spots from the $\{110\}$ planes were observed with a tendency to stretch out to the nearby $\langle 210 \rangle$ orientation with time, Shimizu et al. [7], consistent with results from a $\langle 111 \rangle$ oriented cathode [8]. Similar experiments with LaB$_6$ tips in various orientations under 10^{-8} Torr were interpreted to indicate that $\{111\}$ faces have a low emission current and the $\{210\}$ faces a higher emission current than the $\{100\}$ and $\{110\}$ faces, Oshima et al. [9].

The observation of thermionic emission from (110) faces that was higher by one order of magnitude than of (100) faces reported by Schmidt et al. [10] was attributed to an enhancement effect by oxygen contamination rather than to be characteristic of clean LaB$_6$ faces, Shimizu et al. [7].

The angular emission distribution, the crossover intensity distributions (in the focal plane of the electron optical system), and stability with time are important parameters for electron guns. They were unpredictable and very unstable for cathodes of sintered LaB$_6$, see for example Ahmed, Broers [11]. The influence of the crystallographic orientation of the cathode rod, cone angle, tip radius, and other parameters has been thoroughly studied, Frosien et al. [12], Yamabe et al. [13], Furukawa et al. [14 to 16], Takigawa et al. [17 to 19], Hagiwara et al. [20], Kato et al. [21], Shimizu et al. [22], Hohn et al. [23]; see the papers for figures and further data. For single-crystal cathodes the crossover intensity and angular emission pattern have a characteristic symmetry which is directly related to the orientation of the cathode. In recent investigations the $\langle 100 \rangle$ direction was preferred as rod axis. Cone-shaped tips with a cone angle of 90° and a tip radius of 2 to 5 µm represent a currently applied standard for application in scanning electron microscopes, Hohn [1]. Crystal orientation and cone angle should be selected in such a manner that low work-function planes such as $\{100\}$, $\{110\}$, and $\{210\}$ contribute to the cone surface. Acute-angle cathodes with small tip radii are to be preferred for electron beam systems requiring high spatial resolution. Obtuse-angle cathodes with large radii are more suitable for homogeneous illumination, Frosien et al. [12].

Further improvement (brightness up to $\sim 5 \times 10^6 \, A \cdot cm^{-2} \cdot sr^{-1}$ at 20 kV, 200 µA, and 1900 K, from a figure in [1]) can be obtained by faceted grinding of low work function planes [12], referring to Hohn [24], see [1]. A flat top of the tip for more uniform emission was described by Yamabe et al. [13], similarly by Takigawa, Sasaki [17]. This preshaping can also reduce the effects of faceting of the emitter tips owing to anisotropic evaporation during longtime operation that was frequently observed [15, 17, 20, 21]. For data on the cathode brightness and the increase with tip temperature for LaB_6 cathodes with different orientations and tip shapes see Hohn [1], Frosien et al. [12], Furukawa et al. [15, 16], Takigawa, Sasaki [17], Shimizu et al. [22], and Noack et al. [25].

For a study of the electric field distribution of a LaB_6 electron emitter see Tagawa et al. [26], for a theoretical study of the space charge limited LaB_6 electron gun see Takaoka, Ura [27].

Electron Energy Distribution

The energy width ΔE of thermionic electrons from sintered LaB_6 cathodes was reported to increase from ~ 0.5 eV (for a cathode current I extrapolated to zero) to 2.6 eV for $I = 400$ µA; cathode temperature $T = 1670$ K, acceleration voltage was 21.5 kV, Vogt [28]; $\Delta E = 0.56$ eV at $I = 2.7$ µA and $T = 1848$ K, Loeffler [from 3] and $\Delta E = 0.95$ eV at $I = 25$ µA, Pfeiffer [from 3].

Effect of Impurities or Additives on Emission Properties

Survey

There are numerous reports, see for example Storms [29], that the thermoemissive properties of LaB_6 cannot be improved by additives, see p. 249. It is undesirable to have inclusions of any second-phase material in single-crystal LaB_6 cathodes, since they contribute to surface nonuniformity and may affect electron emission spatial distribution, Davis et al. [30]. This is consistent with the fact that pure or single-crystal LaB_6 cathodes are preferred in recently designed and studied electron sources, see [1, 21, 31, 32]. Even though increased emission currents or lower work functions due to the presence of some additives were sometimes reported for sintered LaB_6. (This inconsistency might be explained by assuming that the LaB_6 cathodes used for comparison did not represent the optimum state of the LaB_6 emission properties.)

Sintered LaB_6 cathodes were much more affected by poor vacuum, leading to unstable emission current than were single-crystal cathodes, Furukawa et al. [33]. LaB_6 cathodes must be activated prior to use to purify the emitter surface, see for example Hosoki et al. [34] or p. 238. The emissive properties may also be degraded by other agents, for example oxygen, which is one of the most active of the poisoning agents, Gallagher [35]. In general, the effect of residual gases on the cathode properties depends on the stability of the solid compounds formed by chemical reaction on the cathode surface, Avdienko, Malev [36]. Unfavorable conditions with respect to poisoning can, to some degree, be compensated by higher operation temperature, Futamoto et al. [37].

Gases and Gaseous Impurities

An argon atmosphere did not poison the LaB_6 cathode at pressures below 10^{-2} Torr (1.3 Pa), Gallagher [35], Avdienko, Malev [36].

In the presence of hydrogen below $\sim 10^{-3}$ Torr (0.13 Pa) invariant (1673 K) or slightly enhanced (1473 K) thermionic emission from a sintered cathode was found by Gallagher [35], confirmed by Avdienko, Malev [36]. In contrast, the emission from a single-crystal cathode decreased at hydrogen pressures above 10^{-4} Torr hydrogen. The decrease was least at 1873 K and more pronounced at 1573 K, Futamoto et al. [37], similarly Buckingham [38]. The emission current from the spherical end of single crystals hardly changed on admission of 1×10^{-6} Torr of hydrogen, however, the emission was shifted from (210) to (310) planes; with time, the (110) spots appeared, Shimizu et al. [8].

The LaB$_6$ cathode was quite resistant to poisoning by nitrogen (no effect at 1573 K). Poisoning was more pronounced at lower temperature, 1473 and 1373 K, Gallagher [35], in agreement with Avdienko, Malev [36] and Carter, Wood [39].

Oxygen is a very effective poison for the LaB$_6$ cathode, Buckingham [38], Gallagher [35], Futamoto et al. [37]. For results of studies on the change of the work function of low-index planes, see pp. 243/4. In the study of Futamoto et al. [37] the decreasing effect on the emission current from a single-crystal cathode was observed at lower oxygen pressure (one to two orders of magnitude) than in the study of Gallagher [35] on sintered LaB$_6$. This was tentatively attributed to differences in the surface composition, possibly due to carbon [37]. The onset of the decreasing effect at a cathode temperature of 1573 K was at $\sim 10^{-5}$ Pa ($\sim 0.75 \times 10^{-7}$ Torr) of oxygen. The emission current had a minimum at roughly the tenfold pressure and increased again for higher pressures. An analogous behavior but shifted to higher pressure was found at higher temperature [37]. For a theoretical model of the LaB$_6$ decomposition in oxidizing atmosphere under thermocathode conditions, see Baranov, Petrosov [40].

The effect of water vapor on the emission current was roughly comparable to that of oxygen, Futamoto et al. [37]. For a cathode temperature of 1673 K a poisoning pressure of $\sim 10^{-4}$ to 10^{-3} Torr was evaluated by Avdienko, Malev [36].

Carbon containing gases: The emission current at cathode temperatures of 1673 to 1873 K decreased in the presence of CH$_4$, pressure range 10^{-4} Pa ($\sim 0.75 \times 10^{-6}$ Torr); CH$_4$ was found to crack slowly and to leave a carbon layer, Futamoto et al. [37]. Acetone or ethyl alcohol at 1×10^{-6} Torr had only very slight effects but heavier hydrocarbons caused poisoning at even lower pressure ($\sim 10^{-7}$ Torr). When CO$_2$ was present the poisoning by the hydrocarbons was diminished, Avdienko, Malev [36].

Solid Impurities or Additives

Degradation of the emissive properties by formation of a carbon layer was recognized as the decisive effect of carbon-containing gases, Futamoto et al. [37], Avdienko, Malev [36]. Carbon on the surface is very stable against thermal desorption but may be removed by oxidation, Swanson et al. [41] or ion etching, Berrada et al. [42].

For the effect on the work function of caesium adsorbed on single-crystal faces [43, 44], see p. 244.

Alloys of LaB$_6$ with TiC, TiB$_2$, and NbC have work functions virtually identical to LaB$_6$, Samsonov et al. [45]. Attempts to change the work function by dissolution of ZrB$_2$ in LaB$_6$ failed, Storms [29]. Older studies, on the other hand, found that Ti added to the LaB$_6$ before sintering resulted in an emission current twice that of pure LaB$_6$. This was interpreted by a formation of a La metal film on the cathode surface, Kudintseva et al. [46]. Also, addition of 10% Ta was found to improve the emission properties and the resistance to thermal shock, Vikhrev et al. [47].

Addition of MoB or MoB$_2$ did not change the work function of LaB$_6$, Storms [29]; similarly stated for addition of WC, Samsonov et al. [45]. Consistently, the addition of 1 or 20% of Mo to LaB$_6$ powder for the preparation of emissive coatings by spark alloying did not significantly alter the thermionic emission of the coatings. Also the substrate, Mo or W, had no significant effect. However, at cathode temperatures above 1400 K the cathodes lost activity. Only small amounts of LaB_6 phase were present in the coatings, which consisted mainly of lower borides of Mo or W, Podchernyaeva et al. [48]. Poisoning of the cathode by Mo vapor was reported by Pelletier, Pomot [49], see also [30]. Older papers reported improved emission properties on Mo addition (10%), Vikhrev et al. [47] or a decreased work function for LaB_6 +10% W, Kudintseva et al. [50], cf. [46].

Re was found to act as a truly inert support for LaB_6 coatings. After activation, a coalesced layer of LaB_6 was found nearest to the Re surfaces, but this coalesced range did not extend to the cathode surface, Ford, Lichtman [51]; improved emission properties, however, were reported by Vikhrev et al. [47].

Addition of Co (10%) and Ni (1 and 30%) to LaB_6 or electric spark alloying of LaB_6 powder with a steel substrate gave emission properties close to that of LaB_6. The LaB_6 coatings on steel lost their emissive properties at lower temperatures than coatings on Mo or W, Podchernyaeva et al. [48]. Addition of 0.3 or 0.7% Ni was reported to lower the work function by ~0.1 eV, Samsonov et al. [45].

References:

[1] Hohn, F. J. (Scanning Electron Microsc. **1985** 1327/38).
[2] Crawford, C. K. (Scanning Electron Microsc. **1979** 19/30).
[3] Nakagawa, S. (JEOL News E 16 [1978] 2/8; C.A. **89** [1978] No. 51874).
[4] Broers, A. N. (J. Appl. Phys. **38** [1967] 1991/2).
[5] Broers, A. N. (J. Sci. Instrum. [2] **2** [1969] 273/6).
[6] Vogel, S. F. (Rev. Sci. Instrum. **41** [1970] 585).
[7] Shimizu, R.; Onoda, H.; Hagiwara, H.; Ishii, S. (J. Appl. Phys. **52** [1981] 6316/21).
[8] Shimizu, R.; Onoda, H.; Hashimoto, H.; Hagiwara, H. (J. Appl. Phys. **55** [1984] 1379/87).
[9] Oshima, C.; Aono, M.; Tanaka, T.; Kawai, S.; Shimizu, R.; Hagiwara, H. (J. Appl. Phys. **51** [1980] 1201/6).
[10] Schmidt, P. H.; Longinotti, L. D.; Joy, D. C.; Ferris, S. D.; Leamy, H. J.; Fisk, Z. (J. Vac. Sci. Technol. **15** [1978] 1554/60).

[11] Ahmed, H.; Broers, A. N. (J. Appl. Phys. **43** [1972] 2185/92).
[12] Frosien, J.; Lischke, B.; Kerner, R. (Microcircuit Eng. 84, Berlin 1984 [1985], pp. 125/30; C.A. **103** [1985] No. 225672).
[13] Yamabe, M.; Furukawa, Y.; Inagaki, T. (J. Vac. Sci. Technol. [2] A **2** [1984] 1361/4).
[14] Furukawa, Y.; Yamabe, M.; Inagaki, T. (Fujitsu Sci. Tech. J. **20** [1984] 241/57; C.A. **102** [1985] No. 14459).
[15] Furukawa, Y.; Yamabe, M.; Inagaki, T. (J. Vac. Sci. Technol. [2] A **1** [1983] 1518/21).
[16] Furukawa, Y.; Yamabe, M.; Itoh, A.; Inagaki, T. (J. Vac. Sci. Technol. **20** [1982] 199/203).
[17] Takigawa, T.; Sasaki, I. (Proc. Electrochem. Soc. **83**-2 [1983] 135/48; C.A. **98** [1983] No. 226383).
[18] Takigawa, T.; Sasaki, I.; Meguro, T.; Motoyama, K. (J. Appl. Phys. **53** [1982] 5891/7).
[19] Takigawa, T.; Yoshii, S.; Sasaki, I.; Motoyama, K.; Meguro, T. (Japan. J. Appl. Phys. **19** [1980] L537/L540).
[20] Hagiwara, H.; Hiraoka, H.; Terasaki, R.; Ishii, M.; Shimizu, R. (Scanning Electron Microsc. **1982** 473/83).

[21] Kato, T.; Shigetomi, A.; Watakabe, Y.; Hagiwara, H.; Hiraoka, H. (J. Vac. Sci. Technol. [2] B **1** [1983] 100/6).

[22] Shimizu, R.; Shinike, T.; Tanaka, T.; Oshima, C.; Kawai, S.; Hiraoka, H.; Hagiwara, H. (Scanning Electron Microsc. **1979** 11/8).

[23] Hohn, F. J.; Chang, T. H. P.; Broers, A. N.; Frankel, G. S.; Peters, E. T.; Lee, D. W. (J. Appl. Phys. **53** No. 3, Pt. 1 [1982] 1283/96).

[24] Hohn, F. J. (U.S. 267319 [1981] and U.S. 267320 [1981] from [12]).

[25] Noack, M. A.; Gibson, E. D.; Verhoeven, J. D. (J. Appl. Phys. **51** [1980] 5566/7).

[26] Tagawa, M.; Takenobu, S.; Ohmae, N.; Umeno, M. (Appl. Phys. Letters **50** [1987] 545/6).

[27] Takaoka, A.; Ura, K. (Optik **75** [1986] 20/5).

[28] Vogt, R. V. (Optik **36** [1972] 262/7).

[29] Storms, E. K. (J. Appl. Phys. **54** [1983] 1076/81).

[30] Davis, P. R.; Swanson, L. W.; Hutta, J. J.; Jones, D. L. (J. Mater. Sci. **21** [1986] 825/36).

[31] Lazorenko, V. I.; Lotsko, D. V.; Platonov, V. F.; Kovalev, A. V.; Galasun, A. P.; Matvienko, A. A.; Klinkov, A. E. (Poroshkovaya Metal. **1987** No. 3, pp. 51/7; Soviet Powder Met. Metal Ceram. **26** [1987] 229/33).

[32] Leung, N. K. (Vacuum **36** [1986] 865/7).

[33] Furukawa, Y.; Yamabe, M.; Ishizuka, T.; Inagaki, T. (J. Appl. Phys. **52** [1981] 533/4).

[34] Hosoki, S.; Yamamoto, S.; Hayakawa, K.; Okano, H. (Japan. J. Appl. Phys. **13** Suppl. 2, Pt. 1 [1974] 285/8).

[35] Gallagher, H. E. (J. Appl. Phys. **40** [1969] 44/51).

[36] Avdienko, A. A.; Malev, M. D. (Vacuum **27** [1977] 283/88).

[37] Futamoto, M.; Nakazawa, M.; Usami, K.; Hosoki, S.; Kawabe, U. (J. Appl. Phys. **51** [1980] 3869/76).

[38] Buckingham, J. D. (Brit. J. Appl. Phys. **16** [1965] 1821/32).

[39] Carter, A. F.; Wood, G. P. (MEMO-2-16-59L [1959] 1/12; C.A. **1959** 17674).

[40] Baranov, V. I.; Petrosov, V. A. (Pis'ma Zh. Tekhn. Fiz. **12** [1986] 1045/8; Soviet Tech. Phys. Letters **12** [1986] 432/3).

[41] Swanson, L. W.; Gesley, M. A.; Davis, P. R. (Surf. Sci. **107** [1981] 263/89).

[42] Berrada, A.; Mercurio, J.-P.; Etourneau, J.; Alexandre, F.; Theeten, J. B.; Duc, T. M. (Surf. Sci. **72** [1978] 177/88).

[43] Davis, P. R.; Chambers, S. A.; Swanson, L. W. (Proc. Intersoc. Energy Convers. Eng. Conf. **15** [1980] 2327/30; C.A. **94** [1981] No. 124554).

[44] Gorodetskii, D. A.; Shchudlo, Yu. G.; Vas'ko, A. A.; Koshelyuk, A. S.; Mel'nik, Yu. P. (Poverkhnost **1985** No. 3, pp. 123/6; Phys. Chem. Mech. Surf. **4** [1986] 877/84).

[45] Samsonov, G. V.; Kondrashov, A. I.; Okhremchuk, L. N.; Podchernyaeva, I. A.; Fomenko, V. S. (Poroshkovaya Metal. **1976** No. 4, pp. 89/91; Soviet Powder Met. Metal Ceram. **15** [1976] 315/7).

[46] Kudintseva, G. A.; Kuznetsova, G. M.; Ni'kulov, V. V. (Radiotekhn. Elektron. **12** [1967] 857/61; Radio Eng. Electron. Phys. [USSR] **12** [1967] 798/802).

[47] Vikhrev, Yu. I.; L'vov, G. V.; Savchenko, V. P.; Fekhretdinov, F. A. (Izv. Leningr. Elektro-tekhn. Inst. No. 104 [1971] 132/6; C.A. **80** [1974] No. 88532).

[48] Podchernyaeva, I. A.; Siman, N. I.; Verkhoturov, A. D.; Chiplik, V. N.; Morozov, V. Ya.; et al. (Poroshkovaya Metal. **1984** No. 2, pp. 50/3; Soviet Powder Met. Metal Ceram. **23** [1984] 132/5).

[49] Pelletier, J.; Pomot, C. (Appl. Phys. Letters **34** [1979] 249/51).

[50] Kudintseva, G. A.; Kuznetsova, G. M.; Mamedov, F. G.; Meerson, G. A.; Tsarev, B. M. (Izv. Akad. Nauk SSSR Neorgan. Materialy **4** [1968] 49/53; Inorg. Materials [USSR] **4** [1968] 38/42).

[51] Ford, R. R.; Lichtman, D. (J. Appl. Phys. **44** [1973] 4378/80).

32.1.8.3.4.9.3.4 Secondary Electron Emission

The yield of secondary electrons from sintered LaB_6 was above $\sigma = 0.8$ for primary beam energies of 500 to 2500 eV with a maximum of $\sigma = 0.9$ to 1 at $\sim$1000 eV. There was no influence of sample temperature, 300, 873, and 1073 K, Lenk [1].

On H^+- and He^+-bombarded LaB_6 (ion energy 20 to 27 keV, doses $\leqq (2 \text{ to } 4) \times 10^{17} \text{ cm}^{-2}$), the coefficient of secondary electron emission, $\gamma = 3.12 \pm 0.1$, was determined, Lesnyakov [2].

References:

[1] Lenk, R. (Czech. J. Phys. **6** [1956] 625/6).
[2] Lesnyakov, G. G. (Poverkhnost **1989** No. 5, pp. 147/50 from C. A. **111** [1989] No. 47 492).

32.1.8.3.4.9.3.5 Field Electron Emission

Field Emission Microscopy (FEM)

Needle-like LaB_6 crystals (Al-flux growth) were investigated by field emission microscopy (FEM). For the FEM images see the paper. To avoid effects of surface contaminants, the emitters were heated to 1650 K at pressures lower than 1×10^{-9} Torr ($\hat{=}$ 133 nPa) and then field evaporated, see p. 253. FEM images were obtained after 2 min at a certain temperature starting from 77 K up to 1800 K with a measurement of the total field emission current I as a function of the applied voltage V. For emitter temperatures up to about 800 K the increasing slope of the Fowler-Nordheim (FN) plots indicated an increase of the average work function with increasing temperature; no significant change in the FEM image (up to $\sim$725 K) was found relative to patterns of the just H_2-field evaporated emitter surface. This together with unchanged FIM patterns (up to $\sim$1000 K, cf. p. 254) indicated that La or B diffusion had no consequences below 1000 K; the increase of the work function was attributed to local surface atom rearrangement. This range of temperature is characterized by the boron-rich surface generated during purification by H_2-field evaporation, Gesley, Swanson [1].

Above about 800 K the average work function decreased with a sharp drop at 1355 K. The FEM pattern at 910 K shows enhanced electron emission from the {110} regions. In this range, surface reconstruction due to thermal equilibration takes place which finally reduces the average work function by $\sim$2.3 eV to the value characteristic of the lanthanum-rich annealed surface. In the range 1355 to 1770 K the FEM images show a dramatic change. In the upper part of the temperature range, the bright emission observed for the {210} planes proceeds to the {310} planes. Thus, the large reduction of the FN slope (representing the average work function) in this range must be attributed to the contribution of the {310} planes which exhibit the lowest work function for the thermally annealed end-form; a constant average emitter radius was inferred from the unchanged best-image voltage, Gesley, Swanson [1], see also Gesley et al. [2]. The enhanced field emission in the (211) region reported previously by Swanson et al. [3], Futamoto et al. [4], and Shimizu et al. [5] was shown to be due to heating in the presence of hydrocarbons [1]. The FEM study by Futamoto et al. [4] on thermally cleaned LaB_6 tips (by electro-etching from Al flux-grown crystals) under 2×10^{-7} Pa ($\hat{=} 1.5 \times 10^{-9}$ Torr) showed dark areas for the planes (100), {110}, and {111} and the brightest areas corresponded to {210} or {310} after heating the tip to between 1673 to 1873 K. The pattern was supposed to represent the pure LaB_6 surface; other temperatures gave irregular patterns. Above about 1073 K, surface diffusion occurred with formation of low-index planes [4]. Bright spots corresponding to {310} planes were also observed by Shimizu et al. [5] by FEM on a [100] oriented tip after heating to 1773 K for several minutes.

Field emitters of remelted LaB$_6$ were studied and revealed properties as emitters superior to tungsten, Elinson, Kudintseva [6], see also [7 to 9], similarly stated by Windsor [10].

Energy Distribution of Field Emitted Electrons

The spread of the total energy distribution was measured for a [100] oriented LaB$_6$ tip at current densities from 81×10^3 to 3.9×10^5 A/cm^2 between 1000 and 1500 K by a retarding field energy analyzer. The spread had a maximum at a certain temperature; it was wider than from cold field emitting or thermal emitting cathodes. The maximum value of the half-width was about half of that of tungsten, Zaima et al. [11]. The results were consistent with theoretical curves by Adachi et al. [12].

References:

[1] Gesley, M. A.; Swanson, L. W. (Surf. Sci. **146** [1984] 583/99).
[2] Gesley, M. A.; Davis, P. R.; Swanson, L. W. (Proc. 29th Intern. Field Emiss. Symp., Goeteborg 1982, pp. 121/7; C.A. **100** [1984] No. 78327).
[3] Swanson, L. W.; Gesley, M. A.; Davis, P. R. (Surf. Sci. **107** [1981] 263/89).
[4] Futamoto, M.; Hosoki, S.; Okano, H.; Kawabe, U. (J. Appl. Phys. **48** [1977] 3541/6).
[5] Shimizu, R.; Kataoka, Y.; Tanaka, T.; Kawai, S. (Japan. J. Appl. Phys. **14** [1975] 1089/90).
[6] Elinson, M. I.; Kudintseva, G. A. (Radiotekhn. Elektron. **7** [1962] 1511/8; Radio Eng. Electron. Phys. [USSR] **7** [1962] 1417/23).
[7] Elinson, M. I.; Gor'kov, V. A.; Kudintseva, G. A. (Proc. Symp. Electron Vac. Phys., Balatonfoldvar, Hung., 1962 [1963], pp. 151/76; C.A. **65** [1966] 4791).
[8] Elinson, M. I.; Vasil'ev, G. F. (Radiotekhn. Elektron. **2** [1957] 348/50; Radio Eng. Electron. Phys. [USSR] **2** No. 3 [1957] 126/9).
[9] Elinson, M. I.; Vasil'ev, G. F. (Radiotekhn. Elektron. **3** [1958] 945/53; Radio Eng. Electron. Phys. [USSR] **3** No. 7 [1958] 122/34).
[10] Windsor, E. E. (Proc. Inst. Elec. Eng. [London] **116** [1969] 348/50).

[11] Zaima, S.; Sase, M.; Adachi, H.; Shibata, Y.; Oshima, C.; Tanaka, T.; Kawai, S. (J. Phys. D **13** [1980] L47/L49).
[12] Adachi, H.; Shibuya, Y.; Hariu, T.; Shibata, Y. (J. Phys. D **10** [1977] L113/L115).

32.1.8.3.4.9.4 Field Evaporation

Since the vaporization energies of La and B from stoichiometric LaB$_6$ are equal ([1], see p. 90), and under conditions of uniform work functions and field strength, the difference between the activation energies for field evaporation of La and B is the difference between their first ionization potentials, 2.68 eV. Thus La will be field-evaporated at a much higher rate than B from the thermally equilibrated surface, both in the presence of noble gases or in vacuum. The evaporation field was reduced by hydrogen, presumably by boron hydride formation. Thus field evaporation of LaB$_6$ under vacuum or in presence of a noble gas occurred at 420 MV/cm, as found by FEM and FIM investigations on a needle-like LaB$_6$ crystal, whereas in the presence of H$_2$ it occurred already at 220 MV/cm, Gesley, Swanson [2].

A layer-by-layer analysis of the LaB$_6$ (001) face was carried out at 20 K under a vacuum of $\sim 10^{-8}$ Pa by using an atom-probe field ion microscope to clarify the field evaporation process on the (001) plane. The surface atoms were "pulse-field-evaporated" by voltage pulses, $V_{pulse} = 2060$ V, superimposed on the d.c. voltage, $V_{d.c.} = 16116$ V. The results indicated that field evaporation of La occurs only at the edges of the (001) plane, together with boron atoms, whereas other boron atoms evaporate continuously from the flat (001) surface planes. The evaporation field for La is lower than that of B. Only single ions evaporate (La^{3+}, B^+, B^{2+}, and some B^{3+}), Murakami et al. [3], cf. [4, 5].

From the analogous study with a pulsed field at 20 K in the presence of hydrogen (8.9×10^{-4} Pa $\cong 6.7 \times 10^{-6}$ Torr), Murakami et al. [6], it was inferred that in the hydrogen FIM image (see below) the La surface atoms make a great contribution. In mass spectra of the field evaporation, the ratio of B to La atoms was 1.5 in the presence of H_2 compared to 4.1 under vacuum; the ratio of boron hydride ions to boron ions was 0.26. The effect of H_2 on the field evaporation was explained as follows: Hydrogen reacts preferentially with boron surface atoms to form boron hydride molecules on the surface. This formation reduces the evaporation field on the boron atom sites and hence, by d.c. field evaporation of the hydride, also the apparent ratio of the number of boron atoms to the La atoms. Lanthanum atoms also form hydride molecules induced by the pulse-field, but these immediately dissociate into La^{3+} and neutral H atoms [6].

Field evaporation studies in the presence of N_2 and O_2 ($\sim 5 \times 10^{-5}$ Torr) indicate chemisorption mainly on the boron sites. Boron nitride, boron oxide, and La oxide ions were observed, Nakamura et al. [7].

References:

[1] Storms, E.; Mueller, B. (J. Phys. Chem. **82** [1978] 51/9).
[2] Gesley, M. A.; Swanson, L. W. (Surf. Sci. **146** [1984] 583/99).
[3] Murakami, K.; Adachi, T.; Kuroda, T.; Nakamura, S.; Komoda, O. (Surf. Sci. **124** [1983] L25/L30).
[4] Murakami, K.; Adachi, T.; Komoda, O.; Kuroda, T.; Nakamura, S. (Proc. 29th Intern. Field Emiss. Symp., Goeteborg 1982, pp. 257/64; C.A. **100** [1984] No. 59182).
[5] Murakami, K.; Adachi, T.; Kuroda, T.; Nakamura, S.; Komoda, O. (Shinku **26** [1983] 461/5; C.A. **99** [1983] No. 111324).
[6] Murakami, K.; Adachi, T.; Kuroda, T.; Nakamura, S. (Surf. Sci. **176** [1986] 327/35).
[7] Nakamura, S.; Ng, Y. S.; Tsong, T. T.; McLane, S. B., Jr. (Surf. Sci. **87** [1979] 656/64).

32.1.8.3.4.9.5 Field Ion Microscopy (FIM)

Needle-like LaB_6 crystals (Al-flux growth) were studied by field ion microscopy (FIM), see the figures in the paper, Gesley, Swanson [1], cf. [2]; for a parallel study by field emission microscopy and the discussion of the results for both, see p. 252. The image gas pressure was typically 1×10^{-4} Torr ($\cong 13$ mPa). Best image field (BIF) was near 220 MV/cm for H_2 image gas, it was constant at all emitter annealing temperatures. FIM patterns were recorded at 77 K starting from the clean, H_2 field-evaporated surface, see p. 253, with heating of the LaB_6 emitter for 2 min to temperatures up to 1800 K. The FIM patterns up to ~ 1000 K showed no significant change relative to that of the H_2 field-evaporated surface. At 1295 K the {110} planes are no longer resolved. In the range 1355 to 1770 K the FIM patterns show a growth of the {111}, {110}, {112} planes, and to some extent of {100} [1], for earlier data see also Swanson et al. [3]. A field ion microscopic study on Al flux-grown crystals etched to fine tips was made with He, Ne, and H_2 as imaging gases at cell voltages of 9.2 kV (He) to 4.8 kV (H_2). FIM patterns were stable with He and Ne; with H_2 slow etching of the LaB_6 tip occurred. The atomic arrangement was clearly resolved on the top (001) layers in the Ne FIM pattern at ~ 70 K. From this, the presence of only one kind of surface atom, La or B, was inferred (other arguments indicating La), Futamoto, Kawabe [4].

An FIM pattern with Ne as image gas was obtained at 78 K, it was suggested that only boron atoms are shown in this pattern; with H_2, no stable cubic structure was formed because of continuous evaporation of boron hydrides [5], see also preliminary results in Nakamura et al. [6]. FIM studies with H_2 showed that the surface La atoms make a great contribution to a hydrogen gas ion image of the (001) plane [7].

FIM studies at 77 K of vapor-deposited LaB₆ on tungsten with He (10^{-5} Torr $\triangleq 1.3$ mPa) as imaging gas indicated a reproducible epitaxial layer of LaB₆ on W {001} and {111} planes. The best image voltage (BIV) decreased from 2.5 kV for the randomly deposited LaB₆ to 2.3 kV after heating for 30 s at 1200 K; the actually imaged species were thought to be La atoms, Joag et al. [8], for earlier reports see also [9 to 11].

Yet there are some strong arguments that only boron atoms show up in a noble gas ion image of the LaB₆ (001) plane [5, 6], cf. [1] (and not lanthanum ions [4, 8, 12]), while lanthanum atoms contribute to the image formation in the presence of hydrogen [7], cf. [6].

References:

[1] Gesley, M. A.; Swanson, L. W. (Surf. Sci. **146** [1984] 583/99).
[2] Gesley, M. A.; Davis, P. R.; Swanson, L. W. (Proc. 25th Intern. Field Emiss. Symp., Goeteborg 1982, pp. 121/7; C.A. **100** [1984] No. 78327).
[3] Swanson, L. W.; Gesley, M. A.; Davis, P. R. (Surf. Sci. **107** [1981] 263/89).
[4] Futamoto, M.; Kawabe, U. (Surf. Sci. **93** [1980] L117/L123).
[5] Murakami, K.; Adachi, T.; Kuroda, T.; Nakamura, S.; Komoda, O. (Surf. Sci. **124** [1983] L25/L30).
[6] Nakamura, S.; Ng, Y. S.; Tsong, T. T.; McLane, S. B., Jr. (Surf. Sci. **87** [1979] 656/64).
[7] Murakami, K.; Adachi, T.; Kuroda, T.; Nakamura, S. (Surf. Sci. **176** [1986] 327/35).
[8] Joag, D. B.; Kanitkar, P. L.; Kanitkar, M. M. (Bull. Mater. Sci. **6** [1984] 573/7).
[9] Kanitkar, P. L.; Dharmadhikari, C. V.; Joag, D. S.; Shukla, V. N. (J. Phys. D **9** [1976] L165/L166).
[10] Dharmadhikari, C. V.; Joag, D. S.; Kanitkar, P. L. (Phys. Status Solidi A **42** [1977] K99/K101).

[11] Dharmadhikari, C. V.; Joag, D. S.; Shukla, V. N.; Kanitkar, P. L. (Proc. Nucl. Phys. Solid State Phys. Symp. C **19** [1976] 506/9).
[12] Futamoto, M.; Aita, T.; Kawabe, U. (Japan. J. Appl. Phys. **14** [1975] 1263/6).

32.1.8.3.4.9.6 Uses

Lanthanum hexaboride is a valuable thermionic emitter of electrons. It is used in a wide variety of devices including X-ray sources, electron-beam laser-pump systems, thermionic energy converters, ion beam sources, halogen atomic beam detectors, negative ion surface ionizers, and many high-brightness electron-beam devices. The latter uses include scanning electron microscopes, transmission microscopes, electron probes, scanning Auger systems, and electron lithography systems. The basic advantage of LaB₆ and other boride cathodes lies in their high electron emission per gram of material lost to evaporation, see, e.g., Crawford [1], who mainly describes electron-beam devices; some relevant data are recorded in Section 32.1.8.3.4.9.3.3, p. 247. Use of LaB₆ in a thermionic energy converter was reported by Gun'ko et al. [2]. A design of a microfocus X-ray tube with a LaB$_6$ cathode is described by Aoki, Sakayanagi [3].

Large area cathodes capable of high current densities were designed from polycrystalline sintered LaB$_6$. For example, an indirectly heated LaB$_6$ disk was used in a plasma generator (some 10^{-4} Torr of H$_2$ or Ar); it was capable of emitting up to 600 A of electrons ($\triangleq$ 20 A/cm^2), Goebel et al. [4], see also Goebel et al. [5]. Directly heated hairpin-shaped LaB$_6$ cathodes were developed that use a tapered filament to provide a uniform filament temperature; also a directly heated cylindrical LaB$_6$ cathode for a plasma generator was developed, Leung et al. [6 to 8].

An early design of an ion source from helium or hydrogen was given by Paderno et al. [9]. A long-pulse ion source for neutral beam injections was developed with filament-free LaB$_6$ multicathodes, Kubota et al. [10, 11]. The good ionization properties of LaB$_6$ for halogens, especially of iodine, were used for intense halogen negative ion sources with a porous LaB$_6$ disk as cathode, Pelletier et al. [12], see also [13]. Other designs for negative ion sources, mainly from halogens are described in [14 to 19].

Additional applications: Resistor materials from mixtures of LaB$_6$ and calciumalumino-borate with a constantly small temperature coefficient between 223 and 423 K were described by Baudry et al. [20], Smirnov, Makhmudbekov [21]. LaB$_6$ resistor elements for use in thermal printer heads are described by Shinmi et al. [22].

References:

[1] Crawford, C. K. (Scanning Electron Microsc. **1979** No. 1, pp. 19/30).
[2] Gun'ko, V. M.; Kucherov, R. Ya.; Oganezov, Z. A.; Tskhakaya, V. K.; Yarygin, V. I. (Poverkhnost **1983**, No. 2, pp. 71/9 from C.A. **98** [1983] No. 201329).
[3] Aoki, S.; Sakayanagi, Y. (Japan. J. Appl. Phys. **20** [1981] 2419/20).
[4] Goebel, D. M.; Hirooka, Y.; Sketchley, T. A. (Rev. Sci. Instrum. **56** [1985] 1717/22).
[5] Goebel, D. M.; Crow, J. T.; Forrester, A. T. (Rev. Sci. Instrum. **49** [1978] 469/72).
[6] Leung, K. N. (Vacuum **36** [1986] 865/7).
[7] Pincosy, P. A.; Leung, K. N. (Rev. Sci. Instrum. **56** No. 5, Pt. 1 [1985] 655/8).
[8] Leung, K. N.; Moussa, D.; Wilde, S. B. (Rev. Sci. Instrum. **57** [1986] 1274/6).
[9] Paderno, Yu. B.; Romanyuk, L. I.; Fomenko, V. S. (Ukr. Fiz. Zh. **8** [1963] 707/8).
[10] Kubota, Y.; Uramoto, J.; Miyahara, A. (Japan. J. Appl. Phys. **21** [1982] 164/7).

[11] Kubota, Y.; Uramoto, J.; Miyahara, A. (IPPJ-513 [1981] 1/17; C.A. **95** [1981] No. 71718).
[12] Pelletier, J.; Pomot, C.; Cocagne, J. (J. Appl. Phys. **50** [1979] 4517/23).
[13] Pelletier, J.; Pomot, C. (Proc. 13th Intern. Conf. Phenom. Ionized Gases, Berlin 1977, Vol. I, pp. 15/6; C.A. **89** [1978] No. 69374).
[14] Chu, A.; Chang, I. D. (J. Spacecr. Rockets **19** [1982] 284/6; C.A. **97** [1982] No. 94955).
[15] Vosicki, B.; Björnstad, T.; Carraz, L. C.; Heinemeier, J.; Ravn, H. L. (Nucl. Instrum. Methods Phys. Res. **186** No. 1/2 [1981] 307/13).
[16] Shmid, M.; Engler, G.; Yoresh, I.; Skurnik, E. (Nucl. Instrum. Methods Phys. Res. **186** No. 1/2 [1981] 349/51).
[17] Dong, W. D.; Kilpatrik, W. D.; Teem, J. M.; Zuccaro, D. E. (Progr. Astronaut. Aeron. **9** [1963] 269/89; C.A. **62** [1965] 6332).
[18] Persky, A.; Greene, E. F.; Kuppermann, A. (J. Chem. Phys. **49** [1968] 2347/57).
[19] Mårtenson, B. M.; Wilhelmsson, S. O. (Intern. J. Mass Spectrom. Ion Processes **67** [1985] 179/89).
[20] Baudry, H.; Monneraye, M.; Ortega, F. (Fr. Demande 2397704 [1978/79] 1/13; C.A. **91** [1979] No. 185583).

[21] Smirnov, M. A.; Makhmudbekov, I. B. (Izv. Vysshikh Uchebn. Zavedenii Neft Gaz **16** [1973] 99/102; C.A. **79** [1973] No. 109218).

[22] Shinmi, A.; Hara, T.; Fukui, M.; Shirato, Y.; Hajimoto, Y. (Japan. Kokai Tokkyo Koho 79-63294 [1979] 1/8; C.A. **91** [1979] No. 132125).

32.1.8.3.4.10 Optical Properties

Color

LaB$_6$ crystals show a typical violet, Mamedov et al. [1], or red-violet color, Betsch, White [2]; powders are reddish purple when dry and deep red when moist, Lafferty [3]. After brief etching, crystals (from Al flux) became bright purple and gradually turned blue-violet on long exposure to air, Zhukova et al. [4].

A change of color from purple for stoichiometric to blue for boron-rich samples was observed by Johnson, Daane [5], Ermakov et al. [6], cf. [1]. Motojima et al. [7]. The composition at the color change is given as roughly LaB$_{6.07}$, McKelvy et al. [8]. Excess of lanthanum apparently had no effect on the (violet) color [6].

Reflectivity. Optical Constants

The optical constants of LaB$_6$ were evaluated from the reflectivity of bulk samples and of thin films, which also allowed transmission measurements.

The normal reflectivity was determined at photon energies between 0.05 and 40 eV on large crystals or from mosaics of up to 60 single-crystal platelets from Al flux crystal growth, Shelykh et al. [9], Gurin et al. [10]. The reflectivity R shows a deep minimum (R<5%) at 2.1 eV and maxima (R = 20 to 30%) at 3.5, 5.5, 7.5, 9.4, 11.6, and 15 eV, see **Fig. 85**, p. 258. By a Kramers-Kronig analysis the refractive index n, the extinction coefficient k, the real part ε' and the imaginary part ε'' of the dielectric function, and the energy loss function $\varepsilon''/[(\varepsilon')^2 + (\varepsilon'')^2] = -\mathrm{Im}\,\varepsilon^{-1}$ were calculated; they are included in Fig. 85, p. 258. (For data evaluated directly from ellipsometry see p. 259.) The maximum of the energy loss function is at 1.98 eV. At energies lower than about 2 eV the optical properties are characteristic of free electrons; at higher energies there are interband transitions. The energy dependence of the absorption coefficient indicated a forbidden interband transition at E = 2.1 eV with an effective electron number of 0.7 per La atom. The dielectric constant $\varepsilon_\infty = 1 + \varepsilon'_b = 4.6$ (see p. 82) was derived by considering the effect of interband transitions in addition to the conduction electrons. Therefore the energy of the plasma frequency ω_p (= 4.3 eV) did not correspond with the frequency of the reflection minimum [10]. The maximum in the energy loss function at 27 to 30 eV and the structure in the range 7 to 10 eV were tentatively related to plasma vibrations of the valence electrons; the difference from the calculated energy of about 20 eV was said to be due to interband transitions. The peak at about 19 eV coincides with the calculated frequency of surface plasmons to within 10% [9]. For the refractive index n and the extinction coefficient k see also Fig. 34 and Fig. 35, respectively, on p. 81, and for the reflectivity see Fig. 36 and Fig. 37, p. 82.

The reflectivity of a single crystal at 300 K for photon energies between 0.002 and 5 eV and the optical conductivity (from a Kramers-Kronig analysis) could not be fitted by the simple Drude expression. Introduction of an energy-dependent relaxation time with four adjustable parameters in addition to the plasma energy $\hbar\omega_p = 2.75$ eV gave good agreement, Kwon et al. [29]. Other experimental studies of the optical properties: reflectivity from 400 to 50000 cm^{-1} ($\hat{=}$ 0.05 to 6.2 eV), Ivanchenko et al. [11, 12]; reflectivity from 100 to 50000 cm^{-1} (0.01 to 6.2 eV), Bessaraba et al. [13]; reflectivity from 0.05 to 6 eV, Henrion, Örtel [14]; reflectivity from 1 to 6 eV, Niemyski, Kierzek-Pecold [15]; reflectivity from wavelength $\lambda = 20$ to 0.3 μm ($\hat{=}$ 0.06 to

4.1 eV), Kauer [16]; reflectivity and transmission from $\lambda = 2.5$ to $0.2\,\mu m$ ($\triangleq 0.5$ to 6.2 eV), Peschmann et al. [17]. The data on the energy E in eV of the reflectivity minimum and results of the Kramers-Kronig analysis, i.e. high-frequency dielectric constant ε_∞, plasma frequency ω_p, relaxation times τ_{opt} from the relation between ε' and $\varepsilon'' \cdot \omega$, and τ_p from the half-width of the peak in the energy loss function, and relative effective mass of the conduction electrons m^*/m_0 are listed in the following table (z.r. = zone refined):

sample	$E(R_{min})$ in eV	ε_∞	$\hbar\omega_p$ in eV	τ_{opt} in 10^{-15} s	τ_p in 10^{-15} s	m^*/m_0	$N/(m^*/m_0)$ in 10^{22} cm^{-3}	Ref.
z.r.	2.08	13.5	2.01	—	4.1	0.35[a]	3.96	[11]
fused	2.1	12.0	1.91	1.5	6.6	0.44	3.2	[12]
fused	2.1	13.5	1.93	—	2.66	0.33	—	[13]
z.r.	2.10	15.5	—	—	—	0.33[a]	4.2	[15]
sintered	2.1	15.6	—	—	—	0.32	—	[16]
film[b]	1.98	12.0	1.69	—	1.47	0.56	—	[13]
film	1.93	12.0	1.69	0.7	1.6	0.56	—	[12]

[a] Calculated from $N/(m^*/m_0)$ using $N = 1.4 \times 10^{22}$ cm^{-3}. – [b] Thickness 0.4 to $2\,\mu m$.

From the zero crossing of ε_f' (contribution of the conduction electrons, see p. 82) and dielectric losses, $\hbar\omega_p = 4.3$ eV, $\tau_{opt} = 30 \times 10^{-15}$ s, $\tau_p = 20 \times 10^{-15}$ s, and $m^*/m_0 = 1.04$ were derived by [10], cf. [9].

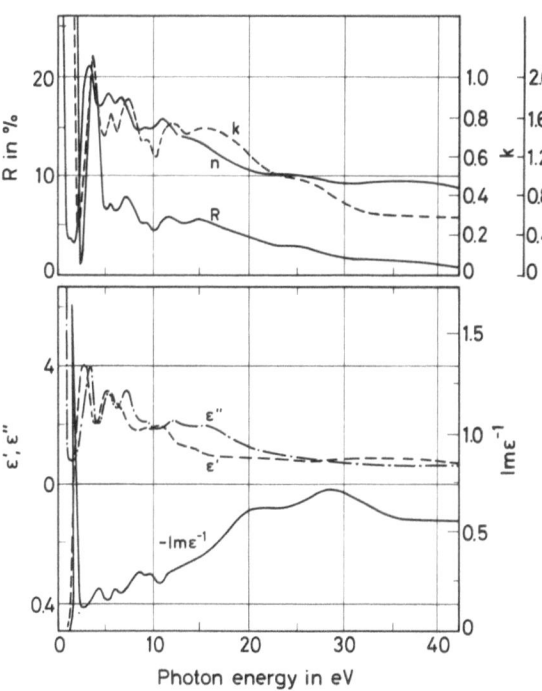

Fig. 85. Reflectivity R, refractive index n, extinction coefficient k, real (ε') and imaginary (ε'') part of the dielectric function, and the energy loss function ($-\mathrm{Im}\,\varepsilon^{-1}$) of LaB$_6$ as a function of the photon energy.

The charge carrier mobility was evaluated as $\mu_{opt} = 12.3$ and 2.0 cm$^2 \cdot$V$^{-1} \cdot$s^{-1} for the fused and thin film sample, respectively [13].

There is only a minor temperature shift of the plasma reflection edge for temperatures from 295 to 1080 K (displacement to higher energy), but a pronounced levelling of the edge, Kauer [16].

More recently the optical properties of the (001) single-crystal surface were investigated using spectroscopic ellipsometry in the range 0.6 to 6.0 eV. By means of this method, the dielectric constants ε' and ε'' can be calculated directly (without KK analysis), see **Fig. 86**. The plasmon was found at $\hbar\omega_p = 2.0$ eV and the scattering time is 4.3×10^{-15} s. The comparison of the experimental optical conductivity σ with that evaluated from band structure calculations, see **Fig. 87**, indicates that the intraband (Drude) contribution to σ is only important at $E < 1$ eV; the shoulder at 1.6 eV (corresponding with the theoretical peak at 2.2 eV) stems from transitions at the top of the boron p complex to the empty La d level. The experimental feature at 4.0 eV originates from the same initial states, but the final states are here the higher lying empty f levels of lanthanum; it corresponds with the theoretical feature which starts at 4 eV [18].

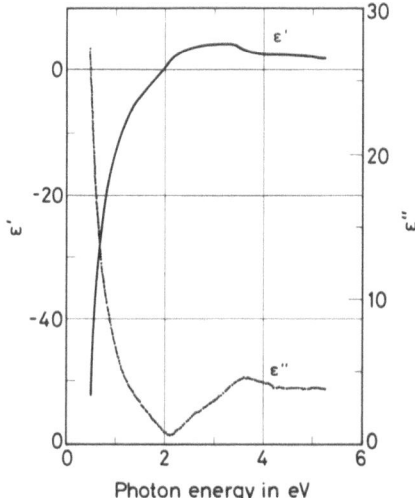

Fig. 86. Real part (ε') and imaginary part (ε'') of the dielectric function of LaB$_6$.

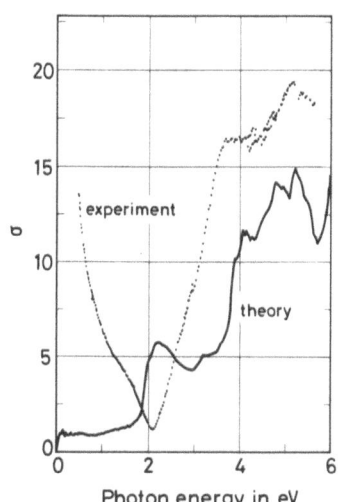

Fig. 87. Experimental and theoretical optical conductivity of LaB$_6$, see text.

In agreement with the interpretation given above (by [9, 10]), Ivanchenko et al. [11], cf. [12], also interpreted the reflectivity as due to the quasi-free conduction electrons below about 2 eV, whereas the interband transitions start at 2.1 eV; maxima in ε'', k, σ, and $-\text{Im}\varepsilon^{-1}$ vs. hν were attributed to transitions from the valence band to maxima in the density of state of the conduction band at 3.6 to 3.9 and 4.9 to 5.1 eV above E_F, i.e. from boron p states to La 5d states [13].

Deviations from the stoichiometric composition induce only slight changes in the position of the reflection minimum E and the plasma frequency ω_p, whereas the constant ε_∞ was significantly different. Differences between the studies of different authors were attributed to different compositions or surface states [11]. For the reflection spectrum (0.05 to 6 eV) of an Ar$^+$ implanted crystal, see [14].

The analysis of reflectivity and transmission data of various films on pyrex or vitreous silica from electron-beam evaporation and sputtering showed the optical properties to deviate from values extrapolated from the bulk properties for films of the same thickness even after annealing at 1273 K. Perturbations by interstitials, vacancies, and possibly by impurities reduce the mobility and relaxation time of the conduction electrons, Peschmann et al. [17]; see the paper for plots of the optical constants n, k, and the energy loss function of LaB_6 films. The reflection minima of sputtered LaB_6 films on glass or vitreous silica (thickness from 0.05 to 0.66 µm) were observed at 2.0 to 2.2 eV, Winsztal et al. [19]. Reflection minima around 600 to 700 nm (2.1 to 1.8 eV) were evaluated from transmission spectra of films (thickness 10 to 1000 nm) in the spectral range 800 to 330 nm (1.55 to 3.8 eV), Paderno et al. [20]. Differently, only a diffuse plasma edge, shifted to $(22 \text{ to } 24) \times 10^3 \text{ cm}^{-1}$ (≈ 2.9 eV), was observed for a 0.3 µm thick LaB_6 film on the Si(111) face. This was attributed to deviations in the composition and to strain in the lattice; a thicker film, 1.5 µm, showed the plasma edge at 16300 to 16500 cm^{-1} ($\triangleq 2.03$ eV), Shaginyan et al. [21].

The reflectivity data on sintered LaB_6 samples, indicating semiconductor properties by Tsarev, Illarionov [22], are apparently superseded.

Emissivity

The spectral emissivity e of LaB_6 at a wavelength $\lambda = 650$ nm between 1600 and 2100 K was described by $e = 1.2144 - 2.467 \times 10^{-4} \cdot T_s$ with T_s the observed apparent surface temperature in K. No effect of surface roughness or composition (purple or blue La–B phases) was found within experimental error, Storms [23]. At temperatures below 1600 K, $e = 0.82$ was assumed, whereas above this temperature the relation from [23] was applied by Swanson et al. [24]. A value $e = 0.7$ at $\lambda = 655$ µm was given by Kudintseva, Tsarev [25]. A value e close to 1 was found appropriate rather than $e = 0.8$ for pyrometric thermometry in the range 1373 to 1623 K, Schmidt et al. [26]

Additional Data

For X-ray emission and absorption spectra, see p. 227. For Raman spectra see p. 195. For piezoreflectivity see p. 83.

Crystals of Ca-doped LaB_6 are laser-active (p-n junction laser) at $\lambda = 930$ nm on excitation at a current density of 6000 A/cm^2, Vickery [27].

A surface luminescence ($0.43 \lesssim \lambda \lesssim 0.53$ µm) was observed for H$^+$- and He$^+$-bombarded LaB_6 at the ion energies 20 to 26 keV for doses $\leqq (2 \text{ to } 4) \times 10^{17} \text{ cm}^{-2}$. A memory effect with respect to the surface state during irradiation was found, Lesnyakov [28].

References:

[1] Mamedov, F. G.; Meerson, G. A.; Zhuravlev, N. N.; Umanskii, Ya. S. (Izv. Akad. Nauk SSSR Neorgan. Materialy **3** [1967] 950/6; Inorg. Materials [USSR] **3** [1967] 851/6).

[2] Betsch, R. J.; White, W. B. (Proc. 12th Rare Earth Res. Conf., Vail, Colo., 1976, Vol. 2, pp. 534/41).

[3] Lafferty, J. M. (J. Appl. Phys. **22** [1951] 299/309).

[4] Zhukova, T. B.; Korsukova, M. M.; Nardov, A. V.; Gurin, V. N. (Izv. Akad. Nauk SSSR Neorgan. Materialy **17** [1981] 353/4).

[5] Johnson, R. W.; Daane, A. H. (J. Phys. Chem. **65** [1961] 909/15).

[6] Ermakov, S. V.; Mamedov, F. G.; Meerson, G. A.; Tsarev, B. M. (Izv. Akad. Nauk SSSR Neorgan. Materialy **3** [1967] 808/12; Inorg. Materials [USSR] **3** [1967] 722/6).

[7] Motojima, S.; Takahashi, Y.; Sugiyama, K. (J. Cryst. Growth **44** [1978] 106/9).

[8] McKelvy, M. J.; Eyring, L.; Storms, E. K. (J. Phys. Chem. **88** [1984] 1785/90).

[9] Shelykh, A. I.; Sidorin, K. K.; Karin, M. G.; Bobrikov, V. N.; Korsukova, M. M.; Gurin, V. N.; Smirnov, I. A. (J. Less-Common Metals **82** [1981] 291/6).

[10] Gurin, V. N.; Korsukova, M. M.; Karin, M. G.; Sidorin, K. K.; Smirnov, I. A.; Shelykh, A. I. (Fiz. Tverd. Tela [Leningrad] **22** [1980] 715/20; Soviet Phys.-Solid State **22** [1980] 418/21).

[11] Ivanchenko, L. A.; Paderno, Yu. B.; Pilyankevich, A. N.; Bekenev, V. L.; Koval'chuk, V. V.; Onipko, A. F.; Perepelitsa, N. I.; Sichkar, V. V.; Chernenko, L. I. (Izv. Akad. Nauk SSSR Neorgan. Materialy **16** [1980] 1551/5; Inorg. Materials [USSR] **16** [1980] 1056/9).

[12] Ivanchenko, L. A.; Paderno, Yu. B.; Pilyankevich, A. N. (Poroshkovaya Metal. **1978** No. 8, pp. 38/48; Soviet Powder Met. Metal Ceram. **17** [1978] 602/9).

[13] Bessaraba, V. I.; Ivanchenko, L. A.; Paderno, Yu. B. (J. Less-Common Metals **67** [1979] 505/9).

[14] Henrion, W.; Örtel, G. (Phys. Halbleiteroberfläche **19** [1988] 161/5 from C.A. **110** [1989] No. 201889).

[15] Niemyski, T.; Kierzek-Pecold, E. (J. Cryst. Growth **3/4** [1968] 162/5).

[16] Kauer, E. (Phys. Letters **7** [1963] 171/3).

[17] Peschmann, K. R.; Calow, J. T.; Knauff, K. G. (J. Appl. Phys. **44** [1973] 2252/6).

[18] van der Heide, P. A. M.; ten Cate, H. W.; ten Dam, L. M.; de Groot, R. A.; de Vroomen, A. R. (J. Phys. F **16** [1986] 1617/23).

[19] Winsztal, S.; Majewska-Minor, H.; Wisniewska, M.; Niemyski, T. (Mater. Res. Bull. **8** [1973] 1329/35).

[20] Paderno, Yu. B.; Ivanchenko, L. A.; Bessaraba, V. I.; Vereshchak, V. M. (Poroshkovaya Metal. **1975** No. 6, pp. 106/8; Soviet Powder Met. Metal Ceram. **14** [1975] 515/6).

[21] Shaginyan, L. R.; Chernyaev, V. N.; Kondrashin, A. A.; Bessaraba, V. I. (Poroshkovaya Metal. **1981** No. 9, pp. 88/91; Soviet Powder Met. Metal Ceram. **20** [1981] 659/61).

[22] Tsarev, B. M.; Illarionov, S. V. (Poroshkovaya Metal. **1962** No. 6, pp. 85/8; Soviet Powder Met. **1962** 468/70).

[23] Storms, E. K. (J. Appl. Phys. **50** [1979] 4450).

[24] Swanson, L. W.; Gesley, M. A.; Davis, P. R. (Surf. Sci. **107** [1981] 263/89).

[25] Kudintseva, G. A.; Tsarev, B. M. (Radiotekhn. Elektron. **3** [1958] 428/30; Radio Eng. Electron. Phys. [USSR] **3** [1958] 182/5).

[26] Schmidt, P. H.; Longinotti, L. D.; Joy, D. C.; Ferris, S. D.; Leamy, H. J.; Fisk, Z. (J. Vac. Sci. Technol. **15** [1978] 1554/60).

[27] Vickery, R. C. (U.S. 3340108 [1963/67] 1/3; C.A. **68** [1968] No. 74014).

[28] Lesnyakov, G. G. (Poverkhnost **1989** No. 5, pp. 147/50 from C.A. **111** [1989] No. 47492).

[29] Kwon, Y. S.; Kimura, S.; Nanba, T.; Kunii, S.; Ikezawa, M.; Suzuki, T.; Kasuya, T. (J. Phys. Colloq. [Paris] **49** [1988] C8-737/C8-738).

32.1.8.3.4.11 Surface Spectroscopy and Surface Structure

For studies concerning the surface electron states, see p. 222.

Surface Spectroscopy

In order to clarify the origin of the low work functions of some LaB$_6$ surfaces, the surface structure of clean low-index planes was studied in a vacuum of 10^{-7} to 10^{-8} Pa by low-energy electron diffraction (LEED), Auger electron spectroscopy (AES), also angle-resolved (ARAES), angle-resolved X-ray photoelectron spectroscopy (ARXPS), and by ion scattering spectroscopy (ARISS) using He$^+$ ions of 1 keV at ~ 0.05 μA.

The surfaces, generally cut from zone-refined samples, were cleaned, for example, by several cycles of bombardment with Ar ions (1 keV, ~1 µA/cm²) and electron-beam heating from behind the sample to ca. 1573 to 1773 K, Aono et al. [1], or by heating to 1400 K in 10^{-4} Pa of oxygen (elimination of carbon as CO) and subsequent heating to above 1700 K (desorption of oxygen), Swanson et al. [2]. LEED patterns of the (100) and (110) surfaces can be recorded up to ~1473 K, above this temperature the brightness of the sample prevents further measurements, see Aono et al. [3], Nishitani et al. [4]. The table on p. 263 assembles results on LEED and gives a survey on the other studies performed (E_p = primary electron energy, ϑ = polar angle of emission direction with respect to surface normal; Φ = azimuthal angle; the various ratios La/B mean intensity ratios; Auger transitions include La(78 eV)NOO, La(625 eV)MNN, and B(179 eV)KLL). For additional data see pp. 226, 228.

Surface Structure

Fig. 88 from Gesley, Swanson [10] shows the surface structure of the four clean surface planes at room temperature, as proposed in the above-mentioned studies:

The (100) surface is unreconstructed and the outermost layer consists of La atoms, based on ARXPS [1, 3] and ARAES, Chambers, Swanson [6].

The (110) surface has the relaxed and reconstructed c(2 × 2) structure with the La atoms displaced by 1.66 Å toward the surface, by ARXPS. This interpretation is supported by the temperature dependence of the LEED spot size in contrast to that of the surface electronic state at ca. 2 eV below the Fermi level, also by the LEED spot size at high temperatures and the ARXPS peak intensity ratio La 4d/B 1s at high polar angle, which both decrease in the order (100) > (110) > (111) surfaces, Nishitani et al. [4]. For an early, different interpretation of the c(2 × 2) pattern see Bas et al. [7]. An unreconstructed nonpolar (110) surface corresponding to the LEED (1×1) pattern at room temperature, observed at high primary electron energies by Oshima et al. [8] and Swanson et al. [2], does not explain the work function of this surface relative to that of (100) and (111) [2]. On the other hand, in the case of the reconstructed (110) and (111) surfaces shown in Fig. 88, significant surface ionic charge modifications are predicted if the work function involves dipole contributions from both the surface and the bulk, Watson, Perlman [11].

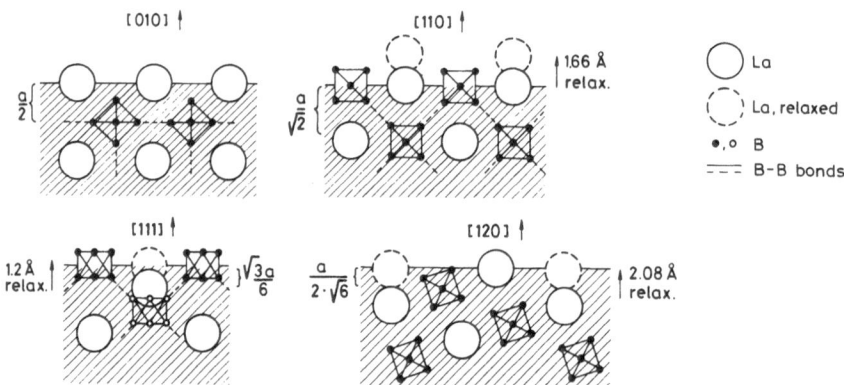

Fig. 88.   Side view of the (010), (110), (111), and (120) faces of LaB₆. Dashed circles represent the proposed displacements of La atoms.

plane	E_p in eV	LEED pattern	T in K	comments and other studies	Ref.
(100)	130	1×1	300	ARXPS: La 4d/B 1s = f(θ) at Φ = 0° and 45°	[1]
	128	1×1	300	AES: La(78)/B = const. for sputter-cleaned surface with subsequent heating (≦1973 K)	[5]
	13	1×1	300 to 1473	LEED: spot size is constant up to ~1273 K, increases rapidly near 1373 K, seems to become independent of T≳1473 K	[3]
				AES: $B(KLL)/La(N_{4,5}O_{2,3}O_{2,3})$ = const. for T = 300 to 1773 K; B 1s/La 4d and B 1s/La 4p = const., T≦1773 K	[6]
	—	—	—	ARAES: polar and azimuthal profiles of the $La(M_5N_{4,5}N_{4,5})$ and $B(KL_1L_1)$ Auger intensities; advantages and disadvantages of ARAES in comparison to AES and ARXPS	
(110)	39	c(2×2)	300	—	[7]
	13	c(2×2) / 1×1	300 to ~1123 / ≳1123 to 1373	LEED: spot size is const. up to ca. 1073 K, then increases and becomes const. again ≳1273 K	[4]
				ARXPS: La 4d/B 1s = f(θ) at various values of Φ	
(111)	23	1×1	300 to 1573	LEED: spot size is const. up to ca. 1473 K, then increases slightly	[4]
				ARXPS: La 4d/B 1s = f(θ) at various values of Φ	
				ARISS: the intensity of the La peak reveals some shadowing effect at the impact angle of 20°	
(100), (110), (111)	100	1×1	300	AES: see p. 264	[2]
	113 to 120	1×1	300	—	[8]
(120)	32	1×1	300	ARXPS: La 4d/B 1s = f(θ) for analyzer rotating in the (001) or (2̄10) plane	[9]

The (111) surface has the relaxed (1×1) structure with the La atoms displaced toward the surface, near the plane that contains the center of the surface boron octahedra. The displacement by 1.2 Å agrees with ARXPS and predicts a total shadowing effect in ARISS at an impact angle of 10° [4]. The ARXPS data are also consistent with a reconstructed nonpolar (111) surface formed by a displacement of originally outermost La atoms into the bulk. For both models of reconstruction, significant surface ionic charge modifications are predicted [11], see p. 262.

The type of reconstruction of the (120) surface in Fig. 88, p. 262, is based on LEED and ARXPS data by Oshima et al. [9], the amount of relaxation (2.08 Å) of La atoms is derived from relations presumed to exist between the value of the work function on different crystal faces and surface dipoles [10]; see p. 240.

The surface composition of the clean LaB_6 (100) surface is apparently independent of the temperature between room temperature and ~1773 K according to the constant ratio of the heights of the B (KLL) versus the La ($N_{4.5}O_{2.3}O_{2.3}$) Auger peaks, Aono et al. [3], see also Goldstein, Szostak [5] for heating up to 1973 K, and Futamoto et al. [12] for temperatures between 1673 and 1973 K after annealing for 10 h at 1973 K; similarly Swanson et al. [2], see below. Variations of the intensity ratio of La ($\leqq 95$ eV) and B Auger lines in polycrystalline material during cleaning by ion bombardment (Ar$^+$, 200 eV, 4 µA/cm^2) and thermal treatment operations (298 or 773 K) were attributed to diminuation in La at the surface during cleaning and again approaching stoichiometric composition during heating, Berrada et al. [13]. A La-deficient surface of crystals with bulk compositions $LaB_{5.86}$ and $LaB_{5.74}$ after a short initial thermal cleaning above 1700 K is inferred from the observed intensity ratio of La (NOO) and B (KLL) Auger peaks. The La (MNN) peak intensity remained mostly unchanged from temperature variations. The surface composition again becomes richer in La during heating. Reproducible results are obtained after several hours heating at ~1800 K [2]. Similar studies indicate an increase of the La concentration at the surface of in situ cleaved LaB_6 (100) on annealing at 1700 K [10]. In general the dependence of the apparent surface composition on the bulk stoichiometry and the thermal history could be accounted for [2] within the model of Storms, Mueller [14, 15], see below. For more details see the paper [2].

From the ratio of XPS intensities of the La 4d and B 1s peaks an atomic ratio B/La = 4.0 was inferred for (100), Aono et al. [16]. A surface ratio B/La = 2.3 to 2.6 was derived after short thermal cleaning at 1700 K, Swanson, Dickinson [17] and B/La = 2.0, Korsukova, Gurin [18], both from AES data on (100).

A comparison of the various high- and low-temperature AES intensity ratios of La and B peaks of at ~1800 K thermally equilibrated surfaces indicates a general increase of the ratio B/La in the sequence (100) < (110) < (111), Swanson et al. [2].

The surface composition of sintered lanthanum boride samples was derived from measurements of the vapor composition above these samples in the temperature range between 1400 and 2100 K. The samples were prepared by arc melting mixtures of B and La with ratios between 5.3 and 8.7, Storms, Mueller [14], see also [15]. The surface composition was found close to the congruently vaporizing composition (CVC), which is at B/La = 6.034 at 1500 K and at B/La = 6.042 at 1700 K. For interior ratios B/La = 5.3 to 8.58, the surface composition at 1500 K changes just slightly from B/La = 6.021 to 6.100 [14], see also [15, 19]. The results were criticized by Nordine, Schiffman [20] as only inappropriately allowing for reactions of the effusion cell, see p. 205.

References:

[1] Aono, M.; Oshima, C.; Tanaka, T.; Bannai, E.; Kawai, S. (J. Appl. Phys. **49** [1978] 2761/4).
[2] Swanson, L. W.; Gesley, M. A.; Davis, P. R. (Surf. Sci. **107** [1981] 263/89).

[3] Aono, M.; Nishitani, R.; Oshima, C.; Tanaka, T.; Bannai, E.; Kawai, S. (J. Appl. Phys. **50** [1979] 4802/7).

[4] Nishitani, R.; Aono, M.; Tanaka, T.; Oshima, C.; Kawai, S.; Iwasaki, H.; Nakamura, S. (Surf. Sci. **93** [1980] 535/49).

[5] Goldstein, B.; Szostak, D. J. (Surf. Sci. **74** [1978] 461/78).

[6] Chambers, S. A.; Swanson, L. W. (Surf. Sci. **131** [1983] 385/402).

[7] Bas, E. B.; Hafner, P.; Klauser, S. (Proc. 7th Intern. Vacuum Congr., Vienna 1977, Vol. 2, pp. 881/4; C.A. **88** [1978] No. 95206).

[8] Oshima, C.; Bannai, E.; Tanaka, T.; Kawai, S. (J. Appl. Phys. **48** [1977] 3925/7).

[9] Oshima, C.; Aono, M.; Tanaka, T.; Nishitani, R.; Kawai, S. (J. Appl. Phys. **51** [1980] 997/1000).

[10] Gesley, M.; Swanson, L. W. (Surf. Sci. **146** [1984] 583/99).

[11] Watson, R. E.; Perlman, M. L. (Surf. Sci. **122** [1982] 371/82).

[12] Futamoto, M.; Nakazawa, M.; Usami, K.; Hosoki, S.; Kawabe, U. (J. Appl. Phys. **51** [1980] 3869/76).

[13] Berrada, A.; Mercurio, J.-P.; Etourneau, J.; Alexandre, F.; Theeten, J. B.; Duc, T. M. (Surf. Sci. **72** [1978] 177/88).

[14] Storms, E. K.; Mueller, B. A. (J. Appl. Phys. **50** [1979] 3691/8).

[15] Storms, E.; Mueller, B. (J. Phys. Chem. **82** [1978] 51/9).

[16] Aono, M.; Tanaka, T.; Bannai, E.; Kawai, S. (Appl. Phys. Letters **31** [1977] 323/5).

[17] Swanson, L. W.; Dickinson, T. (Appl. Phys. Letters **28** [1976] 578/80).

[18] Korsukova, M. M.; Gurin, V. N. (Curr. Topics Mater. Sci. **11** [1984] 390/439, 428/9).

[19] Storms, E. K. (J. Appl. Phys. **52** [1981] 2961/5).

[20] Nordine, P. C.; Schiffman, R. A. (High Temp. Sci. **20** [1985] 1/20).

32.1.8.3.4.12 Chemical Behavior

Behavior against Hydrogen, Deuterium

LaB$_6$ powder is catalytically active in the hydrogen/deuterium equilibration after heating to ~1100 K in a vacuum greater than 10^{-6} Torr (~130 µPa). The reaction was studied in the range from 146 to 179 K and pressures up to 40 Torr (5.3 kPa). The thermal desorption spectrum of hydrogen (LaB$_6$ activated at 1373 K, exposed to H$_2$ at room temperature, cooled to 78 K) shows three peaks around 120 K (β_1), 160 K (β_2), and 340 K (α). The desorption spectrum of deuterium is similar. The activation energy for the desorption of the β_1 hydrogen is $E_A = 19.7 \pm 3$ kJ/mol, in good agreement with $E_A = 20.9 \pm 0.8$ kJ/mol for the equilibration. The β_1 hydrogen was adsorbed dissociatively and is responsible for the reaction. HD formation was shown to proceed via the Bonhoeffer-Farkas mechanism even at low temperature. La vacancies were thought to be responsible for the catalytic effect because less than 1% of the surface was active. For the effect of pre-adsorbed CO see the paper, Nagaki et al. [1].

Behavior against Oxygen

Adsorption and Desorption

1 L (Langmuir) = 10^{-6} Torr·s.

Oxygen adsorption on single-crystal faces was studied under UHV with oxygen pressures of, for example, ~1×10^{-9} Torr (130 nPa), Nishitani et al. [2, 3], Gorodetskii et al. [4, 5], up to 1×10^{-7} Torr (13 µPa), Goldstein, Szostak [6], using low-energy electron diffraction (LEED)

[2 to 8], UPS [2, 3], Auger electron emission (AES) [4, 6, 8, 9], and angle-resolved Auger electron emission (ARAES), Chambers, Swanson [10], together with work function measurements. For a study of thermal desorption of oxygen from polycrystalline LaB_6, see also Tkach et al. [11].

On the (100) Face. At room temperature, adsorption of oxygen gave LEED patterns up to exposures of several hundred L. First adsorption mainly on boron sites up to exposures of 1.4 L ($\triangleq$ relative coverage $\Theta = 0.5$) and then mainly on La sites was deduced from the change of the work function Φ and the photoelectron spectra. The oxygen sticking coefficient initially increases from zero up to ~1 for an exposure of 1.4 L and decreases again, becoming close to zero at exposures higher than ~5 L [2]. Saturation is reached at exposures above ~6 L; two oxygen atoms per unit cell are adsorbed on the surface [3]. Adsorption of oxygen on B and La sites was confirmed by Chambers, Swanson [10] using ARAES, but in contrast, these authors inferred an oxygen density of only 1 per unit cell [9] in agreement with Goldstein, Szostak [6]; both studies suggested preferred oxygen positions above the boron octahedra with participation of the surface La atoms in the bonding of oxygen. A poor LEED pattern was obtained on room temperature adsorption (in contrast to [2]) indicating an amorphous state; the pattern became clear after heating the sample at 1273 or 1523 K [6]. A clear c(2×2) LEED pattern on the (100) face, observed at an oxygen exposure (~10 L [5]) that caused an increase of Φ by 1.1 eV, was interpreted by ordered adsorption of ~1.5 oxygen atoms per surface unit cell (with all boron sites occupied), Gorodetskii et al. [4].

During thermal desorption of oxygen from LaB_6 (100) a sequence of different LEED patterns occurred, which were attributed to different degrees of the desorption process: a (2×1) structure between 900 and 1200 K was assigned to a degree of coverage $\Theta = 0.5$; a (3×1) and a (3×3) structure between 1200 and 1300 K to $\Theta = 0.33$ and 0.11, respectively; a (4×4) structure at above 1300 K to $\Theta = 0.06$, Gorodetskii et al. [4].

Measurements of the thermal desorption of oxygen from (100), both on pre-adsorbed oxygen layers in vacuum and under steady state conditions in fixed O_2 pressure revealed several species as desorption products; the desorption energies (in eV) were 4.3 ± 0.2 for LaO, 3.6 ± 0.1 and 4.3 ± 0.1 for BO, 3.3 ± 0.1 and 3.5 ± 0.1 for B_2O_2, and 3.4 ± 0.1 for B_2O_3, Davis, Chambers [12].

On the Faces (110), (111), and (120). On oxygen exposure, the LEED pattern of the clean (110) face disappears and a (1×1) pattern appears for exposures between 0.33 and ~4 L, Nishitani et al. [3], consistent with observations by Bas et al. [8]. The oxygen atoms are supposed to be adsorbed mainly by surface boron atoms but not at a particular site [3]. Saturation is reached between ~6 L (from figures in [3, 9]) and 16 L [8], and an oxygen density of 2 per unit cell is assumed [3, 8]. In contrast only one adsorption site per unit cell was assumed by Chambers et al. [9]. The initial sticking coefficient on (110) is large [3], a value of 0.42 is calculated by [8].

On the (111) face, oxygen is supposed to be adsorbed mainly to the surface boron octahedra for exposures up to ~2 L; at higher exposures oxygen is also adsorbed on other sites, voids surrounded by three surface La atoms and three surface boron octahedra. At exposures of 2 L, the amount of oxygen on the surface is about half of the saturation value at about 100 L, which corresponds to two oxygen atoms per unit cell [3].

On the (120) face, the oxygen exposure of 15 L saturates the increase in the work function Φ and makes the LEED pattern disappear, Oshima et al. [7].

On flash desorption of the oxygen-covered (110) face, only LaO (with a desorption energy of 4.2 eV) and BO species (3.4, 3.9, and 4.1 eV) were found in mass spectra, Bas et al. [8]. Heating to between 1500 and 1650 K under oxygen (~10^{-6} Torr (~130 µPa), exposures up to 5×10^4 L) gave restructuring of the (110) and (111) surface to form (100) faces, Klauser, Bas [13]. Under poor vacuum, single-crystal surfaces transformed to a polycrystalline appearance during 2 d at 1823 K, Oshima et al. [14], see also [15].

Reactions with O$_2$ and Air

LaB$_6$ is oxidized in air at room temperature, forming a thin film (20 to 30 Å) of La$_2$O$_3$ and a boron oxide. No oxidation was observed below 873 K under a vacuum of 10^{-8} to 10^{-9} Torr ($\hat{=}10^{-6}$ to 10^{-7} Pa). However, after several hours above 873 K under a vacuum of 10^{-7} to 10^{-8} Torr, the formation of La$_2$O$_3$ and a boron oxide could be established on the surface, Berrada et al. [16].

Oxidation of LaB$_6$ single crystals in air was studied by thermogravimetry and differential thermal analysis in the range 933 to 1473 K. On isothermal heating, surficial oxidation started at 1173 K; at 1273 K the oxidation rate was slightly increased, La$_2$O$_3$ deposits were observed, but the weight gain saturated within about 2 h. The oxidation isotherm at 1373 K started with an initial smooth weight increase, which became much steeper after $\sim$2 h. This behavior was attributed to the formation of a protective B$_2$O$_3$ film. At 1473 K there was a strong increase in weight. On heating at a rate of 15 K/min, oxidation started at $\sim$1333 K, Paderno et al. [17]. The oxidation of the single crystals started at 100 K lower than observed earlier on a hot-pressed sample by Voitovich, Pugach [18]. This was interpreted by an initially higher oxygen content in the hot-pressed sample, which was said to block further attack of oxygen on the LaB$_6$ grains. The oxidation kinetics were described by mixed rate laws that were due to the complex phase composition [17]. Previous thermogravimetric and X-ray studies of the LaB$_6$-surface oxidation in air at 1073 and 1273 K were expressed by the reaction: $4\,LaB_6 + 21O_2 \rightarrow 4\,LaBO_3 + 10\,B_2O_3$. The amount μ of LaB$_6$ oxidized per unit surface area follows the law $\mu = a \cdot t^k$ where t is the time of oxidation and $k = 0.75$ at 1073 K and 0.95 at 1273 K, Hsu, Kuo [19].

The oxidation products were mainly B$_2$O$_3$, La$_2$O$_3$, but also BN, B, and some LaB$_4$ according to X-ray and metallographic analyses [18]. Differential thermoanalyses on LaB$_6$ powders with heating in air at 20 K/min showed peaks at 673 to 773 K and at 1018 to 1073 K, the size of both depended on the particle size. The first effect was due to surface oxidation. According to X-ray and crystal-optical analysis the scale from oxidation up to $\sim$988 K consisted of particles of B$_2$O$_3$, B$_2$O$_3 \cdot$ mH$_2$O, and La(OH)$_3$; at 1123 K no B$_2$O$_3$, but La(OH)$_3$ and La$_2$O$_3 \cdot 3\,$B$_2$O$_3$ ($\hat{=}$ La(BO$_2$)$_3$) was detected, Tel'nikov et al. [20]; for crystal-optical investigations of oxidized LaB$_6$ see also the previous paper of Lugovskaya [21].

Oxidation in oxygen, studied between 973 and 1473 K, gave a scale of mainly LaB$_4$ and B$_2$O$_3$ between 1073 and 1273 K and of LaB$_4$, B$_2$O$_3$, and La$_2$O$_3$ above 1273 K. The oxidation rate depended strongly on the oxygen pressure and sample porosity, it followed a parabolic rate from 973 to 1473 K, Lavrenko et al. [22].

Behavior against Additional Gases

Reactions with F$_2$/H$_2$F$_2$

The stability of 35 materials including LaB$_6$ against a hydrogen-fluorine flame, "HF torch" was tested. A protective fluoride film formed rapidly on LaB$_6$, which was the best of the tested materials (CaB$_6$, La$_2$O$_3$, Si$_3$N$_4$, Ni, and others), Holcombe et al. [23].

Reactions with CO and CO$_2$

The thermal desorption spectrum of CO on a clean LaB$_6$ surface (adsorption at room temperature) shows a single peak at $\sim$340 K and a broad peak above 1000 K, Nagaki et al. [1].

The reaction of LaB_6 with CO_2 at 773 to 1273 K was studied by thermogravimetry, X-ray, and chemical analyses. LaB_6 is stable below 973 K. Above this temperature, the main reaction follows the equation $2LaB_6 + 21CO_2 \rightarrow La_2O_3 \cdot 3B_2O_3 + 3B_2O_3 + 21CO$ and to a less extent, $4LaB_6 + 21CO_2 \rightarrow 2La_2O_3 \cdot 3B_2O_3 + 6B_2O_3 + 21C$. The fraction of reaction R changes with the time t according to the rate law $1 - (1 - R)^{\frac{1}{3}} = k_1 \cdot t$ at low temperatures and to $[1 - (1 - R)^{\frac{1}{3}}]^2 = k_2 \cdot t$ at high temperatures. Weight-loss curves and log k vs. 1/T plots are given in the paper, Xu et al. [24].

Behavior against Carbon

LaB_6 and graphite do not significantly interact below 2323 K, Rabenau et al. [25], cf. also Colton [26], Crawford [27]. LEED patterns of the surface carbon layer, formed by carbon diffusion from the bulk, showed a twelvefold symmetry and were explained by a patchwork of hexagonal nets consistent with the graphite (0001) layer. Heating at 1200 to 1900 K resulted in changes of the properties that were interpreted as a transition to C–La and C–B bonds; the hexagonal nets were no longer present, Oshima et al. [28].

Behavior against Metals

The caesium adsorption on the LaB_6 (100) face at 77 K was studied by LEED, and two structures with a square unit cell and ad-atom concentrations of 2.9×10^{14} and 4.0×10^{14} cm^{-2} were observed in the first monolayer. Further deposition led to a hexagonal structure with a concentration of 4.0×10^{14} cm^{-2} in each layer, Gorodetskii et al. [29]. For the adsorption of Cs and the Cs + oxygen co-adsorption on a (100) face at room temperature, see Gorodetskii et al. [5] and on a polycrystalline sample see Gorodetskii et al. [4]. The adsorption of Cs and the co-adsorption properties of Cs and oxygen were studied by AES on (100) and (110) faces. Precoverage of oxygen increases the ionic character of the Cs bond, Chambers et al [9]. The adsorption of Cs at room temperature is apparently limited to a single layer. Thermal desorption of Cs adsorbed on the clean (100) face yields activation energies of 1.2 and 1.5 eV, which, on oxygen pre-adsorption, shift to 1.4 and 1.7 eV. The terminal coverage Cs desorption energy increased from 2.3 to 3.0 eV [30].

Wetting and contact interaction of drops of liquid Fe with sintered LaB_6 samples were studied under excess pressure of 3×10^4 Pa of He at 1813 and 1923 K. Dimples under the drops after 20 min contact indicated vigorous chemical reaction. Electron probe microanalysis of a gray phase at the melt-substrate interface was explained by the formation of a Fe solid solution in LaB_6, Yupko et al. [31]. Similar studies of the wetting behavior of LaB_6 by Co and Ni melts in the temperature range 1373 to 1948 K showed dissolution of LaB_6 in the liquid phase with formation of $LaM_{12}B_6$, M = Co, Ni in the metal drops, and solid solutions of Co or Ni in LaB_6, Yupko et al. [32]. The reaction intensity increases in the order Fe, Co, Ni [31]. Addition of 1 to 2% of Ni to LaB_6 powder as sintering activator, followed by hot pressing at 2000 to 2350 K, led to the formation of $LaNi_{12}B_6$, LaB_4, and Ni_3B at the LaB_6 grain boundaries. The effect of 0.3 to 2 wt% Ni on the microstructure is shown in figures in the paper, Kondrashev et al. [33].

Wetting and contact interactions of LaB_6 with liquid Cu at 1583 ± 10 K were studied by Verkhovodov et al. [34], of Cu and a Cu–(5 to 15%)Ni alloy by Yupko et al. [35], and of LaB_6 with Cu–Bi and Cu–Sb alloys between 1370 and 1770 K by Yupko et al. [36].

Solubility

LaB_6 is very soluble in nitric acid and is insoluble in water, hydrochloric acid, aqueous KOH, aqueous NaOH, and alcohol, Schmidt et al. [37].

The dissolution rate of hot-pressed or single-crystalline LaB$_6$ rods in aqueous nitric acid with <30% HNO$_3$ is low; ~0.02 mm/min for rods with 1 mm diameter is read from a figure in the paper. Above 30 wt%, the dissolution rate increases to ~0.8 mm/min at 50 wt% HNO$_3$. A LaB$_6$ anode (Pt cathode) can be dissolved in nitric acid with 10 wt% HNO$_3$ in a controllable way at potentials of 1.6 to 2.5 V, Aida et al. [38].

LaB$_6$–MB$_2$ Systems

A survey on the LaB$_6$–MB$_2$ systems with M = Ti, Zr, Hf, V, Nb, Ta, and Cr revealed that LaB$_6$ and MB$_2$ form quasi-binary eutectics. The eutectic temperature increases and the MB$_2$ contents of the eutectic decrease with increasing MB$_2$ stability (melting point, heat of formation), Ordan'yan [39].

References:

[1] Nagaki, T.; Inoue, Y.; Kojima, I.; Yasumori, I. (J. Phys. Chem. **84** [1980] 1919/25).

[2] Nishitani, R.; Kawai, S.; Iwasaki, H.; Nakamura, S.; Aono, M.; Tanaka, T. (Surf. Sci. **92** [1980] 191/200).

[3] Nishitani, R.; Oshima, C.; Aono, M.; Tanaka, T.; Kawai, S.; Iwasaki, H.; Nakamura, S. (Surf. Sci. **115** [1982] 48/60).

[4] Gorodetskii, D. A.; Tskhakaya, V. K.; Shchudlo, Yu. G.; Yarygin, V. I.; Yas'ko, A. A. (Izv. Akad. Nauk SSSR Ser. Fiz. **46** [1982] 1224/9; Bull. Acad. Sci. USSR Phys. Ser. **46** No. 7 [1982] 6/10).

[5] Gorodetskii, D. A.; Koshelyuk, A. S.; Nonik, V. P.; Tskhakaya, V. K.; Shchudlo, Yu. G.; Yarygin, V. I.; Yas'ko, A. A. (Poverkhnost **1983** No. 10, pp. 79/82; Phys. Chem. Mech. Surf. **2** [1985] 2991/9).

[6] Goldstein, B.; Szostak, D. J. (Surf. Sci. **74** [1978] 461/78).

[7] Oshima, C.; Aono, M.; Tanaka, T.; Nishitani, R.; Kawai, S. (J. Appl. Phys. **51** [1980] 997/1000).

[8] Bas, E. B.; Hafner, P.; Klauser, S. (Proc. 7th Intern. Vacuum Congr., Vienna 1977, Vol. 2, pp. 881/4).

[9] Chambers, S. A.; Davis, P. R.; Swanson, L. W.; Gesley, M. A. (Surf. Sci. **118** [1982] 75/92).

[10] Chambers, S. A.; Swanson, L. W. (Surf. Sci. **131** [1983] 385/402).

[11] Tkach, A. V.; Paderno, Yu. B.; Paderno, V. N.; Lazorenko, V. I. (Stud. Surf. Sci. Catal. **23** [1985] 221/2).

[12] Davis, P. R.; Chambers, S. A. (Appl. Surf. Sci. **8** [1981] 197/205).

[13] Klauser, S. J.; Bas, E. B. (Appl. Surf. Sci. **3** [1979] 356/63).

[14] Oshima, C.; Aono, M.; Tanaka, T.; Kawai, S.; Shimizu, R.; Hagiwara, H. (J. Appl. Phys. **51** [1980] 1201/6).

[15] Oshima, C.; Bannai, E.; Tanaka, T.; Kawai, S. (J. Appl. Phys. **48** [1977] 3925/7).

[16] Berrada, A.; Mercurio, J.-P.; Etourneau, J.; Alexandre, F.; Theeten, J. B.; Duc, T. M. (Surf. Sci. **72** [1978] 177/88).

[17] Paderno, Yu. B.; Pugach, E. A.; Lazorenko, V. I.; Galasun, A. P.; Lavrinenko, L. N.; Filipchenko, S. I. (Poroshkovaya Metal. **1984** No. 4, pp. 74/8; Soviet Powder Met. Metal Ceram. **23** [1984] 316/9).

[18] Voitovich, R. F.; Pugach, E. A. (Poroshkovaya Metal. **1973** No. 2, pp. 71/5; Soviet Powder Met. Metal Ceram. **12** [1973] 145/8).

[19] Hsu, H.-P.; Kuo, K.-H. (Hua Hsueh Hsueh Pao **31** [1965] 271/6; C.A. **64** [1966] 3001).

[20] Tel'nikov, E. Ya.; Vinitskii, I. M.; Lugovskaya, E. S.; Rud', B. M. (Izv. Akad. Nauk SSSR Neorgan. Materialy **24** [1988] 602/5; Inorg. Materials [USSR] **24** [1988] 498/501).

[21] Lugovskaya, E. S. (Poroshkovaya Metal. **1982** No. 6, pp. 58/61; Soviet Powder Met. Metal Ceram. **21** [1982] 473/5).

[22] Lavrenko, V. A.; Glebov, L. A.; Lugovskaya, Y. S.; Frantsevich, I. N. (Oxid. Metals 7 [1973] 131/9).

[23] Holcombe, C. E., Jr.; Weber, G. W.; Kovach, L. (Am. Ceram. Soc. Bull. **58** [1979] 1185/8, 1192).

[24] Xu, X. (Hsu, H.-P.); Peng, M.; Gao, J. (Huaxue Xuebao **40** [1982] 233/42; C. A. **97** [1982] No. 84096).

[25] Rabenau, A.; Kauer, E.; Klotz, H. (Colloq. Int. Centre Natl. Rech. Sci. [Paris] No. 157 [1967] 495/8).

[26] Colton, E. (Proc. 12th Rare Earth Res. Conf., Vail, Colo., 1976, Vol. 2, pp. 1026/31).

[27] Crawford, C. K. (Scanning Electron Microsc. **1979** No. 1, pp. 19/30).

[28] Oshima, C.; Bannai, E.; Tanaka, T.; Kawai, S. (Japan. J. Appl. Phys. **16** [1977] 965/9).

[29] Gorodetskii, D. A.; Shchudlo, Yu. G.; Yas'ko, A. A.; Koshelyuk, A. S.; Mel'nik, Yu. P. (Poverkhnost **1985** No. 3, pp. 123/6; Phys. Chem. Mech. Surf. **4** [1986] 877/84).

[30] Chambers, S. A.; Davis, P. R.; Swanson, L. W. (Surf. Sci. **118** [1982] 93/102).

[31] Yupko, V. L.; Verkhovodov, P. A.; Morozov, V. V.; Besov, A. V.; Shlyuko, V. Ya. (Poroshkovaya Metal. **1984** No. 1, pp. 60/2; Soviet Powder Met. Metal Ceram. **23** [1984] 57/9).

[32] Yupko, V. L.; Verkhovodov, P. A.; Morozov, V. V.; Besov, A. V.; Shlyuko, V. Ya. (Poroshkovaya Metal. **1981** No. 3, pp. 64/8; Soviet Powder Met. Metal Ceram. **20** [1981] 207/10).

[33] Kondrashev, A. I.; Paderno, V. N.; Martynenko, A. N.; Paderno, Yu. B.; Smirnov, V. P. (Poroshkovaya Metal **1983** No. 6, pp. 96/102; Soviet Powder Met. Metal Ceram. **22** [1983] 506/12).

[34] Verkhovodov, P. A.; Kuz'mina, T. I.; Levchenko, G. V.; Luban, R. B.; Yupko, V. L. (Poroshkovaya Metal. **1979** No. 8, pp. 59/63; Soviet Powder Met. Metal Ceram. **18** [1979] 559/62).

[35] Yupko, V. L.; Levchenko, G. V.; Verkhovodov, P. A.; Luban, R. B.; Yadova, O. V. (Poroshkovaya Metal. **1981** No. 7, pp. 68/72; Soviet Powder Met. Metal Ceram. **20** [1981] 498/501).

[36] Yupko, V. L.; Levchenko, G. V.; Luban, R. B.; Kryzhanovskaya, R. I. (Poroshkovaya Metal. **1986** No. 11, pp. 64/9; Soviet Powder Met. Metal Ceram. **25** [1986] 918/22).

[37] Schmidt, P. H.; Longinotti, L. D.; Joy, D. C.; Ferris, S. D.; Leamy, H. J.; Fisk, Z. (J. Vac. Sci. Technol. **15** [1978] 1554/60).

[38] Aida, T.; Futamoto, M.; Kawabe, U. (Japan. J. Appl. Phys. **18** [1979] 1393/4).

[39] Ordan'yan, S. S. (Izv. Akad. Nauk SSSR Neorgan. Materialy **24** [1988] 235/8; Inorg. Materials [USSR] **24** [1988] 173/6).

32.1.8.3.5 LaB$_6$–MB$_6$ Systems

32.1.8.3.5.1 The LaB$_6$–"ScB$_6$" System

The existence of pure ScB$_6$ is questionable, see p. 126. Single phase solid solutions in the La$_{1-x}$Sc$_x$B$_6$ system were observed only with x=0.1, samples with x>0.1 consisted of a multiphase system, as shown by X-ray diffraction studies of polycrystalline samples, Samsonov et al. [1, 2], Kondrashov [3]. Earlier the existence of homogeneous solid solutions with x≦0.5 was assumed by Tsolovski, Peshev [4]. The samples were obtained by the borothermal reduction of the oxide mixtures in vacuum for 1 h (2 h by [3]). Compacted samples were made by vacuum hot pressing with subsequent annealing for 10 h under argon [1 to 3]. The reduction was carried out at 2123 K by [1], at 2073 K by [3]; hot-pressing temperatures ranged from 2103 to 2373 K [1]; annealing temperature was 1773 K [1, 3]; annealing at 1.5 atm ($\triangleq$ 0.15 MPa) argon pressure was published in [3].

Lattice constants of La$_{1-x}$Sc$_x$B$_6$ solid solutions were found to be independent of x and equal to that of LaB$_6$ [1 to 4].

The microhardness H$_\mu$ = 2994 ± 217 kg/mm^2 ($\triangleq$ 29.4 ± 2 GPa) under a load of 100 g was measured for La$_{0.9}$Sc$_{0.1}$B$_6$ [1, 2].

Thermionic electron emission measurements were performed on single phase La$_{1-x}$Sc$_x$B$_6$ (and multiphase samples with x > 0.1) up to about 1900 K. The emission of La$_{0.9}$Sc$_{0.1}$B$_6$ is unstable and the work function is greater than that of LaB$_6$, for details and figures, see the paper [1, 2]. The results of previous thermionic emission measurements from about 1200 to about 1580 K (see figures in the paper) of [4] were questioned by [1].

References:

[1] Samsonov, G. V.; Kondrashov, A. I.; Okhremchuk, L. N.; Podchernyaeva, I. A.; Siman, N. I.; Fomenko, V. S. (Poroshkovaya Metal. **1977** No. 1, pp. 21/8; Soviet Powder Met. Metal Ceram. **16** [1977] 16/22).

[2] Samsonov, G. V.; Kondrashov, A. I.; Okhremchuk, L. N.; Podchernyaeva, I. A.; Siman, N. I.; Fomenko, V. S. (J. Less-Common Metals **67** [1979] 415/8).

[3] Kondrashov, A. I. (Poluch. Issled. Svoistv. Nov. Mater. Mater. 10th Nauchn. Konf. Aspir. Molodykh Issled. Inst. Probl. Materialoved. Akad. Nauk Ukr. SSR, Kiev 1976 [1978], pp. 48/52; C.A. **91** [1979] No. 111144).

[4] Tsolovski, I. A.; Peshev, P. D. (Dokl. Bolg. Akad. Nauk **25** [1972] 209/12; C.A. **77** [1972] No. 25711).

32.1.8.3.5.2 The LaB$_6$–YB$_6$ System

Formation. Preparation

Metallographic investigations and X-ray diffraction studies showed the existence of a continuous series of solid solutions La$_{1-x}$Y$_x$B$_6$, Samsonov et al. [1], see also Fisk et al. [2]. The samples were prepared by borothermal reduction of mixtures of the oxides at ~1873 K, Fisk et al. [2], or at 2073 K for the La-rich compositions and at 1973 K for the Y-rich compositions, Samsonov et al. [1], see also Sobczak, Sienko [3]. In contrast, arc melting of the mixture of the two hexaborides with the nominal composition La$_{0.5}$Y$_{0.5}$B$_6$ gave two cubic phases corresponding to values x = 0.15 ± 0.05 and x = 1, as well as tetraboride contamination. The observed immiscibility may be due to the morphology of the liquidus and solidus curves in the ternary phase diagram, favoring nearly complete fractionation [2].

Crystals were grown from the oxides plus boron in an Al flux, see p. 179, but were mixed with crystals of YB$_4$. The compositions obtained, x = 0.02, 0.06, and 0.53, were within 15 mol% of the initial ratio of the oxides, Olsen, Cafiero [4].

Properties

The change of the lattice parameters with the composition of the La$_{1-x}$Y$_x$B$_6$ solid solutions between a = 4.156 Å for LaB$_6$ and a = 4.102 Å for YB$_6$ is reasonably well described by Vegard's law, Fisk et al. [2], see also a figure in the papers of Samsonov et al. [1], Kovenskaya et al. [8]; a = 4.1552 Å at x = 0.02 and 4.1526 Å at x = 0.06 were given by Olsen, Cafiero [4]. Figures in [8] show a maximum at x = 0.8 for the expansion coefficient α (300 to 1200 K) of $\sim 6.6 \times 10^{-6}$ K^{-1} and minima at x = 0.8 for the microhardness (~ 22 GPa) and the melting point (~ 2800 K); the Debye temperature increases smoothly with x.

The electrical resistivity, in $\mu\Omega \cdot$cm, of sintered La$_{1-x}$Y$_x$B$_6$ samples at room temperature increases from $\varrho = 12.0$ at x = 0 to a maximum of 54.7 at x = 0.8 and decreases to $\varrho = 38.7$ at x = 1, Kovenskaya et al. [8], see also [5]. A maximum at x = 0.8 was also observed for the effective electron mass (1.15 m$_0$) and minima at x = 0.8 for Hall coefficient ($- 5.89 \times 10^{-10}$ m^3/C), Seebeck coefficient ($- 4.91\,\mu$V/K), carrier concentration (1.06×10^{22} cm^{-3}), carrier mobility (10.8 cm$^2 \cdot$V$^{-1} \cdot$s^{-1}), and Fermi energy (1.49 eV); additional values are tabulated [8].

The solid solutions La$_{1-x}$Y$_x$B$_6$ are superconducting at lower transition temperatures T$_c$ than YB$_6$. The following critical temperatures were reported: T$_c$ = 1.6 $\pm$ 0.1 K at x = 0.5 and T$_c$ = 4.4 $\pm$ 0.1 K at x = 0.6, Sobzcak, Sienko [3]; T$_c$ = 4, 5.8, and 6.8 K for x = 0.67, 0.80, and 0.91, respectively (read from a figure), Fisk et al. [2]; T$_c$ = 6.0 $\pm$ 0.1 K for YB$_6$ [3].

After annealing at 1800 K, polycrystalline La$_{1-x}$Y$_x$B$_6$ samples were stable electron emitters. The thermionic work function (presumably Φ_{eff}) increases strongly with the yttrium content and, for a given composition, it increases slightly with the temperature according to $\Phi_T = \Phi_0 + d\Phi/dT \cdot T$ with $\Phi_0 = 2.66$, 2.77, 3.64, and 3.33 eV, $d\Phi/dT = 2.0 \times 10^{-4}$, 1.6×10^{-4}, 4.0×10^{-5}, and 4.5×10^{-4} eV/K, for x = 0, 0.1, 0.5, and 0.8, respectively, measured in the ranges (in K): 1500 to 1675, 1435 to 1720, 1625 to 1910, and 1480 to 1775, respectively, Samsonov et al. [1], see also [6]. Addition of 0.5% Y$_2$O$_3$ to LaB$_6$ increases the work function slightly, Shlyuko et al. [7].

References:

[1] Samsonov, G. V.; Kondrashov, A. I.; Okhremchuk, L. N.; Podchernyaeva, I. A.; Siman, N. I.; Fomenko, V. S. (Poroshkovaya Metal. **1977** No. 1, pp. 21/8; Soviet Powder Met. Metal Ceram. **16** [1977] 16/22).

[2] Fisk, Z.; Lawson, A. C.; Fitzgerald, R. W. (Mater. Res. Bull. **9** [1974] 633/6).

[3] Sobzcak, R. J.; Sienko, M. J. (J. Less-Common Metals **67** [1979] 167/71).

[4] Olsen, G. H.; Cafiero, A. V. (J. Cryst. Growth **44** [1978] 287/90).

[5] Samsonov, G. V.; Neshpor, V. S.; Serebryakova, T. I. (Inzh. Fiz. Zh. **1959** No. 2, pp. 118/20; N.S.A. **13** [1959] No. 20926).

[6] Samsonov, G. V.; Kondrashov, A. I.; Okhremchuk, L. N.; Podchernyaeva, I. A.; Siman, N. I.; Fomenko, V. S. (J. Less-Common Metals **67** [1979] 415/8).

[7] Shlyuko, V. Ya.; Morozov, V. V.; Besov, A. V.; Chernyak, L. V.; Guzenko, L. V. (Poroshkovaya Metal. **1974** No. 7, pp. 29/33; Soviet Powder Met. Metal Ceram. **13** [1974] 544/6).

[8] Kovenskaya, B. A.; Kondrashov, A. I.; Dudnik, E. M.; Kolotun, V. F. (Tugoplavkie Soedin. Redkozemel. Met. Mater. 3rd Vses. Semin., Novosibirsk 1977 [1979], pp. 36/9; C.A. **93** [1980] No. 60 231).

32.1.8.3.5.3   The LaB$_6$–YB$_6$–"ScB$_6$" System

Polycrystalline single phase La$_{1-x}$Sc$_{0.5x}$Y$_{0.5x}$B$_6$ solid solutions with x = 0.1 and 0.2 were prepared by the borothermal reduction of corresponding metal oxides under 10^{-4} Torr ($\cong$13 mPa) at 2073 K for 1 h. Hot-pressed samples were annealed under vacuum for 10 h at 1773 K at 1.5 atm ($\cong$0.15 MPa) argon pressure, Kondrashov et al. [1], Kondrashov [2]. Products with x > 0.2 consisted of a multiphase system [2]. Solid solutions with x = 0.1 and 0.2 crystallize in the CaB$_6$ structure (see p. 33). The lattice parameters decrease slightly with increasing x [1], see a figure in [2].

Thermionic electron emission was measured between ~1300 and ~1900 K at a residual pressure of 10^{-7} to 10^{-8} Torr ($\triangleq$13 to 1.3 μPa). The work function increases with increasing x, and follows the equation $\Phi_T = \Phi_0 + T \cdot d\Phi/dT$ (between ~1500 and ~1900 K) with $\Phi_0 = 2.48$ and 2.59 eV and (both samples) $d\Phi/dT = 3.5 \times 10^{-4}$ eV/K for x = 0.05 and 0.1, respectively. The passivating effect of Sc on the thermionic emission properties of LaB$_6$ might be connected with a preferential buildup of Sc atoms on the surface during heating and an increased activation energy for the diffusion of La in the presence of Sc [1]. Φ(T) plots are shown for x = 0.0, 0.1, and 0.2 in [1] and in Samsonov et al. [3].

References:

[1] Kondrashov, A. I.; Siman, N. I.; Podchernyaeva, I. A. (Poroshkovaya Metal. **1977** No. 8, pp. 62/4; Soviet Powder Met. Metal Ceram. **16** [1977] 619/21).

[2] Kondrashov, A. I. (Poluch. Issled. Svoistv Nov. Mater. Mater. 10th Nauchn. Konf. Aspir. Molodykh Issled. Inst. Probl. Materialoved. Akad. Nauk Ukr. SSR, Kiev 1976 [1978], pp. 48/52; C.A. **91** [1979] No. 111144).

[3] Samsonov, G. V.; Kondrashov, A. I.; Okhremchuk, L. N.; Podchernyaeva, I. A.; Siman, N. I.; Fomenko, V. S. (J. Less-Common Metals **67** [1979] 415/8).

32.1.8.3.6 LaB$_9$ (?)

The formation of blue LaB$_9$ from LaB$_6$ and B at 2280 K is concluded from thermal and metallographic investigations, Storms, Mueller [1]. However, on the basis of more recent investigations on samples prepared by arc melting a mixture of the elements with an overall composition of LaB$_9$ it is concluded that a phase near LaB$_9$ exists only between 2280 and 1900 K or less. The blue boride observed at room temperature is not LaB$_9$, but has the composition LaB$_{6.13 \pm 0.03}$ due to random lanthanum vacancies in the LaB$_6$ structure. This is concluded from electron microscopy, X-ray analysis, and NMR studies, McKelvy et al. [2]. For an assumed formation of LaB$_9$ by ion bombardment of LaB$_6$, see Krizhanovskii et al. [3].

References:

[1] Storms, E.; Mueller, B. (J. Phys. Chem. **82** [1978] 51/9).

[2] McKelvy, M. J.; Eyring, L.; Storms, E. (J. Phys. Chem. **88** [1984] 1785/90).

[3] Krizhanovskii, V. I.; Kuzmichev, G.; Levchenko, V.; Luban, R. B.; Shendakov, A. I. (Poroshkovaya Metal. **1981** No. 9, pp. 73/5; C.A. **95** [1981] No. 195943).

32.1.8.3.7 LaB$_{12}$ (?)

Attempts were made to prepare LaB$_{12}$ in a vacuum at 2073 K according to the reactions $La_2O_3 + 3B_4C + 12B \rightarrow 2LaB_{12} + 3CO$ and $La_2O_3 + 24B + 3C \rightarrow 2LaB_{12} + 3CO$. However, the compound did not form. The product obtained was LaB$_6$ with some B content, Neshpor [1]. This was confirmed by Zhuravlev et al. [2]. Earlier the formation of blue LaB$_{12}$ was assumed by evaporation of La on heating LaB$_6$ in a vacuum to ~2273 K, Spedding et al. [3]. See also p. 162.

References:

[1] Neshpor, V. S. (Vysokotemp. Metallokeram. Mater. **1962** 96/101; C.A. **58** [1963] 6436).

[2] Zhuravlev, N. N.; Manelis, R. M.; Gramm, N. V.; Stepanova, A. A. (Poroshkovaya Metal. **1967** No. 2, pp. 95/101; Soviet Powder Met. Metal Ceram. **1967** 158/62).

[3] Spedding, F. H.; Goetz, C. A.; Banks, C. V.; et al. (ISC-1116 [1959] 1/69, 14; N.S.A. **13** [1959] No. 18894).

Physical Constants and Conversion Factors

Avogadro constant N_A (or L) = 6.02214×10^{23} mol^{-1}
Faraday constant   F = 9.64853×10^4 C/mol
molar gas constant   R = 8.31451 J·mol^{-1}·K^{-1}
molar volume (ideal gas) V_m = 2.24141×10^1 L/mol
(273.15 K, 101325 Pa)

Planck constant   h = 6.62608×10^{-34} J·s
elementary charge   e = 1.60218×10^{-19} C
electron mass m_e = 9.10939×10^{-31} kg
proton mass m_p = 1.67262×10^{-27} kg

1 kg = 2.205 pounds
1 m = 3.937×10^1 inches = 3.281 feet
1 m^3 = 2.642×10^2 gallons (U.S.)
1 m^3 = 2.200×10^2 gallons (Imperial)

Force	N	dyn	kp
1 N	1	10^5	1.019716×10^{-1}
1 dyn	10^{-5}	1	1.019716×10^{-6}
1 kp	9.80665	9.80665×10^5	1

Pressure	Pa	bar	kp/m^2	at	atm	Torr	lb/in^2
1 Pa = 1 N/m^2	1	10^{-5}	1.019716×10^{-1}	1.019716×10^{-5}	9.86923×10^{-6}	7.50062×10^{-3}	1.450378×10^{-4}
1 bar = 10^6 dyn/cm^2	10^5	1	1.019716×10^4	1.019716	9.86923×10^{-1}	7.50062×10^2	1.450378×10^1
1 kp/m^2 = 1 mm H$_2$O	9.80665	9.80665×10^{-5}	1	10^{-4}	9.67841×10^{-5}	7.35559×10^{-2}	1.422335×10^{-3}
1 at (technical)	9.80665×10^4	9.80665×10^{-1}	10^4	1	9.67841×10^{-1}	7.35559×10^2	1.422335×10^1
1 atm = 760 Torr	1.01325×10^5	1.01325	1.033227×10^4	1.033227	1	7.60×10^2	1.469595×10^1
1 Torr = 1 mm Hg	1.333224×10^2	1.333224×10^{-3}	1.359510×10^1	1.359510×10^{-3}	1.315789×10^{-3}	1	1.933678×10^{-2}
1 lb/in^2 = 1 psi	6.89476×10^3	6.89476×10^{-2}	7.03069×10^2	7.03069×10^{-2}	6.80460×10^{-2}	5.17149×10^1	1

Work, Energy, Heat

Work, Energy, Heat	J	kW·h	kcal	Btu	eV
1 J = 1 W·s = 1 N·m = 10^7 erg	1	2.778×10^{-7}	2.39006×10^{-4}	9.4781×10^{-4}	6.242×10^{18}
1 kW·h	3.6×10^6	1	8.604×10^2	3.41214×10^3	2.247×10^{25}
1 kcal	4.1840×10^3	1.1622×10^{-3}	1	3.96566	2.6117×10^{22}
1 Btu (British thermal unit)	1.05506×10^3	2.93071×10^{-4}	2.5164×10^{-1}	1	6.5858×10^{21}
1 eV	1.602×10^{-19}	4.450×10^{-26}	3.8289×10^{-23}	1.51840×10^{-22}	1

$$1 \text{ cm}^{-1} = 1.239842 \times 10^{-4} \text{ eV}$$
$$1 \text{ hartree} = 27.2114 \text{ eV}$$
$$1 \text{ Hz} = 4.135669 \times 10^{-15} \text{ eV}$$
$$1 \text{ eV} \cong 23.0578 \text{ kcal/mol}$$

Power

Power	kW	hp	kp·m·s^{-1}	kcal/s
1 kW = 10^3 J/s	1	1.35962	1.01972×10^2	2.39006×10^{-1}
1 hp (horsepower, metric)	7.3550×10^{-1}	1	7.5×10^1	1.7579×10^{-1}
1 kp·m·s^{-1}	9.80665×10^{-3}	1.333×10^{-2}	1	2.34384×10^{-3}
1 kcal/s	4.1840	5.6886	4.26650×10^2	1

References:

International Union of Pure and Applied Chemistry, Manual of Symbols and Terminology for Physicochemical Quantities and Units, Pergamon, London 1979; Pure Appl. Chem. **51** [1979] 1/41.

The International System of Units (SI), National Bureau of Standards Spec. Publ. 330 [1972].

Landolt-Börnstein, 6th Ed., Vol. II, Pt. 1, 1971, pp. 1/14.

ISO Standards Handbook 2, Units of Measurement, 2nd Ed., Geneva 1982.

Cohen, E. R., Taylor, B. N., Codata Bulletin No. 63, Pergamon, Oxford 1986.

Key to the Gmelin System
of Elements and Compounds

System Number	Symbol	Element
1		Noble Gases
2	H	Hydrogen
3	O	Oxygen
4	N	Nitrogen
5	F	Fluorine
6	**Cl**	**Chlorine**
7	Br	Bromine
8	I	Iodine
8a	At	Astatine
9	S	Sulfur
10	Se	Selenium
11	Te	Tellurium
12	Po	Polonium
13	B	Boron
14	C	Carbon
15	Si	Silicon
16	P	Phosphorus
17	As	Arsenic
18	Sb	Antimony
19	Bi	Bismuth
20	Li	Lithium
21	Na	Sodium
22	K	Potassium
23	NH_4	Ammonium
24	Rb	Rubidium
25	Cs	Caesium
25a	Fr	Francium
26	Be	Beryllium
27	Mg	Magnesium
28	Ca	Calcium
29	Sr	Strontium
30	Ba	Barium
31	Ra	Radium
32	**Zn**	**Zinc**
33	Cd	Cadmium
34	Hg	Mercury
35	Al	Aluminium
36	Ga	Gallium

HCl

$ZnCl_2$

System Number	Symbol	Element
37	In	Indium
38	Tl	Thallium
39	Sc, Y La−Lu	Rare Earth Elements
40	Ac	Actinium
41	Ti	Titanium
42	Zr	Zirconium
43	Hf	Hafnium
44	Th	Thorium
45	Ge	Germanium
46	Sn	Tin
47	Pb	Lead
48	V	Vanadium
49	Nb	Niobium
50	Ta	Tantalum
51	Pa	Protactinium
52	**Cr**	**Chromium**
53	Mo	Molybdenum
54	W	Tungsten
55	U	Uranium
56	Mn	Manganese
57	Ni	Nickel
58	Co	Cobalt
59	Fe	Iron
60	Cu	Copper
61	Ag	Silver
62	Au	Gold
63	Ru	Ruthenium
64	Rh	Rhodium
65	Pd	Palladium
66	Os	Osmium
67	Ir	Iridium
68	Pt	Platinum
69	Tc	Technetium[1]
70	Re	Rhenium
71	Np,Pu...	Transuranium Elements

$CrCl_2$

$ZnCrO_4$

Material presented under each Gmelin System Number includes all information concerning the element(s) listed for that number plus the compounds with elements of lower System Number.

For example, zinc (System Number 32) as well as all zinc compounds with elements numbered from 1 to 31 are classified under number 32.

---

[1] A Gmelin volume titled "Masurium" was published with this System Number in 1941.

A Periodic Table of the Elements with the Gmelin System Numbers is given on the Inside Front Cover